Protected Areas of the World

A review of national systems

Volume 1: Indomalaya, Oceania, Australia and Antarctic

Published by: IUCN, Gland, Switzerland and Cambridge, UK,
with the financial support of The British Petroleum Company p.l.c.

Prepared by: The World Conservation Monitoring Centre, Cambridge, UK

A contribution to GEMS – The Global Environment Monitoring System

WORLD CONSERVATION
MONITORING CENTRE

Citation: IUCN (1992). *Protected Areas of the World: A review of national systems. Volume 1: Indomalaya,
Oceania, Australia and Antarctic.* IUCN, Gland, Switzerland and Cambridge, UK. xx + 352 pp.

ISBN: 2-8317-0090-6

Printed by: Page Bros (Norwich) Ltd, UK

Cover photos: Horton Plains National Park, Sri Lanka: Michael J. Green
South Adelaide Island, Antarctic Peninsula: R.K. Headland
Islands of the Palau Archipelago, Republic of Palau, North Pacific Ocean: WWF/D. Faulkner
Nambung's Pinnacles, limestone formation, Western Australia: WWF/Fredy Mercay

Produced by: IUCN Publications Services Unit, Cambridge, UK, on desktop publishing equipment purchased
through a gift from Mrs Julia Ward

Available from: IUCN Publications Services Unit,
219c Huntingdon Road, Cambridge, CB3 0DL, UK

The designations of geographical entities in this book, and the presentation of the material, do not imply the expression
of any opinion whatsoever on the part of IUCN, WCMC or BP concerning the legal status of any country, territory, or
area, or of its authorities, or concerning the delimitation of its frontiers or boundaries.

IUCN – THE WORLD CONSERVATION UNION

Founded in 1948, IUCN – The World Conservation Union – is a membership organisation comprising governments, non-governmental organisations (NGOs), research institutions, and conservation agencies in over 100 countries. The Union's mission is to provide leadership and promote a common approach for the world conservation movement in order to safeguard the integrity and diversity of the natural world, and to ensure that human use of natural resources is appropriate, sustainable and equitable.

Several thousand scientists and experts from all continents form part of a network supporting the work of its Commissions: threatened species, protected areas, ecology, environmental strategy and planning, environmental law, and education and communication. Its thematic programmes include forest conservation, wetlands, marine ecosystems, plants, the Sahel, Antarctica, population and natural resources, and Eastern Europe. The Union's work is also supported by 12 regional and country offices located principally in developing countries.

WCMC – THE WORLD CONSERVATION MONITORING CENTRE

The World Conservation Monitoring Centre (WCMC) is a joint venture between the three partners who developed the World Conservation Strategy: IUCN – The World Conservation Union, UNEP – United Nations Environment Programme, and WWF – World Wide Fund For Nature (formerly World Wildlife Fund). Its mission is to support conservation and sustainable development through the provision of information on the world's biological diversity.

WCMC has developed a global overview database that includes threatened plant and animal species, habitats of conservation concern, critical sites, protected areas of the world, and the utilisation and trade in wildlife species and products. Drawing on this database, WCMC provides an information service to the conservation and development communities, governments and United Nations agencies, scientific institutions, the business and commercial sector, and the media. WCMC produces a wide variety of specialist outputs and reports based on analyses of its data.

Protected Areas of the World

A review of national systems

Volume 1: Indomalaya, Oceania, Australia and Antarctic

Compiled by the World Conservation Monitoring Centre

in collaboration with

The IUCN Commission on National Parks and Protected Areas

for the

IVth World Congress on National Parks and Protected Areas,
Caracas, Venezuela, 10-21 February 1992

with the support of

The British Petroleum Company p.l.c.

IUCN – The World Conservation Union

November 1991

TABLE OF CONTENTS

AUSTRALASIA

THE ANTARCTIC AND SOUTHERN OCEAN ISLANDS

———————

FOREWORD

The inspirational and aesthetic values of fine examples of the beauty and bounty of nature lay behind the establishment of many national parks and other types of protected areas. More recently there has been increasing recognition of the range of the value of protected areas and of their contribution to meet the needs of society by conserving the world's natural and physical resources. These values range from protection of representative samples of natural regions and the preservation of biological diversity, to the maintenance of environmental stability in surrounding country. Protected areas can also facilitate complementary rural development and rational use of marginal lands, and provide opportunities for research and monitoring, conservation education, and recreation and tourism.

Over the past thirty years, since the *First World Conference on National Parks* was held in Seattle, Washington (1962), our view of the world, and our impact on the world, has changed significantly. Throughout this time, and despite the mounting pressures of expanding human populations, the number of protected areas established has continued to rise. Since the centennial of national parks was commemorated at the time of the *Second World Conference on National Parks* at Yellowstone and Grand Teton, Wyoming in 1972, the "human" element of protected areas has come more and more into focus. They are no longer seen as being "locked up" or "set apart". Rather, they are seen as being integral to strategic approaches to resource management, a concept enshrined in the *World Conservation Strategy* (1980) based on managing natural areas to support development in a sustainable way.

The fundamental contribution of protected areas to sustainable management was reaffirmed by participants at the *World Congress on National Parks* held in Bali, Indonesia (1982), and for the last decade the *Bali Action Plan* has focused attention on a range of actions necessary for promoting and supporting protected areas. These actions were further focused in regional action plans subsequently developed by members of the IUCN Commission on National Parks and Protected Areas, covering the Afrotropical, Indomalayan, Neotropical and Oceanian regions.

More recently, two significant, and widely accepted documents have stressed the very vital roles that protected areas play. The report of the *World Commission on Environment and Development* was published in 1987, and more recently a new strategy *Caring for the World* was launched in 1991. This latter strategy, which has its roots in the *World Conservation Strategy*, clearly identifies the functions and benefits of protected area systems, what they safeguard, and why they are important for development opportunities.

Many countries have declared extensive systems of protected areas, and are continuing to develop and expand them. The systems and the sites they contain vary considerably from one country to another, depending on national needs and priorities, and on differences in legislative, institutional and financial support. Consequently, protected areas have been established under many different national designations to provide for a spectrum of management objectives, ranging from total protection to sustainable use: from strict nature reserves to lived-in landscapes.

IUCN – The World Conservation Union has been involved in protected areas issues for many years, and has published a significant body of information on the subject. The IUCN Commission on National Parks and Protected Areas was set up both to ensure that the appropriate expertise was available to advise the Union, and to bring together professionals to share information and experience. IUCN and CNPPA have together had a very strong hand in developing the programme for the *IV World Congress on National Parks and Protected Areas* in Caracas, Venezuela (1992).

For more than 10 years, IUCN and CNPPA have worked closely with what is now the World Conservation Monitoring Centre, to help in building an information resource on protected areas. The information is of value to the Commission in developing its own programmes, in identifying priorities, and for a wide range of other purposes such as supporting international initiatives in World Heritage, wetlands and biosphere reserves. It is also important to both IUCN and the Commission that such information is made available to others, so that the roles and values of protected areas are more widely recognised, appreciated and respected.

The three volume *Protected Areas of the World: A review of national systems* is being published for the World Parks Congress by WCMC and IUCN in cooperation with British Petroleum, and aims to provide a standard format "overview" of the world's protected area systems. While this product has gaps, and no doubt inaccuracies, it does illustrate very clearly the range of protected areas activities around the world, and gives an indication of the protected areas estate under the stewardship of our managers. This product, in combination with the protected areas reviews being prepared for the Congress by the CNPPA Regional Vice-Chairs, will also provide a benchmark against which to measure our achievements over the next decade.

P.H.C. (Bing) Lucas
Chair
IUCN Commission on National Parks and Protected Areas

INTRODUCTION

Participants at the *Third World National Parks Congress* held in Bali, Indonesia, in 1982, clearly recognised that the availability of comprehensive, good-quality information on the world's protected areas was essential to a wide range of international organisations, governments, protected area managers, voluntary bodies and individuals. Such information is a prerequisite for assessing the coverage and status of protected areas from regional and global perspectives, and is key to the development of regional and global priorities and strategies. Monitoring protected areas is vital to ensure that those areas allocated to conserve the world's natural resources meet the needs of society.

The World Conservation Monitoring Centre (WCMC) is expanding its capabilities as an international centre for information on the conservation of biological diversity. Working closely with the IUCN Commission on National Parks and Protected Areas (CNPPA), WCMC continues to compile an extensive database on the world's protected areas, which is being used more and more frequently as a source of information.

One result of WCMC's work as an information centre is the ability to draw material together into publications which provide background information on protected areas and protected area systems. At the previous congress in 1982, two publications from the protected areas database were available, the *1982 UN List*, and the *IUCN Directory of Neotropical Protected Areas*. Since then, the Centre has collaborated with CNPPA and others on a wide range of publications, including two subsequent *UN Lists* in 1985 and 1990, directories of protected areas for Africa, Oceania, South Asia, and the mountains of central Asia, and various publications on eastern Europe. A full list of publications on protected areas (including those published by others with information provided by WCMC) is available from the Centre.

The present work, *Protected Areas of the world: A review of national systems*, is the first attempt by WCMC to compile a world-wide survey of protected area systems. The book is organised into national (or occasionally sub-national) accounts, each comprising a description of the national protected areas system, accompanied by a summary list and map of protected areas. The book is divided into three volumes, with volume one covering the Indomalayan, Oceanian, Australian and Antarctic realms, volume two the Palaearctic and Afrotropical realms, and volume three the Nearctic and Neotropical realms.

Publication of such a book serves two purposes. First, it provides extensive background information on the protected area systems of the world, relevant to several plenary sessions and workshops at the *IV World Congress on National Parks and Protected Areas*. In particular, it is a contribution to the third plenary session *The Contributions of Protected Areas to Sustaining Society: A Global Review*. Secondly, and perhaps more significantly, it is also part of the process of information collection and verification. Feedback from protected areas professionals, and others familiar with protected areas, is therefore both welcomed and encouraged, because only by a continual process of review and update can we present a true picture.

Jeremy Harrison
World Conservation Monitoring Centre

ACKNOWLEDGEMENTS

Preparation of a directory of this magnitude is only achieved through a tremendous amount of effort and cooperation. Over the years, protected areas professionals throughout the world have reviewed or compiled material for us, or provided new information. Quite simply, without their cooperation this book could not have been completed, and we greatly appreciate their support.

This assistance has been facilitated in part by the IUCN Commission on National Parks and Protected Areas, and the support of the Commission and its past and present officers is gratefully acknowledged. In particular we would like to thank the present chair P.H.C. (Bing) Lucas of New Zealand, and his predecessor Harold Eidsvik of Canada. Work has also been supported by the staff of the IUCN Protected Areas Unit, and in particular James Thorsell and Jeffrey McNeely.

This particular publication is the product of two projects. The *British Petroleum Company plc* has provided support for the preparation of part of the text and maps (for those areas not covered by the second mentioned project), and have provided funds for publishing the book and distributing it at the World Parks Congress in Caracas, Venezuela. At the same time, the *British Overseas Development Administration* has supported review of information on tropical and sub-tropical countries, as part of a project contributing to the FAO Forest Resources Assessment 1990. Thanks are due to both organisations for their support.

A number of past and present staff of WCMC have been involved in preparing this directory, which includes material published in several earlier directories. Compilation of country accounts has been the responsibility of: Patricia Almada-Villela, Daphne Clark, Graham Drucker, Harriet Gillett, Michael Green, Donald Gordon, Jeremy Harrison, Zbigniew Karpowicz, Sara Oldfield, James Paine and Mark Spalding. Assistance with the preparation of maps has been provided by Mike Adam, Clare Billington, Simon Blythe and Gillian Bunting.

Notwithstanding the significant contributions of the many individuals who have provided information to WCMC and CNPPA, errors and omissions must remain the responsibility of the compilers.

This directory is not intended to be a final statement but a review of the world's protected area systems. If WCMC is to continue to carry out its mission, there is a continual need to maintain and update this information as national protected areas systems change and as more information becomes available. Therefore, with this directory goes a plea for corrections, comments and additional material to help WCMC carry out its mission as effectively as possible. By the same token, the information that WCMC collects and manages is available to others to support their work and programmes.

World Conservation Monitoring Centre
219 Huntingdon Road, Cambridge, CB3 ODL, United Kingdom
Tel: (44) 223 277314
Fax: (44) 223 277136
Tlx: 817036 SCMU G

MANAGING INFORMATION ON PROTECTED AREAS AT WCMC

Many individuals and organisations need basic information on protected areas systems, lists of protected areas with certain features, or analyses of protected areas statistics, yet it is unlikely that they will have the time or resources to collect, compile and analyse all of the information for themselves. Such information also needs to be kept up-to-date, as properties are added or extended, and as legislation or administrative regimes change. Users may also require details about the major protected areas within national systems, such as physical features, vegetation and fauna, or on other aspects such as management status and constraints.

It is to meet these needs that the WCMC Protected Areas Data Unit (PADU) was founded. This service enables users to obtain quickly information on protected areas from a single source, be it for purposes of analysis and assessment, or as briefing material. It is not intended that this service should by-pass any need for users to contact or visit the relevant national authorities for such information, but use of PADU's resources enables users to be well informed prior to making such approaches and in a better position to ask the right questions when so doing.

Institutional background

IUCN – The World Conservation Union has been closely involved in protected areas issues for many years. As early as 1960, it established a Commission on National Parks and Protected Areas (CNPPA) to serve as the "leading international, scientific and technical body concerned with the selection, establishment and management of national parks and other protected areas". CNPPA has always emphasised the need for information on which to base effective conservation planning and management, and has been very active in collecting and disseminating information on protected areas.

As the world's network of protected areas has expanded and its management improved, information on national protected areas systems and individual protected areas has proliferated. This led CNPPA to set up PADU in 1981 to manage this increasing volume of information. Establishment of this Unit was supported by the United Nations Environment Programme (UNEP), as part of its Global Environment Monitoring System (GEMS). Originally part of the IUCN Conservation Monitoring Centre, PADU is now a unit within the World Conservation Monitoring Centre (WCMC), restructured in July 1988 and jointly managed by IUCN, the World Wide Fund for Nature (WWF) and UNEP.

Objectives

WCMC aims to provide accurate up-to-date information on protected area systems of the world for use by its partners (IUCN, WWF and UNEP) in the support and development of their programmes, and by other international bodies, governmental and non-governmental organisations, scientists and the general public. Such information covers the entire spectrum of protected areas, from national parks and sanctuaries established under protected areas legislation or customary regimes to forest reserves created under forestry legislation. It also includes privately-owned reserves in which nature is protected.

Specific objectives are to:

– maintain a comprehensive and up-to-date database of the world's protected areas;
– compile definitive, standard-format accounts summarising national protected areas systems;
– hold maps of protected areas systems and digitise them;
– compile definitive, standard-format accounts covering individual protected areas, particularly the major properties in tropical countries and those of international importance;
– accumulate current and historical information on protected areas; and
– provide support to regional and international activities, programmes and conventions relating to protected areas.

Information capture, management and compilation

Information is collected from official sources, namely national agencies responsible for administering protected areas, and other sources through a global network of contacts ranging in profession from policy-makers and administrators to land managers and scientists. It is also obtained from published and unpublished literature. Regional CNPPA meetings and other relevant scientific and technical meetings provide valuable opportunities for making new contacts and collecting fresh information. This material in itself is a major asset of the Centre.

Information, ranging from books, reports, management plans, scientific papers, maps and correspondence, is stored as hard copy in manual files. Basic data on individual protected areas are extracted and, after verification, entered in a protected areas database, which currently holds some 26,000 records. This computerised database can be used for generating lists of protected areas meeting pre-defined criteria, together with summary statistics, as well as performing more complex tasks. In addition, maps of protected areas are gradually being digitised, using a Geographic Information System, in order to generate computerised graphic output.

The information is also used to produce accounts of protected areas systems and individual protected areas. These accounts are compiled according to standard formats developed over the years by WCMC in collaboration with CNPPA.

Dissemination of information

In keeping with its primary objective, WCMC aims to make available good quality information on protected areas to a wide range of users, including international organisations, governments, protected area managers, conservation organisations, commercial companies involved in natural resource exploitation, scientists, and the media and general public. Information may be provided or consulted by arrangement.

Material may be prepared under contract: for example, WCMC regularly provides UNEP with summary data on protected areas for its biennial *Environmental Data Report*. WCMC is experimenting with providing outside users with direct access to its protected areas database. Trials have been ongoing with the US National Park Service since 1986 and it is hoped to be able to extend this service to other users shortly.

Compiled information is periodically published in the form of regional or thematic directories and lists. Directories comprise sections on individual countries, each with a protected areas *system information sheet*, a *list* of protected areas and accompanying *location map*, and a series of *site information sheets* covering at least the more important properties. Prior to releasing or publishing documents, draft material is circulated for review by relevant government agencies and experts to help ensure that compiled information is accurate and comprehensive.

Major lists and directories published to date are as follows:

- *United Nations List of National Parks and Protected Areas* (1982, 1985, 1990)

- *IUCN Directory of Neotropical Protected Areas* (1982)
- *IUCN Directory of Afrotropical Protected Areas* (1987)
- *IUCN Directory of South Asian Protected Areas* (1990)
- *Protected Areas in Eastern and Central Europe and the USSR* (1990)
- *IUCN Directory of Protected Areas in Oceania* (1991)
- *Nature Reserves of the Himalaya and the Mountains of Central Asia* (1992)

- *Information System: Biosphere Reserves: Compilation 4* (1986)
- *Biosphere Reserves: Compilation 5* (1990)

- *Directory of Wetlands of International Importance* (1987, 1990)
- *Protected Landscapes: Experience around the World* (1987)

In addition, numerous draft directories, reports papers and reviews have been produced. A list of these is available from WCMC.

WCMC also disseminates information through the *CNPPA Newsletter* and *Parks* magazine. In the case of the latter, WCMC has assumed responsibility for compiling *Clipboard* in which world news on protected areas is featured.

Special services

WCMC has a very close working relationship with CNPPA. While the Commission provides expert advice and support through its network of members, WCMC supports many of the Commission's activities through provision of technical information. WCMC has a particular responsibility for managing information on natural properties designated under international conventions and programmes, namely the *Convention concerning the Protection of the World Cultural and Natural Heritage* (World Heritage Convention), *Convention on Wetlands of International Importance especially as Waterfowl Habitat* (Ramsar Convention), and the Unesco *Man and the Biosphere Programme*. Thus, WCMC cooperates closely with the Division of Ecological Sciences, Unesco, in maintaining information on biosphere reserves and World Heritage sites accorded by the MAB Secretariat and World Heritage Committee, respectively. Likewise, it works closely with the Ramsar Bureau with respect to managing information on Ramsar wetlands.

The rest of the World Conservation Monitoring Centre

Protected areas is only one aspect of the programme of the World Conservation Monitoring Centre, which also covers information on plant and animal species of conservation concern, important natural habitats and sites of high biological diversity, wildlife utilisation, and the international trade in wildlife.

To monitor the impact of man on nature is a major task. This requires close collaboration between agencies, and between agencies and individuals, and the development and exchange of information. WCMC acts both as an information centre, and as a facilitator of information management and exchange. WCMC has now embarked on an ambitious programme to promote improvements in the availability of information, and to develop its database capabilities and information services. Information on the distribution and status of the world's protected areas is an essential component of this programme.

COUNTRY ACCOUNTS: GUIDELINES TO THEIR CONTENTS

In general, there is an account for each country, divided up into a series of sections with standard headings. The following notes summarise the type of information included in each section where it is available. In certain cases, accounts have been prepared for areas which are parts of countries, usually where the area concerned is geographically separate from the "parent" country.

Country

Full name of country or political unit, as used by the United Nations (United Nations *Terminology Bulletin* on Names of Countries and Adjectives of Nationality).

Area

Total area according to the latest volume of the *FAO Production Yearbook* prepared by the Statistics Division of the Economic and Social Policy Department, FAO, unless otherwise stated (with full reference). Terrestrial and marine components are distinguished, if appropriate.

Population

Total population and its mean annual rate of growth according to the latest issue of *World Population Prospects*, published by the United Nations Population Division. Year of census or estimate is indicated in parentheses. If another source has to be used, it is cited.

Economic Indicators

Gross domestic product and gross national product per capita in US dollars (or net material product in the case of centrally planned economies), with year in parentheses. These figures are according to the latest issue of *National Accounts Statistics: Analysis of Main Aggregates* (prepared by the United Nations Statistical Office) and *The World Bank Atlas*.

Policy and Legislation

Information on aspects of the constitution that are relevant to nature conservation and protected areas.

Summary of national policies that relate to nature conservation, particularly with respect to the protection of ecosystems. This may include reference to policies relating to environmental impact assessments, and national/regional conservation strategies.

Brief chronological account of past and present national legislation and traditions that relate to the establishment of the protected areas system, with names (in English), dates and numbers of acts, decrees and ordinances. Legislation covering forestry and other resource sectors is included, in so far as it provides for protected areas establishment. Procedures for the notification and declassification of protected areas are summarised.

Outline of legal provisions for administering protected areas

National designations of protected areas are cited and their range of provisions outlined. Their legal definitions, together with the names of the authorities legally responsible for their administration, are summarised in an Annex (see below).

Reviews of protected areas policy and legislation are noted, with any identified deficiencies in prevailing provisions highlighted.

International Activities

Participation in international conventions and programmes (World Heritage and Ramsar conventions, MAB Programme, UNEP Regional Seas Programme) and regional conventions and agreements (such as the African, ASEAN and Berne conventions, the FAO Latin American/Caribbean Technical Cooperation Network, South Asian Cooperative Environmental Programme and the South Pacific Regional Environment Programme) relevant to habitat protection is summarised.

Outline of any international, multilateral and bilateral cooperative programmes or transfrontier cooperative agreements relevant to protected areas.

Administration and Management

All authorities responsible for the administration and management of protected areas are named and described, with a brief history of their establishment, administrative organisation, staff structure, budget and any training programmes. Authorities responsible for different types of protected areas are clearly distinguished.

Outline of the role of any advisory boards

Cooperative agreements between management authorities and national or foreign universities and institutes, with details of any research underway or completed.

Names and brief details of non-governmental organisations concerned with protected areas. Reference to any national directories of voluntary conservation bodies is included.

Effectiveness of protected areas management is noted where information has been provided. Attention is drawn to any sites registered as threatened under the World Heritage Convention, or by the IUCN Commission on National Parks and Protected Areas.

Systems Reviews

Short account of physical features, biological resources, and land use patterns (with percentages if available), including the extent and integrity of major ecosystems.

Brief review of the development of nature conservation programmes, so far as it relates to the establishment and expansion of the national protected areas network. Emphasis is given to any systems reviews or comprehensive surveys of biological resources, with details of major recommendations arising from such studies.

Threats to the protected areas system beyond the control of the management agencies are outlined.

Other relevant information

Tourism and other economic benefits of the protected areas system, if applicable

Other items, as appropriate

Addresses

Names and addresses (with telephone, telex and fax numbers, and cable) of authorities responsible for administering protected areas. Names are given in the original language or transliterated, with English translation in brackets as appropriate, and followed by the title of the post of the chief executive.

Names and addresses (with telephone, telex and fax numbers, and cable) of non-governmental organisations actively involved in protected areas issues. Names are given in the original language or transliterated, with English translation in brackets as appropriate, and followed by the title of the post of the chief executive,

References

Key references (including all cited works) to the protected areas system, in particular, and nature conservation, in general, are listed.

ANNEX
Definitions of protected area designations, as legislated, together with authorities responsible for their administration.

The annex includes the following sections:

Title: Name and number of law in the original language or transliterated, with the English translation underneath, as appropriate.

Date: Day, month and year of enactment, followed by dates of subsequent major amendments

Brief description: Summary of main provisions (often this is stated at the beginning of the legislation)

Administrative authority: Name of authority responsible for administering the law, given in the original language or transliterated, with the English translation underneath as appropriate. This is followed by the title of the post of the chief executive.

Designations: National designation of protected area in the original language or transliterated, followed in brackets by the English translation as appropriate. For each designation this would be followed by: definition of designation (if given in legislation), summary of activities permitted or prohibited, outline of penalties for offences, and, where relevant, reference to subsequent legislation relating to the original law.

Source: This may be "original legislation", "translation of original legislation" or a referenced secondary source.

MAPS and LISTS

The descriptive sections are followed by lists of protected areas, and maps showing their location. In most cases, the lists comprise all of those areas qualifying for inclusion in IUCN management categories I-VIII, which have an area of over 1,000 hectares. However, forest and hunting reserves qualifying for IUCN Management Category VIII have been omitted, largely because our information is not comprehensive. Also, size has been ignored for island nations. Note that in certain cases, nationally designated areas (such as some national parks) will not appear in the lists, as they do not meet the criteria. World Heritage sites, biosphere reserves and Ramsar sites are also listed.

Categories and management objectives of protected areas

I   *Scientific Reserve/Strict Nature Reserve*: to protect nature and maintain natural processes in an undisturbed state in order to have ecologically representative examples of the natural environment available for scientific study, environmental monitoring, education, and for the maintenance of genetic resources in a dynamic and evolutionary state.

II   *National Park*: to protect natural and scenic areas of national or international significance for scientific, educational and recreational use.

III *Natural Monument/Natural Landmark*: to protect and preserve nationally significant natural features because of their special interest or unique characteristics.

IV *Managed Nature Reserve/Wildlife Sanctuary*: to assure the natural conditions necessary to protect nationally significant species, groups of species, biotic communities, or physical features of the environment where these require specific human manipulation for their perpetuation.

V *Protected Landscape or Seascape*: to maintain nationally significant natural landscapes which are characteristic of the harmonious interaction of man and land while providing opportunities for public enjoyment through recreation and tourism within the normal life style and economic activity of these areas.

VI *Resource Reserve*: to protect the natural resources of the area for future use and prevent or contain development activities that could affect the resource pending the establishment of objectives which are based upon appropriate knowledge and planning.

VII *Natural Biotic Area/Anthropological Reserve*: to allow the way of life of societies living in harmony with the environment to continue undisturbed by modern technology.

VIII *Multiple-Use Management Area/Managed Resource Area*: to provide for the sustained production of water, timber, wildlife, pasture, and outdoor recreation, with the conservation of nature primarily oriented to the support of economic activities (although specific zones may also be designed within these areas to achieve specific conservation objectives).

Abridged from IUCN (1984). Categories and criteria for protected areas. In: McNeely, J.A. and Miller, K.R. (Eds), *National parks, conservation, and development. The role of protected areas in sustaining society*. Smithsonian Institution Press, Washington. Pp. 47-53

INTERNATIONALLY DESIGNATED SITES

There are two international conventions and one international programme that include provision for designation of internationally important sites in *any* region of the world. These are the World Heritage Convention, the Ramsar (Wetlands) Convention, and the Unesco Man and the Biosphere (MAB) Programme. While there is a wide range of other international conventions and programmes, these cover only regions, or small groups of countries.

Both World Heritage sites and Ramsar sites must be nominated by a State that is party to the relevant convention. While there is an established review procedure for World Heritage sites (and nomination is no guarantee of listing), all nominated Ramsar sites are placed on the List of Wetlands of International Importance. Biosphere reserves are nominated by the national MAB committee of the country concerned, and are only designated following review and acceptance by the MAB Bureau.

Each Contracting Party to the Ramsar (Wetlands) Convention is obliged to nominate at least one wetland of international importance. However, a country can be party to the World Heritage Convention without having a natural site inscribed on the List, and may participate in the MAB programme without designating a biosphere reserve.

World Heritage Sites

The Convention Concerning the Protection of the World Cultural and Natural Heritage was adopted in Paris in 1972, and came into force in December 1975. The Convention provides for the designation of areas of "outstanding universal value" as World Heritage sites, with the principal aim of fostering international cooperation in safeguarding these important areas. Sites, which must be nominated by the signatory nation responsible, are evaluated for their World Heritage quality before being inscribed by the international World Heritage Committee. Only natural sites, and those with mixed natural and cultural aspects, are considered in this publication.

Article 2 of the World Heritage Convention considers as natural heritage: natural features consisting of physical and biological formations or groups of such formations, which are of outstanding universal value from the aesthetic or scientific point of view; geological or physiographical formations and precisely delineated areas which constitute the habitat of threatened species of animals and plants of outstanding universal value from the point of view of science or conservation; and natural sites or precisely delineated areas of outstanding universal value from the point of view of science, conservation or natural beauty. Criteria for inclusion in the list are published by Unesco.

The following States Party to the Convention lie at least partially within the regions covered by this volume:

Australia
Bangladesh
Fiji
France
India
Indonesia
Lao People's Democratic Republic
Malaysia
Maldives
Nepal
New Zealand
Pakistan
Philippines
Sri Lanka
Thailand
United Kingdom
United States of America
Viet Nam

The following natural World Heritage sites lie within the regions covered by this volume:

Australia
Australian East Coast Rainforest Parks
Great Barrier Reef
Kakadu National Park
Lord Howe Island Group
Tasmanian Wilderness
Uluru (Ayers Rock) National Park
Wet Tropics of Queensland
Willandra Lakes Region

India
Kaziranga National Park
Keoladeo National Park
Manas National Park
Nanda Devi National Park
Sundarbans National Park

Nepal
Royal Chitwan National Park
Sagarmatha National Park

New Zealand
South West New Zealand (Te Wahipounamu)
Tongariro National Park

Sri Lanka
Sinharaja Forest Reserve

United Kingdom
Henderson Island

United States of America
Hawaii Volcanoes National Park

Ramsar Sites

The Convention on Wetlands of International Importance Especially as Waterfowl Habitat was signed in Ramsar (Iran) in 1971, and also came into force in

December 1975. This Convention provides a framework for international cooperation for the conservation of wetland habitats. The Convention places general obligations on contracting party states relating to the conservation of wetlands throughout their territory, with special obligations pertaining to those wetlands which have been designated to the "List of Wetlands of International Importance".

Each State Party is obliged to list at least one site. Wetlands are defined by the convention as: areas of marsh, fen, peatland or water, whether natural or artificial, permanent or temporary, with water that is static or flowing, fresh, brackish or salt, including areas of marine waters, the depth of which at low tide does not exceed six metres.

The following States Party to the Convention lie at least partially within the regions covered by this volume:

Australia
France
India
Nepal
New Zealand
Norway
Pakistan
South Africa
Sri Lanka
United Kingdom
United States of America
Viet Nam

The following wetlands which lie within the region have been included in the List of Wetlands of International Importance:

Australia
Apsley Marshes
Barmah Forest
Bool and Hacks Lagoons
Cape Barren Is. East Coast Lagoons
Cobourg Peninsula
Coongie Lakes
Corner Inlet
Eighty-mile Beach
Forrestdale and Thomsons Lakes
Gippsland Lakes
Gunbower Forest
Hattah-Kulkyne Lakes
Hosnie's Spring (Christmas Island)
Jocks Lagoon
Kakadu (Stage II)
Kakadu (Stage I)
Kerang Wetlands
Kooragang Nature Reserve
Lake Crescent (northwestern corner)
Lake Albacutya
Lake Warden System
Lake Toolibin
Lakes Argyle and Kununurra
Little Waterhouse Lake

Logan Lagoon
Lower Ringarooma River
Macquarie Marshes Nature Reserve
Moulting Lagoon
Ord River Floodplain
Peel-Yalgorup System
Pittwater-Orielton Lagoon
Port Phillip Bay (western shoreline)/Bellarine Peninsula
Riverland
Roebuck Bay
Sea Elephant Conservation Area
The Coorong and Lakes Alexandrina and Albert
Towra Point Nature Reserve
Vasse-Wonnerup System
Western District Lakes
Western Port

India
Chilka Lake
Harike Lake
Keoladeo National Park
Loktak Lake
Sambhar Lake
Wular Lake

Nepal
Koshi Toppu

New Zealand
Farewell Spit
Firth of Thames
Kopuatai Peat Dome
Waituna Lagoon
Whangamarino

Pakistan
Drigh Lake
Haleji Lake
Kandar Dam
Khabbaki Lake
Kheshki Reservoir
Kinjhar (Kalri) Lake
Malugul Dhand
Tanda Dam
Thanadarwala

Sri Lanka
Bundala Sanctuary

Viet Nam
Red River Estuary

Biosphere Reserves

The designation of biosphere reserves differs somewhat from that of either of the previous designations in that it is not made under a specific convention, but as part of an international scientific programme, the Unesco Man and the Biosphere Programme. The objectives of a network of biosphere reserves, and the characteristics which biosphere reserves might display, are identified in various documents, including the Action Plan for Biosphere Reserves (Unesco, 1984).

Biosphere reserves differ from World Heritage and Ramsar sites in that they are designated not exclusively for protection of unique areas or significant wetlands, but for a range of objectives which include research, monitoring, training and demonstration, as well as conservation. In most cases the human component is vital to the functioning of the biosphere reserve, something which is not always true for either World Heritage or Ramsar sites.

The following biosphere reserves are located within the region:

Australia
Croajingolong
Danggali Conservation Park
Fitzgerald River National Park
Hattah-Kulkyne National Park/Murray-Kulkyne Park
Kosciusko National Park
Macquarie Island Nature Reserve
Prince Regent River Nature Reserve
Southwest National Park
Uluru (Ayers Rock-Mount Olga) National Park
Unnamed Conservation Park of South Australia
Wilson's Promontory National Park
Yathong Nature Reserve

France
Atoll de Taiaro

Indonesia
Cibodas Biosphere Reserve
    (Gunung Gede-Pangrango)
Gunung Leuser National Park
Komodo National Park
Lore Lindu National Park
Siberut Nature Reserve
Tanjung Puting National Park

Pakistan
Lal Suhanra National Park

Philippines
Palawan Biosphere Reserve
Puerto Galera Biosphere Reserve

Sri Lanka
Hurulu Forest Reserve
Sinharaja Forest Reserve

Thailand
Hauy Tak Teak Reserve
Mae Sa-Kog Ma Reserve
Sakaerat Environmental Research Station

United States of America
Hawaii Islands Biosphere Reserve

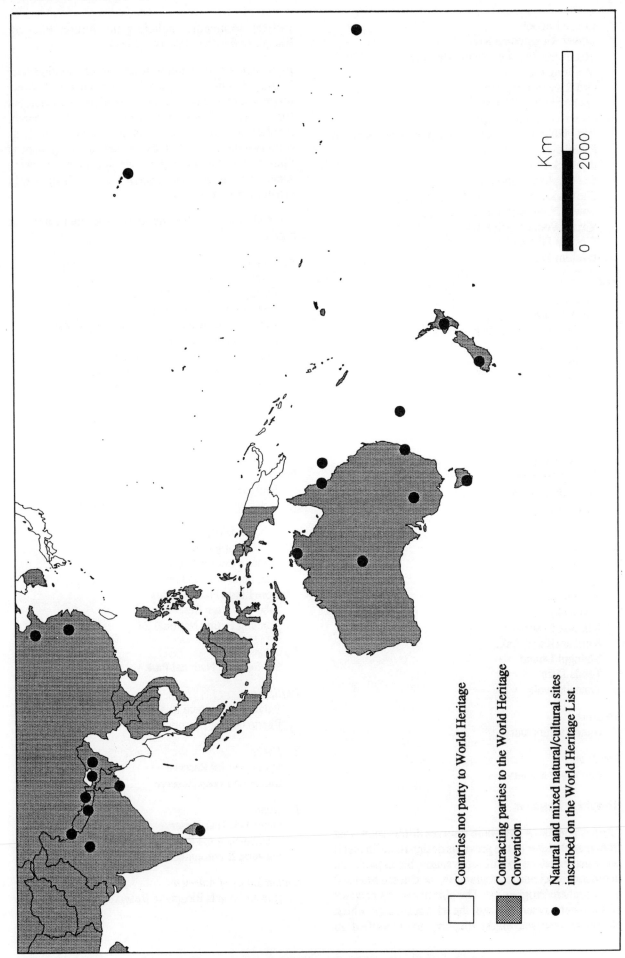

Internationally Designated Sites – World Heritage Convention

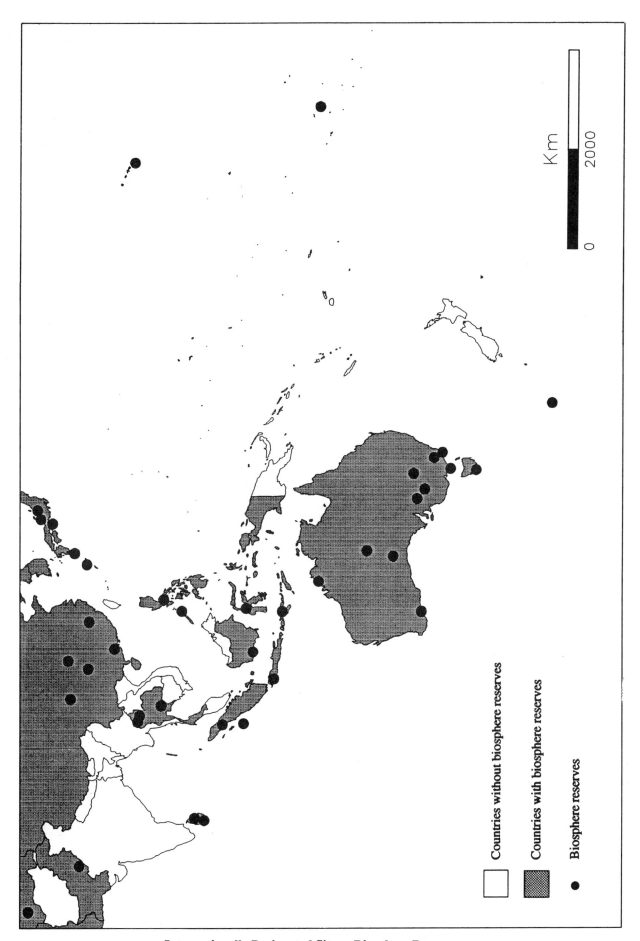

Internationally Designated Sites – Biosphere Reserves

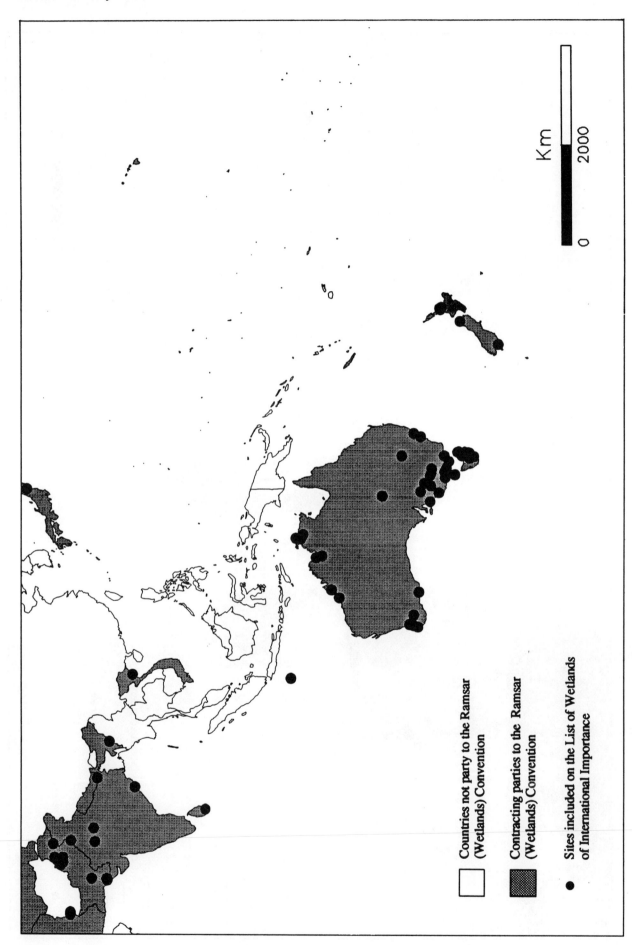

Internationally Designated Sites – Ramsar (Wetlands) Convention

Indomalaya

BANGLADESH

Area 144,000 sq. km

Population 114,800,000 (1990)
Natural increase: 2.5% per annum

Economic Indicators
GNP: US$ 170 per capita (1988)

Policy and Legislation Environmental policy in Bangladesh is based on the following three broad principles: precautionary, whereby harm to the environment is avoided; originator, whereby the costs of ameliorating damage to the environment are borne by those responsible; and cooperation, whereby relevant bodies are involved in planning for environmental protection (Rahman, 1983).

The need for an explicit national policy on environmental protection and management has been repeatedly highlighted (BARC, 1987), and is presently under consideration by the government. Objectives of such a policy will be as follows: to create, develop, maintain and improve conditions under which man and nature can thrive in productive and enjoyable harmony with each other; to fulfil the social, economic and other requirements of present and future generations; and to ensure the attainment of an environmental quality that is conducive to a life of dignity and well-being (Rahman, 1983).

Environmental impact assessment for anticipating adverse impacts has not yet been incorporated into the development planning process, nor is it a mandatory requirement of project-approving agencies. According to government policy, sanctioning agencies should ensure that project proposals contain adequate environmental safeguards but, in practice, this is not strictly followed (BARC, 1987).

Bangladesh has completed the first phase of a national conservation strategy aimed at integrating conservation goals with national development objectives and overcoming identified obstacles to sustainable development (BARC, 1987). Some twenty sectors in the current Third Five-Year Plan are identified for critical analysis during a second phase, including the conservation of genetic resources, and wildlife management and protected areas. The Bangladesh Agricultural Research Council, Ministry of Agriculture is the lead agency for the implementation of Phase II which began in October 1989.

There is no national wildlife conservation policy. The Bangladesh Wildlife (Preservation) Order, 1973, promulgated under Presidential Order No. 23 on 27 March 1973 and subsequently enacted and amended in two phases as the Bangladesh Wildlife (Preservation) (Amendment) Act, 1974, provides for the establishment of national parks, wildlife sanctuaries, game reserves and private game reserves (see Annex). Under Article 23, wildlife sanctuaries enjoy a greater degree of protection than national parks. For example, entry or residence, introduction of exotic or domestic species of animals and lighting of fires is prohibited in wildlife sanctuaries, but not national parks. No specific rules are detailed for game reserves. The Article makes provision, however, for the government to relax any of these prohibitions for scientific, aesthetic or other exceptional reasons, and to alter the boundaries of protected areas. Under Article 24, provision is made for the establishment of private game reserves upon application by the landowner. The owner of a private game reserve may exercise all the powers of an officer provided under the Act. Proposals are being drawn up to strengthen the existing legislation, largely through raising fines and terms of imprisonment for offences.

Conservation, use and exploitation of marine resources are provided for under the Territorial Water and Maritime Zones Act, 1974. According to provisions in this Act, conservation zones may be established to protect marine resources from indiscriminate exploitation, depletion or destruction. At present, there is no legal provision for the management of coastal zones.

The Forest Act, 1927, enables the government to declare any forest or waste land to be reserved forest or protected forest (see Annex). Activities are generally prohibited in reserved forests; certain activities, such as removal of forest produce, may be permitted under license in protected forests, while others, such as quarrying of stone and clearing for cultivation, may be prohibited. The rights of government to any land constituted as reserved forest may be assigned to village communities, with conditions for their management prescribed by government. Such forests are called village forests. Under the Forest (Amendment) Ordinance, 1989, penalties for offences committed within reserved and protected forests have been increased from a maximum of six months imprisonment and a fine of Tk 500 to five years imprisonment and a Tk 5,000 (US$ 1,700) fine. In accordance with the National Forest Policy, adopted in 1979, effective measures will be taken to conserve the natural environment and wildlife resources. The Policy does not, however, deal explicitly with the need to set aside special areas as protected forests, as distinct from productive forests, to preserve genetic diversity and maintain ecological processes within the context of sustainable development (BARC, 1987).

Other environmental legislation less specifically related to protected areas is reviewed elsewhere (DS/ST, 1980; Rahman, 1983).

International Activities Bangladesh is party to the Convention concerning the Protection of the World Cultural and Natural Heritage (World Heritage

Convention) which it accepted on 3 August 1983. No natural sites have been inscribed to date. Bangladesh participates in the Unesco Man and the Biosphere Programme. Apart from a couple of reserved forests proposed as candidate sites by the Bangladesh MAB National Committee in the late 1970s, there does not appear to have been any significant development in recent years. A proposal to become a party to the Convention on Wetlands of International Importance especially as Waterfowl Habitat (Ramsar Convention) was submitted to the erstwhile Ministry of Agriculture and Forestry by the Forest Department and awaits approval. It is proposed to nominate the Sundarbans mangrove forests as a wetland of international importance, in partial fulfilment of the requirements of the Convention (Rahman and Akonda, 1987).

Administration and Management Wildlife conservation, including the management of protected areas, is the responsibility of the Forest Directorate within the new Ministry of Environment and Forests formed in 1989. Previously, the Forest Directorate came under the Ministry of Agriculture and Forests while the former Department of Environmental Pollution Control, concerned largely with environmental pollution, was under the Ministry of Local Government and Rural development.

In 1976 a Wildlife Circle was established within what was then known as the Forest Department, with specific responsibility for wildlife matters under the charge of a Conservator of Forests responsible directly to the Chief Conservator of Forests. A $13.3 million scheme, entitled "Development of Wildlife Management and Game Reserves", was incorporated within the country's First Five-Year Plan, but reduced to $92,000 in the subsequent Two-Year Approach Plan (Olivier, 1979). The Wildlife Circle was subsequently abolished in June 1983, allegedly in the interests of economy and following the recommendations of the Inam Commission. The post of Conservator of Forests (General Administration and Wildlife) remains but the incumbent has many other administrative duties unrelated to wildlife. Following its general down-grading within the Forest Department, wildlife conservation has become the theoretical responsibility of the various divisional forest officers (Blower, 1985; Husain, 1986). Separate staff are deployed for protection purposes in a number of national parks and wildlife sanctuaries (Sarker and Fazlul Huq, 1985).

The Bangladesh Wildlife (Preservation)(Amendment) Act, 1974, also provides for the establishment of a Wildlife Advisory Board, which was set up in 1976 under the chairmanship of the Minister of Agriculture. The Board is supposed to approve important wildlife management decisions and directives (Olivier, 1979). Although it still exists, it does not appear to be a dynamic force (Blower, 1985; BARC, 1987).

In view of the low priority accorded to protected areas, a Task Force was formed by the Ministry of Agriculture

in 1985 to identify institutional and other measures needed to improve current provisions for wildlife conservation. Recommendations of the Task Force, submitted to the government in July 1986, await approval by the competent authority. They include a plan to immediately revive the erstwhile Wildlife Circle, review Phase II of the Wildlife Development Project and secure protection of 5% of the total land area of the country for conservation purposes (Rahman and Akonda, 1987).

The principal non-governmental conservation organisations within the country are the Society for Conservation of Nature and Environment (SCONE), which is mainly concerned with environmental pollution, and the Wildlife Society of Bangladesh. Pothikrit, based in Chunati, and Polli Unnayan Sangstha (POUSH), founded in 1984, are both involved in promoting the adoption of sound management practices in and around protected areas. Their efforts are presently focused on Chunati Wildlife Sanctuary and Teknaf Game Reserve. IUCN (The World Conservation Union) has a project office in Dhaka.

Given that wildlife resources are vested largely in reserved forests, their conservation has in the past been diametrically opposed to forest management practices. Few, if any, protected areas are effectively managed and protected. Lack of personnel trained in wildlife conservation is a further handicap (Gittins and Akonda, 1982; Khan, 1985; Olivier, 1979). The very low priority apparently now accorded to wildlife conservation is reflected in the recent abolition of the Wildlife Circle, the reassignment of staff to normal duties, the lack of any separate financial provision within the Forest Directorate's budget and the now moribund Wildlife Advisory Board (Blower, 1985).

Systems Reviews Some 80% of Bangladesh is lowland, comprising an alluvial plain cut by the three great river systems (Ganges-Padma, Brahmaputra- Jamuna and Meghna) that flow into the Bay of Bengal. Typically, at least one-half of the land is inundated annually, with one-tenth subject to severe flooding. The entire flood plain was well-vegetated, but much of the forest has been replaced by cultivations and plantations in recent decades due to mounting pressure from human populations. Here, the only extensive tract of forest remaining is the Sundarbans. Hills are confined chiefly to the east and south-east, notably the Chittagong Hills where forest cover is among the most extensive in the country.

According to the 1987 Statistical Yearbook of Bangladesh, forests cover 2.1 million hectares or 14.7% of total land area, but this represents neither the area under forest nor that under the control of the Forest Department (Rashid, 1989). In 1980, Gittins and Akonda (1982) estimated remaining natural forest to be 4,782 sq.km (3.3%) and scrub forest 9,260ha (6.5%). Actual forest cover is presently estimated to be 1 million hectares or 6.9% of total land area, a reduction of more than 50% over the past 20 years (WRI/CIDE, 1990).

The major forest types are mangrove, moist deciduous or sal *Shorea robusta*, restricted to the Madhupur Tract and northern frontier with Meghalaya, and evergreen forests found in the eastern districts of Sylhet, Chittagong and Chittagong Hill Tracts. A small amount of freshwater swamp occupies the basins of the north-east region.

Wetlands, variously estimated as covering between seven and eight million hectares or nearly 50% of total land area, support a variety of wildlife, as well as being of enormous economic importance (Scott, 1989).

The only known coral reef is around Jinjiradwip (St Martin's Island) in the Bay of Bengal. It is reputed to be a submerged reef but little is known about it (UNEP/IUCN, 1988).

Conservation efforts began in 1966, prior to independence, when the government of Pakistan invited the World Wildlife Fund to assess its wildlife resources and recommend measures to arrest their depletion. Two expeditions were mounted (Mountfort and Poore, 1967, 1968) and the severity of the situation confirmed, whereupon the government was urged to appoint its own Wildlife Enquiry Committee. The committee was established in 1968 and by 1970 had drafted a report. That part relating to East Pakistan was published as a separate report (Government of East Pakistan, 1971). Considerable progress was made with the establishment of several protected areas (Mountfort, 1969), research undertaken on the Sundarbans tiger population of East Pakistan (Hendrichs, 1975), and technical input from UNDP/FAO (Grimwood, 1969). Then, in 1971, came the War of Liberation which inevitably disrupted subsequent progress. In spite of political instability, however, the Bangladesh Wildlife (Preservation) Order was promulgated in 1973 and an ambitious programme of wildlife management developed, followed by the formation of a Wildlife Circle in 1976 and further technical assistance from UNDP/FAO (Olivier, 1979). Economic constraints, however, have subsequently been responsible for the loss of much of this initiative (Blower, 1985).

The existing system of protected areas has been reviewed recently (Green, 1989). It is not comprehensive, having been established with little regard to ecological and other criteria, and falls well below the target of 5% recommended by the erstwhile Ministry of Agriculture Task Force. Some effort has been made to include representative samples of the major habitats but, for example, marine and freshwater areas have been largely neglected (Gittins and Akonda, 1982; Khan, 1985; Olivier, 19 9; Rahman and Akonda, 1987). Priorities to develop the present network of protected areas are identified in the *IUCN Systems review of the Indomalayan Realm* (MacKinnon and MacKinnon, 1986) and further recommendations are made in the

Corbett Action Plan (IUCN, 1985), many of which are based on earlier recommendations by Olivier (1979). More recently, wetlands of conservation value have been identified (Scott, 1989). Of outstanding importance is the need to prepare a plan for the development of the country's protected areas network.

Addresses

Office of the Chief Conservator of Forests, Conservator of Forests (General Administration and Wildlife), Bana Bhawan, Gulshan Road, Monakhali, DHAKA 12 (Tel: 2 603537; Cable: FORESTS)

Forest Directorate, Chief Conservator of Forests, Ministry of Environment and Forests, Bana Bhawan, Gulshan Road, Monakhali, DHAKA 12 (Cable: FORESTS)

IUCN (The World Conservation Union), Country Representative, 35 B/2 Indira Road, Dhaka 1215 (Tel: 2 815061; FAX: 2 813466; Tlx: 671054 FRC BJ)

Polli Unnayan Sangstha, 43 New Eskaton Road, DHAKA (Tel: 2 402801/ 406628; Tlx: 642639 OCNBJ)

Pothikrit, CHUNATI VILLAGE, Chittagong District

The Society for Conservation of Nature and Environment, Secretary General, 146 Shanti Nagar, DACCA 17 (Tel: 2 409119; Cable: ENVIRON DHAKA)

Wildlife Society of Bangladesh, General Secretary, c/o Department of Zoology, University of Dhaka, DHAKA 1000

References

BARC (1987). National conservation strategy for Bangladesh. Draft prospectus (Phase I). Bangladesh Agricultural Research Council/IUCN, Gland, Switzerland. 154 pp.

Blower, J.H. (1985). *Sundarbans Forest Inventory Project, Bangladesh. Wildlife conservation in the Sundarbans.* Project Report No. 151. ODA Land Resources Development Centre, Surbiton, UK. 39 pp.

DS/ST (1980). Draft environmental profile on Bangladesh. Science and Technology Division, Library of Congress. Washington, DC. 98 pp.

Gittins, S.P. and Akonda, A W. (1982). What survives in Bangladesh? *Oryx* 16: 275-281.

Government of East Pakistan (1971). Report of the Technical Sub-committee for East Pakistan of the Wildlife Enquiry Committee. Dacca.

Green, M.J.B. (1989). Bangladesh: an overview of its protected areas system. World Conservation Monitoring Centre, Cambridge, UK. 63 pp.

Grimwood, I.R. (1969). Wildlife Conservation in Pakistan. *Pakistan National Forestry Research and Training Project.* Report No. 17. FAO, Rome. 31 pp.

Hendrichs, H. (1975). The status of the tiger *Panthera tigris* (Linne, 1758) in the Sundarbans mangrove forest (Bay of Bengal). *Säugetierkundliche Mitteilungen* 23: 161-199.

Husain, K.Z. (1986). Wildlife study, research and conservation in Bangladesh. *Eleventh Annual Bangladesh Science Conference* Section 2: 1-32.

IUCN (1985). *The Corbett Action Plan for protected areas of the Indomalayan Realm*. IUCN, Gland, Switzerland and Cambridge, UK. 23 pp.

Khan, M.A.R. (1985). Future conservation directions for Bangladesh. In: Thorsell, J.W. (Ed.), *Conserving Asia's natural heritage*. IUCN, Gland, Switzerland. Pp. 114-122.

MacKinnon, J. and MacKinnon, K. (1986). *Review of the protected areas system in the Indo-Malayan Realm*. IUCN, Gland, Switzerland Cambridge, UK. 284 pp.

Mountfort, G. (1969). Pakistan's progress. *Oryx* 10: 39-43.

Mountfort, G. and Poore, D. (1967). The conservation of wildlife in Pakistan. World Wildlife Fund, Morges, Switzerland. Unpublished report. 27 pp.

Mountfort, G. and Poore, D. (1968). Report on the Second World Wildlife Fund Expedition to Pakistan. World Wildlife Fund, Morges, Switzerland. Unpublished. 25 pp.

Olivier, R.C.D. (1979). *Wildlife conservation and management in Bangladesh*. UNDP/FAO Project No. BGD/72/005. Forest Research Institute, Chittagong. 121 pp.

Rahman, S. (1983). Country monograph on institutional and legislative framework on environment, Bangladesh. UN/ESCAP and Government of Bangladesh. 76 pp.

Rahman, S.A. and Akonda, A.W. (1987). Bangladesh national conservation strategy: wildlife and protected areas. Department of Forestry, Ministry of Agriculture and Forestry, Dhaka. Unpublished report. 33 pp.

Rashid, H. Er (1989). Land use in Bangladesh: selected topics. Bangladesh Agriculture Sector Review. UNDP Project No. BGD/87/023. Pp. 106-155.

Sarker, N.M. and Fazlul Huq, A.K.M. (1985). Protected areas of Bangladesh. In: Thorsell, J.W. (Ed.), *Conserving Asia's natural heritage*. IUCN, Gland, Switzerland. Pp. 36-38.

Scott, D.A. (Ed.) (1989). *A directory of Asian wetlands*. IUCN, Gland, Switzerland and Cambridge, UK. 1,181 pp.

UNEP/IUCN (1988). *Coral reefs of the world. Volume 2: Indian Ocean, Red Sea and Gulf*. UNEP Regional Seas Directories and Bibliographies. IUCN, Gland, Switzerland and Cambridge, UK/UNEP, Nairobi, Kenya. 440 pp.

WRI/CIDE (1990). Bangladesh environment and natural resource assessment. Draft for review. World Resources Institute/Centre for International Development and Environment, Washington DC. 86 pp.

ANNEX
Definitions of protected area designations, as legislated, together with authorities responsible for their administration

Title: Bangladesh Wildlife (Preservation) (Amendment) Act

Date: 1974

Brief description: Provides for the preservation, conservation and management of wildlife in Bangladesh

Administrative authority: Forest Directorate, Ministry of Environment and Forests

Designations:

National park A comparatively large area of outstanding scenic and natural beauty, in which the protection of wildlife and preservation of the scenery, flora and fauna in their natural state is the primary objective, and to which the public may be allowed access for recreation, education and research.

Hunting, killing or capturing any wild animal within a national park or one mile (1.6km) of its boundaries, causing any disturbance (including firing of any gun) to any wild animal or its breeding place, felling, tapping, burning or in any other way damaging any plant or tree, cultivation, mining or breaking up any land, and polluting water flowing through a national park are not allowed. Such prohibitions may be relaxed for scientific purposes, aesthetic enjoyment of the scenery or any other exceptional reason.

Construction of access roads, rest houses, hotels and public amenities should be planned so as not to impair the primary objective of the establishment of a national park.

Wildlife sanctuary An area closed to hunting and maintained as an undisturbed breeding ground, primarily for the protection of wildlife including all natural resources such as vegetation, soil and water.

Entry or residence, cultivation, damage to vegetation, killing or capturing wild animals within one mile (1.6km) of its boundary, introduction of exotic or domestic species of animals, lighting of fires, and pollution of water are not allowed, but any of these prohibitions may be relaxed for scientific reasons, or for the improvement or aesthetic enjoyment of the scenery.

Game reserve An area in which the wildlife is protected to enable populations of important species to increase. Capture of wild animals is prohibited.

Hunting and shooting may be allowed on a permit basis.

Private game reserve Area of private land set aside by the owner for the same purpose as a game reserve. On application by the owner, such an area may be notified as a private game reserve.

The owner shall exercise all the powers of an officer under this Act.

Source: Original legislation

Title: Forest Act

Date: 1927

Brief description: An Act to consolidate the law relating to forests, the transit of forest produce and the duty leviable on timber and other forest produce.

Administrative authority: Forest Directorate, Ministry of Environment and Forests

Designations:

Reserved forest Any forest land or wasteland belonging to the government, or to which it has proprietary rights, may be constituted a reserved forest subject to completion of notification and settlement procedures provided under the Act.

Prohibited activities include: making fresh clearings or breaking up land for cultivation; kindling or carrying fire; trespass and cattle grazing; felling or otherwise damaging any tree; quarrying stone, burning lime or charcoal; removing forest produce; and hunting, shooting, fishing, trapping and poisoning water.

Village forest Any land constituted as reserved forest that has been assigned to a village community by the government.

Rules for regulating the provision of timber, other forest produce or pasture to the community, and their duties for protecting and improving such forest may be prescribed by the government.

All provisions of the Act relating to reserved forest apply to village forest, in so far as they are consistent with the rules.

Protected forest Any forest land or wasteland not included in a reserved forest and belonging to the government, or to which it has proprietary rights, may be declared a protected forest provided that the nature and extent of rights of government and of private persons in or over such land have been recorded.

Any trees or class of trees may be reserved; any portion of forest may be closed for up to 30 years; and quarrying of stone, burning of lime or charcoal, collection and removal of any forest produce, and breaking up or clearing of any land for any purpose may be prohibited.

Rules may be made to regulate collection and removal of forest produce, granting of licences to inhabitants of nearby settlements to remove forest products for domestic consumption, granting of licences for commercial extraction of forest products, clearing or breaking up of land for cultivation or other purposes, and the protection from fire of timber lying in such forests and of trees reserved under the Act.

Source: Original legislation

5

SUMMARY OF PROTECTED AREAS

Map ref.	*National/international designation* Name of area	IUCN management category	Area (ha)	Year notified
	National Parks			
1	Bhawal	V	5,022	1982
3	Madhupur	V	8,436	1982
	Wildlife Sanctuaries			
6	Chunati	IV	7,764	1986
7	Pablakhali	IV	42,087	1983
8	Rema-Kalenga	IV	1,095	1981
9	Sundarbans East	IV	5,439	1977
10	Sundarbans South	IV	17,878	1977
11	Sundarbans West	IV	9,069	1977
	Game Reserve			
12	Teknaf	VIII	11,615	1983

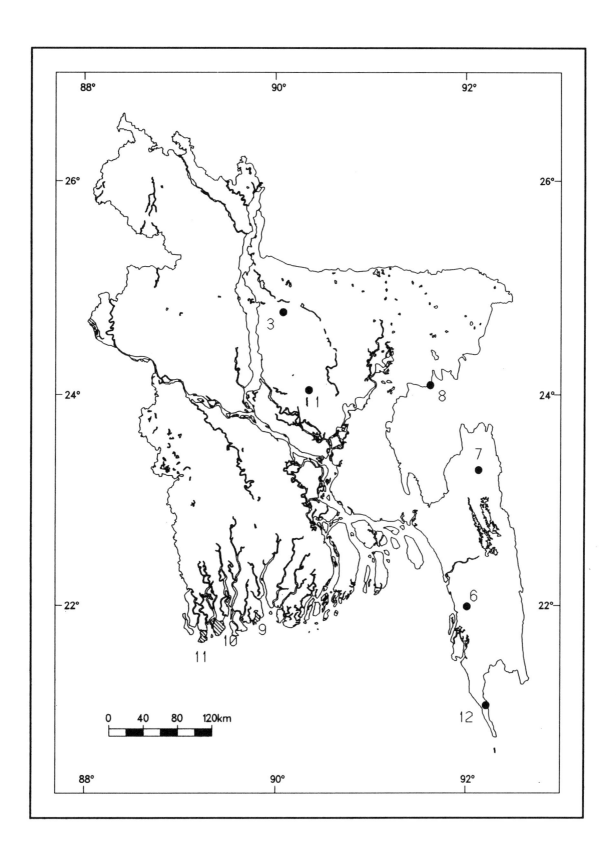

Protected Areas of Bangladesh

BHUTAN

Area 46,620 sq. km

Population 1,600,000 (1990)
Natural increase: 2.1% per annum

Economic Indicators
GNP: US$ 150 per capita (1988)

Policy And Legislation Government policy on environmental conservation is strong, with emphasis consistently given to nature conservation and careful management of natural resources. National development plans have stressed the potential for ecological damage from exploitation of the nation's natural resources, particularly its forests (Blower, 1989; World Bank, 1988). The need to formulate a national conservation strategy has been recognised and a proposal submitted to His Majesty's Government by Danida is under consideration (Tenzing, 1989).

The existing National Forest Policy of 1974 emphasises the importance of maintaining adequate forest cover, with a minimum of 60%, in order to prevent soil erosion and to maintain climatic equilibrium. It recognises the problems caused by grazing and shifting cultivation, and the need to regulate both practices. A new National Forest Policy was prepared in 1985 at the express command of His Majesty the King, but this has yet to be adopted. The new policy lays even greater stress on conservation, its basis being that the nation's forest resources should be regarded more in terms of their conservation value and less as a source of revenue. Prescriptions include: designation of all forest land above 2,700m or on slopes exceeding 60° as protection forest; establishment of a protected areas network (including biosphere reserves) to conserve representative samples of the diverse fauna and flora in their pristine state; control of shifting cultivation and its prohibition on slopes of 45° and more; and the total banning of grazing in forests reserved for protection or conservation (Blower, 1986, 1989).

The Bhutan Forest Act of 1969 is the only legislation covering environmental conservation. Under this Act, all forested land other than any privately owned, is declared as government reserved forest. Activities prohibited within reserved forest are annexed. The maximum penalty for any offence under the Forest Act is one month's imprisonment or a fine of Nu. 200/- (US$ 13) or both. There is no specific provision in the Forest Act for the establishment or management of any other category of protected area, although it is mentioned that "nothing shall be done to fell or damage trees or clear forests in the area of a national park or game sanctuary or the shooting grounds of His Majesty the King." Protected areas, other than reserved forests, have been established by notification, notably No. TIF-11/74 of 1 November 1979 under which three wildlife sanctuaries, one game reserve, one national park and

three reserved forests were designated. A further six sites were declared under Notification No. TIF/FAO/111-8/83/7049. The provisions of the Forest Act apply to these areas (see Annex), together with additional restrictions. These include: prohibition of entry except for Bhutanese officials or visitors with written permission from the Divisional Forest Officer; felling of trees or cutting of other vegetation, except under the provision of a Forest Department Working Plan; no use of land for agricultural, horticultural or other purposes; and no grazing by domestic cattle without permission from the Forest Department. Penalties prescribed for infringements are up to six months' imprisonment or a fine of up to Nu. 1,000/- (US$ 65).

While the present forestry legislation covers many of the essential requirements for conservation, there are serious omissions with respect to such matters as the criteria for different categories of protected area and procedures for their establishment and management. New legislation entitled Bhutan Wildlife (Protection) Act, based on the Indian Wild Life (Protection) Act, was drafted in 1985 but it was considered to be unnecessarily lengthy and complicated. Blower (1986) recommended that new conservation legislation be formulated to provide the basis for an effective conservation programme. This should take the form of a basic enabling act with more emphasis on the broader aspects of environmental conservation, rather than merely on the protection of wildlife and control of hunting. A new Forest and Nature Conservation Act has since been prepared which will replace the Forest Act of 1969. The new law expands on the forestry policy to include related aspects of wildlife and biological diversity (Adams, 1989). It was due to have been presented to the National Assembly in 1988 (H. Wollenhaupt, pers. comm., 1988).

International Activities Bhutan is not as yet party to any international convention concerned with protecting natural areas, such as the Convention Concerning the Protection of the World Cultural and Natural Heritage (World Heritage Convention) and Convention on Wetlands of International Importance especially as Waterfowl Habitat (Ramsar Convention), nor does it participate in the Unesco Man and the Biosphere Programme.

Administration and Management The Forest Department, under the Ministry of Agriculture, is responsible for the management of reserved forests in particular. It is headed by a Director-General and divided into various functional divisions (e.g. planning, management) at its headquarters in Thimphu and a number of territorial divisions. These coincide with the administrative districts, or Dzongkhags, and are headed by a divisional forest officer (Blower, 1989). Forestry has a recent origin in Bhutan, beginning in 1952 with the establishment of the first administrative unit at Samchi.

Further divisions were established at Sarbhang in 1961 and Thimphu in 1967, but funding was very limited until the Third Development Plan (1971-1976) when forest development activities gathered momentum. A forest guard school was established at Kalikhola in 1971, later shifted to Taba in 1977 and upgraded in 1982 for training foresters. Officials and rangers are trained in India (Tenzing, 1989).

Nature conservation is the responsibility of the Wildlife Division established within the Forest Department in 1984. The Division consists of two wildlife circles, each under the charge of a deputy director. The Northern Wildlife Circle, with its headquarters at Thimphu, is nominally responsible for the whole of northern Bhutan including the vast Jigme Dorji Wildlife Sanctuary. With a staff of only one forest ranger and three guards, this is obviously an impossible task. The Southern Wildlife Circle, based at Sarbhang, is responsible for southern Bhutan, including the management of 10 protected areas. Staff include two forest rangers, nine foresters and 36 guards under the charge of a deputy director. The budget for the Northern and Southern Wildlife Circles in 1988/1989 was Nu. 420,000 (US$ 27,300) and Nu. 1,708,000 (US$ 110,000), respectively. In addition, WWF has contributed US$ 300,000 for the development of Manas Wildlife Sanctuary over a three-year period (Blower, 1989).

The Royal Society for the Protection of Nature is the first non-governmental conservation organisation in the country, established in 1987 with assistance from WWF. Its principal aim is to promote conservation and wise management of natural resources through raising public awareness, instituting programmes and acting as an information centre. Due to the Society's efforts, two areas (Phobjikah Valley and Bomdiling) have been declared by the government as sanctuaries for cranes (Adam, 1989; Bunting, 1989).

The Forest Department is short of trained personnel and this has led to a reduction of field staff in its Wildlife Division from 66 in 1986 to 53 in 1989. The Wildlife Division is so inadequately staffed as to be virtually ineffective as far as the country as a whole is concerned (Blower, 1989).

Systems Reviews Bhutan is a small kingdom in the Eastern Himalaya similar in size to Switzerland, but with a much wider altitudinal range (200m to over 7,500m) and only one-fifth of the population density. There has been almost no industrial development in the country: about 95% of the population is primarily dependent on agriculture and animal husbandry. The Himalayan chain runs along the northern border and the interior of the country is made up of a series of six major north-south-aligned mountain ranges. The largest of these, the Black Mountains, rise to nearly 5,000m and form a substantial physical barrier between eastern and western Bhutan. Four of the seven river valleys merge to form the Manas and all of them flow southwards across the plains of West Bengal and Assam into the Brahmaputra.

The enormous altitudinal range and varied climatic conditions are reflected in the country's great ecological diversity, ranging from tropical moist deciduous forest along the southern foothills, through extensive temperate broad-leaved and coniferous forests across the middle of the country, to alpine scrub and meadows up to the permanent snowline to the north.

Bhutan's most valuable natural resources are its forests and its major river systems. Most of the original forest remains. Analysis of LANDSAT 2 imagery for 1978 shows that some 53% of Bhutan is forested, of which 19% is broad-leaved evergreen forest and 34% coniferous and deciduous. The remaining landcover comprises snow/water/scree (19%) and pasture/scrub/arable (28%) (Sargent, 1985; Sargent *et al.*, 1985). This is lower than the official estimate of 64% forest cover (Negi, 1983), which is based on visual inspection of LANDSAT images without recourse to objective ground surveys (Sargent *et al.*, 1985). There was extensive commercial exploitation of forest resources up until 1979, when logging operatives were nationalised and severe restrictions imposed on the export of timer in the interests of sound forestry management and ecological stability (World Bank, 1984, 1986).

The conservation importance of major rivers (Torsa/Ammo Chu, Paidak/Wong Chu, Sankosh/Mo Chu and Manas) is reviewed by Scott (1989). Rivers are generally rocky and fast-flowing, with marshes restricted to flat valley bottoms in the inner valleys. Most marshes have been drained for agricultural purposes but some of those remaining are internationally important for black-necked crane.

Isolated for centuries by its remote geographic location and, latterly, by its resistance to outside influence, Bhutan has maintained a relatively pristine environment along with a strong cultural heritage. Following its membership of the United Nations in 1971, a more open foreign policy has emerged but, acutely aware of mistakes made in neighbouring countries and elsewhere, the government has proceeded cautiously with its development programme. Recognising the need to promote economic growth while sustaining the natural resource base, the government has maintained a strong traditional conservation ethic as the basis of its forest and other policies (Bunting, 1989; Tenzing, 1989). In the case of tourism, for example, numbers of foreign visitors are strictly limited to minimise erosion of the Buddhist culture (Hickman and Edmunds, 1988; Singh, 1989).

Bhutan's oldest protected area is Manas, maintained as a royal hunting reserve for many years prior to being notified a wildlife sanctuary in 1966 and more recently (1988) upgraded to a national park. The bulk of the protected areas network, covering nearly 19% of the country, was established in 1974 and subsequently expanded by a further 2% in 1984. The entire north of Bhutan, comprising nearly 17% of the total area, is protected within the 790,495ha Jigme Dorji Wildlife Sanctuary. While such provisions are impressive,

Sanctuary. While such provisions are impressive, exceeding those of all other countries in South Asia and many elsewhere, the protected areas system is unevenly distributed, with inadequate representation across the middle of the country. Moreover, the relative conservation value of protected areas varies enormously, as does the effectiveness of their protection (Blower, 1985). The only areas considered to be under any form of effective management in 1986 were Manas and the adjacent Nangyal Wangchuk (now combined within Royal Manas National Park), and Mochu Wildlife Sanctuary (Blower, 1986). These deficiencies are being addressed, partly through various internationally assisted development projects. WWF is presently financing a cooperative nature conservation programme to the extent of Nu. 9,120,000 for the period 1988-1993 (Bunting, 1989; Tenzing, 1989). This includes assistance for the institutional development of the Wildlife Division and infrastructural support for Royal Manas National Park. Under the UNDP/FAO Integrated Forest Management and Conservation Project (1987-1991), priorities for nature conservation have been identified, including the strengthening of the protected areas system through the establishment of two large protected areas in the middle of the country (Blower, 1989). As part of the Forestry II Development Project financed by World Bank, there are proposals to strengthen the protected areas system, particularly with regard to the middle sector of the country, and to provide technical assistance for the development of a national conservation strategy (C.W. Holloway, pers. comm., 1987).

Bhutan's natural resources are becoming increasingly threatened. While less pronounced than in other parts of the Himalaya, there is substantial evidence that uplands in Bhutan are being degraded at accelerating rates (Denholm, 1990; Thinley, 1989). The main conservation problem is the conversion of forests to other forms of land use as a result of human settlement, high domestic consumption of fuelwood and timber, shifting cultivation, overgrazing and encroachment, all of which reflect the rising human population (Blower, 1985; Jackson, 1981; Mahat, 1985; Sargent, 1985). Forests are grazed by excessive numbers of domestic livestock and are burnt, while the wildlife is declining due to habitat destruction, grazing competition with domestic livestock and, in some southern areas, organised poaching (Blower, 1985). The southernmost forest belt has been almost completely cleared for human settlement (Mahat, 1985). People are concentrated in the fertile valleys and, in the south-western foothills, at densities approaching an upper limit given present production methods, which are unlikely to change in the near future (Jackson, 1981).

Addresses

Northern Wildlife Circle (Deputy Director of Wildlife), Forest Department, Thimphu (Cable: BHUFOREST; Tel: 22452; Fax: 22395)

Forest Department (Director-General of Forests), Ministry of Agriculture & Forests, Royal Government of Bhutan, PO Box 130, Thimphu (Cable: BHUFOREST; Tel: 22503; Fax: 22395)

Royal Society for the Protection of Nature (President), Thimphu (Tel: 22056; FAX: 22578)

References

Adams, J. (1989). Bhutan: right from the start. *World Wildlife Fund Letter* 1989(6): 1-8.

Blower, J.H. (1985). Nature conservation and wildlife management in Bhutan. FAO, Rome. Unpublished report. 23 pp.

Blower, J.H. (1986). *Nature conservation in Bhutan: project findings and recommendations*. Project HDP/BHU/83/022. FAO, Rome. 55 pp.

Blower, J.H. (1989). *Nature conservation in northern and central Bhutan*. Project BHU/85/016. FAO, Rome. 48 pp.

Bunting, B. (1989). A strategy for environmental conservation in Bhutan: a WWF/RGOB cooperative programme. *Tiger Paper* 16(4): 5-12.

Denholm, J. (1990). Bhutan must protect its green health. *Himal* 3(1): 24.

Hickman, K. and Edmunds, T.O. (1988). Tourism in Bhutan: "The serpent in paradise". *The Geographical Magazine* 60(11): 18-23.

Jackson, P. (1981). Conservation in Bhutan. Unpublished report. 15 pp.

Mahat, G. (1985). Protected areas of Bhutan. In: Thorsell, J.W. (Ed.), *Conserving Asia's natural heritage*. IUCN, Gland, Switzerland. Pp. 26-29.

Sargent, C. (1985). The forests of Bhutan. *Ambio* 14: 75-80.

Sargent, C., Sargent, O., and Parsell, R. (1985). The forests of Bhutan: a vital resource for the Himalaya? *Journal of Tropical Ecology* 1: 265-286.

Scott, D.A. (Ed.) (1989). *A directory of Asian wetlands*. IUCN, Gland, Switzerland and Cambridge, UK. 1,181 pp.

Singh, M.M. (1989). Controlled growth in Bhutan. *Himal* 2(3): 11.

Tenzing, D. (1989). Forestry in Bhutan: policies and programmes. *Forest News* 3(4): 5-9.

Thinley, S. (1989). Upland conservation in Bhutan. *Forest News* 3(4): 10-15.

World Bank (1984). Bhutan, development in a Himalayan kingdom. World Bank, Washington, DC. (Unseen)

World Bank (1986). Bhutan Forestry II Development Project. Preparation Mission Report. FAO/World Bank Co-operative Programme Investment Centre, FAO, Rome. (Unseen)

World Bank (1988). Bhutan, development planning in a unique environment. Report No. 7189-BHU. World Bank, Washington, DC. (Unseen)

ANNEX
Definitions of protected area designations, as legislated, together with authorities responsible for their administration

Title: The Bhutan Forest Act

Date: 1 November 1969

Brief description: To amend the law relating to forests, forest produce and the duty leviable on timber and other forest produce

Administrative authority: Forest Department (Director-General of Forests)

Designations:

Reserved forest Any land under forest to which no person has acquired a permanent, heritable and transferable right of use and occupancy is declared as government reserved forest.

Prohibited activities include: any fresh clearing or breaking up of land for cultivation or other purpose; burning or leaving a fire unattended; felling; girdling; tapping, lopping or otherwise injuring any tree; quarrying of minerals, rocks and sand; poisoning water; hunting and fishing, or setting traps or snares; grazing cattle in new plantations, regeneration areas, catchments reserved for supply of drinking water and hydroelectric projects, and such areas as may be restricted by His Majesty's Government.

Shifting cultivation is allowed in areas where it was practised prior to issue of this Act, but this concession may be withdrawn if highways or public property are endangered. Fresh clearance for shifting cultivation is strictly prohibited.

All forest operations are prohibited within catchments that supply water to townships or are sites of hydroelectric projects.

Only His Majesty the King of Bhutan may grant a "special permit for any forest produce".

Rights and concessions of the local people include: cattle grazing (except in areas defined above), subject to payment of taxes; collection of timber for domestic consumption on payment of royalties; collection of firewood for domestic consumption from dead, dying and fallen trees (or from thinnings and cuttings if such firewood is not available); and collection of leaf-litter, boulders, stones and sand for domestic consumption provided their removal does not accelerate erosion.

Source: Original legislation

SUMMARY OF PROTECTED AREAS

Map ref.	National/international designation Name of area	IUCN management category	Area (ha)	Year notified
	National Parks			
2	Royal Manas	II	65,800	1988
	Wildlife Reserves			
3	Dungsum	IV	18,000	1984
4	Mochu	IV	27,843	1984
	Wildlife Sanctuaries			
6	Jigme Dorji	IV	790,495	1974
7	Neoli	IV	4,000	1984
	Reserved Forests			
8	Khaling	VIII	23,569	1974
9	Pochu	VIII	14,193	1974
10	Sinchula	VIII	8,000	1984

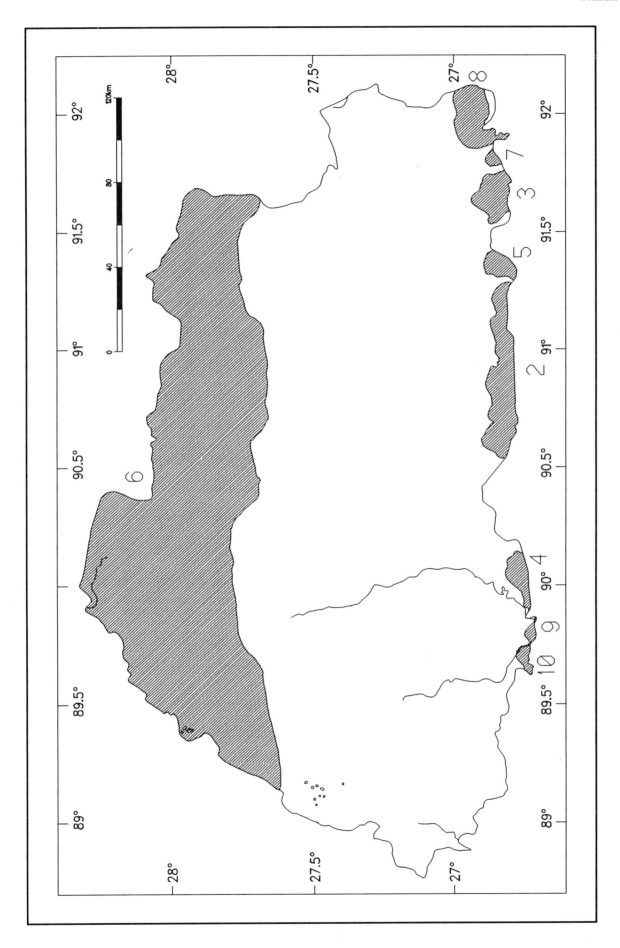

Protected Areas of Bhutan

BRUNEI DARUSSALAM

Area 5,765 sq. km

Population 300,000 (1990)
Natural increase: 2.5% per annum

Economic Indicators
GNP: US$ 15,390 per capita (1987)

Policy and Legislation The Wildlife Protection Act, 1978 provides for the establishment of wildlife sanctuaries (see Annex). A Wildlife Conservation Enactment has been drafted as part of a Special Report on Wildlife Conservation and Management prepared for the Negara Brunei Darussalam Master Plan 2000 to provide, *inter alia*, for the establishment of national parks, wildlife sanctuaries and nature reserves (Farmer *et al.*, 1986).

The Forest Act, 1934 (together with the Forest Rules, 1955 and 1975) provides for the establishment of forest reserves. Land within forest reserves is allocated to one of four categories: protection forest; conservation area; recreational area and production forest (see Annex).

For some time, the Brunei Museum has used the Antiquities and Treasure Trove Enactment, 1967 to designate certain of Brunei's islands (namely the Pelong Rocks and Pulau Punyut) as historical sites, with a view to protecting their fauna and flora. In addition, Pulau Siarau has been unofficially designated a conservation area by the Brunei Museum for some years, in order to afford some protection to the proboscis monkey and flying foxes. Several other islands in Brunei Bay have received partial protection on the pretext of preserving historical sites and antiquities. Although not officially gazetted, Tasek Merimbun is *de facto* Brunei's first national park. It is managed jointly by the Tutong District Office, under whose jurisdiction the area falls, and the Brunei Museum which has been responsible for preliminary scientific studies and management plans.

Other relevant legislation includes the Fisheries Enactment, 1972 under which designated areas may be closed to fishing and other forms of exploitation.

International Activities Brunei does not participate in any international conventions or programmes concerned with protected areas, such as the Convention concerning the Protection of the World Cultural and Natural Heritage (World Heritage Convention), the Convention on Wetlands of International Importance especially as Waterfowl Habitat (Ramsar Convention), and the Unesco Man and the Biosphere Programme.

Field studies have been undertaken in Batu Apoi Forest Reserve as part of the Brunei Rainforest Project 1991-1992, an international cooperative programme involving the Royal Geographic Society (UK) and the University of Brunei, with a view to establishing a field studies centre and to the possible redesignation of this area as a national park (Macklin, 1991).

Administration and Management The Director of the Brunei Museum is responsible for wildlife conservation in the country, including the administration and enforcement of the Wild Life Protection Enactment, 1978. The Forestry Department, headed by the Director of Forestry of the rank of Conservator of Forests, is responsible for the administration and management of forest reserves. A Parks and Conservation Department has been established within the Forestry Department. The Director of the Fisheries Department is responsible for the enforcement of the Fisheries Enactment, 1972 including the protection of living marine resources.

Non-governmental conservation organisations include the Brunei Nature Society, based in Bandar Seri Begawan and the Panaga Natural History Society, based in Seria and run under the auspices of the Brunei Shell Petroleum Company Sdn Bhd.

Systems Reviews Brunei is a small country on the north-west coast of Borneo. The state is divided into two parts by Limbang District in the Malaysian State of Sarawak. The 130km coastline bordering the South China Sea consists of high-profile sandy beaches with a complex estuarine mangrove and mudflat zone in the north-east. In the western part of the country, the alluvial and often swampy coastal plain is backed by low hills, with further swamps inland. Most of the interior is below 90m, rising to almost 400m in the extreme west. The eastern part comprises a swampy coastal plain rising gradually through low hills to mountainous terrain inland. The main mountain range along the border with Sarawak rises to 1,850m (Bukit Pagon).

The natural vegetation throughout Brunei is tropical evergreen rain forest. Forest covers 4,690 sq. km (81% of total land area), of which 22% is secondary forest and plantations and 59% primary forest. Just over half of the primary forest is mixed dipterocarp, one quarter is peat swamp forest and the remainder either swamp forest, heath or montane forest (Bennett, 1991). The mangroves on the Brunei coast probably represent the largest remaining intact mangroves in northern Borneo. Together with those in neighbouring countries in Brunei Bay, they comprise one of the largest tracts of relatively undisturbed mangroves in eastern Asia. Mangrove resources are exploited for various purposes but to a lesser extent than in other countries in the region (de Silva, 1988). The main areas of peat swamp forest are along the basin of the Belait River in western Brunei. Substantial areas of seasonally flooded peat swamp forest occur in the middle reaches of the Tutong River. All of these swamp forests are still in almost pristine condition (Scott, 1989). The dipterocarp forest, which covers most of the country, gives way to montane forest from about 700m upwards. In the upper Temburong area, where the land rises above 1,500m, this in turn gives way to montane vegetation with stunted, gnarled trees covered with mosses (Bennett, 1991). These forests have not been widely exploited, because most people live along the coast, and most of the country's development and economy has been centred around hydrocarbon fossil fuels (Farmer *et al.*, 1986). Timber extraction for local consumption is allowed, under strict control by the Forest Department, but clear felling is prohibited (Bennett, 1991), and no timber is exported (Mittermeier, 1981).

Little coral occurs due to the turbid nature of Brunei's coastal waters and its sedimentary and mangrove-fringed coastline (UNEP/IUCN, 1988). The total extent of coralline areas within the territorial waters of Brunei is about 45 sq. km out of a continental shelf area of 9,390 sq. km. These areas are not heavily fished nor used for recreation and tourism (de Silva, 1988).

All notified protected areas are forest reserves. Although these have an important role in nature conservation, their original purpose was to protect forest resources for commercial exploitation. Forest reserves cover about 3,190 sq. km (55% total land area), of which a quarter is considered unexploitable or has been allocated for protection or conservation purposes (Farmer *et al.*, 1986). In 1986 the Town and Country Planning Department commissioned a special report on wildlife conservation and management (Farmer *et al.*, 1986) as an addendum to the Negara Brunei Darussalam Master Plan 2000. Although this report contains recommendations for the designation of specific protected areas and draft legislation, there is little evidence as yet of its implementation. In 1989, the Forest Department formulated proposals for forest development and conservation in a Forestry Strategic Plan. This Plan may prove to be one of the major vehicles for consolidating the nation's protected area system, as well as maintaining the sustained use of forest for production purposes. The situation is somewhat confused, however, as both the Brunei Museum and the Town and Country Planning Department of the Ministry of Development also claim responsibility for planning and managing protected areas other than forest reserves, which are under the Forest Department (A.S.D. Farmer, pers. comm., 1991).

In general, the major habitat types are well represented within the protected areas system (MacKinnon and MacKinnon, 1986), except swamp forest (Bennett, 1991). It has been recommended, however, that the forest reserve network be developed, with boundaries of certain reserves altered and new reserves established (Anderson and Marsden, 1988). Additional recommendations include the establishment of Batu Apoi Forest Reserve as a national park (Bennett *et al.*, 1984; Farmer *et al.*, 1986; IUCN, 1985). Six key critical sites merit priority attention and continued protection: the primary inland forests of Ulu Temburong (Batu Apoi); the peat swamp forests of the Belait River system

(Ulu Mendaram); the mangroves of Brunei Bay; Tasek Merimbun (an area of freshwater and peat swamps); the Bukit Batu-Sungei Ingei area, which is contiguous with Gunung Mulu National Park in Sarawak; and the coastal kerangas forests (Bennett, 1991).

Addresses

Brunei Museum (Director), Kota Batu, Bandar Seri Begawan

Forest Department (Director of Forestry), Ministry of Industry and Primary Resources, Bandar Seri Begawan 2067 (Tel: 2 22450/22687; Fax: 2 41012)

Fisheries Department (Director), Ministry of Development, PO Box 2161, Bandar Seri Begawan (Tel: 2 42067/44131)

Town and Country Planning Department, Ministry of Development, PO Box 2204, Bandar Seri Begawan (Tel: 2 24591)

Brunei Nature Society, PO Box 2241, Bandar Seri Begawan

Panaga Natural History Society, c/o Brunei Shell Petroleum Company Sdn Bhd, Seria (Tel: Seria 2623)

References

Anderson, J.A.R. and Marsden, D. (1988). Brunei forest resources and strategic planning study. 2 volumes. Report to the Government of His Majesty the Sultan and Yang Di-Pertuan of Negara Brunei Darussalam. Unpublished.

Bennett, E. (1991). Brunei. In: N.M.Collins, J.A.Sayer, T.C.Whitmore (Eds), *The conservation atlas of tropical forests. Asia and the Pacific.* The Macmillan Press Ltd, London. Pp. 98-102.

Bennett, E.L., Caldecott, J.D. and Davison, G.W.H. (1984). A wildlife study of Ulu Temburong, Brunei. Forest Department, Kuching and Universiti Kebangsaan, Malaysia. Unpublished report. 61 pp.

Farmer, A.S.D., Caldecott, J.O., Phillips, A., Prince, G. and Thomson, N. (1986). *Negara Brunei Darussalam masterplan. Special report: wildlife conservation and management.* 4 volumes. Huszar Brammah and Associates/Department of Town and Country Planning, Bandar Seri Begawan.

IUCN (1985). *The Corbett Action Plan for protected areas of the Indomalayan realm.* IUCN, Gland, Switzerland and Cambridge, UK. 23 pp.

Mittermeier, R.A. (1981). Brunei protects its wildlife. *Oryx* 16: 67-70.

MacKinnon, J. and MacKinnon, K. (1986). *Review of the protected areas system in the Indo-Malayan realm.* IUCN, Gland, Switzerland, and Cambridge, UK/UNEP, Nairobi, Kenya. 284 pp.

Macklin, D. (1991). Joining hands in jungle science. *Geographical Magazine* 63(1): 44-46.

Scott, D.A. (Ed.) (1989). *A directory of Asian wetlands.* IUCN, Gland, Switzerland and Cambridge, UK. 1181 pp.

de Silva, M.W.R.N. (1988). The coastal environmental profile of Brunei Darussalam. *Tropical Coastal Area Management* 3: 1-4.

UNEP/IUCN (1988). *Coral reefs of the world. Volume 2: Indian Ocean, Red Sea and Gulf.* UNEP Regional Seas Directories and Bibliographies. IUCN, Gland, Switzerland and Cambridge, UK/UNEP, Nairobi, Kenya. 440 pp.

ANNEX
Definitions of protected area designations, as legislated, together with authorities responsible for their administration

Title: Wildlife Protection Act

Date: 1978

Brief description: Provides inter alia for the establishment of wildlife sanctuaries by decree of the Sultan

Administrative authority: Brunei Museum (Director)

Designations:

Wildlife sanctuary May include the whole part of a forest reserve or a protected forest, in which case nothing herein contained shall prohibit or restrict the management of the forest reserve or protected forest.

Source: Original legislation

Title: Forest Enactment

Date: 1934

Brief description: Provides for the protection, management and exploitation of forest reserves, and all associated forest produce (including guano of birds and bats, peat, rock, sea sand, river sand, sea shells, shell sand, surface soil, trees and all their parts or produce, silk, cocoons, honey and wax, edible birds' nests, timber, firewood, charcoal, getah, getah taban leaves, wood oil, bark, extracts of bard, damar and attap)

Administrative authority: Forest Department (Director of Forestry)

Designations:

Conservation forest To conserve adequate areas of all existing principal terrestrial habitats and ecosystems

Production forest To provide stands of timber of economic potential for exploitation on a sustainable basis

Protection forest To safeguard water catchment areas and very steep terrain against erosion

Recreation forest To provide areas suitable for recreation

Source: Original legislation

SUMMARY OF PROTECTED AREAS

Map ref.	*National/international designation* Name of area	IUCN management category	Area (ha)	Year notified
	Forest Reserves			
1	Andulau (Conservation)	IV	1,309	1940
2	Batu Apoi Conservation)	IV	46,210	1991
3	Labi Hills (Conservation/Protection)	IV	64,283	1947
4	Ladan Hills (Protection)	IV	10,565	1950

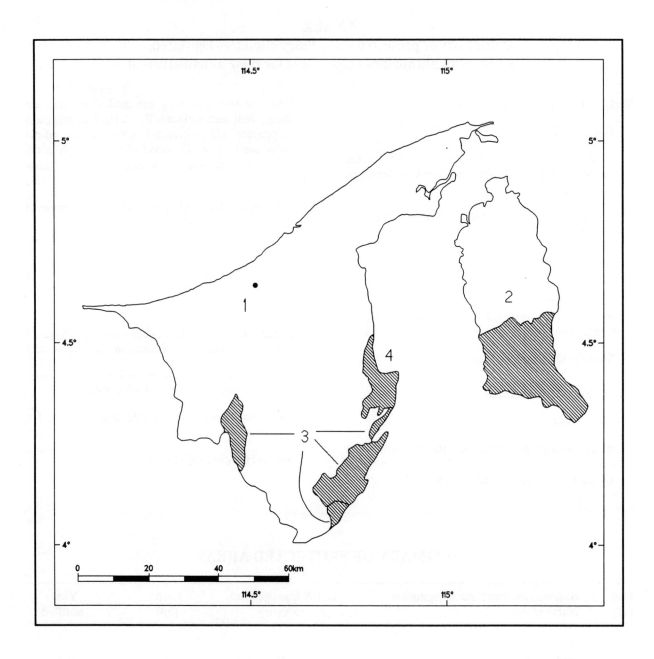

Protected Areas of Brunei Darussalam

CAMBODIA

Area 181,035 sq. km

Population 8,246,000 (1990)
Natural increase: 2.20% per annum

Economic Indicators
GNP: Not available
GDP: Not available

Policy and Legislation There is currently no protected areas legislation. During the French colonial period a number of wildlife reserves were established and, although these sites are still recognised by the current government, they are not properly mapped, demarcated, protected or managed (MacKinnon and MacKinnon, 1986). Royal Ordinance No. 24 of 26 January 1940 regulated hunting and afforded protection to a number of species. Prakas No. 194 (1960) was a temporary measure forbidding the hunting of large wild animals, whilst in the same year a number of faunal reserves were established (Anon., 1968). However, these regulations have lapsed with subsequent changes in government. There are currently 172 production forest reserves covering 3,875,000ha, six forest reserves for wildlife protection and a single national park of 10,717ha. In addition, there are 1,080 sites which encircle ancient monuments (Sarun, 1985). However, the legal basis of these sites is not known.

International Activities Cambodia does not participate in any international conventions or programmes concerned with protected areas, such as the Convention Concerning the Protection of the World Cultural and Natural Heritage Convention (World Heritage Convention), the Convention on Wetlands of International Importance Especially as Waterfowl Habitat (Ramsar Convention) and the Unesco Man and the Biosphere (MAB) Programme.

In 1986 Cambodia signed international agreements with IUCN and her neighbours Laos and Vietnam to cooperate in projects to conserve kouprey *Bos sauveli* (E) and to protect waterbirds in the Mekong wetlands. This may lead to transfrontier reserves comprising contiguous territory in all three countries (MacKinnon and Stuart, 1989)

Administration And Management Prior to 1975, wildlife protection was the responsibility of the Water, Forest and Hunting Service of the Ministry of Agriculture. In 1971 the Service had a budget of US$ 4,200 for fauna protection and US$ 5,000 for reafforestation (McNeely, 1975). At present, the Forestry Department includes divisions of conservation, silviculture, reafforestation and plantations, forest management, timber technology and forest research (Sarun, 1985). However, the number of professional staff in all fields, including forestry, was severely depleted during the 1975-1979 premiership of Pol Pot.

A lack of professional staff thereafter hampered forestry and conservation (MacKinnon and MacKinnon, 1986). During the initial post-war reorganisation the Directorate of Forests and Hunting (Direction des Forêts et Chasse) was not operationally effective due to lack of management, training and support staff. However, accelerated training of new staff, promotion of forest officers and forest-hunting staff has brought the number of personnel to 235.

Systems Reviews Cambodia is situated in tropical Mainland South-east Asia, straddling the Mekong River between Thailand and Viet Nam, with Laos to the north-east. The greater part of Cambodia comprises the plain of the lower Mekong valley, with the western slopes of the Annamite Chain to the east lying along the international border with Viet Nam, and the isolated highlands of the Elephant and Cardamom mountains in the west. The country has a short coastline of only 435km. The climate is dominated by the south-west and north-east monsoons; the south-west monsoon lasting from May to October and the north-east from November to March. The mean annual rainfall is 1200mm-1875mm, but rainfall of up to 3000-4000mm may be experienced in the south-west.

The Mekong River runs southwards across the plains, its delta lying in Viet Nam. Part of the western plains are occupied by the shallow Tonlé Sap (Great Lake), which flows into the Mekong throughout most of the year, but which usually floods back from the main river during the rainy season, becoming a vast storage reservoir. The Cardamom Mountain range dominates the south-west, rising to 1,563m. In the south and south-east are low plains bordering the Mekong River, extensive areas of which are seasonally flooded. North of Tonlé Sap, the area leading to the border with Laos and Thailand consists of rolling savanna country with some open grassland areas of deciduous forest. A review of wetlands important for conservation is given in Scott (1989).

The most recent information on forest resources derives from FAO/UNEP (1981) which was published in FAO (1988). Closed broadleaved forests were estimated to cover an area of 71,500 sq. km. A 1990 estimate indicates 66,500 sq. km of forest cover remains. The 1981-1985 rate of deforestation is estimated at 300 sq. km annually and most forest is lost to shifting cultivation (Collins *et al.*, 1991).

Cambodia suffered prolonged periods of warfare in the 1970s, during which agricultural output fell and extensive areas of forest were destroyed by bombing and defoliation. From 1970 to 1975 a civil war was fought and the country slipped from a position of agricultural surplus to extensive famine conditions (Shawcross, 1979). In 1975 the victorious Khmer Rouge instituted a highly disciplined regime in which contacts with the outside world were severely curtailed, foreigners were expelled and towns and cities were forcibly evacuated. Until defeat in 1979 by the Vietnamese, the Khmer Rouge implemented a genocidal policy towards professional cadres, former officials and others. Consequently, all forestry and conservation activities ceased and there was a severe shortage of professional staff in all fields (MacKinnon, 1986).

The principal activities of the Directorate of Forests and Hunting are: the management and control of forests; operation of saw mills and workshops at the industrial level for export, and at the craft level for local consumption, as well as the transport of wood and wood products; and reforestation in the security zones of non-forested provinces with the establishment of experimental nurseries for forest plants. In addition to professional staff, local labour is used for maintaining nurseries and plantations. Wide-ranging reforestation is planned in other areas, based on the experience gathered from the experimental stations. The privatisation of saw mills and forest exploitation is envisaged in order to achieve effective reforestation, by releasing the Directorate of Forests and Hunting to focus its resources on protection and conservation of existing forests, and on silviculture. The Directorate also organises "tree" and "nature" days which it uses to promote environmental awareness (C. Sarun, pers. comm., 1991).

Projected activities of the Directorate for the period 1990-2000 will focus on the restoration of the five large forested regions which constitute the national forest resource. These activities will include: inventory and mapping of the areas concerned; surveys of areas that could be exploited; reforestation of areas for the production of veneer and paper and timber for both export and local consumption; boundary demarcation; commercial enrichment planting; and the development of protection, recreation and wildlife forest reserves. Development of wildlife reserves will ensure provision of water sources and habitats. Attention will also be given to the development of protected areas for the conservation of Cervidae and Bovidae species, including kouprey *Bos sauveli* (E) (C. Sarun, pers. comm., 1991).

Occupation by Viet Nam resulted in international isolation, with an embargo on aid and development assistance, and civil war continued in several areas. Although the Vietnamese forces withdrew by September 1990, hostilities continued between the administration in Phnom Penh and the three major guerilla groupings (Paxton, 1990). An United Nations plan, including a truce, the demobilisation of all four disputing factions, and an extensive role for the UN in the country's administration prior to elections, was due to be discussed during August 1991 at a meeting of the Cambodian Supreme National Council (Anon., 1991)

Addresses

Directorate of Forest and Hunting, Ministry of Agriculture, Phnom Penh

References

Anon. (1968). Conservation in Cambodia. In Talbot, L.M. and Anon. (1991). Phnom Penh prepared to compromise for peace. *Financial Times* 19 August. P. 4.

Talbot, M.H. (Ed.). *Conservation in tropical South-east Asia.* IUCN, Morges. Pp. 456-459.

Collins, N.M., Sayer, J.A. and Whitmore, T.C. (Eds) (1991). *The conservation atlas of tropical forests: Asia and the Pacific.* Macmillan Press Ltd, London. 256 pp.

Davis, S. D., Droop, S. J. M., Gregerson, P., Henson, L., Leon, C.J., Lamlein Villa-Lobos, J., Synge, H. and Zantovska, J. (1986). *Plants in danger: what do we know?* IUCN, Cambridge. P. 201

FAO (1981). *Tropical forest resources assessment project: forest resources of Tropical Asia.* FAO, Rome. 475 pp.

FAO (1988). An interim report on the state of forest resources in the developing countries. FAO, Rome. 18 pp.

MacKinnon J. (1986). Bid to save the kouprey. *WWF Monthly Report*: 91-97.

MacKinnon, J.R. and MacKinnon, K. (1986). *Review of the protected areas systems in the Indo-Malayan Realm.* IUCN, Cambridge.

MacKinnon, J.R. and Stuart, S.N. (Eds) (1989). The kouprey: an action plan for its conservation. Prepared by the Species Survival Commission of IUCN and WWF. IUCN, Gland, Switzerland and Cambridge, UK. 20 pp.

McNeely, J.A. (1975). Draft report on wildlife and national parks in the Lower Mekong Basin. United Nations Economic and Social Commission for Asia and the Pacific. Committee for the Coordination of Investigations of the Lower Mekong Basin.

Paxton, J. (Ed.) (1985). *The Statesman's yearbook.* The Macmillan Press Ltd, London. P. 259

Paxton, J. (Ed.) (1990). *The Statesman's yearbook.* The Macmillan Press Ltd, London. P. 261

Scott, D. A. (Ed.) (1989). *A directory of Asian wetlands.* Compiled for WWF/IUCN/ICBP/IWRB. IUCN, Gland, Switzerland and Cambridge, UK. Pp. 795-807.

Shawcross, W. (1979). *Sideshow: Kissinger, Nixon and the destruction of Cambodia.* London. Deutsch. 467 pp.

Sarun, C. (1985). Intervention de la délégation du Kampuchea. In: Thorsell, J.W. (Ed.), *Conserving Asia's natural heritage.* IUCN, Gland, Switzerland and Cambridge, UK. 251 pp.

INDIA

Area 3,166,830 sq. km

Population 853,400,000 (1990)
Natural Increase: 2.1% per annum

Economic Indicators
GNP: US$ 330 per capita (1988)

Policy and Legislation A commitment to protect and enhance the environment is enshrined within India's Constitution (Forty-Second Amendment) Act, 1977, as follows:

"The State shall endeavour to protect and improve the environment and to safeguard the forests and wildlife of the country." (Article 48A); and

"It shall be the duty of every citizen of India ... (g) to protect and improve the natural environment including forests, lakes, rivers and wildlife, and to have compassion for living creatures." (Article 51A).

There is no provision, however, which enables the Union to enact legislation pertaining to environmental issues that is uniformly applicable to all states and union territories. In addition to the separate federal and state jurisdiction, there exists a Concurrent List of legislative powers which includes *inter alia* forests and the protection of wild animals and birds. The Concurrent List gives over-riding power to the Union but executive authority lies with the state governments (Dwivedi and Kishore, 1984). Among the recommendations of the Tiwari Committee, a high-powered committee appointed by the government in February 1980 to suggest administrative and legislative reforms to improve environmental protection in the country, was the introduction of environment protection in the Concurrent List of the Constitution. The constitutional directives have provided a strong basis for the enactment of legislative measures for environmental protection. The need to integrate environmental considerations with economic development was explicitly articulated for the first time in the 4th Five-Year Plan, 1969-1974 (Biswas and Bannerjee, 1984).

The National Environment Policy envisages conservation and development, as well as equity among the people sharing the environment, but these tend to be mutually incompatible under much of the existing legislation (Singh, 1985). There is no statutory requirement for environmental impact assessment at present but a mechanism has been initiated whereby assessment is an integral part of the planning process, with appraisals of major projects being the responsibility of the Department of Environment, Forests and Wildlife (Dwivedi and Kishore, 1984).

The protection of wildlife has a long tradition in Indian history. Wise use of natural resources was a prerequisite for many hunter-gatherer societies which date back to at least 6000 BC. The most notable of such traditions are sacred groves, totally inviolate to any human interference, and village groves where only limited use by members of the community is permitted. Many of these are still in existence. Extensive clearance of forests accompanied the advance of agricultural and pastoral societies in subsequent millenia, but an awareness of the need for ecological prudence emerged and many so-called pagan nature conservation practices were retained (Gadgil, 1989). Among the earliest provisions for the establishment of protected areas are those codified in the *Arthasashtra, Indica* (321-300 BC), written by Kautilya, reputedly the Prime Minister of King Chandra Gupta Maurya. Prescriptions included rules for the administration and management of forests, and provisions for three classes of forests, namely those reserved for the king, those allocated for ascetics and those for the public which could be used only for hunting purposes. Kautilya is also the first-recorded person to have advocated the creation of "Abhayaranyas", or sanctuaries for wildlife. The following century, during the reign of Emperor Ashoka, the first-recorded conservation measures for wildlife were enacted, and reserves were established for wild animals (Singh, 1986; Mitra, 1989). Hindu, Moslem and, latterly, British rulers continued these traditions in subsequent centuries, setting up reserves for privileged hunting over much of India. As more and more land became settled or cultivated, so these hunting reserves increasingly became refuges for wildlife. Many of these reserves were subsequently declared as national parks or sanctuaries, mostly after Independence in 1947. Examples include Gir in Gujarat, Dachigam in Jammu & Kashmir, Bandipur in Karnataka, Eravikulum in Kerala, Madhav (now Shivpuri) in Madhya Pradesh, Simlipal in Orissa, and Keoladeo, Ranthambore and Sariska in Rajasthan. The fact that the great majority of the Indian population is vegetarian (devout Hindus and Jains) has undoubtedly helped to preserve that part of India's natural heritage which remains today (Singh, 1985; Gadgil, 1989).

Following independence, a number of states (Goa, Haryana, Himachal Pradesh, Jammu & Kashmir, Madhya Pradesh, Maharashtra, Mysore – now Karnataka, Punjab, Rajasthan and Tamil Nadu) enacted wildlife preservation acts, while others (Assam, Uttar Pradesh and West Bengal) continued to enforce the government of India Wild Birds and Animals Protection Act, 1912. National park acts were enacted by a few states but only five national parks were established under these acts, namely Kanha, Bandhavgarh and Shivpuri in Madhya Pradesh, Tadoba in Maharashtra and Hailey (now Corbett) in Uttar Pradesh. The Hailey National Park Act of 1936 was probably the first law in India intended for the exclusive protection of wildlife and its habitat (IBWL, 1970; Kothari *et al.*, 1989).

A National Wildlife Policy for India was first formulated by an Expert Committee of the Indian Board for Wildlife in 1970 (IBWL, 1970). A major aim was to reserve at least 4% of the total land area for wildlife, both plants and animals – an objective which has recently been exceeded. Much of this policy was subsequently enshrined in the Wild Life (Protection) Act, 1972. The Act provides the necessary uniform legislation for the establishment of protected areas and has since been adopted by all states and union territories. Provisions include *inter alia* the constitution of state wildlife advisory boards and the notification of sanctuaries, national parks, game reserves and closed areas by state governments (see Annex). Setting up a sanctuary involves settling all private rights, either allowing them to continue or acquiring them after adequate compensation. Only a completely unencumbered area, in which all rights have become vested in the government, may be declared a national park. Once established, its boundaries may not be altered except through a resolution passed by the state legislature.

The basis to present nature conservation policy in India is the National Wildlife Action Plan (Department of Environment, n.d.). Drawing on the World Conservation Strategy launched by IUCN in March 1980, the Bali Action Plan arising from the 3rd World Parks Congress in October 1982, and the World Charter for Nature proclaimed by the United Nations General Assembly in October 1982, it was adopted by the government of India in October 1983 on the recommendation of the Indian Board for Wildlife. Objectives include the establishment of a representative network of protected areas and development of appropriate management systems (together with the restoration of degraded habitats), and the adoption of a National Conservation Strategy, which is now being formulated.

The Indian Forest Act, first enacted in 1865 and succeeded by a more comprehensive act in 1927, provides significant protection to wildlife through the provision of reserved and protected forests which may be established in any forest or waste lands belonging to the government, or over which the government has proprietary rights (see Annex). Some states enacted their own forest legislation after the National Forest Policy was announced in 1952, while others amended the Act to suit their own requirements. The Act also makes provision for the rights of government over land constituted as reserved or, in the case of a few states, protected forest to be assigned to village communities. The Forest (Conservation) Act was promulgated in 1980 (and later amended in 1988) to stem the indiscriminate diversion of forest land to non-forestry purposes. Under

this Act, no forest land can be de-reserved or diverted to non-forestry purposes without the approval of Central Government. Enforcement of this Act has had a salutary effect, the annual rate of diversion of forests having been reduced from about 140,000ha pre-1980 to 6,500ha in the 1980s (Panwar, 1990). Other initiatives include a moratorium imposed since 1983 on the felling of trees at altitudes of 1,000m and above (Ministry of Environment and Forests, 1985). The 1952 National Forest Policy was superceded by a new National Forest Policy (Resolution No. 3-1/86 FP) on 7 December 1988. Objectives include the maintenance of environmental stability, conserving the nation's natural heritage by preserving remaining natural forests, preventing soil erosion and denudation of catchment areas, and creating a mass people's movement with the involvement of women to achieve such aims and minimise pressure on existing forests. A target has been set for one-third of the total land area of the country to be under forest, as originally stipulated in the 1952 National Forest Policy, but in the hills and mountainous regions the target is two-thirds. In addition, forest management must provide "corridors" to link protected areas and thereby maintain genetic continuity between artificially separated subpopulations of migrant wildlife. Also, full protection of the rights and concessions of tribals and poor people dependent on forests is advocated (Government of India, 1988).

A selected list of other environmental legislation is given in *A Second Citizen's Report* (CSE, 1985). Of particular note is the Environment (Protection) Act, 1986, which provides a focus for environmental issues in the country and plugs loopholes in the existing legislation (Ministry of Environment and Forests, 1987a).

Inadequacies in the existing nature conservation legislation are reviewed by Dwivedi and Kishore (1984) and by Singh (1986). The recognition of only wild animals and birds, without reference to plants, is an important omission from both the Wild Life (Protection) Act and the Constitution. Uniform and comprehensive forest legislation is urgently needed, with emphasis on forest conservation rather than the existing system of resource exploitation. Both acts are currently under revision.

International Activities India ratified the Convention concerning the Protection of the World Cultural and Natural Heritage (World Heritage Convention) on 14 November 1977. Five natural sites have been inscribed on the World Heritage List to date. India acceded to the Convention on Wetlands of International Importance especially as Waterfowl Habitat (Ramsar Convention) on 1 October 1981, at which time two sites were designated as wetlands of international importance. Four more sites were designated on 23 March 1990. Participation in the Unesco Man and Biosphere Programme began in 1972 with the constitution of the Indian National MAB Committee. The Indian Biosphere Reserves Programme will operate within the ambit of existing state and federal legislation; separate legislation for biosphere reserves is not envisaged (Ministry of

Environment and Forests, 1987b). Thirteen potential biosphere reserves have been identified, of which Nilgiri Biosphere Reserve is the first to have been established but has yet to be nominated for inclusion in the international biosphere reserve network.

Administration and Management The Department of Environment, Forests and Wildlife within the Ministry of Environment and Forests was created in September 1985. It serves as the administrative focus within central government for planning, promoting and coordinating environmental and forestry programmes, including the preservation and protection of wildlife and the biosphere reserve programme (Ministry of Environment and Forests, 1987a). Previously, wildlife management was the responsibility of the Forest Department within the Ministry of Agriculture. Following recommendations made by the Tiwari Committee, a separate Department of Environment was constituted on 1 November 1980 to which wildlife management was transferred in September 1982. This Department became part of a new Ministry of Environment and Forests, constituted under Presidential Notification No. 74/2/1/85-Cab. dated 4 January 1985. At that time, the Ministry consisted of two departments, namely Environment and Forests & Wildlife, but these were merged later that year (Government of India, n.d.; Ministry of Environment and Forests, 1986, 1987a). Departments of Environment have also been set up in a number of states (Biswas and Bannerjee, 1984).

Wildlife, together with forestry, has traditionally been managed under a single administrative organisation within the forest departments of each state or union territory, with the role of central government being mainly advisory. There have been two recent developments. First, the Wild Life (Protection) Act has provided for the creation of posts of chief wildlife wardens and wildlife wardens in the states to exercise statutory powers under the Act. This has largely been responsible for the creation of wildlife wings within each state headed by a chief wildlife warden. Under this Act, it is also mandatory for the states to set up state wildlife advisory boards. Secondly, the inclusion of protection of wild animals and birds in the Concurrent List of the Constitution, has provided the Union with some legislative control over the states in the conservation of wildlife (Pillai, 1982). Guidelines specifying that the management of protected areas should be under the remit of the wildlife wings were issued by central government in 1975, but progress in implementing them was slow. This prompted central government to threaten cessation of financial assistance to states which had not transferred protected areas to their respective wildlife wings. The situation has since improved, all states and union territories with national parks or sanctuaries having set up wildlife wings. However, by 1987, three states (Andhra Pradesh, Punjab and Tamil Nadu) had not transferred control over any protected area to their respective wildlife wings, while eight others (Bihar, Gujarat, Karnataka, Maharashtra, Madhya Pradesh,

Orissa, Uttar Pradesh and West Bengal) had transferred only some of their national parks and sanctuaries (Kothari *et al.*, 1989; Ministry of Environment and Forests, 1987a). Management of protected areas in individual states and union territories is summarised by Pillai (1982).

The Indian Board for Wildlife, under the chairmanship of the Prime Minister, is the main advisory body to the government of India on wildlife matters. First constituted in 1952 as the Central Board for Wildlife, it was later redesignated as the Indian Board for Wildlife. Among its various achievements, it has been instrumental in the formulation of the Wild Life (Protection) Act, the establishment of many new protected areas (including tiger reserves), and in the formation of separate departments for wildlife conservation both at the Centre and in the states. State wildlife advisory boards have been constituted under statutory provisions of the Wild Life (Protection) Act to advise state governments (Saharia and Pillai, 1982).

The administration of Project Tiger, initiated as a Central Sector Scheme in 1973, is overseen by a Steering Committee headed by the Minister of State for Environment and Forests. The Director is responsible for coordinating the project within central government. Execution of the project is the responsibility of the chief conservators of forests in the relevant states, with tiger reserves managed by field directors. The project's present status is that of a centrally-sponsored scheme, with costs shared equally between the union and state governments (Panwar, 1982).

Training in wildlife management is undertaken at the Wildlife Institute of India, which became an autonomous institution of the Ministry of Environment and Forests with effect from 1 April 1986. Its objectives include training in protected areas management, research and extension services, building a computerised wildlife information system, and providing advisory services. The Institute offers a one-year post-graduate diploma course for forest officers, a three-month certificate course for forest rangers and an M.Sc. Wildlife Biology course (WII, 1987).

There are many non-governmental organisations involved in nature conservation. The oldest is the Bombay Natural History Society, established in 1883 and currently comprising about 3,000 members. Whereas work undertaken in its early years was concentrated on collecting, identifying and documenting India's flora and fauna, the emphasis has shifted to conservation-oriented research in recent decades, particularly threatened species and habitats. Long-term field studies are based in a number of protected areas, such as Keoladeo National Park (Rajasthan), Mudumalai Sanctuary (Tamil Nadu) and Dalma Sanctuary (Bihar). The Society's *Journal of the Bombay Natural History Society* is widely circulated in India and overseas.

World Wide Fund for Nature-India (formerly World Wildlife Fund-India), established in 1969, has quickly developed to become the largest non-governmental nature conservation organisation in India, with 20 branches and a total staff of about 130. Its activities include ecological research and surveys, policy reviews, conservation projects, nature education and responsibility to the Ministry of Environment and Forests for environmental information relating to federal and state legislatures, NGOs and the media. Two recent initiatives underway are the establishment of the Indira Gandhi Conservation Monitoring Centre and the launch of a Community Biodiversity Conservation Movement.

The Indian National Trust for Art and Cultural Heritage, constituted in January 1984, has rapidly emerged as one of the most progressive and influential conservation bodies in India. It has 150 regional chapters spread over India's 32 states and union territories, the ultimate goal being to establish a chapter in each district. Its aim is to develop an awareness among the public of India's cultural and natural heritage and to promote its conservation. The Trust set up a Natural Heritage Cell in May 1985 which promotes land-use planning and management in areas of critical conservation importance.

The Centre for Science and Environment aims to publicise topical environmental issues, as well as promoting people's participation in environmentally-sound rural development. Its findings are documented in its citizens' reports, two of which have been published to date (CSE, 1982, 1985).

Other national conservation organisations include the Wildlife Preservation Society of India, founded in 1958 and publisher of the journal *Cheetal*, and the Indian Society of Naturalists, which publishes *Environmental Awareness*. Details of some 700 environmental non-governmental organisations can be found in a directory produced by WWF-India.

Protected areas are often poorly managed, with little consideration given to the local people living in and around them (Singh, 1986). The legal, ecological and management status of protected areas has recently been examined by the Environment Studies Division, Indian Institute of Public Administration (Kothari *et al.*, 1989). The study was commissioned by the National Committee on Environmental Planning in 1984 and sponsored by the Ministry of Environment and Forests. The survey shows, for example, that only 40% of 52 national parks and 8% of 209 sanctuaries have completed legal procedures for their establishment. Only 43% of national parks and 28% of sanctuaries surveyed have management plans; in many cases they are cursory documents and have never been approved by the state government. Many of the deficiencies in protected areas management reflect a lack of commitment of resources on the part of state governments. For example, in 1983-84, expenditure on protected areas was 1.5% of forest department budgets. The Environmental Studies Division is currently engaged in a series of in-depth studies of management issues in a selection of India's major protected areas.

Systems Reviews India is a nation of extraordinary diversity, the seventh largest and second most populous in the world. Its relief can be conceptualised in terms of three well-defined regions: the Himalayan mountain system along its northern margin; the Gangetic Plain, which extends some 2,400km from Assam in the east to the Punjab in the west and southwards to the Rann of Kutch in Gujarat; and the Deccan Plateau which is flanked on either side by the Western Ghats and Eastern Ghats (Mani, 1974). Its rich diversity of ecosystems, which range from tropical rain forests to deserts, and from marine and coastal systems to high mountains, support an estimated 5-8% of the world's known flowering plant and animal species, of which a significant proportion are endemic (Gadgil and Meher-Homji, 1986a). Important centres of biological diversity, particularly for plants, are the Western Ghats, north-eastern India and the Andaman and Nicobar Islands (Nayar, 1989).

Forest once covered most of India but much of it has been destroyed or severely degraded as a result of human population pressures, particularly in the fertile lowlands which are among the most densely populated areas in the world. For example, 4.1 million hectares of forest were cleared mainly for agriculture between 1951 and 1980 (Singh, 1986; Vedant, 1986). Probably less than 1% of the total land area is covered by primary forest (Mani, 1974). Forests are estimated to have covered 64.01 million hectares in 1985-1987, or 19.5% of total land area comprising 11.5% dense forest (at least 40% crown density), 7.8% open forest (at least 10% crown density) and 0.1% mangrove forest (FSI, 1989).

The total area of wetlands (excluding rivers) in India is 58,286,000ha, or 18.4% of the country, 70% of which comprises areas under paddy cultivation. A total of 1,193 wetlands, covering an area of 3,904,543ha, were recorded in a preliminary inventory coordinated by the Department of Science and Technology, of which 572 were natural. In a recent review of India's wetlands, 93 are identified as being of conservation importance (Scott, 1989).

Coral reefs occur along only a few sections of the mainland, principally the Gulf of Kutch, off the southern mainland coast, and around a number of islands opposite Sri Lanka. This general absence is due largely to the presence of major river systems and the sedimentary regime on the continental shelf. Elsewhere, corals are also found in the Andaman, Nicobar and Lakshadweep groups, although their diversity is reported to be lower than in south-east India (UNEP/IUCN, 1988).

Historically, conservation in India stems mainly from the creation of large forest reserves in the late 19th and early 20th centuries to safeguard timber, soil and water resources. Superimposed on this network of reserved forests has been a much smaller number of national parks and sanctuaries where the value of the biological resource has persuaded authorities to reduce the level of forest product utilisation (Rodgers, 1985). Both the adoption of a National Policy for Wildlife Conservation in 1970 and the enactment of the Wild Life (Protection) Act in 1972 lead to significant growth in the protected areas network, from 5 national parks and 60 sanctuaries in 1960 to 69 and 410, respectively, in 1990 (Panwar, 1990). The network was further strengthened by a number of national conservation projects, notably Project Tiger, initiated on 1 April 1973 by the Government of India with support from WWF (IBWL, 1972; Panwar, 1982), and the Crocodile Breeding and Management Project, launched on 1 April 1975 with technical assistance from UNDP/FAO (Bustard, 1982). Project Tiger has been acclaimed as an internationally outstanding conservation success story. The number of tiger reserves has increased from 9 (covering a total area of 13,723 sq.km) at the time of its launch to 18 (covering 28,017 sq.km) by 1990 (Panwar, 1990). Its achievements and shortcomings are reviewed by Panwar (1984) and Singh (1986). The government of India subsequently initiated a Snow Leopard Conservation Scheme along the lines of Project Tiger, but with the emphasis on resolving conflicts between wildlife and resident human populations without having to relocate villagers from within protected areas (Ministry of Environment and Forests, 1987a). This has yet to be implemented.

In fulfillment of one of the major objectives of the National Wildlife Action Plan (Department of Environment, n.d.), the existing protected areas system has been reviewed and plans formulated for a comprehensive network which covers the full range of biological diversity in the country (Rodgers and Panwar, 1988). In mid-1987, there were 426 national parks and sanctuaries covering a combined area of 109,652 sq.km, or 3.3% of the country. Major gaps in the network include inadequate representation (1%) of the following biotic provinces: Ladakh, South Deccan, Gangetic Plains, Assam Hills and Nicobars. Recommendations in the systems plan bring the total number of protected areas to 651, covering 151,342 sq. km or 4.6% of the country. Particular emphasis is given to protecting sites of high species diversity and endemism, as well as ecologically fragile areas. This plan for a national network of protected areas has been accepted by central government and commended to the states for implementation. Proposals in the plan supercede previous recommendations emanating from the *Corbett Action Plan* (IUCN, 1985) and the *IUCN systems review of the Indomalayan region* (MacKinnon and MacKinnon, 1986). They also endorse the earlier work of Gadgil and Meher-Homji (1986b), in which representation of the main vegetation types of India within the protected areas network is assessed. A number of states are now implementing many of the recommendations made in the systems plan, to the extent that total coverage by national parks and sanctuaries is nearly 4%. There are financial provisions under the 8th Five-Year Plan (1991-1995) to enhance the protected areas network in accordance with the systems plan, and to improve management of protected areas and promote ecodevelopment in areas surrounding them, with emphasis on at least 20 important national parks and sanctuaries (Panwar, 1990).

Wildlife conservation in India has met with tremendous success but protected areas management is beset with problems of inadequate fund allocation, a reluctance on the part of the states to establish national parks and sanctuaries because the land is lost forever for other uses (moreover, industries are not permitted within 30km of the boundary of a sanctuary), insufficient magisterial powers for wildlife staff to deal with poachers, difficulties of communication in often remote areas, and lack of trained manpower at lower levels (Chandha, 1989).

Other Relevant Information

The following resumé of states and union territories is based on a review by Pillai (1982), unless otherwise indicated. Conservation priorities for individual states and union territories are identified elsewhere by Rodgers and Panwar (1988).

Andaman and Nicobar Islands Union Territory (8,249 sq. km) There are more than 325 islands (21 inhabited) in the Andamans and more than 24 (13 inhabited) in the Nicobars, covering some 6,408 sq. km and 1,841 sq. km, respectively (Saldanha, 1989). Based on Landsat imagery for 1985-1987, the total area of forest is estimated to be 7,624 sq. km (FSI, 1989). Notified forests cover 7,144 sq. km. In the case of Little Andaman, reserved forest covers 92% (671 sq. km) of the island (Saldhana, 1989). Much of the forest, perhaps 70% in the case of the Andamans, is degraded, secondary growth. The forests are unique in almost the whole of South-east Asia, with a protective belt of mangroves surrounding the islands, or in some areas giant evergreen forest growing right down to the sea. Inland from the mangrove and littoral forests are evergreen dipterocarp forests, within which occur patches of moist deciduous and semi-evergreen forest. Higher slopes and ridges support moist or dry deciduous forests (Whitaker, 1985). The total area of mangroves is estimated to be 1,150 sq. km, of which 500 sq. km occur in the Andamans (Saldanha, 1989).

The Wild Life (Protection) Act, 1972 and wildlife rules have been enforced since 1973 and 1974, respectively. The Chief Conservator of Forests is designated as the Chief Wildlife Warden. A Wildlife Advisory Board has been constituted. The status of conservation areas is reviewed by Nair (n.d.) and Whitaker (1985).

Andhra Pradesh State (276,820 sq. km) Forests cover 47,911 sq. km (FSI, 1989) and are mostly of the tropical dry deciduous type.

The Wild Life (Protection) Act, 1972 has been enforced in the state since 1973 and the rules framed under this

Act have been operative since 1974. A State Wildlife Advisory Board was constituted in 1976, but this body has not been very active. A separate wildlife wing under an Additional Chief Conservator of Forests (Chief Wildlife Warden) has been created. The management of sanctuaries has not yet been transferred to the wildlife wing. For implementation of the Wild Life (Protected) Act, 1972, all territorial divisional forest officers have been delegated the powers of wildlife wardens. The appointment of honorary wildlife wardens has also been initiated to augment conservation efforts.

Local non-governmental organisations include the Andhra Pradesh Natural History Society.

Arunachal Pradesh State (83,578 sq. km) Forests cover 68,763 sq. km (FSI, 1989). According to Mehta (n.d.), the total forest area is 51,540 sq. km, of which 12,606 sq. km is reserved forest, 251 sq. km *anchal* reserved forest, and 7 sq. km protected forest.

The Wild Life (Protection) Act, 1972 came into force on 15 May 1973 (Notification No. G.S.R. 272(E)). There are six wildlife divisions under a Conservator of Forests, *ex officio* Chief Wildlife Warden (Pillai, 1982). Each national park and sanctuary is under the charge of a Divisional Forest Officer (Mehta, n.d.).

Assam State (78,525 sq. km) Forests cover 26,058 sq. km (FSI, 1989) and are mostly of the tropical evergreen type.

Little was done to protect wildlife and its habitat in Assam prior to 1930, other than designate Kaziranga and Manas as game sanctuaries to protect dwindling populations of Indian rhinoceros. Measures had already been taken to protect the elephant, in accordance with the Elephant Preservation Act, 1879, and in the early 1930s the Elephant Hunting Rules became law. Later, in 1938, the new Shooting Rules for the Preservation of Wild Life in Reserved Forests came into force, with limits and royalties set for each type of game shot by licence-holders. During World War II, large numbers of military were stationed all over the province, with consequent indiscriminate shooting of wildlife. Early in 1949, at the invitation of the Assam Government, a survey of the status of the wildlife was carried out and recommendations made by the Bombay Natural History Society (Gee, 1950).

The Assam National Park Act was introduced in 1969, prior to the adoption of the Wild Life (Protection) Act, 1972, which has been enforced since 1976. There are two wildlife divisions, each under a Deputy Conservator of Forests, headed by a Conservator of Forests. The Chief Conservator of Forests is designated as Chief Wildlife Warden for the purpose of implementing the Wild Life (Protection) Act. A State Wildlife Advisory Board was constituted in 1977. Honorary wildlife wardens have been appointed. Assam has its own State Department of the Environment.

Bihar State (173,875 sq. km) Forests cover 26,934 sq. km (FSI, 1989). Moist deciduous and dry deciduous forest are predominant.

The Wild Life (Protection) Act, 1972 has been enforced since 1 February 1973 (Notification No. GSR 40(E)). The Chief Wildlife Warden is an officer of the rank of Conservator of Forests working within the Forest Department.

Chandigarh Union Territory (114 sq. km) Forests cover 8 sq. km (FSI, 1989).

The Wild Life (Protection) Act, 1972 has been enforced since 1977. A Wildlife Advisory Board has been constituted. The role of Wildlife Warden is assumed by a Forest Officer.

Dadra and Nagar Haveli Union Territory (491 sq. km) Forests cover 205 sq. km (FSI, 1989).

The Wild Life (Protection) Act, 1972 has been enforced since 1973. There is no separate wildlife wing but a Deputy Conservator of Forests functions as the Chief Wildlife Warden. There are no protected areas.

Daman and Diu Union Territory (112 sq. km) Forests cover 2 sq. km (FSI, 1989). No protected areas exist at present, but there are plans to develop a sanctuary for tourism puposes (Rodgers and Panwar, 1988).

Delhi Union Territory (1,485 sq. km) Forests cover 22 sq. km (FSI, 1989).

The Wild Life (Protection) Act, 1972 has been enforced since 1 June 1973 (Notification No. GSR 299 (E)). Wildlife rules were enforced in the same year. There is no separate wildlife wing. The Deputy Commissioner of Delhi is designated as the Chief Wildlife Warden.

Goa State (3,698 sq. km) Forests cover 1,300 sq. km (FSI, 1989).

The Wild Life (Protection) Act, 1972 and wildlife rules have been enforced in the state since 1973 and 1977, respectively. The Conservator of Forests is designated as the Chief Wildlife Warden. A Wildlife Advisory Board has been constituted.

Gujarat State (195,985 sq. km) Forests cover 11,670 sq. km (FSI, 1989). Mostly open dry deciduous and tropical thorn forests are predominant.

Notified forests extend over 19,657 sq. km of the state (Ministry of the Environment and Forests, 1986). Alarmed by the rapid denudation of the state's forests, the Forest Department began a social forestry programme in 1970 to meet fuel and fodder requirements. Its success led to the formulation of a community forest project in 1980 with backing from the World Bank (Ranganathan, 1981).

The Wild Life (Protection) Act, 1972 has been enforced since 1 February 1973 (Notification No. GSR 42(E)). The Gujarat Wild Life (Protection) Rules were framed in 1974 and these are under revision. A State Wild Life Advisory Board was constituted in 1973. A separate wildlife wing, now headed by a special Chief Conservator of Forests, was created in 1970. Honorary wildlife wardens have been appointed in each district. In addition, all district superintendents of Police have been appointed wildlife wardens to help implement wildlife protection measures. Training courses in wildlife management are run from the Gujarat Forest Rangers' College, Rajpipla (Rashid, n.d.).

Local non-government organisations include the Wild Life Preservation Society of Bhavnagar.

Haryana State (44,220 sq. km) Forests cover only 563 sq. km (FSI, 1989). Sal and tropical dry deciduous forests predominate. A pilot project to control flash-flooding and erosion in the Siwalik Hills, through construction of check-dams, has met with such success that it is being extended to hundreds of villages in the area, with financial support from central government and the World Bank (Sharma, 1986).

The Wildlife Life (Protection) Act, 1972 and wildlife rules have been enforced from 12 March 1973 (Notification No. GSR 63 (E)) and 1974, respectively. The wildlife wing is headed by an Additional Chief Conservator of Forests who is the Chief Wildlife Warden. A State Wildlife Advisory Board has been constituted and is an active body. No honorary wildlife wardens have been appointed to date. All divisional forest officers with territorial responsibilities have been appointed wildlife wardens.

Himachal Pradesh State (55,673 sq. km) Forests cover 13,377 sq. km (FSI, 1989) and are mostly of the sub-tropical, temperate and sub-alpine types. Reserved forest covers 1,896 sq. km and protected forest 33,350 sq. km (of which 31% is demarcated and the rest undemarcated) (DFFC, 1990). From an examination of forest department records, it seems likely that forest cover has not changed greatly over the past 50 years, despite periods of rapid forest destruction. More likely to account for accelerating siltation rates and flooding along Himalayan rivers is the destruction of the forest understorey as a result of overgrazing by domestic livestock (Gaston, 1983).

The Department of Forest Farming and Conservation is headed by a Principal Chief Conservator of Forests. In 1990 the total staff complement was 6,627, of which 255 were gazetted officers. Total revenue in 1986-1987 was Rs 213.7 million and expenditure Rs 356.4 million (DFFC, 1990).

The Wildlife (Protection) Act, 1972 has been enforced since 2 April 1973 (Notification No. G.S.R. 19(E)). The State Wildlife Advisory Board was constituted in 1975 but it is not active. A separate wildlife wing, headed by a Chief Wildlife Warden of the rank of Additional Chief Conservator of Forests, has been created but it is not fully fledged because the administration of protected areas is still vested with territorial staff. There are four wildlife

divisions, each headed by a divisional forest officer (Arya, n.d.). Revenue in 1987-1987 totalled Rs 29,000 and expenditure Rs 10.5 million, or 2.9% of total expenditure within the Department (DFFC, 1990). The government of Himachal Pradesh has banned the commercial felling of trees within national parks and 21 of 29 sanctuaries. Legal procedures have been completed in the case of only two sanctuaries (Bandli and Shikari Devi). Neither of the two national parks have been finally notified. Further details of the management status of protected areas are given by Singh *et al.* (1990).

Jammu & Kashmir State (138,942 sq. km) Forests, predominantly of temperate and sub-tropical types, cover 20,182 sq. km or 15% of the state (52% if the cold deserts of Ladakh and Zanskar are excluded) (J&K Forest Department, 1987, 1989).

The Forest Department, created in 1891, is headed by a Chief Conservator of Forests. The total staff complement in 1986 was 5,920 (234 gazetted officers and the rest ungazetted). Revenue in 1986-1987 totalled Rs 400.0 million and expenditure Rs 217.5 million (J&K Forest Department, 1987, 1989).

Jammu & Kashmir has enacted separate legislation known as the Jammu & Kashmir Wildlife Act, 1978, modelled on the Wild Life (Protection) Act, 1972, which has been enforced since late January 1979. Full details of the Act are given by Ganhar (1979). Legislation also exists for the establishment of biosphere reserves. A Wildlife Advisory Board has been constituted under the provisions of the Act. A separate wing of the Forest Department, known as the Directorate of Wildlife Protection, came into existence in 1978. This was upgraded to departmental status in 1982 with the establishment of the Department of Wildlife Protection (Bacha, 1986). This is headed by a Chief Wildlife Warden of the rank of Chief Conservator of Forests. There are three wildlife divisions at present, each managed by a Deputy Conservator of Forests. The conservation importance of Ladakh, India's largest district and administratively part of Jammu & Kashmir, was first officially recognised in 1978, since when a network of conservation areas has been identified and given protected status (Bacha, 1985).

Karnataka State (191,775 sq. km) Karnataka was formerly the princely state of Mysore which became part of the Indian Union in 1956. The state was renamed Karnataka in 1973.

Forests cover 32,100 sq. km (FSI, 1989) Tropical evergreen, moist deciduous and dry deciduous forests are predominant.

A State Wild Life Advisory Board was originally constituted in 1952, and in 1963 the Mysore Wild Animals and Wild Birds Act was enacted to establish a uniform code for wildlife (Spillet, 1968). The Wild Life (Protection) Act, 1972 has been enforced since 1973. There is a wildlife wing headed by a Chief Wildlife Warden of the rank of Additional Chief Conservator of

Forests. Protected areas are managed by territorial staff within the forest department. Project Tiger has now been brought under the control of the Chief Wildlife Warden. There are proposals to reorganise the wildlife wing with the creation of three wildlife divisions. Steps are also being taken to appoint honorary wildlife wardens within each district (Chief Wildlife Warden, 1985). In a recent review of the status of wildlife within the state, protection status is considered to be inadequate in most sanctuaries and in one of the three national parks (Karanth, 1987). Karnataka has its own State Department of the Environment.

Local non-governmental organisations include: Coorg Wildlife Society, Madikeri; Life Environmental Awareness Foundation, Bangalore; Wildlife Association of South India, Bangalore; Youth for Conservation, Bangalore; and Wildlife Preservation Group, Bangalore (Chief Wildlife Warden, 1985).

Kerala State (38,870 sq. km) Forests cover 10,149 sq. km (FSI, 1989). Forests are mostly tropical evergreen and moist deciduous types. In the past, there had been large-scale destruction of tropical evergreen forests to raise plantations and set up hydroelectric projects. The controversy over the proposed Silent Valley Hydroelectric Project became a national and international issue (Variava, 1983). The project was finally dropped in deference to the sentiments of the then Prime Minister, Mrs Indira Gandhi and subsequently, in 1984, the area was included within Silent Valley National Park.

The Wild Life (Protection) Act, 1972 has been enforced in the state since 1 June 1973 and the rules since 1979. Wildlife conservation gained momentum after the launch of the National Action Plan, marked by the establishment of a wildlife advisory board which met for the first time in January 1984. A separate wildlife wing, headed by a Chief Wildlife Warden of the rank of Chief Conservator of Forests, was created on 1 March 1985. Management plans were due to be completed for all protected areas by 1990, in accordance with central government policy.

Local non-governmental organisations include the Kerala Sastra Sahitya Parishad (Kerala Science & Literature Society) and Prakrithi Samrekshua Samithi (Environmental Conservation Society) in Trivandrum, and the Society for Environmental Education in Kerala (SEEK) in Cannanore.

Lakshadweep Union Territory (32 sq. km) The Lakshadweep Archipelago comprises 36 island, ten of which are inhabited. There is virtually no natural vegetation remaining, most islands having been planted with coconuts (Saldanha, 1989; Rodgers and Panwar, 1988).

The Wild Life (Protection) Act, 1972, has been enforced since 1973. There are no existing protected areas.

Madhya Pradesh State (442,840 sq. km) Forests cover 133,191 sq. km (FSI, 1989) and are mainly of sal, teak and mixed types. Environmentalists are extremely concerned about the potential loss of forests from submergence and resettlement planned under the Narmada Valley Project, described by the World Bank as "the largest river basin population resettlement to date" (Jackman, 1989). The amount of forest to be submerged is estimated to be 3,500 sq. km, but this may well be an underestimate (Kothari and Singh, 1988).

The Wild Life (Protection) Act, 1972 has been enforced since 25 January 1973 (Notification No. GSR 28 (E)) and the wildlife rules since 1974. There were nine sanctuaries originally notified under Forest Rules that were not renotified under the present Act due to their unimportance as conservation areas (Sinha, 1979). A Wildlife Wing, headed by a Chief Wildlife Warden of the rank of Chief Conservator of Forests, was created in 1977. Responsibility for managing protected areas has been transferred to this wing. Whereas directors of national parks report directly to this office, superintendents of sanctuaries work in close collaboration with the respective divisional forest officers (Joshi, 1985). The State Wildlife Advisory Board is very active and has been instrumental in making important decisions for wildlife conservation. Honorary wildlife wardens have been appointed to strengthen conservation efforts. A training school for game guards has been set up in Bandhavgarh National Park.

Maharashtra State (307,760 sq. km) Forests cover 44,058 sq. km (FSI, 1989).

The Wild Life (Protection) Act, 1972 has been enforced since 1 June 1973 (Notification No. GSR 293 (E)) and rules framed under the Act since 6 March 1975. There is an active State Wildlife Advisory Board (Anon., 1985).

Manipur State (22,327 sq. km) Forests cover 15,021 sq. km (Government of Manipur, 1990). This estimate, based on Landsat imagery for 1987-1989, is lower than that of 17,885 sq. km obtained by FSI (1989) using 1985-1987 imagery.

The Wild Life (Protection) Act, 1972 has been enforced since 15 May 1973 (Notification No. GSR 269 (E)) and wildlife rules since 1974. A State Wildlife Advisory Board has been constituted.

Meghalaya State (22,490 sq. km) Forests cover 15,690 sq. km (FSI, 1989) but only 3% of total land area is state-controlled forest, the remaining forest being controlled by district councils, local villages and clans, and under private ownership (Rodgers and Gupta, 1989).

The Wild Life (Protection) Act, 1972 has been enforced in the state since 1976 and wildlife rules since 1977. A State Wildlife Advisory Board has been constituted. There is only one Wildlife Division functioning in the State.

Mizoram State (21,087 sq. km) The state has vast natural forest resources but extensive tracts have been degraded due to shifting cultivation. Forests cover 18,178 sq. km (FSI, 1989). Reserved forests cover about 6,400 sq. km, protected forests (in which utilisation of any land is prohibited and cutting of trees not allowed without permission) 1,447 sq. km, and village safety and supply forests 1,485 sq. km. An important landmark in the protection of forests was the notification of the Inner Line Reserve (1,320 sq. km) in 1877, to which access by outsiders was prohibited (Government of Mizoram, 1989, 1991).

The scientific management of the forest estate was a low priority in the 1970s, as reflected in 1973 by the small staffing levels (155 personnel under a Conservator of Forests) within the former Department of Forests. Considerable progress has been made since the creation of the post of Principal Chief Conservator of Forests in 1987. The Department of Environment and Forests now comprises 10 territorial forest divisions and 6 functional divisions, of which one is the Wildlife Division. The total staff complement in 1990 was 1,238, with 47 in the Wildlife Division. Revenue for the Department in 1989-1990 totalled Rs 7.7 million, of which Rs 2.7 million was collected by the Wildlife Division, and expenditure Rs 67.6 million (Government of Mizoram, 1989, 1991).

The Wild Life (Protection) Act, 1972 has been enforced since 1974. A Wildlife Advisory Board has been constituted and is active. A Wildlife Division was created in 1986 and began functioning the following year. The Conservator of Forests, Northern Circle was appointed Chief Wildlife Warden, and all territorial Divisional Forest Officers as Wildlife Wardens of their respective territories in 1986 (Government of Mizoram, 1991).

Nagaland State (16,525 sq. km) Forests cover 14,356 sq. km. A unique feature is that 88% of the total recorded forest is under private ownership (FSI, 1989). Deforestation is estimated to be currently about 180 sq. km per year (Thakkar, 1987).

The Wild Life (Protection) Act, 1972 has been enforced since the 1970s. There is a State Wildlife Advisory Board. The Conservator of Forests (Wildlife) is designated as the Chief Wildlife. There is only one Wildlife Division.

Orissa State (155,780 sq. km) Forests cover 47,137 sq. km (FSI, 1989). Notified forest totals 57,184 sq. km, of which 27,087 sq. km is reserved forest and the rest mostly protected forest (Department of Forest, Fisheries and Animal Husbandry, pers. comm., 1991).

A number of conservation measures have been introduced in the forestry sector since 1988. These include a ban on felling of trees within five districts, namely Bolangir, Ganjam, Kalahandi, Koraput and Phulbani, and the involvement of village communities in the protection of reserved forests in return for concessions to meet domestic fuelwood and timber requirements (Department of Forest, Fisheries and Animal Husbandry, pers. comm., 1991).

The Wild Life (Protection) Act, 1972 has been enforced in the state since 1974 and Wildlife (Protection) (Orissa) Rules, since 1975. A separate Wildlife Conservation Division created in 1966 now has six wildlife divisions and is headed by a Chief Wildlife Warden of the rank of Additional Chief Conservator of Forests. All divisional forest officers (territorial) have been designated as wildlife wardens for implementing the Act and Rules. The Nandankanan Development Board, constituted by the state government, is responsible for the administration and development of Nandankanan Biological Park. A State Wildlife Advisory Board has been constituted, and a forum of the Orissa Legislative Assembly on Environment also advises the state government on wildlife conservation matters (Padhi, 1985).

Non-governmental organisations registered with the state include the Nature and Wildlife Conservation Society of Orissa and the Orissa Environmental Society (Padhi, 1985).

Pondicherry Union Territory (492 sq. km) There are no forests or protected areas in Pondicherry.

The Wild Life (Protection) Act, 1972 has been enforced since 1975.

Punjab State (50,362 sq. km) Forest covers 1,161 sq. km (FSI, 1989), much of the rest of the land being predominantly under agriculture.

The Wild Life (Protection) Act, 1972 and wildlife rules have been enforced since 1975, prior to which a separate organisation already existed. A State Wildlife Advisory Board has been constituted and is reported to be active.

Rajasthan State (342,215 sq. km) Forests cover 12,966 sq. km (FSI,1989) and range from the tropical semi-evergreen forests of Mt Abu, tropical dry deciduous teak *Tectona grandis* forests of Banswara and tropical dry deciduous *Anogeissus pendula* forests of the Aravalis to the tropical dry thorn forests of the desert zone (Sankala, 1964).

The Wildlife (Protection) Act, 1972 has been enforced since 1973. A separate Wildlife Organisation has been set up, headed by a Chief Wildlife Warden of the rank of Additional Chief Conservator of Forests. A State Wild Life Board was constituted in 1955.

Sikkim State (7,300 sq. km) Forests cover 3,124 sq. km or 43% of total land area (FSI, 1989). Alpine pastures and permanently snow-covered areas occupy a further 30% (Ali, 1981). A disturbing situation is that of the 2,650 sq. km of notified forest, only 1,577 sq. km are reserved forest; the rest is of nebulous legal status (FSI, 1989).

The Wild Life (Protection) Act, 1972, and wildlife rules have been enforced since 1976, following the integration of Sikkim within the Indian Union in 1975. The Chief Conservator of Forests is designated as the Chief Wildlife Warden. A State Wildlife Advisory Board has been constituted and is active on conservation policy matters.

Tamil Nadu State (130,058 sq. km) Forests cover 17,715 sq. km (FSI, 1989). Notified forests extend over 21,938 sq. km (Tamil Nadu Forest Department, 1990).

The Wild Life (Protection) Act, 1972 and wildlife rules have been enforced in the state since 1974 and 1976, respectively. A State Wildlife Advisory Board was constituted in the 1960s. There is a Wildlife Department, headed by a Chief Wildlife Warden of the rank of Additional Chief Conservator of Forests. Five wildlife divisions have been created but dual control still exists between wildlife and forest departments. Tamil Nadu is one of the few governments to appoint hunters as honorary game wardens, thereby involving them in conservation efforts. An historical account of wildlife conservation in the state is given by Davidar (1987).

Locally important conservation organisations include Nilgiri Wildlife and Environment Association, Madras Snake Park, Madras Crocodile Bank, Madras Naturalists Society, Tirunelveli Wildlife Preservation Society and Ramanathapuram Wildlife Society.

Tripura State (10,475 sq. km) Forests cover 5,325 sq. km (FSI, 1989). Only about 8% is dense natural forest, the rest having been much depleted by clearance for shifting cultivation and, recently, for settlement of refugees from Bangladesh (Paxton, 1990).

The Wild Life (Protection) Act, 1972 was enforced on 2 October 1973 (Notification No. GSR 465 (E)). A Conservator of Forests is *ex officio* Chief Wildlife Warden. A State Wildlife Board has been constituted. No honorary wildlife wardens have been appointed.

Uttar Pradesh State (294,411 sq. km) Forests cover 33,844 sq. km (FSI, 1989).

The Wild Life (Protection) Act, 1972 has been enforced since 1 February 1973 (Notification No. GSR 44 (E)) and the wildlife rules since 1974. The Wildlife Preservation Organisation was originally set up in 1956, with the introduction of a Wild Life Preservation Scheme under the Second Five-Year Plan (Srivastava, 1969), and later reorganised in 1958. It is now headed by a Chief Wildlife Warden of the rank of Additional Chief Conservator of Forests. There are five wildlife divisions. Honorary wildlife wardens have been appointed. A State Wildlife Advisory Board has been constituted.

West Bengal (87,852 sq. km) Forests cover 8,394 sq. km (FSI, 1989).

The Wild Life (Protection) Act, 1972 has been enforced since 1 May 1973 (Notification No. GSR 224 (E)). The

wildlife rules were also enforced in 1973. The Chief Conservator of Forests is designated as the Chief Wildlife Warden and wildlife management duties are the responsibility of the territorial staff of the Forest Department. A State Wildlife Advisory Board has been constituted.

Addresses

Department of Environment, Forests and Wildlife (Joint Secretary, Wildlife),

Ministry of Environment and Forests, Paryavaran Bhawan, CGO Complex, Lodi Road, NEW DELHI 110 003 (Tel: 306156I; Tlx: 3163015 WILD IN; Cable: PARYAVARAN, NEW DELHI)

Department of Environment, Forests and Wildlife (Inspector-General of Forests), Ministry of Environment and Forests, Paryavaran Bhawan, CGO Complex, Lodi Road, NEW DELHI 110 003 (Cable: AGRINDIA, NEW DELHI)

Project Tiger (Director), Bikaner House, New Delhi 110011

Wildlife Institute of India (Director), PO New Forest, Dehra Dun 248 006 (Tel. 27021-8, 28760, 27724; Tlx 585238 PRES IN, 585258 FRIC IN; Cable: WILDLIFE)

Chief Wildlife Warden, PO Chatham, PORT BLAIR, Andaman & Nicobar 744 101 (Union Territory)

Conservator of Forests (Wildlife), HYDERABAD, Andra Pradesh 500 004

Government of Arunachal Pradesh (Chief Wildlife Warden), ITANAGAR, Arunachal Pradesh 791 111

Chief Conservator of Forests and Chief Wildlife Warden, Retabari, GUWAHATI, Assam 788 735

Chief Wildlife Warden, PO Hinoo, RANCHI, Bihar 834 002

Divisional Forest Officer and Chief Wildlife Warden, Estate Office, Sector 17, CHANDIGARH 160 017 (Union Territory)

Chief Wildlife Warden, SILVASSA, Dadra and Nagar Haveli 396230 (Union Territory)

ADM and Chief Wildlife Warden, Room 39, 1st Floor, Western Wing, Tis Hazari, DELHI 110 054 (Union Territory)

Wildlife and Parks Division (Conservator of Forests and Chief Wildlife Warden), Junta House, 3rd Floor, PANAJI, Goa 403 001

Chief Conservator of Forests and Chief Wildlife Warden, Kothi Annexe, BARODA, Gujarat 390 001

Chief Wildlife Warden, Kothi 70, Sector 8, PANCHKULA, Haryana 134 109

Department of Forest Farming and Conservation (Chief Conservator of Forests, Wildlife and Chief Wildlife Warden), Talland, SIMLA, Himachal Pradesh 171 002

Department of Wildlife Protection (Chief Wildlife Warden), Tourist Reception Centre, SRINAGAR, Jammu & Kashmir 190 001

Chief Wildlife Warden, 11 Floor, Aranya Bhawan, 18th Cross, Malleshwaram, BANGALORE, Karnataka 560 003

Additional Chief Conservator of Forests and Chief Wildlife Warden, TRIVANDRUM, Kerala 695 001

Administrator, Via Kavaratti, Lakshadweep 673 555 (Union Territory)

Forest Department (Chief Conservator of Forests, Wildlife and Chief Wildlife Warden), 1st Floor, B-Wing, Satpura Bhawan, BHOPAL, Madhya Pradesh 462 001

Chief Wildlife Warden, Nature Conservation, MS, NAGPUR, Maharashtra 440 001

Government of Manipur (Chief Wildlife Warden), PO Sanjenthong, IMPHAL, Manipur 795 001

Government of Meghalaya (Chief Wildlife Warden), Risa Colony, SHILLONG, Meghalaya 793 003

Department of Environment and Forests (Chief Wildlife Warden), AIZAWAL, Mizoram 796 001

Government of Nagaland (Chief Conservator of Forests, Wildlife), DIMAPUR, Nagaland 797 112

Additional Conservator of Forests (Wildlife) and Chief Wildlife Warden, 315 Kharvel Nagar, BHUBANESHWAR, Orissa 751 001

Chief Wildlife Warden, PONDICHERRY 605 001 (Union Territory)

Chief Wildlife Warden, Punjab SCD 2463-64, Sector 22C, CHANDIGARH, Punjab 160 022

Chief Wildlife Warden, Van Bhawan, Bhagwan Das Road, JAIPUR, Rajasthan 302 001

Forest Secretariat (Chief Wildlife Warden), GANGTOK, Sikkim 737 101

Conservator of Forests, 62 James Cottage, TIRUNELVELI 7, Tirunelveli District, Tamil Nadu

Government of Tripura (Chief Conservator of Forests and Chief Wildlife Warden), P O Kunjaban, AGARTALA, Tripura 799 006

Wildlife Preservation Organisation (Chief Wildlife Warden), 17 Rana Pratap Marg, LUCKNOW, Uttar Pradesh 226 001

Office of the Chief Conservator of Forests, West Bengal (Chief Wildlife Warden), P-16 India Exchange Place Extension, New CIT Building, CALCUTTA, West Bengal 700 073

Bombay Natural History Society (BNHS) (Curator), Hornbill House, Shahid Bhagat Singh Road, BOMBAY 400 023 (Tel. 243869, 244085; Cable: HORNBILL)

Centre for Science and Environment (CSE) (Director), F6 Kailash Colony, NEW DELHI (Tel. 6438109)

Indian National Trust for Art and Cultural Heritage (INTACH) (Director-Natural Heritage), 71 Lodi Estate, NEW DELHI 110 003 (Tel. 611362, 618912, 616581)

Indian Society of Naturalists (INSONA), Oza Building, SALATWADA, Baroda 390 001

Wildlife Preservation Society of India (Honorary Secretary), 7 Astley Hall, DEHRA DUN (Tel. 5392)

World Wide Fund For Nature-India (WWF-India) (Secretary General), Secretariat, 172-B Lodi Estate, NEW DELHI 110 003 (Tel. 616532, 693744; Fax 626837)

References

Ali, Mohammad S. (1981). Ecological reconnaissance in eastern Himalaya. *Tiger Paper* 8(2): 1-3.

Anon. (1985). A paper on wildlife and nature conservation measures in Maharastra State. Paper presented at 25th Working Session of IUCN Commission on National Parks and Protected Areas, Corbett National Park, India. 4-8 February 1985. 4 pp.

Arya, S.R. (n.d.). *Status of the wild life protection areas and their management in Himachal Pradesh.* Department of Forest Farming and Conservation, Simla. 7 pp.

Bacha, M.S. (1985). *Ecological cum management plan for Hemis High Altitude National Park, Jammu and Kashmir State. 1985-90.* Department of Wildlife Protection, Srinagar. 31 pp.

Bacha, M.S. (1986). *Snow leopard recovery programme for Kishtwar High Altitude National Park, Jammu and Kashmir State, 1986-87 to 1989-90.* Department of Wildlife Protection, Srinagar. 51 pp.

Biswas, D.K. and Bannerjee, P.K. (1984). Environmental programmes of the Government of India. In: Singh, Shekhar (Ed.), *Environmental policy in India.* Indian Institute of Public Administration, New Delhi. Pp. 97-115.

Bustard, H.R. (1982). Crocodile breeding project. In Saharia, V.B. (Ed.), *Wildlife in India.* Natraj Publishers, Dehra Dun. Pp. 147-163.

Chandha, C.M. (1989). National parks and sanctuaries in India. In: *Proceedings of the International Conference on National Parks and Protected Areas.* 13-15 November 1989, Kuala Lumpur. Department of National Parks, Penninsular Malaysia. Pp.111-114.

Chief Wildlife Warden (1985). Protected area of Karnataka and their management: a status report. Paper presented at 25th Working Session of IUCN's Commission on National Parks and Protected Areas, Corbett National Park, India. 4-8 February 1985. 4 pp.

CSE (1982). *The state of India's environment 1982. A first citizens' report.* Centre for Science and Environment, New Delhi. 189 pp.

CSE (1985). *The state of India's environment 1984-85. A second citizens' report.* Centre for Science and Environment, New Delhi. 393 pp.

Davidar, E.R.C. (1987). Conservation of wildlife in Tamil Nadu. *Journal of the Bombay Natural History Society* 83 (Supplement): 65-71.

Department of Environment (n.d.). *National Wildlife Action Plan.* Government of India, New Delhi. 28 pp.

DFFC (1990). *H.P. statistics 1990.* Department of Forest Farming and Conservation, Simla. 257 pp.

Dwivedi, O.P. and Kishore, B. (1984). India's environmental policies: a review. In: Singh, Shekhar (Ed.), *Environmental policy in India*. Indian Institute of Public Administration, New Delhi. Pp. 47-84.

FSI (1989). *The state of forest report 1989*. Forest Survey of India, Government of India, Dehra Dun. 50 pp.

Gadgil, Madhav (1989). The Indian heritage of a conservation ethic. In: Allchin, B., Allchin, F.R. and Thapar, B.K. (Eds), *Conservation of the Indian Heritage*. Cosmo Publishers, New Delhi. Pp. 13-21.

Gadgil, Madhav and Meher-Homji, V.M. (1986a). Localities of great significance to conservation of India's biological diversity. *Proceedings of the Indian Academy of Sciences (Animal Sciences/Plant Sciences) Supplement* 1986: 165-180.

Gadgil, Madhav and Meher-Homji, V.M. (1986b). Role of protected areas in conservation. In: Chopra, V.L. and Khoshoo, T.N. (Eds) *Conservation for productive agriculture*. Indian Council of Agricultural Research, New Delhi. Pp. 143-159.

Ganhar, J.N. (1979). *The wildlife of Ladakh*. Haramukh Publications, Srinager.

Gaston, A.J. (1983). Forests and forest policy in northwest India since 1800. Paper presented at Bombay Natural History Society Centenary Symposium. Powai, Bombay. 30 pp.

Gee, E.P. (1950). Wildlife reserves in India, Assam. *Oryx* 49: 81-89.

Government of India (n.d.). *Department of Environment: a profile*. Government of India, New Delhi. 23 pp.

Government of India (1988). *National Forest Policy 1988*. Ministry of Environment and Forests, Government of India, New Delhi. 13 pp.

Government of Manipur (1990). Report on land use/land cover, Manipur State (Bishnupur, Chandel, Churachandrpur, Imphal, Senapati, Tamenglong, Thoubal and Ukhrul districts). State Remote Sensing Centre, Government of Manipur, Imphal and National Remote Sensing Agency, Government of India, Hyderabad. Unpublished.

Government of Mizoram (1989). Mizoram forests. *Forest Extension Series* 89(2): 1-35.

Government of Mizoram (1991). *Progress report of forestry in Mizoram 1990*. Department of Environment & Forests, Government of Mizoram, Aizwal. 73 pp.

IBWL (1970). *Wildlife conservation in India*. Report of the Expert Committee. Indian Board for Wildlife, Government of India, New Delhi. 149 pp.

IBWL (1972). *Project Tiger. A planning proposal for preservation of tiger (Panthera tigris tigris Linn.) in India*. Indian Board for Wildlife, Government of India, New Delhi. 114 pp.

IUCN (1985). *The Corbett Action Plan for protected areas of the Indomalayan Realm*. IUCN, Gland, Switzerland and Cambridge, U.K. 23 pp.

Jackman, B. (1989). India be dammed. *Geographical Magazine* 61: 10-14.

J&K Forest Department (1987). A digest of forest statistics 1987. *J&K Forest Record* No. 1. 199 pp.

J&K Forest Department (1989). *Jammu & Kashmir Forest Department. Activities and future strategies*. 24 pp.

Joshi, N.K. (1985). Wildlife conservation in Madhya Pradesh. Paper presented at 25th Working Session of IUCN Commission on National Parks and Protected Areas, Corbett National Park, India. 4-8 February 1985. 6 pp.

Karanth, K.U. (1987). Status of wildlife and habitat conservation in Karnataka. *Journal of the Bombay Natural History Society* 83 (Supplement): 166-179.

Kothari, A., Pande, P., Singh, S., Variava, D. (1989). Management of national parks and sanctuaries in India: a status report. Environmental Studies Division, Indian Institute of Public Administration, New Delhi. 298 pp.

Kothari, A. and Singh, S. (1988). *The Narmada Valley Project: a critique*. Kalpavriksh, Delhi. 24 pp.

MacKinnon, J. and MacKinnon, K. (1986). *Review of the protected areas system in the Indo-Malayan Realm*. IUCN, Gland, Switzerland and Cambridge, UK. 284 pp.

Mani, M.S. (Ed.)(1974). Ecology and biogeography in India. Junk, The Hague.

Mehta, J.K. (n.d.). Status paper on the protected areas and their management in Arunachal Pradesh. Government of Arunachal, Itanagar. Unpublished. 9 pp.

Ministry of Environment and Forests (1985). India. Country Report 1985. Presented to IX World Forestry Congress. Government of India, New Delhi. 53 pp.

Ministry of Environment and Forests (1986). *Annual report 1985-86*. Government of India, New Delhi. 57 pp.

Ministry of Environment and Forests (1987a). *Annual report 1986-87*. Government of India, New Delhi. 73 pp.

Ministry of Environment and Forests (1987b). *Biosphere reserves*. Government of India, New Delhi. 250 pp.

Mitra, D.K. (1989). A note on the ancient history of nature and wildlife conservatuon in India. *Zoo's Print* 4(12): 13.

Nair, S.C. (n.d.). *Natural resources conservation and development in Andaman and Nicobar islands*. Department of Environment, New Delhi. 76 pp.

Nayar, M.P. (1989). *In situ* conservation of wild flora resources. Paper presented at National Symposium on the Conservation of India's Genetic Estate, New Delhi, 3-4 November 1989. 19 pp.

Padhi, S.C. (1985). Wildlife management in Orissa. Paper presented at 25th Working Session of IUCN Commission on National Parks and Protected Areas, Corbett National Park, India. 4-8 February 1985. 4 pp.

Panwar, H.S. (1982). Project Tiger. In: Saharia, V.B. (Ed.), *Wildlife in India*. Natraj Publications, Dehra Dun. Pp. 130-137.

Panwar, H.S. (1984). What to do when you've succeeded: Project Tiger, ten years later. In: McNeely, J.A. and Miller, K.R. (Eds), *National parks, conservation and development: the role of protected areas in sustaining society.* Smithsonian Institution Press, Washington DC. Pp. 183-189.

Panwar, H.S. (1990). Status of management of protected areas in India: problems and prospects (revised). Regional Expert Consultation on Management of Protected Areas in the Asia-Pacific Region, 10-14 December 1990. FAO Regional Office for Asia and Pacific, Bangkok. Unpublished report. 21 pp.

Paxton, J. (Ed.) (1990). *The stateman's year-book 1990-91.* The Macmillan Press, London. Pp. 689-690.

Pillai, V.N.K. (1982). Status of wildlife conservation in states and union territories. In: Saharia, V.B. (Ed.), *Wildlife in India.* Natraj Publishers, Dehra Dun. Pp. 74-91.

Ram, A. (1991). Slippery slope for project controls. *Panoscope* 22: 6-7.

Ranganthan, Shankar (1981). Gujarat Forestry Scheme: the start of something good. *Parks* 5(4): 10-11.

Rashid, M.A. (n.d.). *Wild life and its conservation in Gujarat State.* Forest Department, Gujarat State, Vadodara. Unpublished report. 14 pp.

Rogers, W.A. (1985). Biogeography and protected area planning in India. In: Thorsell, J.W. (Ed.), *Conserving Asia's natural heritage.* IUCN, Gland, Switzerland and Cambridge, UK. Pp. 103-113.

Rodgers, W.A. and Gupta, S. (1989). The Pitcher Plant (*Nepenthes khasiana* HK.F.) Sanctuary of Jaintia Hills, Meghalaya: lessons for conservation. *Journal of the Bombay Natural History Society* 86: 17-21.

Rodgers, W.A. and Panwar, H.S. (1988). *Planning a wildlife protected area network in India.* 2 vols. Project FO: IND/82/003. FAO, Dehra Dun. 339, 267 pp.

Saharia, V.B. and Pillai, V.N.K. (1982). Organisation and legislation. In: Saharia, V.B. (Ed.), *Wildlife in India.* Natraj Publishers, Dehra Dun. Pp. 53-73.

Saldanha, C.J. (1989). *Andaman, Nicobar and Lakshadweep: an environmental impact assessment.* Oxford and IBH Publishing, New Delhi. 114 pp.

Sankhala, K.S. (1964). Wildlife sanctuaries of Rajasthan. *Journal of the Bombay Natural History Society* 61: 27-34.

Scott, D.A. (Ed.) (1989). *A directory of Asian wetlands.* IUCN, Gland, Switzerland and Cambridge, UK. 1,181 pp.

Sharma, Sudhirendar (1986). Sukhomajri – an example of effective conservation. *Environmental Conservation* 13: 76-79.

Singh, Chhatrapati (1985). Emerging principles of environmental laws for development. In: Bandopadhyay, J., Jayal, N.D., Schoettli, V., Singh, Chhatrapati (Eds), *India's environment: crises and responses.* Natraj Publishers, Dehra Dun. Pp. 247-275.

Singh, Samar (1985). Protected areas in India. In: Thorsell, J.W. (Ed.), *Conserving Asia's natural heritage.* IUCN, Gland, Switzerland and Cambridge, UK. Pp. 11-18.

Singh, Samar (1986). *Conserving India's Natural Heritage.* Natraj Publishers, Dehra Dun. 219 pp.

Singh, S., Kothari, A. and Pande, P. (1990). *Directory of national parks and sanctuaries in Himachal Pradesh: management status and profiles.* Environmental Studies Division, Indian Institute of Public Administration, New Delhi. 164 pp.

Sinha, N.K. (1979). List of national parks and wildlife sanctuaries in Madhya Pradesh. *Journal of the Bombay Natural History Society* 75: 469-472.

Spillet, J.J. (1968). A report on wild life surveys in south and west India. *Journal of the Bombay Natural History Society* 65: 296-325.

Srivastava, T.N. (1969). *Wild life of Uttar Pradesh, India.* Wild Life Preservation Organisation, Lucknow. Unpublished report. 21 pp.

Tamil Nadu Forest Department (1990). *The forest cover report.* Government of Tamil Nadu, Madras. 48 pp.

Thakkar, Natwar (1987). Deforestation and Nagaland – a layman's observations. Views from keepers of the forests. Paper presented at Asia-Pacific NGOs Conference on Deforestation and Desertification. New Delhi, 23-25 October 1987. 5 pp.

UNEP/IUCN (1988). *Coral reefs of the world. Volume 2: Indian Ocean, Red Sea and Gulf.* UNEP Regional Seas Directories and Bibliographies. IUCN, Gland, Switzerland and Cambridge, UK/UNEP, Nairobi, Kenya. 440 pp.

Variava, D. (1983). Silent Valley: a case study in environmental education. Paper presented at the Bombay Natural History Society Centenary Symposium, Bombay. 12 pp.

Vedant, C.S. (1986). Comment: afforestation in India. *Ambio* 15: 254-255.

Whitaker, R. (1985). *Endangered Andamans.* Environmental Services Group, WWF-India, New Delhi. 51 pp.

WII (1987). *Annual report 1986-87.* Wildlife Institute of India, Dehra Dun. 58 pp.

ANNEX
Definitions of protected area designations, as legislated, together with authorities responsible for their administration

Title: Wild Life (Protection) Act

Date: 1972, last amended 1987

Brief description: An Act to provide for the protection of wild animals and birds, and related or ancillary matters

Administrative authority: Central Government (Director of Wild Life Preservation), State Government (Chief Wild Life Warden)

Designations:

Sanctuary[1] An area of "adequate ecological, faunal, floral, geomorphological, natural or zoological significance" may be declared a sanctuary for the protection and propagation of its wildlife[2] or environment.

Permission to enter or reside in a sanctuary may be granted by the Chief Wildlife Warden for purposes of photography, scientific research, tourism and transaction of lawful business with any resident. Entry is restricted to a public servant on duty, a person permitted by the Chief Wildlife Warden to reside in a sanctuary or who has any right over immovable property within a sanctuary, a person using a public highway, or dependents of any of the above.

Hunting without a permit, entry with any weapon, causing fire, and using substances potentially injurious to wildlife are prohibited. Fishing and grazing by livestock may be allowed on a controlled basis.

National park[1] An area of "ecological, faunal, floral, geomorphological, or zoological importance" may be declared a national park for the protection, propagation or development of its wildlife or environment once all rights have become vested in the state Government.

No alteration of boundaries may be made except by resolution passed by the state legislature.

Entry, unless used as a vehicle by an authorised person, and grazing of any cattle is prohibited.

Restrictions on entry, in so far as they apply, are the same as those for a sanctuary.

Destruction, exploitation or removal of any wildlife or its habitat is prohibited, except with permission from the Chief Wildlife Warden and provided it is necessary for the improvement and better management of wildlife. Other prohibited activities,

in so far as they apply, are the same as those for a sanctuary.

Game reserve An area in which only licensed hunting is permitted.

Closed area An area closed to hunting for such periods as may be specified in the notification.

Source: Original legislation

[1]State-owned land leased or otherwise transferred to central government may be declared as a sanctuary or national park by the federal authority.

[2]Wildlife is defined in the Act as including any animal, bee, butterfly, crustacean, fish and moth, and aquatic or land vegetation which forms part of any habitat.

Title: Indian Forest Act

Date: 1927, amended 1930,1933, 1948 (Central Legislation)

Brief description: An Act to consolidate the law relating to forests, the transit of forest produce and the duty leviable on timber and other forest produce

Administrative authority: Central Government (Inspector General of Forests) State Government (Chief Conservator of Forests)

Designations:

Reserved forest Any forest land or wasteland belonging to the government, or to which it has proprietary rights, may be constituted a reserved forest once all lands within the proposed forest have become invested in the government.

Prohibited activities include: making fresh clearings or breaking up land for cultivation; kindling or carrying fire; trespass and cattle grazing; felling or otherwise damaging any tree; quarrying stone, burning lime or charcoal; removing forest produce; and hunting, shooting, fishing, trapping and poisoning water.

Village forest Any land constituted as reserved forest that has been assigned to a village community by the state government. Such an assignment may be cancelled.

Rules for regulating the provision of timber, other forest produce or pasture to the community, and their duties for protecting and improving such forest may be prescribed by the state government.

All provisions of the Act relating to reserved forest apply to village forest, in so far as they are consistent with the rules.

Protected forest Any forest land or wasteland not included in a reserved forest and belonging to the Government, or to which it has proprietary rights, may be declared a protected forest provided that the nature and extent of rights of government and any private persons in or over such land have been recorded.

Activities prohibited within reserved forests are subject to regulations in protected forests. In addition, in protected forests, any trees, class of trees or portion of forest may be temporarily closed to all forms of exploitation, including the quarrying of stone and burning of lime.

Source: Original legislation

SUMMARY OF PROTECTED AREAS

Map[†] ref.	*National/international designation* Name of area	IUCN management category	Area (ha)	Year notified
	ANDAMAN AND NICOBAR ISLANDS UNION TERRITORY			
	National Parks			
1	Marine (Wandur)	II	28,150	1983
2	Mount Harriet Island	II	4,622	1979
3	Saddle Peak	II	3,255	1979
	Sanctuaries			
4	Barren Island	IV	810	1977
5	Battimalve Island	IV	223	1985
6	Benett Island	IV	346	1987
7	Bluff Island	IV	114	1987
8	Bondoville Island	IV	255	1987
9	Buchanan Island	IV	933	1987
10	Cinque Island	IV	951	1987
11	Crocodile (Lohabarrack)	IV	10,600	1983
12	Defence Island	IV	1,049	1987
13	East (Inglis) Island	IV	355	1987
14	East Island	IV	611	1987
15	Flat Island	IV	936	1987
16	Interview Island	IV	13,300	1985
17	James Island	IV	210	1987
18	Kyd Island	IV	800	1987
19	Landfall Island	IV	2,948	1987
20	Narcondum Island	IV	681	1977
21	North Reef Island	IV	348	1977
22	Paget Island	IV	736	1987
23	Pitman Island	IV	137	1987
24	Point Island	IV	307	1987
25	Ranger Island	IV	426	1987
26	Reef Island	IV	174	1987
27	Roper Island	IV	146	1987
28	Ross Island	IV	101	1987
29	Sandy Island	IV	158	1987
30	Shearme Island	IV	785	1987
31	Sir Hugh Rose Island	IV	106	1987
32	South Brother Island	IV	124	1987
33	South Reef Island	IV	117	1987
34	South Sentinel Island	IV	161	1977
35	Spike Island-2	IV	1,170	1987
36	Swamp Island	IV	409	1987
37	Table (Delgarno) Island	IV	229	1987
38	Table (Excelsior) Island	IV	169	1987
39	Talabaicha Island	IV	321	1987
40	Temple Island	IV	104	1987
41	Tillanchong Island	IV	1,683	1985
42	West Island	IV	640	1987
	ANDHRA PRADESH STATE			
	Sanctuaries			
43	Coringa	IV	23,570	1978
44	Eturnagaram	IV	81,259	1953
45	Kawal	IV	89,228	1965
46	Kinnersani	IV	63,540	1977
47	Kolleru	IV	90,100	1963
48	Lanjamadugu	IV	3,620	1978

Map[†] ref.	National/international designation Name of area	IUCN management category	Area (ha)	Year notified
49	Manjira	IV	2,000	1978
50	Nagarjunasagar Srisailam	IV	35,689	1978
51	Pakhal	IV	86,205	1952
52	Papikonda	IV	59,068	1978
53	Pocharam	IV	12,964	1952
54	Pranhita	IV	13,603	1980
55	Pulicat	IV	58,000	1976
56	Siwaram	IV	2,992	1978
57	Srivenkateswara	IV	50,700	
	ARUNACHAL PRADESH STATE			
	National Parks			
58	Mouling	II	48,300	1986
59	Namdapha	II	198,524	1983
	Sanctuaries			
60	D'Ering Memorial	IV	19,000	1978
61	Itanagar	IV	14,030	1978
62	Mehao	IV	28,150	1980
63	Pakhui	IV	86,195	1977
	ASSAM STATE			
	National Park			
64	Kaziranga	II	42,996	1974
	Sanctuaries			
65	Barnadi	IV	2,622	1980
66	Dipor Beel	IV	4,000	1989
67	Laokhowa	IV	7,014	1979
68	Manas	IV	39,100	1928
69	Nameri	IV	13,707	1985
70	Orang	IV	7,260	1985
71	Pabha	IV	4,900	
72	Pobitora	IV	3,883	1987
73	Sonai Rupai	IV	17,500	1934
	BIHAR STATE			
	National Park			
74	Palamau	II	21,300	1986
	Sanctuaries			
75	Bhimbandh	IV	68,190	1976
76	Dalma	IV	19,322	1976
77	Gautam Budha	IV	25,948	1976
78	Hazaribagh	IV	18,625	1976
79	Kabar	IV	20,400	1986
80	Kaimur	IV	134,222	1978
81	Koderma	IV	17,795	1985
82	Lawalang	IV	21,103	1978
83	Mahuadaur	IV	6,325	1976
84	Nakti Dam	IV	20,640	1985
85	Palamau	IV	76,700	1976
86	Parasnath	IV	4,923	1984
87	Rajgir	IV	3,584	1978
88	Valmikinagar	IV	46,160	1978

Map[†] ref.	National/international designation Name of area	IUCN management category	Area (ha)	Year notified
	CHANDIGARH UNION TERRITORY			
	Sanctuary			
89	Sukhna	IV	2,542	
	DELHI UNION TERRITORY			
	Sanctuary			
90	Indira Priyadarshini	IV	1,320	
	GOA STATE			
	National Park			
91	Bhagwan Mahavir	II	10,700	1978
	Sanctuaries			
92	Bhagwan Mahavir	IV	14,852	1967
93	Cotigao	IV	10,500	1968
	GUJARAT STATE			
	National Parks			
94	Bansda	II	2,399	1976
95	Gir	II	25,871	1975
96	Great Rann	II	700,000	1990
97	Marine (Gulf of Kutch)	II	16,289	1982
98	Velavadar	II	3,408	1976
	Sanctuaries			
99	Barda	IV	19,931	1979
100	Dumkhal	IV	44,818	1982
101	Gir	IV	115,342	1965
102	Jessore	IV	18,066	1978
103	Kachchh Desert	IV	750,622	1986
104	Marine (Gulf of Kutch)	IV	29,303	1986
105	Nal Sarovar	IV	12,082	1969
106	Narayan Sarovar	IV	76,579	1981
107	Rampura	IV	1,501	1988
108	Ratanmahal	IV	5,565	1982
109	Wild Ass	IV	495,370	1973
	HARYANA STATE			
	Sanctuaries			
110	Bir Shikargah	IV	1,093	1975
111	Chautala	IV	11,396	1987
	HIMACHAL PRADESH STATE			
	National Parks			
114	Great Himalayan*	II	60,561	1984
115	Pin Valley*	II	80,736	1987
	Sanctuaries			
116	Bandli*	IV	3,947	1962
117	Chail*	IV	11,004	1976
118	Churdhar*	IV	5,659	1985
119	Daranghati*	IV	2,701	1962
120	Darlaghat*	IV	9,871	1962
121	Gamgul Siahbehi*	IV	10,546	1949
122	Gobind Sagar*	IV	12,067	1962
123	Kais*	IV	1,220	1954
124	Kalatop & Khajjiar*	IV	3,069	1949

Map[†] ·ref.	National/international designation Name of area	IUCN management category	Area (ha)	Year notified
125	Kanawar*	IV	6,157	1954
126	Khokhan*	IV	1,760	1954
127	Kugti*	IV	33,000	1962
128	Lippa Asrang*	IV	2,953	1962
129	Majathal*	IV	3,164	1962
130	Manali*	IV	3,127	1954
131	Naina Devi*	IV	3,719	1962
132	Nargu*	IV	24,313	1962
133	Pong Dam*	IV	32,270	1983
134	Raksham Chitkul*	IV	3,827	1962
135	Rupi Bhabha*	IV	85,414	1982
136	Sechu Tuan Nala*	IV	65,532	1962
137	Shikari Devi*	IV	7,119	1962
138	Simbalbara*	IV	1,720	1958
139	Talra*	IV	3,616	1962
140	Tirthan*	IV	6,825	1976
141	Tundah*	IV	41,948	1962

JAMMU AND KASHMIR STATE

National Parks

142	Dachigam	II	14,100	1981
143	Hemis	II	410,000	1981
144	Kishtwar	II	31,000	1981

Sanctuaries

145	Baltal	IV	20,300	1987
146	Changthang	IV	400,000	1987
147	Gulmarg	V	18,600	1987
148	Hirapora	IV	11,000	1987
149	Hokarsar	IV	1,000	
150	Karakoram	IV	180,000	
151	Lachipora	IV	8,000	1987
152	Limber	IV	2,600	1987
153	Nandini	IV	3,372	1981
154	Overa	IV	3,237	1981
155	Overa-Aru	IV	42,500	1987
156	Ramnagar	IV	1,290	1981
157	Surinsar-Mansar	IV	3,958	1981
158	Tongri	IV	2,000	

KARNATAKA STATE

National Parks

159	Anshi	II	25,000	1987
160	Bandipur	II	87,420	1974
161	Bannerghatta	II	10,427	1974
162	Kudremukh	II	60,032	1987
163	Nagarahole	II	64,339	1975

Sanctuaries

164	Arabithittu	IV	1,350	
165	Bhadra	IV	49,246	1974
166	Biligiri Ranga Swamy Temple	IV	32,440	1974
167	Brahmagiri	IV	18,129	1974
168	Cauvery	IV	51,051	1987
169	Ghataprabha	IV	2,979	1974
170	Melkote Temple	IV	4,982	1974
171	Mookambika	IV	24,700	1974
172	Nugu	IV	3,032	1974

Map[†] ref.	National/international designation Name of area	IUCN management category	Area (ha)	Year notified
173	Pushpagiri	IV	10,292	1987
174	Ranebennur	IV	11,900	1974
175	Sharavathi Valley	IV	43,123	1974
176	Shettihally	IV	39,560	1974
177	Someshwara	IV	8,840	1974
178	Talkaveri	IV	10,500	1987
179	Tungabadra	IV	22,422	
	KERALA STATE			
	National Parks			
180	Eravikulam	II	9,700	1978
181	Periyar	II	30,500	1982
182	Silent Valley	II	8,952	1980
	Sanctuaries			
183	Aralam	IV	5,500	1984
184	Chimony	IV	10,500	1984
185	Chinnar	IV	9,044	1984
186	Idukki	IV	7,700	1976
187	Neyyar	IV	12,800	1958
188	Parambikulam	IV	28,500	1973
189	Peechi Vazhani	IV	12,500	1958
190	Peppara	IV	5,300	1983
191	Periyar	IV	47,200	1950
192	Shenduruny	IV	10,032	1984
193	Thattekkad Bird	IV	2,500	1983
194	Wynad	IV	34,444	1973
	MADHAYA PRADESH STATE			
	National Parks			
195	Bandhavgarh	II	44,884	1968
196	Indravati	II	125,837	1978
197	Kanger Ghati	II	20,000	1982
198	Kanha	II	94,000	1955
199	Madhav	II	15,615	1959
200	Panna	II	542,666	1981
201	Pench	II	29,286	1977
202	Sanjay	II	193,801	1981
203	Satpura	II	52,437	1981
	Sanctuaries			
204	Achanakmar	IV	55,155	1975
205	Badalkhol	IV	10,435	1975
206	Bagdara	IV	47,890	1978
207	Barnawapara	IV	24,466	1976
208	Bhairamgarh	IV	13,895	1983
209	Bori	IV	51,825	1977
210	Gandhi Sagar	IV	36,862	1974
211	Ghatigaon Great Indian Bustard	IV	51,200	1981
212	Gomarda	IV	27,782	1972
213	Karera Great Indian Bustard	IV	20,221	1981
214	Ken Gharial	IV	4,500	1981
215	Kheoni	IV	12,270	1982
216	Narsingarh	IV	5,719	1974
217	National Chambal	IV	42,300	1978
218	Noradehi	IV	118,696	1975
219	Pachmarhi	IV	46,086	1977
220	Palpur (Kuno)	IV	34,468	1981

Map[†] ref.	National/international designation Name of area	IUCN management category	Area (ha)	Year notified
221	Pamed	IV	26,212	1983
222	Panpatha	IV	24,584	1983
223	Pench	IV	11,847	1977
224	Phen	IV	11,024	1983
225	Ratapani	IV	68,879	1976
226	Sailana	IV	1,296	1983
227	Sanjay (Dubri)	IV	36,459	1975
228	Sardarpur	IV	34,812	1983
229	Semarsot	IV	43,036	1978
230	Singhori (Sindhari)	IV	28,791	1976
231	Sitanadi	IV	55,336	1974
232	Son Gharial	IV	4,180	1981
233	Tamor Pingla	IV	60,852	1978
234	Udanti	IV	24,759	1983

MAHARASHTRA STATE

National Parks

235	Gugamal	II	36,180	1987
236	Nawegaon	II	13,388	1975
237	Pench	II	25,726	1975
238	Sanjay Gandhi	II	8,696	1983
239	Tadoba	II	11,655	1955

Sanctuaries

240	Andhari	IV	50,900	1986
241	Aner Dam	IV	8,294	1986
242	Bhimashankar	IV	13,100	1985
243	Bor	IV	6,110	1970
244	Chandoli	IV	30,900	1985
245	Chaprala	IV	13,500	1986
246	Gautala Autram	IV	26,100	1986
247	Great Indian Bustard	IV	849,644	1979
248	Jaikwadi	IV	34,105	1986
249	Kalsubai Harishchandragad	IV	36,200	1986
250	Katepurna	IV	1,500	
251	Koyna	IV	42,400	1985
252	Malvan	IV	2,912	1987
253	Melghat	IV	159,733	1985
254	Nagzira	IV	15,281	1969
255	Nandur Madmeshwar	IV	10,010	1986
256	Painganga	IV	32,462	1986
257	Phansad	IV	7,000	1986
258	Radhanagari	IV	37,200	1958
259	Sagareshwar	IV	1,100	1985
260	Tansa	IV	30,481	1970
261	Yawal	IV	17,752	1969

MANIPUR STATE

National Parks

262	Keibul Lamjao	II	4,000	1977
263	Siroi	II	4,130	1982

Sanctuaries

264	Yagoupokpi Lokchao	IV	18,480	1989

MEGHALAYA STATE

National Parks

265	Balphakram	II	22,000	1986

Map[†] ref.	National/international designation Name of area	IUCN management category	Area (ha)	Year notified
266	Nokrek	II	6,801	1985
	Sanctuaries			
267	Nongkhyllem	IV	2,900	1981
	MIZORAM STATE			
	Sanctuaries			
268	Dampa	IV	48,000	1985
269	Murlen	IV	5,000	1989
	NAGALAND STATE			
	Sanctuary			
270	Intanki	IV	20,202	1975
	ORISSA STATE			
	National Park			
271	North Simlipal	II	84,570	1978
	Sanctuaries			
272	Balimela	IV	16,000	
273	Balukhand	IV	7,200	1984
274	Bhitar Kanika	IV	17,000	1975
275	Chandaka Dampada	IV	22,000	1982
276	Chilka	IV	1,553	1987
277	Debrigarh	IV	34,690	1985
278	Hadgarh	IV	19,160	1978
279	Kapilasa	IV	12,600	1970
280	Karlapat	IV	25,500	1969
281	Khalasuni	IV	11,600	1982
282	Kondakameru	IV	43,000	
283	Kotgarh	IV	39,950	1981
284	Kuldiha	IV	27,275	1984
285	Lakhari	IV	11,835	1985
286	Mahanadi Baisipalli	IV	16,835	1981
287	Saptasajya	IV	2,000	1970
288	Satkosia Gorge	IV	79,552	1976
289	Simlipal	IV	135,50C	1978
290	Sunabeda	IV	44,213	1983
291	Ushakothi	IV	30,403	1987
	PUNJAB STATE			
	Sanctuaries			
292	Abohar	IV	18,824	1975
293	Harike Lake	IV	4,300	1982
	RAJASTHAN STATE			
	National Parks			
294	Desert	II	316,200	1981
295	Keoladeo	II	2,873	1981
296	Ranthambore	II	39,200	1980
297	Sariska	II	27,380	1982
	Sanctuaries			
298	Bandh Baretha	IV	19,276	1985
299	Bassi	IV	15,290	
300	Bhensrodgarh	IV	22,914	1983
301	Darrah	IV	26,583	1955
302	Jaisamand	IV	5,200	1956

Map[†] ref.	National/international designation Name of area	IUCN management category	Area (ha)	Year notified
303	Jamwa Ramgarh	IV	30,000	1982
304	Jawahar Sagar	IV	10,000	1980
305	Keladevi	IV	67,600	1983
306	Kumbhalgarh	IV	57,826	1971
307	Mount Abu	IV	28,884	1960
308	Nahargarh	IV	5,000	1980
309	National Chambal	IV	28,000	1983
310	Phulwari	IV	51,141	1983
311	Ramgarh Bundi	IV	30,700	1982
312	Sariska	IV	49,200	1958
313	Sawai Mansingh	IV	10,325	1984
314	Shergarh	IV	9,871	1983
315	Sita Mata	IV	42,294	1979
316	Sunda Mata	IV	10,700	
317	Todgarh Raoli	IV	49,527	1983
318	Van Vihar	IV	5,993	1955
	SIKKIM STATE			
	National Park			
319	Khangchendzonga	II	84,950	1977
	Sanctuaries			
320	Fambong Lho	IV	5,176	1984
321	Maenam	IV	3,534	1987
322	Shingba	IV	3,250	1984
	TAMIL NADU STATE			
	National Park			
323	Indira Gandhi	II	11,808	1989
	Sanctuaries			
324	Anamalai	IV	84,935	1976
325	Kalakad	IV	22,358	1976
326	Mudumalai	IV	32,155	1940
327	Mukurthi	IV	7,846	1982
328.	Mundanthurai	IV	56,738	1962
329	Point Calimere	IV	1,726	1967
330	Pulicat	IV	46,102	1980
331	Srivilliputhur	IV	48,520	1988
	TRIPURA STATE			
	Sanctuaries			
332	Gumti	IV	38,954	1988
333	Sepahijala	IV	1,853	1987
334	Trishna	IV	17,056	1987
	UTTAR PRADESH STATE			
	National Parks			
335	Corbett	II	52,082	1936
336	Dudwa	II	49,029	1977
337	Gangotri	II	155,273	1991
338	Govind	II	47,208	1991
339	Nanda Devi	I	63,033	1982
340	Rajaji	II	83,153	1988
341	Valley of Flowers	II	8,950	1982

Map[†] ref.	National/international designation Name of area	IUCN management category	Area (ha)	Year notified
	Sanctuaries			
342	Binsar	IV	4,559	1988
343	Chandra Prabha	IV	7,800	1957
344	Govind Pashu Vihar	IV	48,104	1954
345	Hastinapur	IV	2,073	1986
346	Kaimur	IV	50,075	1982
347	Katarniaghat	IV	40,009	1976
348	Kedarnath	IV	97,524	1972
349	Kishanpur	IV	22,712	1972
350	National Chambal	IV	63,500	1979
351	Ranipur	IV	23,031	1977
352	Sohagabarwa	IV	42,820	1987
353	Sonanadi	IV	30,118	1987
	WEST BENGAL STATE			
	National Parks			
354	Neora Valley	II	8,689	1986
355	Singalila	II	7,860	1986
356	Sundarbans	I	133,010	1984
	Sanctuaries			
357	Buxa	IV	31,452	1986
358	Halliday Island	IV	583	1976
359	Jaldapara	IV	11,563	1941
360	Lothian Island	IV	3,885	1976
361	Mahananda	IV	12,722	1976
362	Sajnakhali	IV	36,234	1976
363	Senchal	IV	3,860	1976
	Ramsar Wetlands			
	Chilka	R	1,553	1987
	Harike Lake	R	4,300	1982
	Keoladeo National Park	R	2,873	1981
	Loktak Lake	R	26,600	1990
	Sambhar Lake	R	24,000	1990
	Wular Lake	R	18,900	1990
	World Heritage Sites			
	Kaziranga National Park	X	42,996	1985
	Keoladeo National Park	X	2,873	1985
	Manas Wildlife Sanctuary	X	39,100	1985
	Nanda Devi National Park	X	63,033	1988
	Sundarbans National Park	X	133,010	1987

[†]Locations of most protected areas are shown on the accompanying maps. More extensive maps showing the locations of national parks and sanctuaries are presented in Rodgers and Panwar (1988).

*Areas are estimated from digitised maps using Autocad 2.6 software (Source: IIPA/Environmental Studies Division) and may differ from official records.

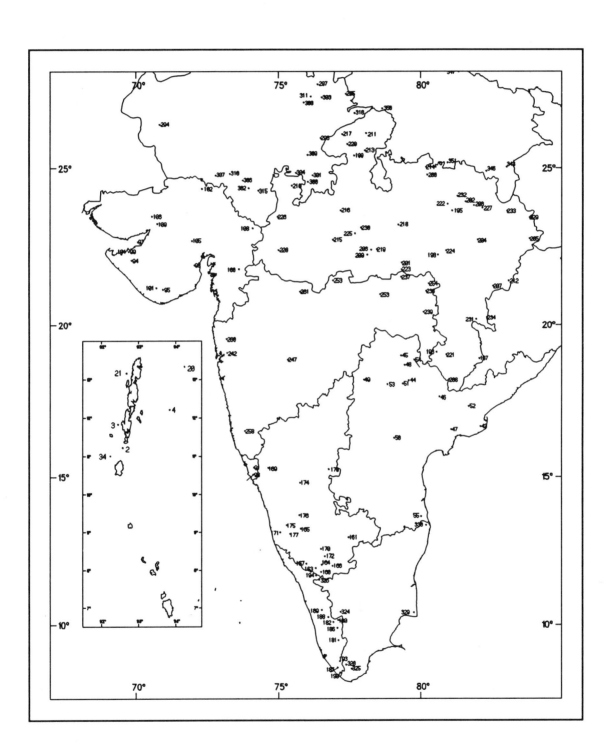

Protected Areas of South India and the Andaman and Nicobar Islands

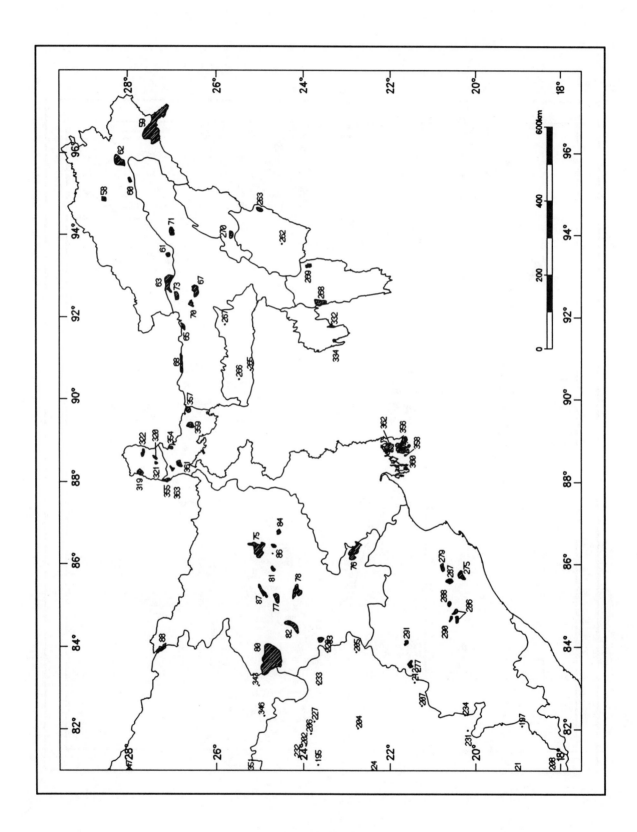

Protected Areas of North-east India

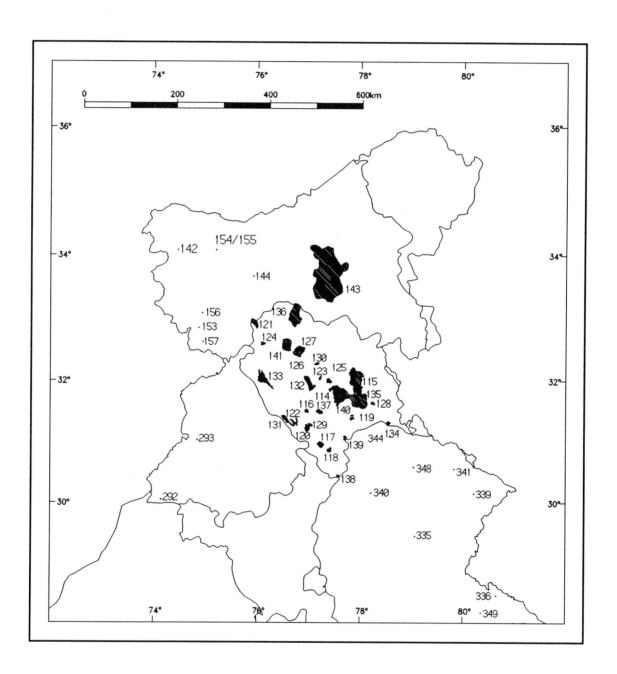

Protected Areas of North-west India

INDONESIA

Area 1,919,445 sq. km

Population 189.4 million (1990)
Natural increase: 1.8% per annum

Economic Indicators
GNP: US$ 430 (1988)

Policy and Legislation Government policy regarding conservation is based upon the desire to promote cultural and economic development of the Indonesian people in harmony with the natural environment, as expressed in the 1945 Constitution. According to this policy, all forms of natural life and examples of all ecosystems have an important role in human welfare and must be preserved for the benefit of present and future generations. This is elucidated in the Basic Environmental Law of 1982 which makes provision for the management of the living environment. It emphasises the importance of forest lands, stating that they form the primary means of maintaining harmony between man and the environment.

Reserves were originally established on the basis of two colonial government ordinances, of which the Nature Protection Ordinance (Natuurbescherminge Ordannantie) of 1941 (Staatsblad No. 167) rescinded the earlier Nature Protection Ordinance of 1932 (Staatsblad No. 17). The now obsolete 1941 Ordinance made provision for nature reserves (Cagar Alam) and game reserves (Suaka Margasatwa) and their management by the Forest Department, under the jurisdiction of the Ministry of Agriculture. The Forest Department was given full ministerial status in 1981 and subsequently assumed full responsibility for the management of the national forest estate.

To overcome a number of weaknesses inherent in the earlier legislation, the Conservation of Living Natural Resources and their Ecosystems Act was passed on 10 August 1990. This rescinds the 1941 Nature Protection Ordinance and distinguishes between nature sanctuaries (nature reserves and game reserves) and nature conservation areas (national parks and grand forest parks). These designations, which are equally applicable to terrestrial or marine reserves, provide differing degrees of legal protection ranging from nature reserves, where no human activities are allowed, to national parks where utilisation of natural resources by local people is permitted within a system of management zoning. The Act also makes provision for existing protected areas to be established as biosphere reserves under the Unesco Man and the Biosphere Programme. Gazettement, alteration and declassification of protected areas is by decree of the Minister of Forestry.

To date there is no clear and comprehensive enunciation of forest policy in Indonesia; it has developed through a series of laws, regulations and decisions on the part of various government departments including the Department of Forestry. The Basic Forestry Act No. 5 of 1967 is the most comprehensive existing statement of government policy and describes the purpose, supply, availability and use of forests in the interests of community development. The Act also provides a number of conservation-related provisions such as protection of flora, fauna, and water catchments (GOI/IIED, 1985). To achieve these objectives, three categories of forest are defined under the Act: production forest (Hutan Produksi), protection forest (Hutan Lindung) and tourist forest, which is subdivided into tourist park (Taman Wisata) and hunting park (Taman Buru). The designations of nature reserve and game reserve are reiterated under the term nature sanctuary (see Annex).

International Activities Indonesia participates in a number of international conservation-related conventions and programmes. It is a party to the Convention concerning the Protection of World Cultural and Natural Heritage (World Heritage Convention) which it accepted on 6 July 1989. Although no sites have yet been inscribed, three protected areas have recently been nominated for inclusion on the World Heritage List. Indonesia is a contracting party to the ASEAN Convention, and has three designated sites to date: Gunung Leuser and Kerinci Seblat in Sumatra, and Lorentz in Irian Jaya. Indonesia also participates in the Unesco Man and the Biosphere Programme. To date six biosphere reserves have been declared. Other international activities include proposals for the cooperative management of a number of reserves that abut international borders with Papua New Guinea and Malaysia (Hadiseputro and Wardojo, 1990).

Administration and Management The first organisation to be given responsibility for establishing and managing protected areas was the Directorate of Nature Conservation and Wildlife (PPA), established in 1971 as a separate unit of the then Directorate General of Forestry, under the Ministry of Agriculture. With elevation of this Directorate to full ministerial status, PPA became the Directorate General of Forest Protection and Nature Conservation (PHPA) within which are four different directorates. Of these, the Directorate of Nature Conservation and the Directorate of National Parks and Recreation Forest are concerned specifically with marine and terrestrial protected areas. Responsibilities of the Directorate of Nature Conservation include overall planning of the terrestrial and marine protected areas network, drafting of conservation legislation and the proposal, establishment and management of individual protected areas. The Directorate of National Parks and Recreation Forest deals specifically with development of the national parks programme (Sumardja and McNeely, 1980). Scientific bodies responsible for screening and approving technical decisions relating to terrestrial and marine conservation are the Centre for Biological Research and Development and the Centre for Oceanography Research and Development, both of which are part of the Indonesian Institute for Sciences (Scott, 1989).

In the field, PHPA operates through the regional structure of the Directorate of Nature Conservation, which consists of eight provincial offices (Balai Konservasi Sumber Daya Alam – Office of Natural Resources Conservation). These work in association with the regional planning boards (BAPPEDA) which are ultimately responsible to the provincial government. They are supported by several regional Sub-Balai (SBKSDA) within each province, with an additional office for each national park which reports directly to the Directorate of National Parks and Recreation Forest.

The work of PHPA has been greatly assisted by a number of international non-governmental and inter-governmental organisations. Of these, the Food and Agriculture Organisation of the United Nations (FAO), in conjunction with the United Nations Development Programme (UNDP), has initiated a number of major projects and programmes. These include two successive UNDP/FAO programmes: INS/73/013 Nature Conservation and Wildlife Management – from 1974 to 1978; and INS/78/061 Development of National Parks – from 1978 to 1982, under which a National Conservation Plan was prepared (FAO, 1981-1982). Under the World Wide Fund for Nature-Indonesia programme, initiated in 1977, WWF has provided field and consultancy staff and assisted in the preparation and implementation of numerous protected areas management plans, a regional strategy for Irian Jaya (Petocz, 1989) and a system plan for the conservation of Indonesia's marine environment (IUCN/WWF, 1984).

Other international non-governmental organisations beginning or extending programmes concerned with the protected areas system include Wildlife Conservation International, Conservation International, The Nature Conservancy, the International Council for Bird Preservation (ICBP) and the Asian Wetland Bureau (AWB), the latter's operations extending from regional and national wetland inventory preparation to detailed studies of protected areas, and preparation of recommendations for their management.

At the national level, non-governmental conservation organisations are increasingly active in lobbying government and mobilising public support for national environmental issues. They frequently play a crucial role by representing local interests and acting as intermediaries between local communities and government. Of the major groups, the Indonesian Environmental Forum (WALHI) was set up during the early 1980s in order to strengthen the environmental programme of Indonesian NGOs and to provide nationwide exposure for their concerns. WALHI today is an umbrella organisation for over 500 Indonesian

NGOs and networks. Of these, the Indonesian Forest Conservation Secretariat (Sekretariat Pelestarian Hutan Indonesia - SKEPHI) coordinates a network of some 54 non-governmental organisations concerned with the conservation of forest resources, maintains contact with the State Ministry of Population and Environment and frequently acts as a pressure group, lobbying the cause of forest conservation. Other Indonesian NGOs include the Natural Resource Conservation Group (Kelompok Pelestarian Sumber Daya Alam – KPSA), which was established in 1983 to function as a mediator between the government and people, in order to facilitate participation in conserving forests and maintaining environmental quality. A list of Indonesian non-governmental organisations can be found in Scott (1989).

Management of the protected areas system is constrained by a number of factors, many of which are outside the PHPA's terms of reference. Despite progressively increasing government budgets and international investment programmes, management has been unable to keep pace with the increased workload demanded by an expanding protected areas network. Major constraints in the field include insufficient trained and motivated personnel and a frequent absence of clearly and accurately defined protected areas boundaries. Management problems are frequently compounded by the conflicting land use objectives of national and local government, the conservation administration, and local people living within and around the reserves themselves. In order to overcome these difficulties, reserve management is becoming increasingly concerned with improving local welfare and recognising the important role local people can play in managing the areas they utilise and perceive of as their own through traditional (adat) rights. In addition, the conservation of marine protected areas presents special difficulties, as PHPA capabilities and training programmes have traditionally been oriented toward the terrestrial environment.

Systems Reviews Spanning two major biogeographic regions, the Oriental and Australian, the Indonesian archipelago consists of approximately 17,000 islands stretching in an east-west direction for 5,200km across the Sunda and Sahul continental shelves. Characterised by an enormously varied physical structure of high mountain ranges, volcanoes, alluvial plains, lakes, swamps and shallow coastal waters, the archipelago exhibits a biological diversity and richness which is without comparison in South-East Asia (FAO, 1981-82; Petocz, 1989; Scott, 1989).

Historically, human settlement has concentrated on the so-called "inner islands" with fertile volcanic soils, particularly Java and Bali and to a lesser extent Lombok. Although the population in Java in 1817 was estimated by Raffles to be only about four million, a tradition of intensive cultivation, with high population densities, has been in evidence for several generations. Introduction of estate crops, especially coffee, tobacco and tea, during the 19th century contributed to the expansion of cultivation into the uplands, destroying the natural forests in these areas. During the last half of this century, increasing population pressure as well as changing economic circumstances have led to the cultivation of many of these former plantations. Today these islands contain more than 64% of the population and produce some 70% of the national food supply (IIED/GOI, 1985).

With the exception of the Lesser Sunda Islands, the natural vegetation of the "outer" islands of Sumatra, Kalimantan, Sulawesi, Moluccas and Irian Jaya consists primarily of tropical moist forest of which Indonesia holds approximately 10% of the world total (Davies *et al.*, 1986). Deciduous monsoon forest occurs in seasonally dry areas, particularly in southern and eastern islands such as the Lesser Sundas and the southern part of Irian Jaya. The vegetation types to the east and west of the "Wallace-line" are divided by a biogeographic boundary that extends from north to south along the Sunda Shelf. Forests on the islands of the shelf itself are principally Malesian and dominated by the commercially important Dipterocarpaceae, while those found to the east have greater affinities with the Australo-Pacific realm and are dominated by mixed tropical hardwood species. Extensive natural wetlands, including many of international importance, are found in the low-lying alluvial plains and basins, flat-bottomed valleys and mangrove estuaries of Sumatra, Kalimantan and Irian Jaya. In addition, Indonesia contains some of the largest artificial wetlands in the world, including millions of hectares of rice paddies and nearly 200,000ha of fish ponds (Scott, 1989). According to IUCN/WWF (1984), Indonesia's marine habitats contain some of the most important coral reef systems in the world, with fringing reef the most common type. These are almost continuous along the coasts of Sulawesi, Maluku, north and west Irian Jaya and the Lesser Sundas and the Sumatran archipelagoes of Mentawi, Belitung and Riau.

Habitat loss has been most acute on the heavily populated central islands of Bali and Java. The remaining natural forests on these islands consist largely of sub-montane and montane forests, the majority of natural lowland vegetation having long since been converted for agriculture. Given the high population density and limited job opportunities for the growing number of landless and unemployed workers, illegal encroachment of forests for arable land, fuel and livestock grazing seems set to continue (GOI/IIED, 1985). Forest degradation in the outer islands, by contrast, is a relatively recent phenomenon, the generally poor, non-volcanic soils of islands such as Kalimantan having discouraged extensive settlement for many centuries. Although valuable timber and other forest products have been exported from these islands since early colonial times, it is only in the past two decades that efforts have been directed towards extensive exploitation of their rich natural resources. An integral part of the development of the outer islands has been the government's programme of transmigration, the primary goal of which was to redistribute some of the population

growth on Java and Bali to the less populous outer islands, thereby relieving pressure on critical watersheds. To avoid repetition of problems associated with poor site planning, the Regional Physical Planning Programme for Transmigration (RePPProT), a joint cooperative venture between the Indonesian and British governments, has been completed in order to provide a natural resources database for transmigration planning. Currently, the transmigration programme is in its second phase and is concerned primarily with upgrading existing settlements rather than developing new sites.

The current protected areas network has its origins in the 19th century under the Dutch colonial administration, when Indonesia's first nature reserve was established at Cibodas, West Java, in 1889. Since the declaration of independence in 1949, by which time over 100 sites had been established, the network has expanded considerably, and now covers 19 million ha, some 10% of the total land area (Hadiseputro and Wardojo, 1990). During this period, a number of systems reviews have been published with recommendations for the further development of the protected areas system.

The first of these, the National Conservation Plan for Indonesia (FAO, 1981-82), outlined a programme for the development of a comprehensive protected areas network covering 11.2 million ha of the terrestrial area. This subsequently formed the basis for the Indonesian section of the IUCN Systems Review of the Indomalayan Realm (MacKinnon and MacKinnon, 1986) and all subsequent reviews. An expanded and updated conservation strategy for Irian Jaya has since been produced by WWF/IUCN (Petocz, 1984, 1989). The establishment of a protected areas network for Indonesia's marine environment was planned by IUCN/WWF (1984), with provision of 10 million ha of marine sites by 1990. Areas of the highest priority for inclusion in the marine reserve network include those which directly safeguard current subsistence and commercial fisheries, threatened resources and the tourist industry (Petocz, n.d.). Recommendations from these studies form the basis of the Corbett Action Plan which was prepared by protected areas managers at the 25th Working Session of the IUCN Commission on National Parks and Protected Areas (IUCN, 1985).

Despite many of the criteria for the design of the protected area system having been met, a number of deficiencies remain. Wetlands in particular have received inadequate coverage in the existing network and are under-represented, partly because of insufficient information regarding their conservation value. A preliminary inventory of wetlands has been compiled by the Asian Wetland Bureau (Silvius *et al.*, 1987), in order to identify sites for incorporation into the protected areas network. A similar situation applies to lowland tropical rain forest, new areas of which will be difficult to allocate to conservation due to their economic value and accessibility (Sumardja and McNeely, 1980). The protection of Indonesia's natural heritage and resources for the benefit of future generations will therefore require innovative conservation measures, possibly outside the existing protected areas network. That these must take account of human needs is increasingly crucial, as many areas of outstanding conservation value are both populated and subject to conflicting land use demands of a growing economy (Leader-Williams *et al.*, 1990).

Addresses

Directorate General of Forest Protection and Nature Conservation, (Director of Nature Conservation), Jalan Ir H Juanda 15, PO Box 133, Bogor 16001 (Tel: 251 24013/23067)

Directorate of National Parks and Recreation Forest, Jalan Ir H Juanda 100, Bogor 16123 (Tel: 251 321014)

Asian Wetland Bureau (National Coordinator), PO Box 254/BOO, Jalan Arzimar III No. 17, Bogor 16001 (Tel: 251 325755; FAX: 251 325755; Cable: INTERWADER)

Indonesian Environmental Forum (WALHI), Jln Penjernihan 1, Komplek Keuangan No. 15, Pejompongan, Jakarta 10210 (Tel: 21 586820; FAX: 21 588416/586181)

Indonesian Forest Conservation Secretariat (SKEPHI), Jalan Ir Suryopranoto No. 8 Lt. IV, Jakarta Pusat

World Wide Fund For Nature-Indonesia, Jalan Ir H Juanda 9, PO Box 133, Bogor 16001 (Tel: 251 327316; FAX: 251 328177)

World Wide Fund For Nature-Indonesia, Jln Pela No. 3, Gandaria Utara, PO Box 29/JKSKM, Jakarta Selatan 12001 (Tel/FAX: 21 769 4726)

References

Davies, S.D., Droop, S.J.M., Gregerson, P., Henson, L., Leon, C.J., Lamlein Villa-Lobos, J., Synge, H. and Zantovska, J. (1986). *Plants in danger: what do we know?* IUCN, Gland, Switzerland and Cambridge, UK. Pp. 173-178.

FAO (1982-83). *A National Conservation Plan for Indonesia.* 8 volumes. UNDP/FAO National Parks Development Project INS/78/061. FAO, Bogor.

Hadiseputro, S. and Wardojo, W. (1990). Status of national parks management in Indonesia. Regional Expert Consultation on management of protected areas in the Asia-Pacific region, 10-14 December. FAO Regional Office, Bangkok, Thailand. 7 pp.

IIED/GOI (1985). *A review of policies affecting the sustainable development of forest lands in Indonesia.* Vol. III. Background paper. International Institute for Environment and Development/Government of Indonesia, Jakarta. 142 pp.

IUCN (1985). *The Corbett Action Plan for protected areas of the Indomalayan Realm.* IUCN, Gland, Switzerland and Cambridge, UK. 23 pp.

IUCN/WWF (1984). *The seas must live: a protected areas system plan for the conservation of Indonesia's marine environment.* 7 vols. IUCN/WWF report. WWF-Indonesia, Bogor. 22 pp.

MacKinnon, J. and MacKinnon, K. (1986). *Review of the protected areas system in the Indo-Malayan Realm.* IUCN, Gland, Switzerland and Cambridge, UK/UNEP, Nairobi, Kenya. Pp. 257-258.

Leader-Williams, Harrison, J. and Green, M.J.B. (1990). Designing protected areas to conserve natural resources. *Scientific Progress Oxford* 74: 189-204.

Petocz, R.G. (n.d.). *Conservation in Indonesia: current status and development of an action strategy.* Report to the World Bank. 57 pp.

Petocz, R.G. (1984). *Conservation and development in Irian Jaya – a strategy for rational resource utilization.* WWF/IUCN Conservation for Development Programme in Indonesia. Prepared for Directorate General of Forest Protection and Nature Conservation, Bogor. 173 pp.

Petocz, R.G. (1989). *Conservation and development in Irian Jaya: a strategy for rational resource utilization.* E.J. Brill, Leiden, the Netherlands. 18 pp.

Scott, D.A. (Ed.) (1989). *A directory of Asian wetlands.* IUCN, Gland, Switzerland and Cambridge, UK. Pp. 981-1103.

Silvius, M.J., Djuharsa, E., Taufik, A.W., Steeman, A.P.J.M. and Berczy, E.T. (1987). *The Indonesian wetland inventory. A preliminary compilation of existing information on wetlands of Indonesia.* PHPA, AWB/Interwader, EDWIN, Bogor. 268 pp.

Sumardja, E.A. and McNeely, J.A. (1980). National parks and protected areas of Indonesia: an up-date. Prepared for the 16th Working Session of the IUCN Commission on National Parks and Protected Areas, Perth, Scotland. Unpublished report. 52 pp.

UNEP (1977). *Nature conservation and wildlife management: Indonesia.* Interim Report INS/73/013. UNEP/FAO, Rome. 133 pp.

UNEP/IUCN (1988). *Coral reefs of the world. Volume 2: Indian Ocean, Red Sea and Gulf.* UNEP Regional Seas Directories and Bibliographies. IUCN, Gland, Switzerland and Cambridge, UK/UNEP, Nairobi, Kenya. Pp. 95-132.

ANNEX
Definitions of protected area designations, as legislated, together with authorities responsible for their administration

Title: Basic Forestry Act

Date: 1967

Brief description: Provides for protection, management and exploitation of forest lands

Administrative authority: Ministry of Forestry

Designations:

Hutan Produksi (production forest) Forest which, because of its natural condition or capacity, can give benefits in the form of timber and other forest products. The removal of forest products is regulated in such a way that it can be continued permanently.

Hutan Lindung (Protection forest) Forest whose natural condition is such that it exerts a good influence upon soil, the surrounding environment and water control, and so must be maintained and protected.

Among forests classified as protective forest, there are some from which, because of their natural condition, products can still be removed within certain limits, without detracting from/diminishing their protection.

Nature Sanctuary

Cagar Alam (Nature reserve) No management or human interference is permitted that changes the character of soil, flora or fauna in any way or affects its pristine condition. Access is for scientific purposes only and is subject to written permission of the Directorate of Forest Protection and Nature Conservation (PHPA).

Suaka Margasatwa (Game reserve) No activities are permitted that damage the flora, fauna or landscape or that could detract from the value of the reserve. Provision is made, however, for hunting, subject to written permission of the Minister of Forestry, and also for development of forest industries subject to a permit issued by the provincial governor for collection of forest produce, grazing of livestock and fishing.

Tourist Forest

Taman Buru (Hunting park) Managed specifically for hunting and fishing.

Taman Wisata (Recreation park) Maintained for outdoor recreation purposes.

Source: FAO, 1977, 1982-83; Scott, 1989

Title: Conservation of Living Resources and their Ecosystems Act

Date: August 1990

Brief description: Concerned with the maintenance of biodiversity and ecosystem function in the context of the sustainable utilisation of living natural resources.

Administrative authority: Directorate General of Forest Protection and Nature Conservation (Director of Nature Conservation)

Designations:

Nature Sanctuary

A specific terrestrial or aquatic area having protection as its main function to preserve biodiversity of plants and animals, as well as their ecosystems which also act as life support systems.

Cagar Alam (Nature reserve) A nature sanctuary which, because of its characteristic plants, animals and/or ecosystems, must be protected and allowed to develop naturally. Activities permitted are research and the development of science, education and other activities protecting breeding stock. Management shall be by the government in an effort to preserve the species diversity of plants and animals and their ecosystems.

Suaka Margasatwa (Game reserve) A nature sanctuary having high species diversity and/or unique animal species, in which the habitat may be managed, in order to assure the continued existence of these species. Management shall be implemented by the government in an effort to preserve the diversity of plant and animal species and their ecosystems.

Biosphere Reserve

An area of unique and/or degraded ecosystems, which needs to be protected and conserved for its research and education value.

Within the framework of international conservation and for those activities defined in Article 17, "sanctuary reserves" and other specified areas can be established as biosphere reserves.

Kawasan Pelestarian Alam (Nature Conservation Area)

A specific terrestrial or aquatic area where the main functions are to protect life support systems, to preserve diversity of plant and animal species, as well as to conserve living natural resources and their ecosystems for sustainable utilisation.

Taman Nasional (National park) A nature conservation area which possesses natural ecosystems, and which is managed through a zoning system for research, science, education, supporting cultivation, recreation and tourism purposes.

Taman Hutan Agung (Grand forest park) A nature conservation area created to provide a collection of indigenous and/or introduced plants and animals for research, science, education, supporting cultivation, culture, recreation and tourism purposes.

Taman Wisata Alam (Nature recreation park) A nature conservation area mainly intended for recreation and tourism purposes.

Source: Translation of original legislation.

SUMMARY OF PROTECTED AREAS

Map[†] ref.	National/international designation Name of area	IUCN management category	Area (ha)	Year notified
	IRIAN JAYA			
	National Parks			
1	Teluk Cenderawasih Marine	II	1,433,000	1990
2	Wasur	II	308,000	1991
	Nature Reserves			
3	Enarotali	I	300,000	1980
4	Gunung Lorentz	I	2,150,000	1978
5	Pegunungan Arfak	I	45,000	
6	Pegunungan Cyclops	I	22,500	1978
7	Pulau Batanta Barat	I	10,000	1981
8	Pulau Biak Utara	I	11,000	1982
9	Pulau Misool	I	84,000	1982
10	Pulau Salawati Utara	I	57,000	1982
11	Pulau Superiori	I	42,000	1982
12	Pulau Waigeo Barat	I	153,000	1981
13	Pulau Yapen Tengah	I	59,000	1982
14	Rawa Biru	I	4,000	1978
	Game Reserves			
15	Pegunungan Jayawijaya	IV	800,000	1981
16	Pulau Anggrameos	IV	2,500	1981
17	Pulau Dolok	IV	600,000	1978
18	Wasur	IV	308,000	1982
	Recreation Park			
19	Teluk Yotefa	V	1,650	1978
	JAVA			
	National Parks			
1	Bali Barat	II	77,727	1982
2	Baluran	II	25,000	1980
3	Bromo-Tengger-Semeru	II	57,606	1982
4	Gunung Gede Pangrango	II	15,000	1980
5	Kepulauan Karimun Jawa Marine	II	111,625	1986
6	Kepulauan Seribu Marine	II	110,000	1982
7	Meru Betiri	II	58,000	1972
8	Ujung Kulon	II	78,359	1980
	Nature Reserves			
9	Arjuno Lalijiwo	I	4,960	1972
10	Batukau I/II/III (Bali)	I	1,762	1974
11	Cibodas-Gunung Gede	I	1,040	1925

Map† ref.	National/international designation Name of area	IUCN management category	Area (ha)	Year notified
12	Gunung Burangrang	I	2,700	1979
13	Gunung Celering	I	1,279	1973
14	Gunung Halimun	I	40,000	1979
15	Gunung Honje	I	10,000	1979
16	Gunung Simpang	I	15,000	1979
17	Gunung Tilu	I	8,000	1978
18	Gunung Tukung Gede	I	1,700	1979
19	Kawah Ijen Ungup-Ungup	I	2,468	1920
20	Kawah Kamojang	I	7,500	1979
21	Laut Pasir Tengger-Gunung Bromo	I	5,287	1919
22	Leuwang Sancang	I	2,157	1978
23	Nusa Barung	I	6,100	1920
24	Nusa Kambangan	I	4,983	1937
25	Pulau Panaitan/Pulau Peucang	I	17,500	1937
26	Pulau Saobi (Kangean Islands)	I	430	1919
27	Pulau Sempu	I	877	1928
28	Ranu Kumbolo	I	1,340	1921
29	Rawa Danau	I	2,500	1921
30	Tangkuban Perahu	I	1,290	1974
	Game Reserves			
31	Bali Barat	IV	19,475	1947
32	Baluran	IV	25,000	1937
33	Banyuwangi Selatan (Blambangan)	IV	62,000	1939
34	Cikepuh	IV	8,128	1973
35	Dataran Tinggi Yang (Yang Plateau)	IV	14,145	1962
36	Gunung Sawal	IV	5,400	1979
37	Meru Betiri	IV	58,000	1972
38	Pulau Bawean	IV	3,832	1979
	Hunting Parks			
39	Maelang	VI	70,000	1966
40	Masigit Kareumbi	VI	12,421	1976
	Recreation Parks			
41	Gunung Gamping	V	1,102	1982
42	Gunung Tampomas	V	1,250	1979
43	Gunung Tangkuban Perahu	V	1,290	1974
	KALIMANTAN			
	National Parks			
1	Gunung Palung	II	90,000	1990
2	Kutai	II	200,000	1982
3	Tanjung Puting	II	300,040	1982
	Nature Reserves			
4	Bukit Baka	I	70,500	1987
5	Bukit Raya	I	110,000	1979
6	Bukit Tangkiling	I	2,061	1977
7	Gunung Palung	I	30,000	1937
8	Gunung Raya Pasi	I	3,742	1978
9	Mandor	I	2,000	1936
10	Muara Kaman Sedulang	I	62,500	1976
11	Padang Luwai	I	5,000	1967
12	Pararawan I,II	I	6,200	1979
13	Sungai Kayan Sungai Mentarang	I	1,600,000	1980
	Marine Nature Reserve			
14	Kepulauan Karimata	I	77,000	1985

Map[†] ref.	National/international designation Name of area	IUCN management category	Area (ha)	Year notified
	Game Reserves			
15	Danau Sentarum	IV	80,000	1985
16	Gunung Palung	IV	60,000	1981
17	Kelumpang-Selat Laut-Selat Sebuku	IV	66,650	1981
18	Kutai	IV	200,000	1971
19	Pleihari Martapura	IV	36,400	1974
20	Pleihari Tanah Laut	IV	35,000	1975
21	Tanjung Puting	IV	300,040	1978
	Marine Parks			
22	Gunung Asuansang	IV	28,000	
23	Pulau Sangalaki	IV	280	1982
	Protection Forests			
24	Bukit Batutenobang	VI	883,000	
25	Bukit Perai	VI	100,000	
26	Bukit Rongga	VI	110,000	
27	Gunung Tunggal	VI	50,830	
	Recreation Parks			
28	Bukit Suharto	V	27,000	
29	Tanjung Keluang	V	2,000	1984
	LESSER SUNDA ISLANDS			
	National Parks			
1	Gunung Rinjani	II	40,000	1990
2	Komodo	II	40,729	1980
	Nature Reserves			
3	Maubesi (West Timor)/NTT	I	1,830	1981
4	Pulau Tujuh Belas (North Flores)	I	11,900	1987
5	Ruteng (Flores Is.)	I	30,000	
6	Tanah Pedauh (Sumbawa Is.)	I	543	1975
7	Wae Wuul Mburak (North Flores)	I	3,000	1985
	Game Reserves			
8	Gunung Wanggameti (Sumba Is.)	IV	6,000	
9	Kateri (West Timor)	IV	4,560	1981
10	Manupau (Sumba Is.)	IV	12,000	
11	Pulau Menipo (West Timor Is.)	IV	3,000	1977
12	Pulau Moyo (Sumbawa)	IV	18,765	1975
13	Pulau Moyo Marine	IV	6,000	
14	Pulau Padar (Komodo Is.)	IV	1,533	1969
15	Pulau Rinca (Komodo Is.)	IV	8,196	1969
16	Wae Wuul Mburak (North Flores)	IV	3,000	1960
	Hunting Parks			
17	Dataran Bena	VI	11,370	1978
18	Tambora Selatan (Sumbawa Is.)	VI	30,000	1978
	Forest Reserve			
19	Gunung Timau FoR (West Timor)	VI	15,000	
	Protection Forests			
20	Gunung Mutis (West Timor)	VI	10,000	
21	Lewotobi (East Flores)	VI	4,200	
22	Manupau (Sumba Is.)	VI	12,000	
23	Selah Legium Complex (Sumbawa Is.)	VI	50,000	
24	Watu Panggota/Bondokapu (Sumba Is.)	VI	4,000	

Map[†] ref.	National/international designation Name of area	IUCN management category	Area (ha)	Year notified
	Recreation Parks			
25	Danau Kelimutu	V	4,984	1984
26	Tuti Adagae (Alor Is.)	V	5,000	1981
	Marine Recreation Park			
27	Pulau Teluk Maumere	V	59,450	1987
	MOLUCCAS			
	National Park			
1	Manusela	II	189,000	1982
	Nature Reserves			
2	Pulau Banda	I	2,500	1977
3	Pulau Seho	I	1,250	1972
4	Wae Mual	I	17,500	1972
5	Wae Nua	I	35,000	1972
	Marine Nature Reserves			
6	Pulau Pombo	I	1,000	1973
7	Sebagian Kep. Aru Bagian Tenggara	I	114,000	1991
	Game Reserves			
8	Pulau Baun	IV	13,000	1974
9	Pulau Kassa	IV	900	1978
10	Pulau Kassa Marine	IV	1,100	1978
11	Pulau Manuk	IV	100	1981
	SULAWESI			
	National Parks			
1	Bunaken Menado Tua Marine	II	89,065	1989
2	Dumoga-Bone	II	300,000	1982
3	Lore Lindu	II	231,000	1982
4	Rawa Aopa Watumohai	II	200,000	
	Nature Reserves			
5	Bantimurung	I	1,018	1980
6	Bulusaraung	I	5,690	1980
7	Dua Saudara	I	4,299	1978
8	Gunung Ambang	I	8,638	1978
9	Gunung Kelabat	I	5,300	1932
10	Karaenta	I	1,000	1976
11	Morowali	I	225,000	1986
12	Paboya	I	1,000	1973
13	Panua	I	1,500	1938
14	Pegunungan Feruhumpenai	I	90,000	1979
15	Pulau Mas Popaya Raja	I	160	1919
16	Tangkoko Batuangus	I	4,446	1919
17	Tangkoko-Dua Saudara	I	8,745	1978
18	Tanjung Api	I	4,246	1977
	Marine Nature Reserves			
19	Arakan Wowontulap	I	13,800	1986
20	Kepulauan Togian	I	100,000	1989
21	Pulau Bunaken	I	75,265	1986
22	Take Bone Rate	I	530,765	1989
	Game Reserves			
23	Bone	IV	110,000	1979
24[†]	Bontobahari	IV	4,000	1980

Map[†] ref.	National/international designation Name of area	IUCN management category	Area (ha)	Year notified
25	Bulawa	IV	75,200	1980
26	Buton Utara	IV	82,000	1979
27	Dumoga	IV	93,500	1979
28	Gunung Manembo-Nembo	IV	6,500	1978
29	Lampoko Mampie	IV	2,000	1978
30	Lombuyan I/II	IV	3,665	1974
31	Lore Kalimanta	IV	200,000	1973
32	Pinjan/Tanjung Matop	IV	1,613	1981
33	Tanjung Batikolo	IV	5,500	1980
34	Tanjung Peropa	IV	38,937	1986
	Hunting Parks			
35	Gunung Watumohai	VI	50,000	1976
36	Karakelang	VI	22,000	1979
	Research Forest			
37	Sungai Camba	IV	1,300	
	Protection Forests			
38	Danau Lindu	VI	31,000	1978
39	Gunung Damar	VI	30,000	1939
40	Gunung Lokon	VI	3,930	1939
41	Gunung Lompobatang	VI	20,000	
42	Gunung Soputan	VI	13,433	1933
43	Lubutodaa & Paguyaman Barat	VI	16,000	1943
44	Pegunungan Latimojong	VI	58,000	1940
45	Tamposo-Sinansajang	VI	15,000	1937
46	Wiau	VI	6,280	1939
	Recreation Parks			
47	Danau Matado/Mahalano	V	30,000	1979
48	Danau Towuti	V	65,000	1979
	SUMATRA			
	National Parks			
1	Barisan Selatan	II	365,000	1982
2	Gunung Leuser	II	792,675	1980
3	Kerinci Seblat	II	1,484,650	1981
4	Way Kambas	II	130,000	
	Nature Reserves			
5	Bukit Tapan	I	66,500	1978
6	Dolok Sibual Bual	I	5,000	1982
7	Dolok Sipirok	I	6,970	1982
8	Gunung Indrapura	I	70,000	1980
9	Indrapura	I	221,136	1929
10	Jantho	I	8,000	1984
11	Kelompok Hutan Bakau Pantai Timur Jambi	I	6,500	1981
12	Krakatau	I	2,500	1919
13	Manua	I	1,500	
14	Pulau Berkeh	I	500	1968
15	Pulau Burung	I	200	1968
16	Rimbo Panti	I	2,830	1934
17	Toba Pananjung	I	1,235	1932
	Game Reserves			
18	Bentayan	IV	19,300	1981
19	Berbak	IV	175,000	1935
20	Bukit Gedang Seblat	IV	48,750	1981

Map† ref.	National/international designation Name of area	IUCN management category	Area (ha)	Year notified
21	Bukit Kayu Embun	IV	106,000	1980
22	Danau Pulau Besar/Danau Bawah	IV	25,000	1980
23	Dangku	IV	29,080	1981
24	Dolok Surungan	IV	23,800	1974
25	Gumai Pasemah	IV	45,883	1976
26	Gunung Raya	IV	39,500	1978
27	Isau-Isau Pasemah	IV	12,144	1978
28	Kappi	IV	8,000	1976
29	Karang Gading & Langkat Timur Laut	IV	15,765	1980
30	Kerumutan	IV	120,000	1979
31	Kluet	IV	23,425	1936
32	Padang Sugihan	IV	75,000	1983
33	Rawas Ulu Lakitan	IV	213,437	1979
34	Sekundur and Langkat (South and West)	IV	218,440	1939
35	Sumatera Selatan	IV	356,800	1935
36	Tai-tai Batti	IV	56,500	1976
37	Way Kambas	IV	130,000	1937
	Marine Park			
38	Pulau Weh	IV	2,600	1982
	Grand Forest Park			
39	Dr. Moch. Hatta	V	70,000	1986
	Hunting Parks			
40	Benakat	VI	30,000	1980
41	Lingga Isaq	VI	80,000	1978
42	Nanuua	VI	10,000	1978
43	Semidang Bukit Kabu	VI	15,300	1973
44	Subanjeriji	VI	65,000	1980
	Protection Forests			
45	Bajang Air Tarusan (Utara)	VI	81,865	
46	Batang Marangin Barat/Menjuta Hulu	VI	64,600	1926
47	Bentayan	VI	19,300	1981
48	Bukit Balairejang	VI	16,700	1926
49	Bukit Balal	VI	13,583	1926
50	Bukit Dingin/Gunung Dempo	VI	38,050	1926
51	Bukit Hitam/Sanggul/Dingin	VI	69,395	1932
52	Bukit Kaba	VI	13,490	1926
53	Bukit Mancung dan Sei Gemuruh	VI	1,500	
54	Bukit Nantiogan Hulu/Nanti Komerung Hulu	VI	36,200	1936
55	Bukit Raja Mandara/Kaur (North)	VI	77,180	1935
56	Bukit Reges/Hulu Sulup	VI	41,060	1926
57	Bukit Sebelah & Batang Pangean	VI	22,803	
58	Dolok Sembelin	VI	33,910	
59	Gunung Betung	VI	22,244	
60	Gunung Merapi	VI	9,670	
61	Gunung Patah/Bepagut/Muara Duakisim	VI	91,655	1936
62	Gunung Sago/Malintang/Karas	VI	5,486	
63	Gunung Singgalang	VI	9,658	
64	Gunung Sumbing/Masurai	VI	300,000	
65	Gurah Serbolangit	VI	9,297	
66	Hulu Bintuanan Complex	VI	76,745	1939
67	Hutan Sinlah	VI	81,000	
68	Kambang/Batanghari I/Bayang	VI	100,000	
69	Krui Utara/Bukit Punggur	VI	34,861	
70	Langsa Kemuning	VI	2,000	
71	Lembah Anai (Extension)	VI	96,002	1922

Map[†] ref.	National/international designation Name of area	IUCN management category	Area (ha)	Year notified
72	Lembah Harau	VI	23,467	1933
73	Maninjau (North and South)	VI	22,106	1920
74	Merangin Barat dan Nunjuta Ulu	VI	64,600	1926
75	Paraduan Gistana & Surroundings	VI	70,000	1936
76	Punguk Bingin	VI	2,400	
77	Sangir Ulu /Batang Tebo/Batang Tabir	VI	61,200	1926
78	Tanggamus	VI	15,660	1941
79	Tangkitebak/Kota Agung Utara/Way Waya	VI	140,600	1941
	Recreation Park			
80	Pulau Weh RP	V	1,300	1982
	Biosphere Reserves			
	Cibodas	IX	14,000	1977
	Gunung Leuser National Park	IX	946,400	1981
	Komodo National Park	IX	30,000	1977
	Lore Lindu National Park	IX	231,000	1977
	Siberut (Tai-tai Batti) Game Reserve	IX	56,500	1981
	Tanjung Puting National Park	IX	205,000	1977

[†]Locations of most protected areas are shown on the accompanying maps.

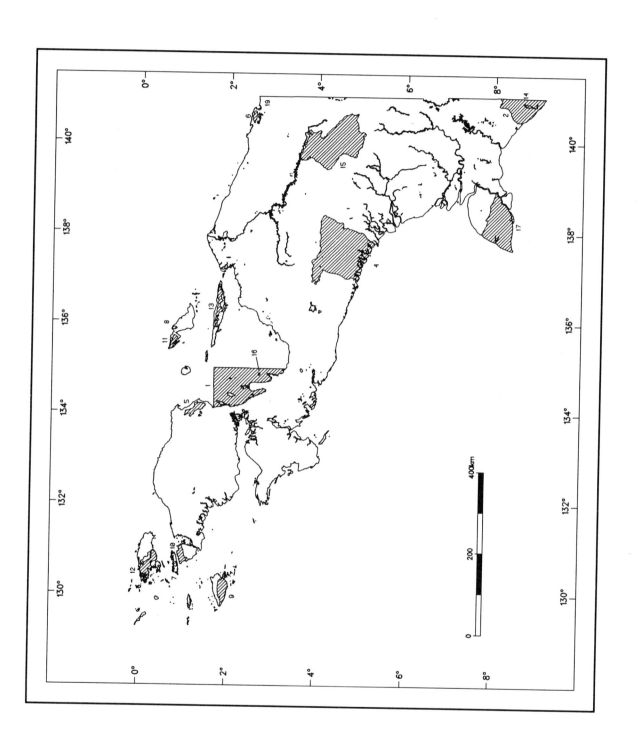

Protected Areas of Irian Jaya

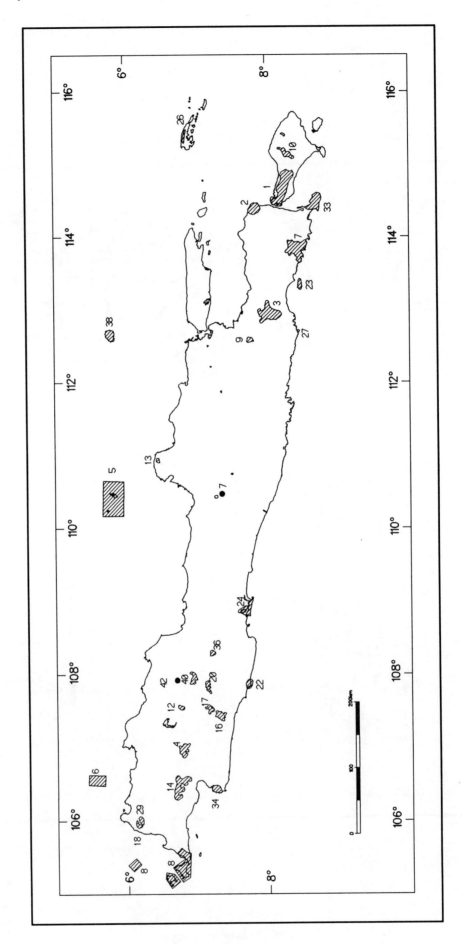

Protected Areas of Java

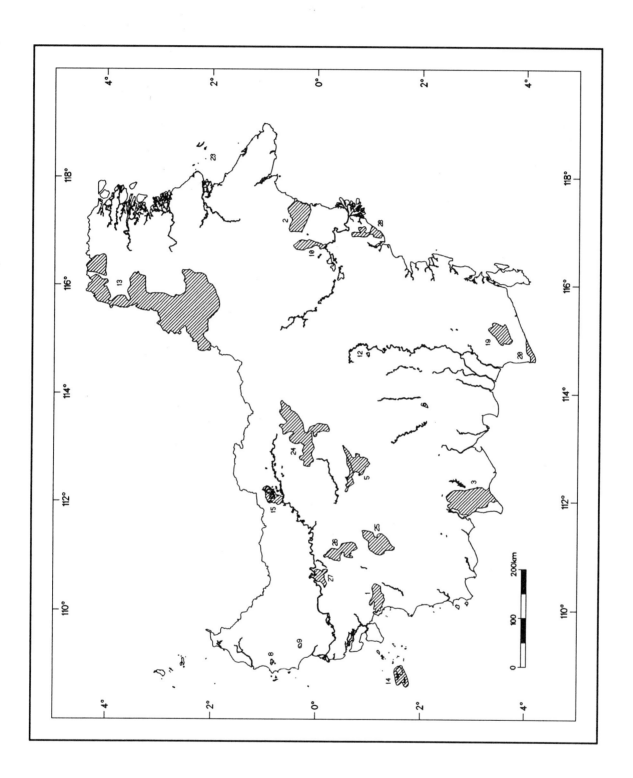

Protected Areas of Kalimantan

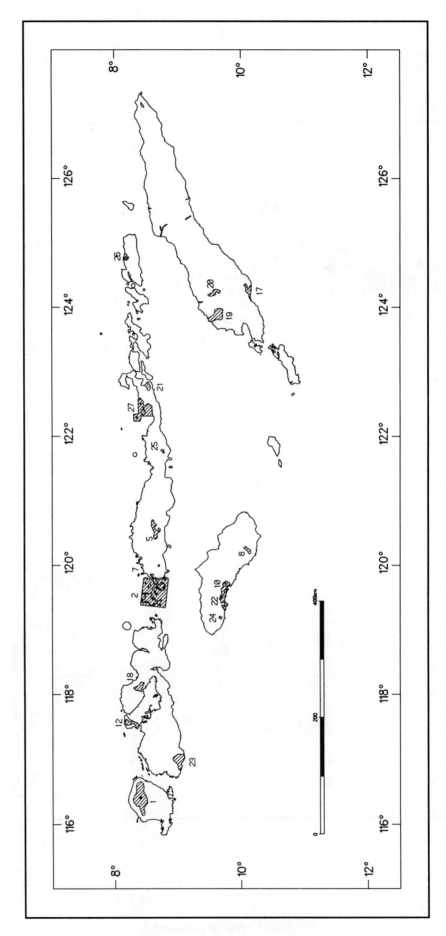

Protected Areas of the Lesser Sunda Islands

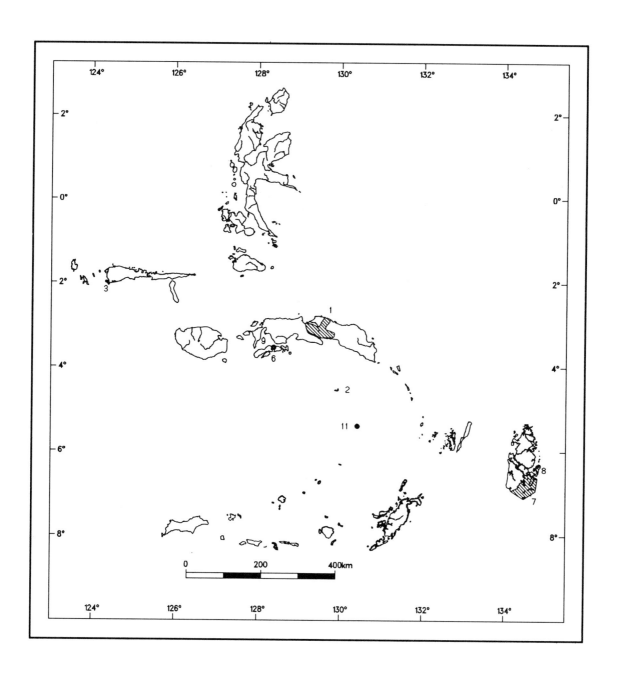

Protected Areas of the Moluccas

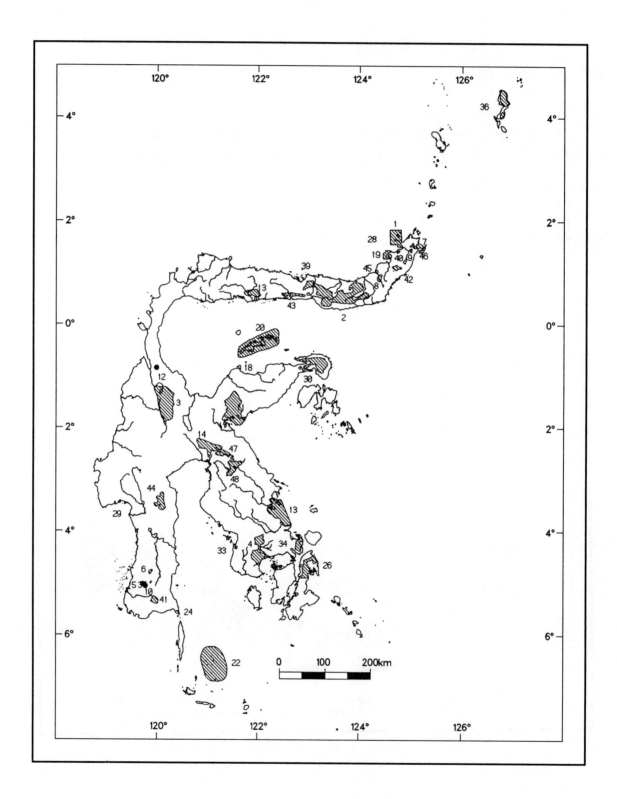

Protected Areas of Sulawesi

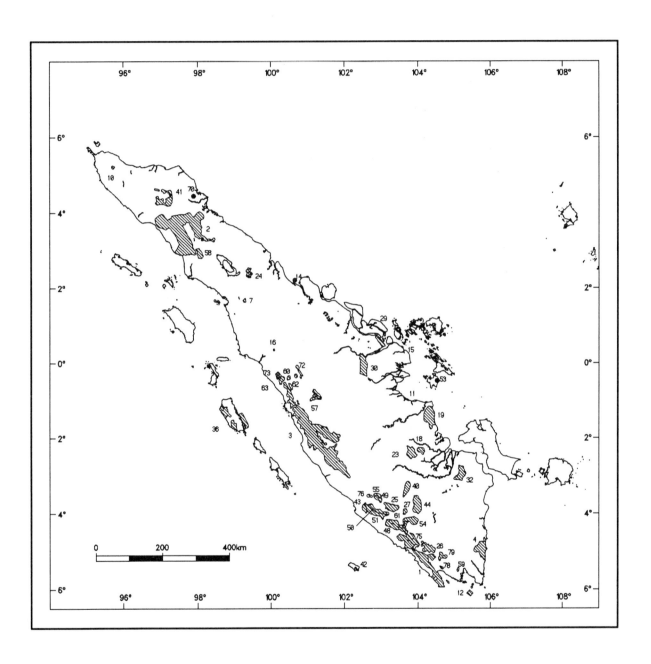

Protected Areas of Sumatra

LAO PEOPLE'S DEMOCRATIC REPUBLIC

Area 236,800 sq. km

Population 4,170,000 (1990)
Natural increase: 2.92% per annum

Economic Indicators
GNP: US$ 171 per capita
GDP: Not available

Policy and Legislation Laos is one of the few countries in Asia that has neither protected areas legislation nor protected areas. A number of central government decrees dating from 1979 deal with various aspects of nature conservation, including forest protection, wildlife trade and hunting and fishing, but these are not comprehensive and have not been effectively implemented (Madar and Salter, 1990). However, a Draft Nature Conservation Act has been formulated and is now under consideration by government, and planning for a representative protected area system is being undertaken with international assistance (see below). A comprehensive forestry policy is also being formulated by government and this will include conservation components.

International Activities Laos ratified the Convention concerning the World Cultural and Natural Heritage (World Heritage Convention) on 20 March 1987. No sites have been inscribed on the World Heritage List. Laos does not participate in the two other major international conventions or programmes concerned with protected areas, namely the Convention on Wetlands of International Importance Especially as Waterfowl Habitat (Ramsar Convention) and the Unesco Man and the Biosphere (MAB) Programme. Laos is also a party to the 1951 International Convention for Plant Protection, the subsequent 1956 Plant Protection Agreement for Southeast Asia and the Pacific, and the 1982 Law of the Sea Convention. Two senior level officials from the Ministry of Agriculture and Forestry attended, as observers, the Fourth Meeting of the Ramsar Parties held in Switzerland in 1990, and accession to the Convention has been recommended (Madar and Salter, 1990).

A large proportion of foreign aid has been dedicated to the mechanisation of forestry, although an early UNDP/FAO forest management project had a small nature conservation and environmental education component (FAO, 1983). International collaboration in these and other projects has been received from UNDP, IUCN and the Swedish International Development Agency (SIDA). SIDA is currently supporting a Forest Resources Conservation Project, for which IUCN is providing technical support. The long-term objectives are: to conserve and achieve sustainable utilisation of the forest and wildlife resources of Laos; to establish a nature conservation capability within the Ministry of Agriculture and Forestry, and to develop programmes for the establishment and management of protected areas, for staff training and for development of an awareness of conservation issues within the population; to promote the integration of conservation considerations into development planning; and to initiate the development of a national conservation strategy (SIDA, 1988). A tentative tripartite agreement with Kampuchea and Viet Nam, to establish a transfrontier reserve for the conservation of kouprey, has been made (MacKinnon and Stuart, 1989), although this has no force in law and has not been followed up (R. Salter, pers. comm., 1991).

A Tropical Forestry Action Plan, with support from the Laotian Ministry of Agriculture and Forestry, UNDP, FAO, Asian Development Bank, the World Bank and SIDA commenced in September 1989; the main report was produced in August 1990, which subsequently received government approval in September 1991 (R. Salter, pers. comm., 1991).

Administration and Management Until recently there has been no central government authority with specific responsibility for nature conservation nor for the establishment and management of protected areas. An administrative unit, established in 1983 to deal with conservation concerns, was split in 1984 into the Environmental Protection Division, which has primary responsibility for watershed management, and the Wildlife and Fisheries Conservation Division (WFCD), which is responsible for wildlife management and for the planning and establishment of a protected areas system. Both are under the Department of Forestry and Environment which is an administrative/technical unit of the central government's Ministry of Agriculture and Forestry. Development of conservation programmes has been constrained by a shortage of qualified staff and limited budgets, but these concerns are now being addressed through support from the Forest Resources Conservation Project (FRCP).

At present, the central WFCD has 10 professional (B.Sc. or equivalent) and 15 technical level staff. Four of the professional staff have attended the Post-Graduate Diploma Course in Wildlife Management at the Wildlife Institute of India; this is the highest level of academic training in conservation so far achieved, although most of the professional and technical staff have also attended various short courses, study tours and technical meetings dealing with conservation subjects.

The national organisational structure is replicated in the provincial governments, although staff with conservation related responsibilities generally also have other duties. Forty provincial forest services staff have attended a two-month "Short Course for Field Staff in Protected Area Management" over the three-year period 1989-91, but as yet none has higher academic qualifications in conservation.

Systems Reviews The original forest cover consisted largely of evergreen and semi-evergreen forests (MacKinnon and MacKinnon, 1986; Salter and Phanthavong, 1989). These comprised dry evergreen forests, which covered much of the mountainous northern part of the country; tropical montane evergreen forests, primarily along the Annamite Mountains and on the Bolovens Plateau; and lowland semi-evergreen forests over the Mekong Plain. Other original vegetation types were tropical montane deciduous forests scattered throughout the north, dry dipterocarp and mixed deciduous forests in the south and on the Mekong Plain, forest on limestone and pine forests in the Annamites and parts of the north, and a small area of subtropical montane forest in the extreme north (along the Chinese border). Seasonally flooded wetlands previously covered large parts of the Mekong Plain, but have now largely been converted to rice cultivation.

The most recent data on forest cover (Lao Forest Inventory and Management Office, 1991), based on 1988-89 SPOT imagery, indicate a current forest cover of 111,816 sq. km (47.2% of land area). An additional 88,051 sq. km (37.2% of area) are classified as potential forest, including bamboo and secondary formations; 15,515 sq. km (6.6% of area) as other wooded areas, primarily savanna and scrub forest; 10,083 sq. km (4.3% of area) as permanent agricultural land; and 11,336 sq. km (4.8% of area) as other non-forest land, including barren areas, grasslands, urban areas and wetlands. The best and most extensive forests are now confined primarily to the southern and central parts of the country, deforestation having been most severe in the north and along the densely settled Mekong Plain. The total current forest area represents an approximately 2% decrease in forest cover from 1981/82 estimates.

Shifting cultivation, fires, uncontrolled hunting and fishing, and unsustainable logging practices all represent serious conservation problems in Laos. Probably the major current cause of actual deforestation is some upland shifting cultivation systems which involve conversion of primary forest areas and continuous, intensive cultivation and repeated fires. Nevertheless, large, sparsely populated areas remain, and these continue to support a fauna of approximately 200 mammal species, including tiger, leopard, two bear species, possibly six deer species, elephant, four species of wild cattle, including kouprey *Bos sauveli* (E), and probably rhinoceros *Didermoceros sumatraensis* (E) and/or *Rhinoceros sondaicus* (E), and more than 600 bird species (FRCP (unpublished data); Interim Mekong Committee, 1978; King and Dickinson, n.d.; King *et al.*, 1975; Lekagul and McNeely, 1988).

Although some forest reserves have been declared in the past, their total extent has been small (less than 1,300 sq. km) and protection ineffective (FAO/UNEP, 1981). The government has recently received assistance, through FRCP, in formulating and implementing a management plan for one of these reserves, the 808ha Houei Nhang Forest Reserve just outside Vientiane

(Salter and Venevongphet, 1988), and in formulating a management plan for the 200,000ha Phou Khao Khouay area (Salter and Phanthavong, 1990).

Planning for the development of a representative protected area system is also proceeding. A recent analysis of needs and priorities (Salter and Phanthavong, 1989) indicated that sufficient forest cover remains for all forest formations to be protected at or about 10% of their original areas, and that for most (including evergreen types) 20% can still be protected. Biophysical profiles of 68 potential protected areas have been developed, and priorities for field surveys and establishment of protective measures have been determined on the basis of size, completeness of original cover, representativeness, regional priorities and degree of threat. Reconnaissance surveys of 28 of these areas were carried out during 1988-91, and four areas totalling over 6,000 sq. km have been selected for the development and implementation of management plans during 1991-95 (Salter, in prep.). An additional four areas covering over 7,000 sq. km have been selected for the development of management plans over the same period, subject to the results of further field surveys.

Address

Wildlife and Fisheries Conservation Division (Director), Department of Forestry and Environment, Vientiane

References

Anon. (1978). *Agriculture in the Lower Mekong Basin.* Committee for the Coordination of Investigations of the Lower Mekong Basin. Draft.

Anon. (1991). Basic statistics about the socioeconomic development in the Lao P.D.R. for 15 years (1975-1990). State Statistical Centre, Ministry of Economy Planning and Finance, Vientiane.

Collins, N.M., Sayer, J.A. and Whitmore, T.C. (Eds) (1991). *The conservation atlas of tropical forests: Asia and the Pacific.* Macmillan Press Ltd, London. 256 pp.

FAO (1983). *Forest management project. Lao People's Democratic Republic. Nature conservation and national parks.* Project LAO/82/006. Final Report. FAO, Vientiane. 59 pp.

FAO/UNEP (1981). *Tropical Forest Resources Assessment Project.* Forest Resources of Tropical Asia. Three Volumes. FAO, Rome.

Interim Committee for Coordination of Investigations of the Lower Mekong Basin (1978). Wildlife and national parks in the lower Mekong basin. Mekong Committee Report, Bangkok.

King, B.F. and Dickinson, E.C. (n.d.). A distribution table of the birds of Southeast Asia. Unpublished MS.

King, B.F., Dickinson, E.C. and Woodcock, M.W. (1975). *A field guide to the birds of Southeast Asia.* Collins, London.

Lao Forest Inventory and Management Office (1991). National Forest Reconnaissance Survey of Lao

P.D.R. Department of Forestry and Environment, Vientiane.

Lekagul, B. and McNeely, J.A. (1988). *Mammals of Thailand*. Second edition. Association for the Conservation of Wildlife, Bangkok.

MacKinnon, J.R. and MacKinnon, K. (1986). *Review of the protected areas system in the Indo-Malayan realm*. IUCN, Gland, Switzerland and Cambridge, UK. 284 pp.

MacKinnon, J.R. (1986). Bid to save the kouprey. *WWF Monthly Report* 1986: 91-97.

MacKinnon, J.R. and Stuart, S.N. (Eds) (1989). The kouprey: an action plan for its conservation. Prepared by the Species Survival Commission of IUCN and WWF. IUCN, Gland, Switzerland and Cambridge, UK. 20 pp.

Madar, Z. and Salter, R.E. (1990). Needs and priorities for conservation legislation in Lao PDR. Forest Resources Conservation Project, Lao/Swedish Forestry Cooperation Programme, Vientiane.

Salter, R.E. (in prep.). Forest Resources Conservation Project, Phase I: 1988-91. Summary report of project activities and recommendations. Part 3: Field surveys (June 1991).

Salter, R.E. and Phanthavong, B. (1989). Needs and priorities for a protected area system in Lao PDR. Forest Resources Conservation Project, Lao/Swedish Forestry Cooperation Programme, Vientiane.

Salter, R.E. and Phanthavong, B. (1990). Phou Khao Khouay Protected Area management plan. Forest Resources Conservation Project, Lao/Swedish Forestry Cooperation Programme, Vientiane.

Salter, R.E. and Venevongphet (1989) (revised). Houei Nhang Forest Reserve management plan 1988-90. Forest Resources Conservation Project, Lao/Swedish Forestry Cooperation Programme, Vientiane.

Sayer, J. (1983). Nature conservation priorities in Laos. *Tigerpaper* 10: 10-14.

SIDA (1988). Lao-Swedish forestry sector programme. Programme documents (1988-1990). Swedish International Development Authority, Stockholm.

MALAYSIA

Area

Peninsular Malaysia:	132,750 sq. km
Sabah:	73,620 sq. km
Sarawak:	123,985 sq. km
Malaysia (total):	330,355 sq. km

Population

Peninsular Malaysia:	14,303,000 (1988)
Sabah:	1,371,000 (1988)
Sarawak:	1,590,000 (1988)
Malaysia (total):	18,400,000 (1991)

Natural increase: 2.26% per annum (1990)

Economic Indicators
GNP: US$ 2,305 per capita (1991)
GDP: US$ 1,590 per capita (1990)

Policy and Legislation

Malaysia comprises 11 states and a federal territory in the Malay Peninsula, together with two other states and a single federal territory on the island of Borneo. Sabah and Sarawak joined the Federation with some constitutional safeguards that give them a greater degree of autonomy than the other states.

The Malaysian constitution reserves rights over land matters to the respective state governments. Thus, the jurisdiction of most land matters is controlled by state decision-making bodies and land ownership is retained by the states. Sabah and Sarawak each have state laws covering forestry, protected areas and wildlife.

Statutes relating to biological resources are in force at both state and federal levels, providing a complexity in approach. Of the many legislative acts in place, only the Environmental Quality Act (1974) and the Fisheries Act (1985) are applied nationally. The latter makes provision for the establishment of marine parks, intended for the conservation of flora, fauna and habitat and the promotion of recreation.

Federal/Peninsular Malaysia Since 1964, land use planning in Peninsular Malaysia has been partly guided by the Land Capability Classification, in which game reserves and protected forest reserves are included in Class V, which also includes land possessing little or no mineral, agricultural or productive forest development potential. The Land Capability Classification Reports need only be taken as advice - there is no enforcement of these overall government policies.

Local authorities have broad powers under Parts VII and XII of the Local Government Act (1976) to establish and manage public places, including parks. These powers can provide for the creation of small protected areas of

natural habitat and intensively managed parks (Scott, 1989).

Conservation is specifically recognised to be an essential element of land use planning under the Town and Country Planning Act (1976). The Act gives certain powers at both state and local levels to require that specific areas of land be conserved in one way or another. However, the form and content of the Town and Country Planning Act adopted by states may differ significantly for the parent Federal Act. Other acts may then be used to delineate an area of land to be conserved and to specify the exact use to which that land should be put (Scott, 1989).

The Fisheries Act (1985) makes provision for the establishment of marine parks and marine reserves by the appropriate Federal Minister, in any Malaysian waters, to afford protection to flora and fauna, and to make provision for scientific study and recreation. However, these reserves may not include any area above the high tide mark and inappropriate management of adjacent land may have serious impacts on such reserves.

The National Parks Act No. 226 of 1980 (amended in 1983) provides for the establishment of national parks and applies to the whole of Malaysia, except for Sabah and Sarawak. The Act, as amended, allows the appropriate Federal Minister to request that any state land be reserved for the purpose of a national park, although this has no legal force. The states have been reluctant to lose control over land ownership and the Act has not been used.

The Protection of Wild Life Act of 1972 is a consolidation of Federal laws relating to wildlife protection in Peninsular Malaysia with provisions for both the protection of fauna and habitats. The Act provides for the establishment, alteration and extinction of wildlife reserves and sanctuaries by the state governments, although all management is in the hands of the federal government.

The National Forestry Act (1984) provides for the State Director of Forests, with the approval of the State Authority, to classify every permanent reserved forest. The following classifications lend themselves particularly to the protection of wildlife habitat: forest sanctuary, virgin jungle reserve, amenity forest, education forest and research forest. However, these classes are not defined in the Act, and do not offer protection to wildlife habitat beyond that offered by forest reserve status (Scott, 1989). State governments have formally agreed to adopt the categories within the National Forestry Act and there are restrictions on use in each category, although these differ slightly from state to state. For example, logging is prohibited in the category "forest reserve for wildlife".

The Aboriginal Peoples Act (1954) provides for the protection, wellbeing and advancement of aboriginal people in Peninsular Malaysia. Since many of these people live near park areas, the provisions of this Act affect park management. The Act states that no land within an aboriginal area can be declared a wildlife reserve or sanctuary.

Some 25 environmentally-related acts are currently being reviewed, in order not only to make stronger legal provisions available, but also to streamline and complement the activities of government departments in relation to the environment. Acts to be reviewed include the Environmental Quality Act (1974), the Town and Country Planning Act (1976), Local Government Act (1976), Land Conservation Act (1960), Forest Enactment (1935)/National Forestry Act (1984), National Land Code (1965), National Parks Act (1980) and the Protection of Wild Life Act (1972) (Shariif, 1991).

Sabah The principal conservation laws in Sabah are the Parks Enactment (1984), Forests (Amendment) Enactment (1984) and the Fauna Conservation Ordinance (1963), amended in 1979.

The Sabah Parks Board of Trustees was constituted with the establishment of Kinabalu "National" Park in 1964 under the National Park Ordinance (1962). The 1962 Ordinance was replaced by the National Parks Enactment (1977), under which several other "national parks" were established. The Parks Enactment (1984) repealed the 1977 Enactment and all five "national" parks existing at that time were reconstituted as state parks to ensure that they remained under state legislation rather federal control. Such areas are intended for both nature conservation and recreation (Ngui, 1990) and may include marine areas that are not covered by the federal Fisheries Act (1985). The 1984 Enactment also described the procedure by which the State Assembly may establish parks, streamlined the control, management and administration of parks and redefined the boundaries of the parks. Land title to the parks is vested in the Board of Trustees for a period of 99 years, free of all liabilities and encumbrances.

The Forests (Amendment) Enactment (1984) was introduced by the Chief Minister on 8 March 1984 and amended the Forest Enactment (1968). It classifies forests on a I-VII scale. Class I is protection forest for soil and water conservation. It is not supposed to be used for commercial timber operations. Class II is commercial forest to be used for commercial timber operations and forest produce collection. Class III is domestic forest for the use of village people for extracting forest produce and occasional small amounts of timber for personal use. Class IV is amenity forest which is often used for trials and research plots and has in many cases been logged in the past. It can be developed for recreational use. Class V is mangrove forest - there are various categories within this definition. Class VI is virgin jungle reserve for genetic conservation. Most jungle reserves are, in fact,

logged but they are supposed to protect the forest completely. Class VII is wildlife reserve. Commercial forest can be selectively logged under licence, but other areas cannot be logged without state authority. Any areas previously constituted as forest reserves, but not included in the schedule under the Forests (Amendment) Enactment (1984), ceased to be forest reserves on 8 March 1984. Legislation regarding existing forest reserves has been strengthened by Section 22 in the amended Act, which states that "No Forest Reserve shall cease to be a Forest Reserve or any portion thereof shall be excised from such a Reserve except by Enactment or except where it is required for conversion to a Park, a Game Sanctuary or a Bird Sanctuary under the law for the time being in force relating thereto". The amended Act also established two new wildlife reserves, Kulamba and Tabin, and transfers the power to excise wildlife reserves from the Chief Minister to the State Assembly.

The Fauna Conservation Ordinance (1963), amended in 1979, makes provision for the Forest Department to take measures to conserve wildlife and to establish protection areas for wildlife (SFD, 1986).

Sarawak The principal protected areas legislation comprises the National Parks Ordinance (1956), the Wildlife Protection Ordinance (1990) and the Forest Ordinance (1954).

The National Parks Ordinance (1956) and amendments (e.g. National Parks (Amendment) Ordinance, 1990) provide for the constitution, maintenance and control of national parks. Such areas are totally protected and aimed at the preservation of "animal and vegetable life ... in a natural state", and are intended to be open to the public for recreation.

The Wildlife Protection Ordinance (1990) makes provision for the gazettement of wildlife sanctuaries, being totally protected areas of natural interest, protected against all forms of exploitation, including tourism. The Ordinance was enacted to provide for the protection of wildlife and wildlife habitat.

International Activities Malaysia became a signatory to the Convention Concerning the World Cultural and Natural Heritage (World Heritage Convention) on 7 December 1988, although no sites have yet been nominated for inscription on the World Heritage list. Research studies have been carried out under the Unesco Man and the Biosphere Programme, although no sites have been designated under the programme. Malaysia does not yet participate in the Convention on Wetlands of International Importance Especially as Waterfowl Habitat (Ramsar Convention),

International assistance has been given by the Asian Development Bank, the International Development Research Centre (Canada), Japanese International Cooperative Agency, OECF, UNDP, UNEP, Unesco and FAO.

Administration and Management

Federal/Peninsular Malaysia The Director General, Department of Wildlife and National Parks (within the Ministry of Science, Technology and Environment), is responsible for managing Taman Negara and for implementing all matters arising from the Protection of Wild Life Act (1972) and the National Parks Act (1980). On the creation of a national park, the Federal Minister appoints a committee which has control over the general management and development of the park. The Committee consists of representatives of government departments considered necessary by the Minister. The Minister has authority to make regulations about activities within the park. However, the Act has not yet been implemented and there are no national parks.

The Forestry Department manages the permanent forest estate. The Federal Fisheries Department is responsible for protection of marine resources throughout the country, including the management of several marine parks. This is complicated by state control over freshwater and coastal matters, requiring the Department to seek either delegation of powers or specific approvals on a case-by-case basis for some parts of its management work.

In addition, numerous sectoral agencies are responsible for natural resource use and management and all states can pass legislation to establish and manage protected areas (WWF-M, 1990).

Sabah The Sabah Parks Board of Trustees, also known as "Sabah Parks", is responsible for the management and administration of state parks in Sabah. Thus, the six state parks all come under the control of the state government, *via* Sabah Parks which is a government agency under the Ministry of Tourism and Environmental Development. The Director of Sabah Parks, appointed by the Board, heads the administration, management and development programmes which are formulated by the Board. The Director is assisted by a Deputy and an Assistant Director, administrative personnel and field staff, totalling some 250 individuals. The administration is based in Kota Kinabalu, and is divided into five divisions, comprising administration, development, enforcement, research and education and regional administration. In addition, three regional offices have been established, namely, Kinabalu Ranau, West Coast/Kota Kinabalu and East Coast/Sandakan, each headed by a park warden. Sabah Parks' primary management responsibility is to ensure that its parks are maintained unimpaired in their natural condition as far as possible. The parks tend to be large enough to maintain viable wildlife populations and the usual approach to conservation management is to allow natural processes to occur unhindered. In certain cases, however, positive management has been implemented, for example for the conservation of slipper orchid and marine turtles. Law enforcement and park protection is also given a high priority, with other government agencies being involved (Alisaputra *et al.*, 1989).

The Department of Wildlife, under the state Ministry of Tourism and Environmental Development, is responsible for the conservation and management of wildlife within the state under the provisions of the Fauna Conservation Ordinance (1963). Amongst other responsibilities, the major role of the Department is the identification and establishment of sanctuaries, but it does not control land that has been set aside for wildlife conservation, a situation that is being reviewed (WWF-M, 1990).

The Forest Department has a responsibility for the conservation of adequate forest estate for recreation, education, research and protection of flora and fauna (Mahedi and Ambu, 1989).

Sarawak The National Parks and Wildlife Office of the Sarawak Forest Department has responsibility for the conservation of wildlife and wildlife habitat under the provisions of the National Parks Ordinance (1956) and the Wildlife Protection Ordinance (1990) (WWF-M, 1990). The number of staff within the Division needs to be increased such that it can properly discharge its responsibilities, and there is insufficient institutional support for the Forest Department itself (Ngui, 1990).

Systems Reviews Unlike any other nation in South-east Asia, Malaysia has a substantial land mass on both the Asian mainland and in the Malay Archipelago. It forms a crescent well over 1,600km long between 1°00-7°00N and 100°-109°E. It occupies two distinct regions, namely the Malay Peninsula, which extends from the Isthmus of Kra to the Singapore Strait, and the north-western part of Borneo (WWF-M, 1990).

Lowland evergreen tropical rain forest is the principal original formation in Peninsular Malaysia on dry land at low altitudes. In the extreme north-west this is replaced by semi-evergreen formations. The rain forest is rich in Dipterocarpacae and may be sub-divided into lowland (below 300m) and hill (300m to 1,300m) forest, on the basis of floristic composition. Along the east coast there remain a few patches of heath forest on recent unconsolidated sands, but most have been degraded to open grasslands or scrub. Widely scattered patches of forest on limestone occur north of Kuala Lumpur, peat swamp and freshwater swamp forests are extensive on both east and west coasts, although most of the latter have been cleared for agriculture, and extensive montane rain forests are found. About 100 years ago rain forests probably covered 90% of the land area, much of it in the lowlands. In 1966 it was estimated that 68% of land area was naturally forested (Lee, 1973). The National Forest Inventory of 1970-1972 showed that this figure had dropped to 83,633 sq. km (63% of land area) (Mohd Darus, 1978). At the end of 1985, forest cover had diminished to 61,870 sq. km (47%) (Ministry of Primary Industries, 1988) and a figure of 57,090 sq. km (43%) has been estimated for 1990 (FAO, 1987). However, much of this remaining cover is disturbed and in 1985 as little as 13,000 sq. km (9.8%) supported intact forest (Collins *et al.*, 1991).

Both Sabah and Sarawak were originally clothed in forest, including lowland evergreen rain forests, peat swamps, heath forests, forests on limestone, a floristically distinct formation on the ultrabasic rock which forms a mountainous arc extending from Mount Kinabalu to the east coast and lower and upper montane forests. According to FAO (1987), the forest estate in Sarawak stood at 84,000 sq. km of broad-leaved forest in 1980 (67.5% of land area) and a predicted 79,639 sq. km (64%) in 1990. In Sabah in 1953 natural forest covered 63,725 sq. km (Fox, 1978) or 86% of land area. Thirty years later the forest cover had diminished to 46,646 sq. km (63%) (Sabah Forest Department, 1984). According to an FAO assessment in 1985, forest cover was 33,130 sq. km (45%) and the prediction was that this would fall to 29,110 sq. km (39%) by 1990 (FAO, 1987).

The wetlands of Malaysia may be divided into ten groups, comprising mangroves, mudflats, nipa swamps, freshwater swamps, peat swamp forest, lakes, oxbow lakes, river systems, marshes and wet rice paddies. These are briefly described in DWNP (1987), and more detailed accounts for some 90 wetland sites important for conservation are also given. This inventory will be revised during 1992 (D. Parish, pers. comm., 1991). An account of the coral reefs is given in UNEP/IUCN (1988), including a general description, an account of reef resources, disturbances and deficiencies, legislation and management and recommendations for further protection; a number of specific sites are described in more detail.

Species conservation dates back to 1924 when Krau Game Reserve was created. The first comprehensive protected areas plan was contained in the Third Malaysian Plan (TMP) for 1976-1980, in which Taman Negara, 22 other wildlife reserves, game reserves and bird sanctuaries were recognised and a further two national parks and 21 other reserves were proposed. The TMP was designed to include representative ecosystems and the major biological communities suggested by the Malayan Nature Society in its Blueprint for Conservation (MNS, 1974). The present record of conservation falls a little short of that proposed in the TMP, with a total of 8,239 sq. km gazetted against the 8,985 sq. km proposed, and the sites recommended in the TMP do not necessarily correspond to those that have been gazetted.

The existing protected areas system in Peninsular Malaysia relies heavily on Taman Negara, one of the largest parks in South-east Asia. However, a number of critical habitats remain under-represented, including peat swamp forests in Pahang and Johor states, mangrove forests and open lake systems. Most importantly, the lowland dipterocarp rain forests are seriously under-represented. Existing "national" parks cover 0.92% and 1.9% of Sarawak and Sabah, respectively. Other categories of existing reserves cover 1.4% in Sarawak, and 4.9% of Sabah. Sarawak has a further 5% in proposed "national" parks and 1.4% in proposed wildlife sanctuaries. In addition, Sabah has 883 sq. km of virgin jungle reserves and about 1,000 sq. km of protection reserved forest that potentially play an important role. Including all categories, 8.3% of Sarawak, and 8.9% of Sabah will be under protection for ecological and biological purposes if all new reserves are gazetted. The protected areas system in both states covers a good representation of the natural communities present. However, there is concern that few of the areas are under complete protection, and they could be subject to disturbance as pressure for land and timber increase. The main short comings are as follows. In Sabah, some of the virgin jungle reserves and other protected areas have been excised, or partially or totally logged, and may be under threat from shifting cultivators and poaching of wildlife. Although the state government passed a bill in 1984 requiring all dereservations to go to the floor of the State Assembly, areas have still been excised from Kinabalu Park. Wildlife sanctuaries in Sarawak had a dual status as part of the Permanent Forest Estate, which gave less protection to biodiversity; the Wildlife Protection Ordinance (1990) has addressed this shortcoming. In Sarawak, great progress has been made in extending and managing the protected areas system in recent years. The State Conservation Strategy, still a confidential document, identified various new areas for protection and good progress has been made, but until the proposed areas are gazetted important omissions remain (Collins *et al.*, 1991).

Each state is responsible for the management of its forests, but it does so under a forest policy that is common. The 1977 Malaysia Forest Policy emphasises that each state should maintain 47% of land area as forest reserves. Some 47,500 sq. km (36% of total land area) of forest are currently allocated to the Permanent Forest Estate within Peninsular Malaysia, of which 28,500 sq. km is managed for production and 19,000 sq. km for protection purposes (Collins *et al.*, 1991). Forested lands outside the Permanent Forest Estate, apart from those already established as conservation areas, are destined to be logged, relogged and may be eventually converted for agricultural use. At present, only about 37% of Sarawak and 45% of Sabah are gazetted under the Permanent Forest Estate, although there are proposals to increase this in Sarawak. In Sabah, 28 protection forest reserves have been gazetted, covering 9% of the Permanent Forest Estate. In Sarawak, "protected forests" cover about 30% of the Permanent Forest Estate but they are set aside by local administration and do not have the same legal status as the protection reserves in Sabah or Peninsular Malaysia. A small number of communal forests have been established for the use of local communities (Collins *et al.*, 1991).

A major cause of forest loss in Sarawak is shifting cultivation, covering over 30,000 sq. km, with as many as 1,500 sq. km cleared annually for hill padi, although some of this may have already been cleared on a previous occasion. Encroachment frequently occurs along logging roads, where access is provided into previously inaccessible forest. There is concern that levels of timber extraction may be unsustainable, although measures are being taken to attempt to improve sustainability (G. Davison, pers. comm., 1991). Shifting cultivation is less of a threat in Sabah, covering only some 15% of total land area, although substantial areas in the lowlands are under threat of conversion. Severe and extensive fires, especially in logged forests, destroyed some 10,000 sq. km of forest in 1983.

World Wide Fund for Nature-Malaysia (WWF-M) is a fully professional, local non-governmental organisation working on a wide range of conservation matters, frequently with government agencies. WWF-M is currently contracted by the government to compile the national conservation strategy. The Malayan Nature Society, which publishes the Malayan Nature Journal and produced a "Blueprint for Conservation in Peninsular Malaysia" in 1974, is the oldest scientific society in Malaysia. Its activities have included the promotion of a protected area at Endau-Rompin, and it is currently developing a "Second Blueprint for Conservation". Sahabat Alam Malaysia (Friends of the Earth Malaysia), the Environmental Protection Society of Malaysia and the Consumers' Association of Penang all maintain more independent positions, frequently critical of the government. The Asian Wetland Bureau, another fully professional, scientific non-governmental organisation which operates a joint programme with the Institute for Advanced Studies, University of Malaya, concentrates on protection and sustainable utilisation of wetland resources. Other organisations include the Sabah Society, Sarawak Nature Society and the Malaysian Society for Marine Sciences (WWF-M, 1990).

Addresses

Federal

Department of Wildlife and National Parks, Km 10 Jalan Cheras, 50664 Kuala Lumpur (Tel: 3 905 2872; FAX: 3 905 2873)

Forestry Department, Jalan Sultan Salahuddin, 50600 Kuala Lumpur

Peninsular Malaysia

Department of Wildlife and National Parks, Km 10 Jalan Cheras, 50664 Kuala Lumpur (Tel: 3 905 2872; FAX: 3 905 2873)

Sabah

Sabah Parks, PO Box 10626, 88806 Kota Kinabalu

Forest Department (Jabatan Perhutanan), Peti Surat 311, 90007 Sandakan (Tel: 660811; FAX: 89 669170; Tlx MA 82016)

Sarawak

Forest Department (Ibu Pejabat Perhutanan), Bangunan Wisma Sumber Alam, Jalan Stadium, Petra Jaya, 93660 Kuching (Tel: 82 442180; FAX: 82 441377)

World Wide Fund For Nature-Malaysia (WWF-M), Locked Bag No. 911, Jalan Sultan PO, 46990 Petaling Jaya (Tel: 3 757 9192; FAX: 3 756 5594)

The Malayan Nature Society, PO Box 10750, 50724 Kuala Lumpur (Tel: 3 791 2185; FAX: 3 791 7722)

Asian Wetland Bureau, Institute for Advanced Studies (IPT), University of Malaya, Lembah Pantai, 59100 Kuala Lumpur (Tel: 3 757 2176/3 756 6624; FAX: 3 757 1225; Tlx: UNIMAL MA 39845)

References

Alisaputra, L.A., Sidek, A.R. and Nais, J. (1989). The management and development of State Parks in Sabah. In: *Proceedings of the International Conference on National Parks and Protected Areas.* Department of Wildlife and National Parks, Peninsular Malaysia, Kuala Lumpur, 13-15 November. Pp. 125-132.

Collins, N.M., Sayer, J.A. and Whitmore, T.C. (Eds) (1991). *The conservation atlas of tropical forests: Asia and the Pacific.* Macmillan Press Ltd, London. 256 pp.

DWNP (1987). *Malaysian wetland directory.* Department of Wildlife and National Parks, Kuala Lumpur. 316 pp.

FAO (1987). Special study of forest management, afforestation and utilization of forest resources in the developing regions. Asia-Pacific Region, Assessment of Forest Resources in Six Countries. *Field Document* 17. FAO, Bangkok. 104 pp

FAO (1987). *Assessment of forest resources in six countries.* FAO, Bangkok. 104pp.

Fox, J.E .D. (1978). The natural vegetation Sabah, Malaysia. The physical environment and classification. *Tropical Ecology* 19: 218-139.

Lee, P.C. (1973). Multi-use management of West Malaysia's forests. *Proceedings of a Symposium on Biological Resources and Natural Development* 93-101. (Unseen)

Mahedi, P.A. and Ambu, L.N. (1989). Management of endangered species in protected areas in Sabah. In: *Proceedings of the International Conference on National Parks and Protected Areas.* Department of Wildlife and National Parks, Peninsular Malaysia, Kuala Lumpur, 13-15 November. Pp. 125-132.

Malayan Nature Society (1974). A blueprint for conservation in Peninsular Malaysia. *Malayan Nature Journal* 27: 1-16.

Mohd Darus, H.M. (1988). Forest conservation, management and development in Malaysia. Forest Department Headquarters, Kuala Lumpur. (Unseen)

Ministry of Primary Resources (1988). *Forestry in Malaysia.* (Unseen)

Muhammad, J. (1980). The national forestry policy (editorial). *Malaysian Forester* 43: 1-6. (Unseen)

Ngui, S.K. (1990). The management status of protected areas in Malaysia. Paper presented at the Regional Expert Consultation on Management of Protected Areas in the Asia-Pacific Region. FAO Regional Office for Asian and Pacific, 10-14 December, Bangkok. 17 pp.

Sabah Forest Department (1984). *Annual report.* State of Sarawak. 31 pp. (Unseen)

Sabah Forest Department (1986). *Forest Department annual report for the year 1986.* 69 pp.

Scott, D.A. (Ed.) (1989). *A directory of Asian wetlands.* IUCN, Gland, Switzerland and Cambridge, UK. 1181 pp.

Shariif, N. (1991). Streamlining environment laws. *New Strait Times.* 28 June. P. 2.

UNEP/IUCN (1988). *Coral reefs of the world. Volume 2: Indian Ocean, Red Sea and Gulf.* UNEP Regional Seas Directories and Bibliographies. IUCN, Gland, Switzerland and Cambridge, UK/UNEP, Nairobi, Kenya. Pp. 95-132.

World Wide Fund for Nature-Malaysia (1990). WWF country plan: Malaysia. Unpublished report. 42 pp.

ANNEX
Definitions of protected area designations, as legislated, together with authorities responsible for their administration

FEDERAL/PENINSULAR MALAYSIA

Title: National Forestry Act

Date: 24 December 1984

Brief description: An Act to provide for the administration, management and conservation of forests and forestry development within the States of Malaysia and for connected purposes.

Administrative authority: State Director of Forestry

Designations:

The State Authority may constitute any land a permanent reserved forest, under the following classification:

Timber production forest under sustained yield
Soil protection forest
Soil reclamation forest
Flood control forest
Water catchment forest
Forest sanctuary for wildlife
Virgin jungle reserved forest
Amenity forest
Education forest
Research forest
Forest for federal purposes

Source: Translation of original legislation

Title: National Parks Act

Date: 1980 (amended 1983)

Brief description: An Act to provide for the establishment and control of National parks and for matter connected therewith. Applies to all Malaysia except Sabah, Sarawak and the state parks of Kelantan, Pahang and Trengganu which together constitute Taman Negara.

Administrative authority: Director General of Wild Life and National Parks, appointed under the Protection of Wild Life Act 1972; National Parks Advisory Council; National Committee.

Designations:

National park The state authority may, on the request of the Minister, reserve any state land within the state, including any marine area, for the purpose of a national park.

The object of the establishment of a national park is the preservation and protection of wild life, plant life and objects of geological, archaeological, historical and ethnological and other scientific and scenic interest and through their conservation and utilisation to promote the education, health, aesthetic values and recreation of the people.

A committee shall be constituted by the Minister for each national park, chaired by the appropriate state secretary.

Title: Protection of Wild Life Act

Date: 1972 (amended February 1976, February 1988)

Brief description: Consolidates the laws relating to and further provides for the protection of wild life and for other purposes connected therewith. The act only applies to West Malaysia. Repealed the Wild Animals and Birds Protection Ordinance 1955.

Administrative authority: Director General of Wildlife and National Parks

Designations:

Wild life reserve Declared by the Ruler, or State Governor, on any state land, which allows licensed hunting, but in which any species may be declared protected from hunting. The Protection of Wild Life (Amendment) Act 1988 stipulates that it is prohibited to disturb, cut or remove vegetation.

Wild life sanctuaries Within which it is prohibited to shoot, kill or disturb any animal, or disturb or remove any vegetation.

Source: Translation of original legislation

SABAH

Title: Parks Enactment

Date: 14 March 1984

Brief description: An Enactment to repeal and re-enact the law relating to the provision and control of national parks in Sabah and to provide for matters incidental thereto and connected therewith so as to make better provisions respecting the constitution, administration, procedure, functions and finance of parks. Repealed the National Parks Enactment (1977).

Administrative authority: Yang di-Pertua Negeri

Designations:

Park Any area of state land so declared under the provisions of Part II of the Act. Any forest reserve declared under the Forest Enactment (1968), or any game sanctuary or bird sanctuary declared under the Fauna Conservation Ordinance (1963) may be converted to a park.

Part III makes provision for the establishment and constitution of a Board of Trustees

Within a park it is forbidden to hunt, damage vegetation, introduce or remove animals or vegetation, remove minerals, archaeological objects etc, erect buildings or clear land, without prejudice to rights gained prior to commencement of the act, and to provisions of any written law relating to mining, prospect for metals or minerals in any park.

Section II (17) constitutes six named parks.

Source: Translation of original legislation

Title: The Forests (Amendment) Act

Date: 8 March 1984

Brief description: An enactment to amend the Forest Enactment, 1968

Administrative authority: Sabah Forest Department

Designations:

Forest reserve
Protection forest (I)
Commercial forest (II)
Domestic forest (III)
Amenity forest (IV)
Mangrove forest (V)
Virgin jungle reserve (VI)
Wildlife reserve (VII)

Source: Translation of original legislation

SUMMARY OF PROTECTED AREAS

Map[†] ref.	*National/international designation* Name of area	IUCN management category	Area (ha)	Year notified
	PENINSULAR MALAYSIA			
	National Park			
1	Taman Negara	II	434,351	1939
	Wildlife Reserves			
2	Bukit Kutu	VIII	1,942	1923
3	Endau-Kluang	VIII	101,174	1933
4	Endau-Kota Tinggi (East)	VIII	7,413	1933
5	Endau-Kota Tinggi (West)	VIII	61,959	1933
6	Krau	IV	53,095	1923
7	Pulau Tioman	VIII	7,160	1972
8	Sungai Dusun	IV	4,330	1964
9	Sungkai	VIII	2,428	1928
	Wildlife Sanctuary			
10	Cameron Highlands	IV	64,953	1962
	Bird Sanctuaries			
11	Bukit Fraser	VIII	2,979	1922
12	Pahang Tua	VIII	1,335	1954
	Park			
13	Templer	V	1,011	1956
	Virgin Jungle Reserves			
14	Berembun	I	1,595	1959
15	Bukit Larut	I	2,747	1962
16	Gunung Jerai	I	1,579	1960
17	Gunung Ledang	I	1,134	1969
	SABAH			
	National Park			
1	Crocker Range	II	139,919	1984
	Wildlife Reserves			
2	Kulamba	VIII	20,682	1984
3	Tabin	VIII	120,521	1984
	Conservation Areas			
4	Danum Valley	VIII	42,755	1983
5	Maliau Basing (Gunung Lotung)	VIII	39,000	1983
	Parks			
6	Bukit Tawau	II	27,972	1979
7	Kinabalu	II	75,370	1964
8	Pulau Penyu (Turtle Islands)	II	1,740	1977
9	Pulau Tiga	II	15,864	1978
10	Tunku Abdul Rahman	II	4,929	1974
	Virgin Jungle Reserves			
11	Brantian-Tatulit	I	4,140	1984
12	Crocker Range	I	3,279	1984
13	Gomantong,Materis, Bod Tai, Keruak, Pangi	I	1,816	1984
14	Kabili Sepilok	I	4,294	1931
15	Kalumpang	I	3,768	1984
16	Lungmanis	I	6,735	1984
17	Madai Baturong	I	5,867	1984

Map[†] ref.	*National/international designation* Name of area	IUCN management category	Area (ha)	Year notified
18	Maligan	I	9,240	1984
19	Mengalong	I	1,008	1984
20	Milian-Labau	I	2,812	1984
21	Pin-Supi	I	4,696	1984
22	Sepagaya	I	4,128	1984
23	Sepilok (Mangrove)	I	1,235	1931
24	Sungai Imbak	I	18,113	1984
25	Sungai Kapur	I	1,250	1984
26	Sungai Lokan	I	1,852	1984
27	Sungai Simpang	I	1,149	1984
28	Tabawan,Bohayan,Maganting, Silumpat Islands	I	1,009	1984
	SARAWAK			
	National Parks			
1	Bako	II	2,728	1957
2	Batang Ai	II	1,000	1991
3	Gunung Gading	II	4,106	1983
4	Gunung Mulu	II	52,865	1974
5	Kubah	II	2,230	1989
6	Lambir Hills	II	6,952	1975
7	Loagan	II	1,000	1991
8	Niah	II	3,140	1974
9	Similajau	II	7,067	1979
	Wildlife Sanctuaries			
10	Lanjak-Entimau	IV	168,758	1983
11	Samunsam	IV	6,092	1979

[†]Locations of most protected areas are shown on the accompanying maps.

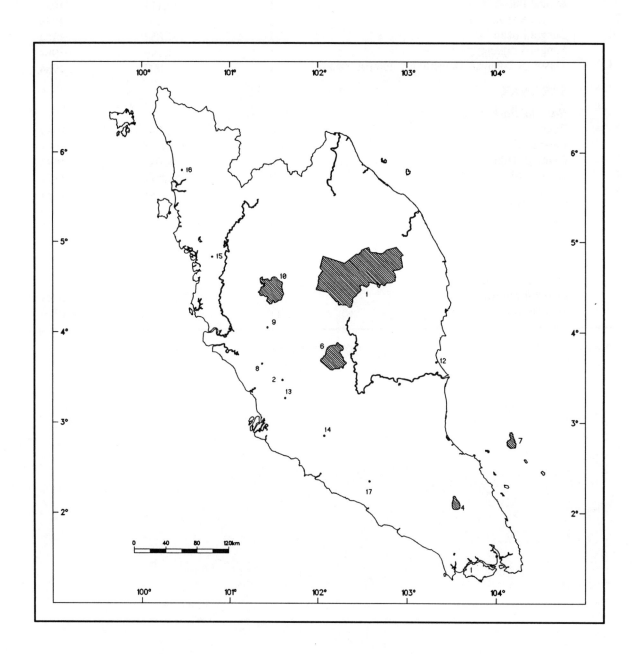

Protected Areas of Peninsular Malaysia (and Singapore)

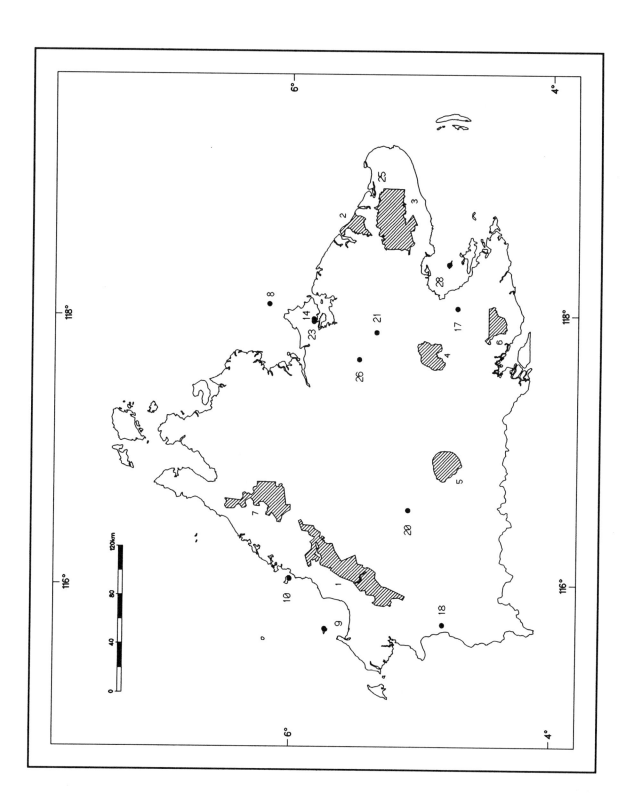

Protected Areas of Sabah

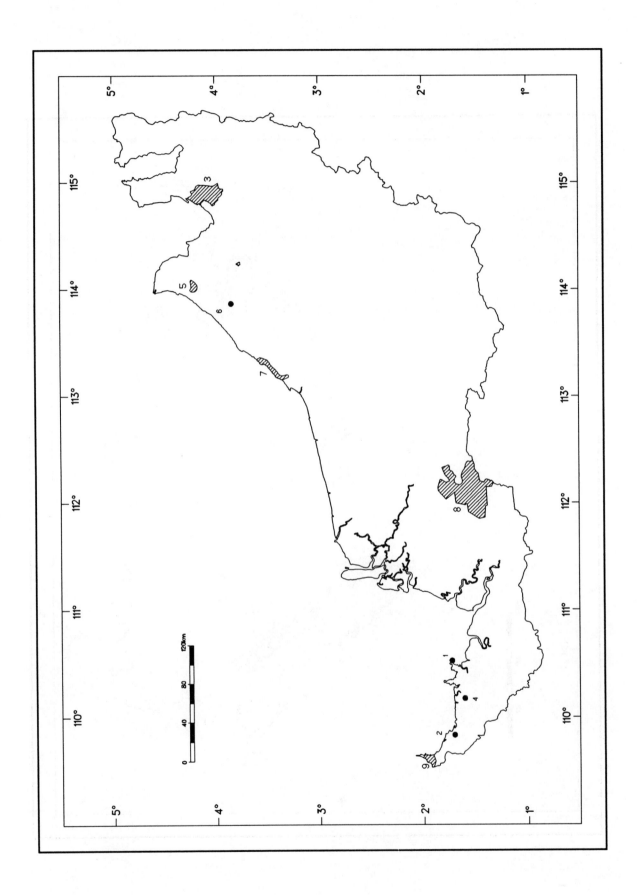

Protected Areas of Sarawak

MALDIVES

Area 298 sq. km

Population 0.2 million (1989)
Natural increase: 3.7% per annum

Economic Indicators
GNP: US$ 300 per capita (1987)

Policy and Legislation There is no legislation which provides for the establishment of protected areas. A draft nature conservation act has been formulated.

Administration and Management A National Environment Council has recently been created and presently functions under the Ministry of Planning and Environment, consisting of representatives from different ministries, such as the Department of Tourism which has a particular interest in ensuring the conservation of natural resources. However, it has no legislative power (UNEP, 1986).

International Activities The Maldives have ratified the Convention concerning the Protection of the World Cultural and Natural Heritage (World Heritage Convention) but no natural sites have been inscribed in the World Heritage list to date. It is not party to the Convention on Wetlands of International Importance especially as Waterfowl Habitat (Ramsar Convention), nor are any biosphere reserves designated under the Unesco Man and the Biosphere Programme.

Systems Reviews The Maldives form the central and largest part of the Laccadive-Chagos Ridge, which extends southwards from India to the centre of the Indian Ocean and consists entirely of a chain of low atolls and associated coralline structures, rising from a submarine ridge which is generally only 270-380m deep but descends to more than 1,000m between the main group of atolls and Suvadiva and Ari in the south. Although the islands cover more than 867km in length, only a small part is dry land with a maximum altitude of no more than 5m above sea level (UNEP/IUCN, 1988). Official sources state that there are 1,190 islands (MPE, *in litt.*, 1991). Most are small: only nine are larger than 2 sq. km, and of these only three exceed 4 sq. km. Ancient rocks are absent from the present geological structure, basic materials being coral sand and coral rock. Soils accordingly are shallow, alkaline, lack significant nutrients and minerals, and have little water retaining ability. Fresh water is found on most islands at a depth of one to three metres in the form of freshwater lenses floating on the salt water below. A number of islands have fresh and brackish lakes (Munch-Petersen, 1985).

Areas with vegetation cover vary from a few sq. m to 6 sq. km (Munch-Petersen, 1985). Within each atoll is a complex of sand banks and reefs, many of which support vegetated islands. There is no available count of reefs and sand banks, partly due to the difficulty in distinguishing between the two, but there are probably some 4,000-6,000 reefs (Kenchington, 1983). Natural vegetation of islands with good to reasonable soils is tropical rain forest. However, none remains undisturbed. Lesser islands have a cover of hardy salt and drought-tolerant bushes and Cyperaceae. Coconuts are the most important crop, palms being cultivated on the greater part of inhabited and uninhabited islands. Other agricultural activity mostly takes place on the permanently inhabited islands (Munch-Petersen, 1985).

Tourism has increased enormously over the last decade, and is the biggest source of foreign exchange. In 1989, there were approximately 140,000 visitors, restricted to 61 tourist resorts (Munch-Petersen, pers. comm., 1990). These resorts are limited by planning regulations to previously uninhabited islands. Despite these careful planning controls, the increase in size of the tourist population, together with the rapidly increasing resident population, has resulted in various ecological problems. Reef destruction to provide building materials takes place to a significant extent in and near the Malé (Kaaf) and Ari (Alif) atolls where the majority of tourist resorts are located. Prior to the 20th century, building in coral rock was limited to mosques, shrines and tombstones, but it is now the main construction material on Malé and tourist resort islands (Brown and Dunne, 1988; Moutou, 1985). A major ecological problem is siltation from the removal and transport of coral rock causing die-offs of reef parts (Munch-Petersen, pers. comm., 1990). Another significant impact of tourism has been the heavy dependence on speed boat travel, which also has a silting impact on live corals, and excursions where anchorages are not controlled with the ensuing destruction of large coral areas (Munch-Petersen, 1983). In the capital, Malé, inadequate waste disposal and fresh water shortage create additional ecological threats (Munch-Petersen, pers. comm., 1990).

Global warming and accompanying rises in sea level pose serious long-term threats (Anon., 1988; Pernetta, 1988; UNEP/IUCN, 1989; Wells and Edwards, 1989).

Proposals for the creation of a nature conservation act to protect fauna and flora and to establish nature reserves are given in an environmental report to the Government of the Maldives (Munch-Petersen, 1983).

Addresses

Environment Research Unit, Ministry of Planning and Environment, Malé (Tlx: 66110 mpdmale mf; FAX: 960 327351)

References

Anon. (1988). A sinking feeling. *The Economist* 1 October. P. 72.

Brown, B.E. and Dunne, R.P. (1988). The environmental impact of coral mining on coral reefs in the Maldives. *Environmental Conservation* 15: 159-165.

Kenchington, R.A. (1983). Report on mission to the Republic of the Maldives. Unesco, Paris. Unpublished. 57 pp.

Moutou, F. (1985). Briefly: the Maldive Islands. *Oryx* 19: 232-233.

MPE (1991) Statistical yearbook of the Maldives. Ministry of Planning and Environment, Malé. (Unseen)

Munch-Petersen, N.F. (1983). *Tourism Development Plan Republic of Maldives*. Final Report, May 1983 (2 volumes). Vol.2: Selected Projects: Project No.5: Protection of Environment and Wildlife Conservation. Kuwait Fund for Arab Economic Development/Dangroup Planning Consultants, Denmark. Pp. 5.1-5.55.

Munch-Petersen, N.F. (1985). Man and reef in the Maldives. Continuity and change. *Proceedings of the Fifth International Coral Reef Congress*, Tahiti 2: 254.

Pernetta, J.C. (1988). The Maldives: present environment and future sea-level rise. *The Siren* 39: 25-26.

UNEP (1986). Environmental problems of the marine and coastal area of the Maldives: national report. *UNEP Regional Seas Reports and Studies* 76. 33 pp.

UNEP/IUCN (1988). *Coral reefs of the world. Volume 2: Indian Ocean, Red Sea and Gulf.* UNEP Regional Seas Directories and Bibliographies. IUCN, Gland, Switzerland and Cambridge, UK/UNEP, Nairobi, Kenya. 440 pp.

Wells, S., and Edwards, A. (1989). Gone with the waves. *New Scientist* 1690: 47-51.

UNION OF MYANMAR
(formerly the Socialist Republic of the Union of Burma)

Area 676,550 sq. km

Population 41,300,000 (1990)
Natural increase: 2.05% per annum

Economic Indicators
GNP: US$ 203 per capita (1988)

Policy and Legislation The 1947 Burmese Constitution, implemented after independence in 1949, defined the state as the "ultimate owner of all lands". Consequently, the state has the right to regulate, alter or abolish land tenures or resume possession of any land for redistribution as it sees fit (Maung, 1961).

Forest policy recognises the basic tenets of conservation, and has three salient principles: the maintenance of environmental stability for the preservation of permanent forest estates; preservation of natural heritage by conserving species and ecosystem diversity and the establishment of a system of protected areas; and ensuring sustainable utilisation of forest resources for the direct benefit of present and future generations (Forest Department, 1991).

Legal protection of natural resources currently rests on two acts, both dating from the pre-World War II colonial period. The 1902 Burma Forest Act repealed all earlier forest acts. This Act allows the Ministry of Agriculture and Forests to establish game sanctuaries and reserved forests on any land at the disposal of the government, and places responsibility for their management and protection on the Forest Department (see Annex). Game sanctuaries were primarily intended to protect hunting stock and the first was established in 1911. The procedure for establishing reserved forests, as laid down in the Act, entails the appointment of a settlement officer to adjudicate in disputes over extant rights and forest use, and makes provision for certain activities, such as agriculture, to continue after designation. Under the Act, wildlife is defined as "forest produce", and local governments are able to issue Game Rules. However, these were not comprehensively formulated until 1927 (Weatherbe, 1940). The application of the Act was complex, and some areas, occupied by hill tribes, were exempt.

The 1902 Forest Act was enhanced by the 1936 Burma Wild Life Protection Act, a consolidation of the earlier Wild Birds and Animal Protection Act of 1912 which was repealed in 1936. Under Sections 26 and 28 of the 1936 Act, the Burma Wildlife Protection Rules were published in the Department of Agriculture and Forests Notification No. 2, dated 2 January 1941 and effective from 11 January 1941. Similar to the 1902 Forest Act, the 1936 Act was not applicable nationwide and certain tribal areas were exempt under the Scheduled Areas

Wildlife Protection Regulation No. 1 of 1941, published by the Defence Department, Political Branch, on 10 February 1941. Tun Yin (1954) details the application of the regulations to specific areas. The 1936 Act makes provision for the establishment of wildlife sanctuaries on any government-owned land or on private land where the owner's consent has been obtained. The Act prohibits all hunting, fishing and wilful disturbance to any animal in sanctuaries, and similar activities in reserved forests have to be licensed (see Annex). In addition, nationwide closed hunting seasons were established and a limited number of species received year-round protection. Although the 1902 Forest Act and the 1936 Wildlife Protection Act theoretically provide protection for wildlife in both reserved forests and in wildlife sanctuaries, neither act includes measures specifically to protect habitat. In 1985, new legislation was proposed which would not only strengthen conservation efforts but also for the first time make provision for the establishment of national parks and nature reserves (FAO, 1985b).

International Activities Myanmar is not yet party to any of the three major international conventions concerned with nature conservation, namely the Unesco Man and the Biosphere Programme, the Convention on Wetlands of International Importance especially as Waterfowl Habitat (Ramsar Convention) and the Convention concerning the Protection of the World Cultural and Natural Heritage (World Heritage Convention).

Administration and Management Responsibility for managing protected areas remains with the Forest Department, which is one of the oldest in Asia. Myanmar is divided into some 40 forest divisions, each of which is supervised by a Divisional Forest Officer; final responsibility rests with the Director-General of Forestry. The Forest Department is responsible for a network of 722 forest reserves, although these are managed primarily for production (FD, 1991). It also manages an Elephant Control Scheme whereby extensive, temporary sanctuaries are established, and elephant capture by the State Timber Corporation is suspended. However, these sanctuaries have no legal status (FAO, 1983). The Forest Department is overshadowed by the politically more influential State Timber Corporation which generates about 25% of the nation's foreign exchange through its monopoly on timber exploitation in reserved forests (Blower, 1985; FAO, 1985a).

A Wildlife Conservation and Sanctuaries Division has been established within the Department, mainly responsible for the management of "national parks" and other protected areas. It has a mandated staff of 2,251, with 498 appointed by 1987; the number of staff currently employed is not available. In addition to responsibilities for protected areas, the Division is concerned with species conservation activities.

Management of wildlife sanctuaries tends to be on an *ad hoc* basis, usually limited to infrequent patrols, and is hampered by inadequate staff, resources, support and relevant infrastructure in the Forest Department. Priorities within the Forest Department have tended to be production oriented, with only modest support for conservation activities. Consequently, there has been a failure to stem both poaching and illegal felling in sanctuaries and reserved forests, some of which have lost their original conservation value (FAO, 1985a; Than, 1989).

Systems Reviews Situated between the Indian subcontinent and the South-east Asian peninsula, Myanmar extends some 2,093km from north to south. Between these extremes there exists an ecological spectrum of almost unique variety, ranging from tropical rain forests and coral reefs in the south to temperate forests of conifers, oaks and rhododendrons in the far north, where snow-capped mountains up to 5,729m high mark the eastern extremity of the Himalaya. High mountain ranges form a continuous barrier along the western border with India and Bangladesh, extending southward parallel with the coast to the Ayeyarwady (Irrawaddy) Delta. In the north-east, the border with China follows the high crest of the Irrawaddy-Salween divide, then bulges out eastward to enclose the ruggedly mountainous Shan Plateau forming the border with Laos and Thailand. Between these mountain barriers to the east and west lies the fertile, heavily-populated basin of the Ayeyarwady, with its largest tributary, the Chindwin, joining from the north-west. Myanmar's other great river, the Salween, flows south through neighbouring Yunnan and then cuts through the Shan Plateau in deep, heavily forested gorges before finally reaching the sea in the Gulf of Martaban. Further south, Tenasserim extends in a long, mountainous arm bordering Thailand down to the Kra Isthmus (Blower, 1982).

The climax vegetation in coastal areas is lowland rain forest, with mangroves and freshwater swamp forest in the Ayeyarwady Delta and flood plain. The Ayeyarwady Basin includes a central dry zone of open, stunted dry deciduous woodland, known as *indaing*. Peripheral to this dry zone are extensive mixed deciduous forests which are of great economic importance as the source of Myanmar's teak and other commercial hardwoods. These are in turn surrounded by a fringe of moist, semi-evergreen and evergreen forest on the semi-circle of higher hills to the west, north and east, merging in the far north with temperate oak and conifer forests and ultimately fir, birch, rhododendron and other sub-alpine vegetation (Blower, 1989). The Forest Department recognises 11 Burma Standard Forest Types, as follows: closed broad-leaf forests, comprising tidal mangroves, beach and dune, swamp, evergreen, mixed deciduous, deciduous dipterocarp, and hill formations; closed

coniferous pine forest; bamboo forest and scrub formations comprising dry scrub and *indaing* scrub (FAO, 1985a).

Results from the UNDP/FAO National Forest Survey and Inventory Project, based on Landsat MSS and RBV imagery for the period 1979-1981, indicate that the total area of closed and degraded forest was 42.3% of total land area. However, for reasons given below, the figure probably exaggerates the current state of Myanmar's forest but more recent imagery is not yet available. Comparison with satellite imagery for 1975 shows that, at a conservative estimate, forest was being depleted at a rate of 6,000 sq. km (2.1%) annually (Allen, 1984). The area of natural forest remaining in 1988/1989 was probably about 245,000 sq. km (35% of land area). However, the extent of closed forest is certainly less than 20% (J. Sayer, pers. comm., 1989). Much of the closed canopy forests are temperate formations in the north, dominated by oak *Quercus* spp., *Castanopsis* spp. and a variety of Ericaceae. Effectively, all forest in the Shan states has been affected by shifting cultivation and is consequently degraded or cleared. There is little intact forest in the Arakan Yoma in the west, with forest on the coastal side degraded to bamboo and only some managed mixed deciduous formations on the eastern side. Conditions in the southern and northern Chin Hills are extremely degraded with only very small islands of natural forest remaining (J. Sayer, pers. comm., 1989). The status and distribution of forests is discussed further in Collins *et al.* (1991).

A current summary of wetlands is given in Scott (1989). With a coastline of 2,278km, several very large estuarine and delta systems and numerous offshore islands, Myanmar possesses a considerable diversity of coastal wetland habitats, including coral reefs, sandy beaches and mudflats. The most extensive wetlands in the interior of the country are the seasonally inundated floodplains of the three main river systems: Ayeyarwady-Chindwin, Sittaung (Sittang) and Salween. These plains have a surface area of some six million hectares during the monsoon season, providing feeding grounds for waterfowl and spawning grounds for fish, notably carp, catfish and perch. The practice of constructing embankments and cultivating flood plains restricts major areas of natural flood plain to the north. Permanent freshwater bodies, including the two main lakes, Inle and Indawngy, cover about 1,300,000ha (Scott, 1989). At least 17 important wetland sites have been identified (Scott, 1989).

The main coral reefs lie in the Mergui Archipelago (Duncan, 1889; Harrison and Poole, 1909). There are no data on the ecology of these reefs but 65 species in 31 genera have been described in a more recent study (Kyi, 1985), suggesting a moderate diversity. It may be assumed from the brief early descriptions and by inference from the better-known islands of adjacent Thailand, that coral reef development in Mergui is appreciable. Rosen (1971) predicts that perhaps 43 or 44 coral genera may be found. There are no known major coral reefs along the mainland coast, although corals have been reported near the mouth of the Bassein River and around Thamihla Kyun (UNEP/IUCN, 1988).

In 1980 the government requested the Food and Agricultural Organisation of the United Nations and the United Nations Development Programme to assist in a joint Nature Conservation and National Parks Project with the Working Peoples Settlement Board. The 1981-1984 FAO/UNDP Nature Conservation and National Parks Project was formulated to conserve natural ecosystems, protect endangered species and develop a system of national parks and nature reserves. Immediate objectives included the development of institutions for conservation, assistance in surveys and feasibility studies for the establishment of national parks and nature reserves, and preparation of management plans and their implementation. A comprehensive set of recommendations was made, covering the following: policy, legislation and organisation; recruitment and training; conservation education; coordination of surveys and planning; establishment, development and management of protected areas; establishing species conservation priorities, law enforcement; control of hunting and capture; control of trade in wildlife and wildlife products; completing natural resource inventories; and obtaining external assistance for a second-phase project (FAO, 1983).

The principal measures required for the planning and implementation of an effective nature conservation programme, and the establishment of a protected areas system, is discussed in some detail in FAO (1985b), drawing on the experience of FAO/UNDP Nature Conservation and National Parks project. The recommendations cover a broad range of topics, namely: policy, legislation and organisation; staff recruitment and training; conservation education; coordination of surveys and planning; establishment, development and management of protected areas; and the establishment of species conservation priorities. Implementation of these recommendations has been slow, with some development of Alaungdaw Kathapa, Hlawaga Wildlife Park and Mount Popa Mountain Park. A further FAO/UNDP project to implement the recommendations was proposed for implementation during 1987-1990, but this was not finalised. Entitled "Support to nature conservation programme", the project was intended to build on the earlier work with the following objectives in view: development and management of protected areas, especially for the benefit of local communities; protection of watersheds, landscapes, representative ecosystems and threatened species; and the strengthening of institutions and administrative capabilities (UNDP, 1985). Funding for this is available but due to political uncertainties the project has not yet been implemented.

At the 25th Working Session of IUCN's Commission on National Parks and Protected Areas, regional field managers developed an action plan for protected areas in the Indomalayan realm (IUCN, 1985). The plan identifies a number of goals for the region, and makes the following specific recommendations for Myanmar: upgrading of Kyatthin Game Sanctuary to a nature reserve and establishment of Thamihla Kyun, South Moscos and Kadonlay Kyun as marine reserves; accession to, and implementation of the World Heritage Convention; exchange of expert staff with national management agencies in Thailand, India, Sri Lanka and Indonesia, to address control and management problems; and promotion of intergovernmental cooperation to implement bilateral management of species (for example, elephant and tiger, which cross the borders with Bangladesh, Bhutan and India) and riverine ecosystems (for example, the Naaf, Mekong and Salween rivers).

Sanctuaries cover only 0.7% of the total land area, which is considered to be an inadequate sample of the nation's natural resources (FAO, 1985a). In contrast, reserved forests, which for conservation purposes are in many respects comparable to wildlife sanctuaries, cover some 100,222 sq. km or 14% of the total land area. Coastal protected areas are limited to Thamihla Kyun and Moscos Islands wildlife sanctuaries, but there is no current legislation for establishing marine protected areas. The most serious omissions from the current protected areas system are lowland evergreen, hill evergreen and semi-evergreen forest (FAO, 1985a; MacKinnon and MacKinnon, 1986) and tidal forest (R.E. Salter, pers. comm., 1987). The proposed Pakchan Nature Reserve, and Natma Taung and Pegu Yomas national parks are intended to rectify much of this (FAO, 1985a; MacKinnon and MacKinnon, 1986). However, MacKinnon and MacKinnon (1986) suggest that even with the designation of the proposed protected areas all vegetation types, with the exception of sub-alpine, will remain threatened. Wetlands are unprotected, even in Wethtigan Wildlife Sanctuary which provides protection for wildlife but not habitat. The proposed Inle Lake and Mong Pai Lake wildlife sanctuaries, and Moyingyi Game Sanctuary are intended to address this omission.

Protected areas are directly threatened by inadequate size, both individually and in aggregate, by failure to provide representative coverage of several important biota, and by weak and poorly-enforced legislation (Blower, 1982). Effective law enforcement and prevention of poaching in reserved forests and game sanctuaries are difficult due to the shortage of Forest Department field staff and to the large numbers of firearms in the hands of military personnel, para-military People's Militia and, in some areas, insurgents (Whitmore and Grimwood, 1976). There is also extensive encroachment in many of the existing forest reserves (J. Blower, pers. comm., 1989). Game sanctuaries only legally protect fauna and not habitat. Many have been seriously damaged and some of the smaller areas, for example Maymyo Game Sanctuary, have little justification for being retained as protected areas (FAO, 1985a). Continuing civil unrest, particularly in more remote regions, largely precludes development of the protected areas system.

Addresses

Wildlife Conservation and Sanctuaries Division (Director), Forest Department, Gyogon, Yangon
Forest Department (Director General), Yangon

References

Allen, P.E.T. (1984). A quick new appraisal of the forest cover of Burma, using Landsat satellite imagery at 1:1,000,000 scale. FAO/UNEP National Forest Survey and Inventory. BUR/79/011. *Technical Note* 11: 1-6.

Anon (1937). Note on the Burma Wild Life Protection Act. *Journal of the Bombay Natural History* 39(3): 606-608.

Blower, J. (1982). Species conservation priorities in Burma. In: Mittermeier, R.A. and Konstant, W.R. (Eds), Species conservation priorities in the tropical forests of South-east Asia. *IUCN SSC Occasional Paper* No. 1: 53-58.

Blower, J. (1985). Conservation priorities in Burma. *Oryx* 19: 79-85.

Blower, J. (1989). Burma: conservation of biological diversity. Draft. World Conservation Monitoring Centre, Cambridge, UK. 13 pp.

Collins, N.M., Sayer, J.A. and Whitmore, T.C. (Eds) (1991). *The conservation atlas of tropical forests: Asia and the Pacific*. Macmillan Press Ltd, London. 256 pp.

Duncan, P.M. (1889). On the Madreporaria of the Mergui Archipelago, collected for the trustees of the Indian Museum by Dr John Anderson FRS, superintendent of the Museum. *Journal of the Linnean Society (Zoology)* 21: 1-25. (Unseen)

FAO (1983). *Summary of currently available information on internationally threatened species in Burma*. Nature Conservation and National Parks Project BUR/80/006. Field Document No. 7. FAO, Rangoon. 76 pp.

FAO (1985a). *Burma: survey data and conservation priorities*. Nature Conservation and National Parks Project BUR/80/006. Technical Report No. 1. FAO, Rome. 102 pp.

FAO (1985b). *Burma: project findings and recommendations*. Nature Conservation and National Parks Project DP/BUR/80/006. Terminal Report. FAO, Rome. 69 pp.

FAO/UNEP (1981). *Tropical forest resources assessment project: forest resources of tropical Asia*. FAO, Rome. 475 pp.

Forest Department (1991). *Forest resources of Myanmar: conservation and management*. Forest Department, Yangon. 13 pp.

Harrison, R.M. and Poole, M. (1909). Marine fauna from Mergui Archipelago, collected by Jas.J. Simpson, M.A., B.Sc. and R.N. Redmose Brown, B.Sc., University of Aberdeen. Madreporaria. *Proceedings of the Zoological Society of London* 1909: 897-912.

IUCN (1985). *The Corbett Action Plan for protected areas of the Indomalayan Realm*. IUCN, Gland, Switzerland and Cambridge, UK. 23 pp.

Kyi, A. (1983). Systematic study of some Scleratinian Corals from Mergui Archipelago of Burma. M.Sc. thesis, Moulmeim Degree College, Moulmeim, Burma. (Unseen)

MacKinnon, J. and MacKinnon, K. (1986). *Review of the protected areas system in the Indo-Malayan Realm*. IUCN, Gland, Switzerland and Cambridge, UK/UNEP, Nairobi, Kenya. 284 pp.

Maung, K. (1961). *Burma's Constitution*. Martinus Nijhoff, The Hague. 340 pp. (Unseen)

Rosen, B.R. (1971). The distribution of reef coral genera in the Indian Ocean. In: Stoddart, D.R. and Yonge, C.M. (Eds), Region Variation in Indian Ocean Coral Reefs. *Symposium of the Zoological Society of London* 28: 263-299.

Scott, D.A. (Ed.) (1989). *A directory of Asian wetlands*. IUCN, Gland, Switzerland and Cambridge, UK. 1,181 pp.

Tun Yin, U. (1954). Wildlife preservation and sanctuaries in the Union of Burma. *Journal of the Bombay Natural History Society* 52: 264-284.

Than, A (1989). A proposal for ecological study and conservation of brow-antlered deer (*Cervus eldi thamin*) in Myanmar (Burma). Unpublished. 19 pp.

UNDP (1985). Support to nature conservation programme: project of The Socialist Republic of the Union of Burma. Revised draft. 16 pp.

UNEP/IUCN (1988). *Coral reefs of the world. Volume 2: Indian Ocean, Red Sea and Gulf*. UNEP Regional Seas Directories and Bibliographies. IUCN, Gland, Switzerland and Cambridge, UK/UNEP, Nairobi, Kenya. 440 pp.

Weatherbe, D'A. (1940). Burma's decreasing wild life. *Journal of the Bombay Natural History Society* 42: 150-160.

Whitmore, T.C. and Grimwood, I.R. (1976). *The conservation of forests, plants and animals in South-east Asia*. Volume 2. Part 1. Continental South East Asia. IUCN, Gland, Switzerland. Unpublished report. 82 pp.

ANNEX
Definitions of protected area designations, as legislated, together with authorities responsible for their administration

Title: The Burma Forest Act

Date: 27 March 1902 (amended 1906, 1912, 1926, 1938 and 1941)

Brief description: An Act to consolidate and amend the law relating to forests, forest produce and the duty leviable on timber. After independence in 1948, the Act was reinstated with only the titles of government and authorities changed (Adaptation of Laws, Order 1948, dated 4 January 1948).

Administrative authority: Forest Department

Designations:

Reserved forest[1] A forest and every part of a forest a) declared to be a reserved forest under the provisions of Section 18 of this Act or the corresponding section of any enactment previously in force in Burma; or b) declared to be a reserved forest under the provisions of any rules in force in Lower Burma previous to 1st July 1882, and brought within the provisions of the Burma Forest Act 1881 by Section 30 of that Act, which shall not, at the time being, have ceased to be a reserved forest under Section 29 of this Act or the corresponding provision of any such enactment or rules.

Prohibited activities include trespass, pasturing, damaging trees, setting fires, quarrying, cultivation, poisoning or dynamiting, hunting, shooting, fishing or setting traps or snares.

Source: Original legislation

Title: Burma Wild Life Protection Act

Date: 1936 (amended 1954)

Brief description: Makes provision for the establishment of sanctuaries (game sanctuaries) on any land at the disposal of the government or, subject to the consent of the owner, any land which is private property. Also provides for the protection of a number of named species outside sanctuaries and reserved forests.

Administrative authority: Wildlife Conservation and Sanctuaries Division, Forest Department

Designations:

Game sanctuary No person is permitted to hunt without the special permission of the local government (which is only granted for scientific purposes or to preserve the balance of animals) or, drive, stampede or wilfully disturb any animal.

Reserved forest No person shall hunt, drive, stampede or wilfully disturb any animal or remove any animal or part of product thereof except under a licence.

Source: Anon (1937)

[1]Forest is classified as follows: commercial reserves, managed for the production of hardwoods for domestic consumption and export; and local supply reserves in close proximity to villages and managed for supply of minor forest products for domestic consumption. Land at the disposal of the state, other than reserved forest, may be constituted as public forest land to meet local requirements for forest products and to discourage encroachment into reserved forest. Timber may also be extracted for commercial purposes from public forest land (Forest Department, 1991).

SUMMARY OF PROTECTED AREAS

Map[†] ref.	National/international designation Name of area	IUCN management category	Area (ha)	Year notified
1	*National Park* Alaungdaw Kathapa	II	160,580	1984
2	*Park* Popa Mountain	V	12,691	1985

[†]Location of the National Park is shown on the accompanying map.

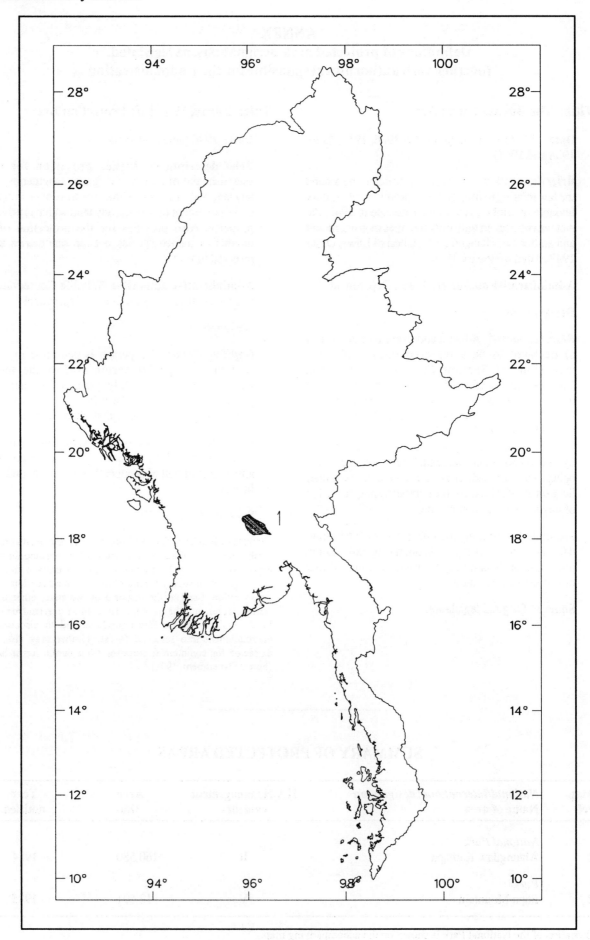

Protected Areas of Myanmar

NEPAL

Area 141,415 sq. km

Population 19,100,000 (1990)
Natural increase: 2.5% per annum

Economic Indicators
GNP: US$ 170 per capita (1988)

Policy and Legislation The new Constitution of the Kingdom of Nepal 2047 (1991) formally recognises the need to preserve the environment and use natural resources wisely. In Chapter 4 it is stated that "The Kingdom of Nepal will give priority to raising public awareness on environmental issues, to mitigating the adverse effects development works have on the environment, and to the conservation of rare fauna and flora." The Constitution makes provision for the formation of committees on Natural Resources and Environmental Conservation by the House of Representatives (Chapter 8).

A National Conservation Strategy for Nepal was completed in 1987 and endorsed as policy in 1988 (HMG Nepal/IUCN, 1983, 1988). Policy resolutions cover the basic requirements of the people, as well as the need to safeguard natural and aesthetic values and to maintain the country's cultural heritage. It is also resolved that a separate body, the National Council for the Conservation of Natural and Cultural Resources, replace the National Commission for the Conservation of Natural Resources and be responsible for implementing the National Conservation Strategy and formulating policy guidelines concerning resource conservation matters. This council has since been formed and represents the most important step to date towards establishing an institutional framework for cooperative environmental management and protection in the country.

Conservation awareness dates back many centuries in Nepalese society. The tradition of preserving large expanses of forest adjacent to places of worship or important sources of water is deep-rooted. In Kathmandu Valley, for example, there are 45 sacred forests ranging in size from one to several thousand hectares which have been preserved by countless generations in accordance with ancient religious traditions (Mansberger, 1990). Various traditional systems of resource administration have also evolved: for example, the *shingo nawa* (forest caretakers) in Sherpa society; the *kipat* system of exclusive and unalienable communal rights over large areas in the Eastern Hills; and the *chitaidar* (local non-official functionaries) responsible for the use of village forests in the 19th century (HMG Nepal/IUCN, 1988). In the first half of this century there was a National Code under which forests in the hills were controlled by village heads and private forests by forest watchers, with the district administrator holding superior authority. Forest clearance was prohibited unless authorised by the government. Traditional forms of resource conservation,

such as *shingo nawa, chitaidari* and *kipat*, disappeared with the handing over of private forests to the state in 1957 under the Forest Nationalisation Act. This Act was introduced to bring forests under management and also to prevent land being converted to agriculture. Increasing pressures on land in the *terai* led to the passing of the Forest Act in 1961 to protect forests, by restricting access to them, and to regulate forest utilisation. State, panchayat, panchayat protected, religious and contract forests are defined under this Act (see Annex). Following recent changes to Nepal's political system, these terms are due to be replaced by national, community plantation, community protected, religious and leased forests, respectively, under new forest legislation which awaits approval by the newly elected government. The Forest Protection Special Act, 1968 provides forest officials with policing and judicial powers. Such measures, introduced to help counter encroachment and wanton destruction of forests, became less applicable with the change in policy towards community forestry and decentralisation. For example, various rules such as the Panchayat Forest Rules, 1978 and Panchayat Forest Protection Rules, 1978 were framed under the Forest Act to give local communities access to or ownership of forest lands to encourage sustained use of such resources. Other forestry-related legislation includes the Soil and Watershed Conservation Act, 1982 which enables the Department of Soil Conservation and Watershed Management to declare, develop and conserve critical watersheds. Under this Act, any area may be designated as a protected watershed area (see Annex). The Act has not yet been applied but two nationally important watersheds are under consideration for designation as protected watershed areas (MFSC, 1988).

Nepal has a well-developed mechanism for formulating and declaring policy through its national five-year plans. Stated policies that affect the forestry sector are more or less adequate, the main problem being translating policy to legislation and in its effective implementation. Moreover, existing laws are not always consistent with current policy. Present forest policy is based on the 1976 National Forestry Plan. Its main objectives are to meet the people's needs for forest products, to maintain and restore the ecological balance through programmes of reforestation and watershed management, and to derive maximum economic gains from forest products. Policies are incorporated within the most recent (seventh) five-year plan. A new forestry sector policy (1989) was formulated under the Master Plan for the Forestry Sector. It was proposed that: forest resources be managed with priority given to products that best contribute to the needs of the people; forest resources be managed according to their ecological capability so as to conserve the forests, soil, water, flora, fauna and scenic beauty, with representative examples of ecosystems unique to

Nepal protected and tourism regulated according to local carrying capacities; and community forestry and the establishment of private forests on leased and private lands be promoted in accordance with the principles of decentralisation policy. Official endorsement and implementation of this new policy was considered to be a priority, requiring extensive reform of existing forest legislation (MFSC, 1988). A revised forestry sector policy (1991) has since been prepared under the Master Plan.

A national conservation programme was initiated by HMG Nepal in 1971. This was given a legal basis following the passing of the National Parks and Wildlife Conservation Act, 2029 in March 1973 which provides for the establishment and administration of protected areas and "the conservation of animals and birds and their habitats." This act supercedes the Wildlife Conservation Act, 2015 (1958) and the Hunting Rules of 1967, under which six royal hunting reserves were established in July 1969. The 1973 Act enables the government to establish any area as a national park, reserve (i.e. controlled natural reserve, wildlife reserve and hunting reserve) and, following an amendment in 1989, conservation area (see Annex). A controlled natural reserve (more commonly referred to as strict nature reserve) is at the protection end of the spectrum, with entry permitted only for scientific study (none has been created to date); a hunting reserve is at the utilisation end and managed for recreational hunting on a sustained yield basis. National parks and wildlife reserves both provide for the conservation of fauna and their habitats, but national parks have a broader emphasis encompassing landscape values. A conservation area provides for a flexible system of resource management through people's participation. It may be managed by the relevant government agency or entrusted to a non-governmental organisation. The government may alienate, transfer ownership or alter the boundaries of national parks, reserves or conservation areas by notification in the *Nepal Gazette*. The various regulations introduced under the Act are the National Parks and Wildlife Protection Regulations, 2030 (1974), Royal Chitwan National Park Regulations, 2030 (1974), Wildlife Reserve Regulations, 2034 (1977), Himalayan National Park Regulations, 2036 (1979) and Khaptad National Park Regulations (1987). Provisions under the Himalayan National Park Regulations include the disposal of rubbish in designated places, prohibition of the use of forest products or their purchase from local residents by visitors, self-sufficiency in fuel for visitors, exemption of park entry fees for pilgrims, and collection of forest products and grazing of livestock by residents in places designated by the warden. Conservation Area Regulations are under preparation. While the Act and accompanying regulations provide considerable discretionary powers to authorised officers, the lack of policy guidelines is a major constraint to achieving effective management of protected areas. A working policy has recently been drafted as part of the Master Plan for the Forestry Sector and represents a guide to the application of the National Parks and Wildlife Conservation Act. In addition to the adoption of a working policy, the Act needs to be amended to strengthen protected areas management by providing for zonation (including the creation of buffer zones), the addition of a new category (biological reserve) of protected area for biologically important areas and wetlands that do not meet national park or wildlife reserve criteria nor need the restrictions on entry of the controlled natural reserve, and income for community development. It is also proposed that the long-term security of protected areas be strengthened by requiring that their alienation or transfer be made subject to special legislation passed through the National Parliament (MFSC, 1988).

Certain other legislation relates to tourism in protected areas. Under the Tourism Act, 2035 (1979), mountaineering expeditions must obtain a permit from the Ministry of Tourism in order to climb listed Himalayan peaks, some of which are in national parks. Similarly, tourists wishing to trek anywhere in Nepal must obtain permission from the Central Immigration Office, Home Ministry, in accordance with the Trekking and River Rafting Regulations, 2041 (1985). Many of the popular trekking routes are in national parks.

International Activities Nepal has entered into a number of obligations and cooperative agreements related to conservation. It is a signatory to the Convention concerning the Protection of the World Cultural and Natural Heritage (World Heritage Convention) which it accepted in 20 June 1978. Two natural sites, Sagarmatha and Royal Chitwan national parks, have been inscribed on the World Heritage List. Nepal acceded to the Convention on Wetlands of International Importance especially as Waterfowl Habitat (Ramsar Convention) on 17 December 1987, at which time Koshi Tappu was added to the List of Wetlands of International Importance established under the terms of the Convention. Nepal participates in the Unesco Man and Biosphere Programme. A National Committee for MAB was established in 1974 under the framework of the Nepal National Committee for Unesco formed in 1971. No biosphere reserves have been established to date. Following initiatives by MAB/Unesco and MAB/Nepal in 1975, the International Centre for Integrated Mountain Development was established in Kathmandu in 1983 following an agreement between HMG Nepal and Unesco signed in 1981. Its primary objectives are "to help promote the development of an economically and environmentally sound mountain ecosystem ...", thereby complementing regional efforts towards conservation. Participating nations are Afghanistan, Bangladesh, Bhutan, Burma, China, India, Nepal and Pakistan (Glaser, 1984; ICIMOD, 1989).

Other regional initiatives concerned with resource conservation in which Nepal participates are the South Asian Cooperative Environmental Programme and the South Asian Association for Regional Cooperation.

Further details are given elsewhere (HMG Nepal/IUCN, 1988).

Nepal and China have both established protected areas on their respective sides of Mount Everest (Sagarmatha/Chomolangma). Management plans are being formulated under cooperative agreements with the Woodlands Mountain Institute in the case of both countries.

Administration and Management A new institutional structure for the Ministry of Forests and Soil Conservation was developed during the formulation of the Master Plan for the Forestry Sector and this is being implemented ahead of legislative reforms under directives issued by His Majesty the King on 1 May 1988 (MFSC, 1988). The focus of the organisational changes is to strengthen the field units in the Department of Forest in order to develop community and private forests based on people's participation and to develop national forests based on management by government agencies. The Department of Forest, one of four departments within the Ministry, and now headed by a Director General, is responsible for protection and utilisation of forest resources. It is split into four divisions responsible for administration, planning, management and community forests, respectively. There are five regional directorates of forests, each headed by a Regional Director, 75 district forest offices, each under a District Forest Officer, and 222 range offices, each under a range officer. In addition, 453 forest service centres are proposed to assist with community forestry. Protection responsibilities are assigned to the armed forest guards in the case of national forests, but in community forests they are the responsibility of the user groups. The total number of approved posts in 1988 was 8,855, of which 1,329 were for armed forest guards.

Wildlife conservation, prior to the National Parks and Wildlife Conservation Act, was the responsibility of the Forest Department, which established wildlife and hunting reserves, issued hunting licences and controlled hunting within forest reserves. A National Parks and Wildlife Conservation Office was set up in July 1972 as a semi-autonomous branch of the Forest Department (FAO, 1980). In 1982, it was upgraded to departmental status within the Ministry of Forests, now the Ministry of Forests and Soil Conservation. The Department is the primary agency for *in situ* conservation of ecosystems and genetic resources. It is headed by a Director General and comprises two divisions (National Parks and Reserves, Planning and Research), three sections (Administration, Financial Administration, Hatisar/ Elephant Camps) and a Central Zoo. Law enforcement within parks and reserves has been the responsibility of the Royal Nepal Army since 1974 (MFSC, 1988). The total number of approved posts for 1990/1991 is 998, of which 595 are field units (but not protection units) responsible for administering parks and reserves. The Department's financial allocation for 1990/1991 is NRs 124.3 million, of which 84% is for protection units (Royal Nepal Army). Revenue totalled NRs 0.6 million

from headquarters and NRs 22.2 million from parks and reserves (B.N. Upreti, pers. comm). In recent years, income generated from tourism, concessions, permits and other sources has consistently exceeded expenditure if the costs of the protection units are excluded. The Smithsonian-Nepal Tiger Ecology Project was launched in 1973 as a joint programme supported by HMG Nepal and the US government. The project was based in Royal Chitwan National Park and in 1984 was succeeded by the Smithsonian-Nepal Terai Ecology Project. Recently, in 1988, the Department of National Parks and Wildlife Conservation signed a 12-year cooperative agreement with the Woodlands Mountain Institute to support the Makalu-Barun Conservation Project (Shrestha *et al.*, 1990).

The Department of Soil Conservation and Watershed Management was established in 1974 in response to a growing awareness of soil conservation and watershed degradation problems. Its objectives are to maintain ecological equilibrium by conserving important watershed areas and by reducing the incidence of natural disasters such as soil erosion, landslides and floods. The Department is a project-based territorial organisation. It is split into Environment and Management and Technology Development divisions, and three sections, with a total complement of 594 staff in 1988, of which 274 were permanent and the rest temporarily assigned to projects (MFSC, 1988).

The Shivapuri Watershed and Wildlife Reserve Board was established in 1975 with the aim of improving the quality and quantity of drinking water in Kathmandu Valley, conserving the natural environment and developing it for tourism. A watershed area of 144 sq. km has been demarcated by a boundary wall and declared a wildlife reserve. The reserve is managed by a committee, members of which include the Director-General of the departments of Forests, National Parks and Wildlife Conservation, and Soil Conservation and Watershed Management. It is planned that management of the reserve should eventually be handed over to the Department of National Parks and Wildlife Conservation (MFSC, 1988).

The shortcomings of the Ministry of Forests and Soil Conservation were assessed as part of the Master Plan process (MFSC, 1988). It is widely accepted that: protective forestry as a general strategy for forest conservation has failed; management of forests located close to farmland should be handed over to the local people; and that only areas which can be legitimately defined and demarcated can be managed successfully by a professional body such as the Department of Forest. Constraints within this Department include a distorted staffing distribution, due primarily to the large number of vacancies still to be filled in the remote areas. The Department of Soil Conservation and Watershed Management is relatively new and lacks the resources to fulfil its mandate. In the long-term, the Department should become economically sustainable. Criteria need to be formulated to identify priority areas and authorities, with a view to optimising the allocation of scarce resources. The Department of National Parks and Wildlife Conservation has been inadequately staffed for the size of its task, a problem exacerbated by secondments, research assignments and overseas fellowships. Many of these constraints have been addressed in the recent organisational reform of the Ministry. Within the Department of National Parks and Wildlife Conservation, however, the assignment of protection responsibilities to the military continues to be a considerable drain on the Department's financial resources. Moreover, this sharing of responsibilities is a constraint to the effective management of protected areas (Upreti, 1990).

The King Mahendra Trust for Nature Conservation is an autonomous non-profit organisation established in October 1983 under the King Mahendra Trust for Nature Conservation Act 2039. The King Mahendra Trust for Nature Conservation Regulations 2041 were published on 15 October 1984. The Trust aims to conserve and manage natural resources in order to improve the quality of life of the human population, complementing the efforts of HMG and foreign agencies. The Trust has been instrumental in the establishment of the Annapurna Conservation Area (pending legal notification) and entrusted with its management. Support for protected areas has also been extended to preliminary surveys of the Barun Valley in cooperation with the Woodlands Mountain Institute, USA (Rana *et al.*, 1986). Major objectives planned for 1988/1989-1991/1992 include the implementation of the conservation area concept in the Annapurna basin and the establishment of the Nepal Conservation Research and Training Centre at Sauraha in Royal Chitwan National Park (KMTNC, 1988).

The Nepal Nature Conservation Society, founded in 1971, encourages local interest in natural history and conservation but is seriously handicapped by lack of financial resources (FAO, 1980). Other non-governmental organisations with a conservation outlook include the Nepal Forum of Environmental Journalists, and Nepal Forestry Association. A small-scale but effective initiative is the Jara Juri programme whereby each year leading efforts to promote resource conservation by an individual or community are formally recognised (HMG Nepal/IUCN, 1988; Pandey, 1988). IUCN (The World Conservation Union) has a project office in Nepal to assist with implementing the Nepal Conservation Strategy.

Systems Reviews Nepal, with its rich biological diversity and spectacular landscape, extends for 800km along the southern slopes of the Himalaya, separating the arid Tibetan Plateau to the north from the fertile Gangetic Plain to the south. More than 80% of the total area is covered by rugged hills and mountains, including Sagarmatha (Mount Everest) and another seven of the world's ten highest peaks. Five physiographic zones can be distinguished: High Himal (23% total area) comprising alpine meadows, rock and ice between the tree line and Great Himalayan divide; High Mountains

(20%) extending from the heavily populated hills of the Middle Mountains to the tree line; Middle Mountains or Middle Hills (30%) of central Nepal; Siwaliks (13%), representing the first and lowest ridges of the Himalayan system and extending from the Gangetic Plain to the Mahabharat Lekh at the southern edge of the Middle Mountains; and the *terai* (14%), a northern extension of the Gangetic Plain (Kenting, 1986; MFSC, 1988). There are four main ecological zones: trans-Himalaya (a small, semi-arid zone north of the main Himalayan axis in Western Nepal), highlands, subtropical/temperate midlands, and tropical lowlands or *terai* (HMG Nepal/IUCN, 1988). The main river systems from west to east are the Mahakali, Karnali, Narayani and Kosi, all of which originate from the Himalaya. Together with other smaller rivers rising in the Mahabharat Lekh and Siwaliks, they contribute up to 40% of the annual flow of the Ganges River and 71% of its dry season flow. Other wetlands include numerous small lakes, reservoirs and village tanks, and a number of large reservoirs under construction in the Gandaki, Bagmati and Karnali river basins (Scott, 1989).

Based on aerial surveys in 1978/1979, it has been estimated that forest (i.e. land with at least 10% tree crown cover) covers 56 million hectares or 38% of the country (most of which is found in the Siwaliks, Middle Mountains and High Mountains). Scrubland accounts for a further 4.7%, cultivations and non-cultivated inclusions 26.8%, grasslands 11.9%, and other lands 18.5%. Much of this forest is in poor conditions with only a scattering of trees: forest cover at 40% tree crown cover is only 28.1%. Furthermore, it is estimated that there has been a 5.7% of loss of forest land during the preceding 14 years, most of which occurred in the *terai* and Siwaliks. Such losses are due to uncontrolled exploitation for fodder, fuelwood, timber and grazing and to conversion for agriculture, exacerbated by the mass migration of people from the Middle Mountains following the eradication of malaria in the lowlands. Although forest cover may not have changed significantly in the Middle Mountains from 1964/1965 to 1978/1979, its quality deteriorated more than anywhere else in the country. Moreover, deforestation has been more extensive in the Middle Mountains (with 41% forest cover) and *terai* (23%) during recent historical times, than in the Siwaliks (76%) and High Mountains (55%) (Kenting, 1986).

Following initiatives in the late-1950s to protect the Indian rhinoceros and its habitat, the need to establish protected areas elsewhere in Nepal was highlighted under the HMG/UNDP/FAO Trisuli Watershed Project (Caughley, 1969). Subsequently, in 1973, HMG embarked on a National Parks and Wildlife Conservation Project with assistance from UNDP and FAO. Its objective was to ensure the more effective conservation and management of Nepal's valuable yet diminishing wildlife resources and associated habitats by establishing a system of national parks and reserves which, in addition to their conservation role, would contribute to the development of the country's economically important tourist industry. Many of Nepal's protected areas were established under this project which ended in 1979. A conservation education programme was included in the project and wildlife staff were trained overseas under this project and a New Zealand Cooperation Project (FAO, 1980). In 1974, Royal Chitwan National Park, Royal Karnali Wildlife Reserve (renamed Royal Bardia Wildlife Reserve) and Royal Sukla Phanta Wildlife Reserve were identified by HMG as important areas for tiger conservation and received substantial support for their development from WWF under the aegis of Operation Tiger. The New Zealand government was instrumental in the establishment of Sagarmatha National Park, providing funds for its development over a six-year period beginning in 1975 (Lucas, 1977). A third two-year HMG/UNDP/FAO National Parks and Protected Areas Management Project was launched in 1986 to strengthen the capability of the Department of National Parks and Wildlife Conservation to effectively manage its protected areas by preparing and implementing management plans and integrating local people into the planning and management process (Heinen *et al.*, 1988). Initiatives are now underway to extend the protected areas network to the Annapurna and Makalu-Barun regions, with particular emphasis on promoting the "conservation area" concept to facilitate people's participation in conserving natural resources (Upreti, 1990).

Nepal has a fairly extensive protected areas network covering 7.7% of total land area. It is in the process of being expanded by a further 2.9%, with the establishment of conservation areas in the Annapurna and Makalu-Barun regions (Sherpa *et al.*, 1986; Shrestha *et al.*, 1990), and a number of earlier proposals for setting up hunting reserves remain outstanding (Wegge, 1976a, 1976b). General recommendations to develop the protected areas network are made in the IUCN *Review of the Protected Areas system in the Indo-Malayan Realm* (MacKinnon and MacKinnon, 1986) and in the *Corbett Action Plan* (IUCN, 1985). A more recent assessment shows that of Nepal's five physiographic zones, the Middle Mountains are poorly represented with only 1.4% protected areas coverage, as compared with at least 4% for all other zones and 17.1% in the case of the High Himal. The limited coverage of the Middle Mountains is improved somewhat by the royal forests of Nagarjun (1,600ha) and Gokarna (250ha), and there are two protected watersheds due to be established in this zone (MFSC, 1988). A more refined review of protected areas coverage of Nepal's forests with respect to breeding birds (Inskipp, 1989) shows that all upper temperate, subalpine and alpine and most tropical forest types are well represented. Tropical evergreen forests, subtropical and lower temperate broad-leaved forests in the far east, and subtropical broad-leaved forests further west are unrepresented or very poorly represented. A high priority for bird conservation is protection of the species-rich forests of

Phulchowki Mountain in Kathmandu Valley, severely threatened by quarrying and removal of fuelwood (Inskipp and Inskipp, 1989), and the Mai Valley in the far east. A comprehensive systems review, covering the full range of habitat types and floral and faunal assemblages, is needed to assess the adequacy of the protected areas system (MFSC, 1988). Policy regarding the selection of additional lands for protection is outlined in the national conservation strategy (HMG Nepal/IUCN, 1988).

Nepal's natural resources are being exploited above their sustainable capacity to meet the increasing needs of a rising human population that is predominantly agrarian and subsistence in nature. The pressure on land and forest resources to meet daily food, fuelwood and fodder requirements inevitably leads to conflicts at the boundaries of protected areas (Upreti, 1985). Major development projects also threaten the integrity of protected areas, as in the case of the proposed irrigation and hydropower projects planned near Royal Chitwan National Park (now listed as a threatened protected area by the IUCN Commission for National Parks and Protected Areas) and Royal Bardia National Park. The need to integrate conservation and development needs is widely recognised and gradually being addressed through, for example, the national planning process and implementation of the national conservation strategy.

Other Relevant Information Protected areas play a very important role in the tourism industry, being a popular destination for mountaineers, trekkers, and those interested in Nepal's wildlife or cultural diversity. The number of visitors to Nepal increased from 45,000 in 1970 to 223,000 by 1986. During this period, the number of tourists who came for trekking and mountaineering rose from 12,600 to 33,600. In 1985, tourism accounted for 48.5% of gross foreign exchange earnings (HMG Nepal/IUCN, 1988). In 1989, protected areas received a total of 84,840 visitors, Annapurna Conservation Area and Royal Chitwan National Park being the most popular destinations.

Addresses

Department of National Parks and Wildlife Conservation (Director General), PO Box 860, Babar Mahal, Kathmandu (Tel: 1 229012/220850/227926; FAX: 1 227675; Tlx: 2567 kmtnc np)

Department of Forest (Director General), Babar Mahal, Kathmandu (Tel: 1 220303/221231)

IUCN (The World Conservation Union), Senior Advisor, PO Box 3923, Kathmandu (Tel: 1 521506; FAX: 1 226820; Tlx: 2566 HOHIL NP)

King Mahendra Trust for Nature Conservation (Secretary), PO Box 3712, Kathmandu (Tel: 1 223229/220109; FAX: 1 226602; Tlx: 2567 kmtnc np; Cable NATRUST)

Nepal Nature Conservation Society (General Secretary), Kathmandu

References

Caughley, G. (1969). *Wildlife and recreation in the Trisuli Watershed and other areas in Nepal.* Report No. 6. HMG/FAO/UNDP Trisuli Watershed Development Project, Kathmandu. 44 pp.

FAO (1980). *National parks and wildlife conservation, Nepal: project findings and recommendations.* UNDP/FAO Terminal Report, Rome. 63 pp.

Glaser, G. (1984). The role of ICIMOD: a presentation of the Centre. In: *Mountain Development: challenges and opportunities.* International Centre for Integrated Mountain Development, Kathmandu. Pp. 59-63.

Heinen, J.T., Kattel, B. and Mehta, J.N. (1988). National park administration and wildlife conservation in Nepal. Draft. Department of National Parks and Wildlife Conservation, Kathmandu. 72 pp.

HMG Nepal/IUCN (1983). *National conservation strategy for Nepal: a prospectus.* IUCN, Gland, Switzerland. 36 pp.

HMG Nepal/IUCN (1988). *Building on success. The National Conservation Strategy for Nepal.* HMG National Planning Commission/NCS for Nepal Secretariat, Kathmandu. 179 pp.

ICIMOD (1989). *The first five years – a summary presentation.* International Centre for Integrated Mountain Development, Kathmandu. 25 pp.

Inskipp, C. (1989). Nepal's forest birds: their status and conservation. *International Council for Bird Preservation Monograph* No. 4. 160 pp.

Inskipp, C. and Inskipp, T. (1989). Pulchowki – hill of flowers. *Oryx* 23: 135-137.

IUCN (1985). *The Corbett Action plan for protected areas of the Indomalayan Realm.* IUCN, Gland, Switzerland and Cambridge, UK. 23 pp.

Kenting (1986). *HMG Nepal/Government of Canada Land Resource Mapping Project.* Kenting Earth Sciences Limited, Kathmandu.

KMTNC (1988). *Strategy for environmental conservation in Nepal.* The initial five-year (1988/89-1992/93) Action Plan of The King Mahendra Trust for Nature Conservation (KMTNC). KMTNC, Kathmandu. 70 pp.

Lucas, P.H.C. (1977). Nepal's park for the highest mountain. *Parks* 2(3): 1-4.

MacKinnon, J. and MacKinnon, K. (1986). *Review of the protected areas system in the Indo-Malayan Realm.* IUCN, Gland, Switzerland and Cambridge, UK/UNEP, Nairobi, Kenya. 284 pp.

Mansberger, J. (1990). Keeping the covenant: sacred forests of Nepal. *The New Road* 12: 2.

MFSC (1988). Master Plan - Forestry Sector Nepal. 13 reports. HMG Nepal/ADB/FINNIDA with Jaakho Poyry/Madecor Consultancy, Kathmandu.

Pandey, S. (1988). Jarajuri – A Nepal NGO with a difference. *Tiger Paper (Forest News)* 15(1): 11-13.

Rana, P.S.J.B., Pandey, N.R. and Mishra, H.R. (1986). An introduction to the King Mahendra Trust for Nature Conservation. *King Mahendra Trust for Nature Conservation Publication Series* No. 1. 35 pp.

Scott, D.A. (Ed.) (1989). *A directory of Asian wetlands.* IUCN, Gland, Switzerland and Cambridge, UK. 1,181 pp.

Sherpa, M.N., Coburn, B. and Gurung, C.P. (1986). *Annapurna Conservation Area, Nepal Operation Plan.* King Mahendra Trust for Nature Conservation, Kathmandu. 74 pp.

Shrestha, T.B., Sherpa, L.N., Banskota, K. and Nepali R.K. (1990). *The Makalu-Barun National Park and Conservation Area management plan.* Department of National Parks and Wildlife Conservation, Kathmandu/Woodlands Mountain Institute, West Virginia, USA. 85 pp.

Upreti, B.N. (1985). The park-people interface in Nepal: problems and new directions. In: McNeely, J.A., Thorsell, J.W. and Chalise, S.R., *People and protected areas in the Hindu Kush - Himalaya.* King Mahendra Trust for Nature Conservation/ International Centre for Integrated Mountain Development, Kathmandu. Pp. 19-24.

Upreti, B.N. (1990). Status of national parks and protected areas in Nepal. *Tiger Paper* 18 (2): 27-32.

Wegge, P. (1976a). *Terai shikar reserves: surveys and management proposals.* Field Document No. 4. FAO/NEP/72/002 Project, Kathmandu. 78 pp.

Wegge, P. (1976b). *Himalayan shikar reserves: surveys and management proposals.* Field Document No. 5. FAO/NEP/72/002 Project, Kathmandu. 96 pp.

ANNEX
Definitions of protected area designations, as legislated, together with authorities responsible for their administration

Title: Forest Act, 2018

Date: 27 December 1961, amended 1963, 1977, 1978

Brief description: To provide for the demarcation and administration of state forests

Administrative authority: Department of Forest, Ministry of Forests and Soil Conservation (Director-General)

Designations:

State forest[1] All forest, inclusive of waste land, streams and ponds, or paths, other than a forest park and Panchayat forest as mentioned in this Act. Designated by the government by notification in the *Nepal Gazette*. No person has any rights within state forests unless provided through contract or permit by the government.

Prohibited activities include deforestation, cultivation, setting fires, grazing, damaging trees, removing stone, manufacturing charcoal or lime etc, and removing forest products.

Any state forest may be declared a forest park.

Panchayat forest[1] State forest, or part thereof, which has been rendered waste or contains only stumps, entrusted to any village Panchayat for reforestation in the interest of the village community.

Under the Panchayat Forest Rules, 1978, ordinarily up to 200 bighas in the *terai* or 2,500 ropanies (125ha) elsewhere shall be maintained as Panchayat forest in each Village Panchayat. The Panchayat is obliged to plant and maintain the forest, and act in accordance with the operational plan of the relevant Forest Division. It is forbidden to sell, mortgage, alienate, reclaim, cultivate or use the land in any manner other than prescribed in the approved plan.

Panchayat protected forest[1] State forest, or part thereof, entrusted to a local Panchayat for its protection and proper management.

Under the Panchayat Protected Forest Rules, 1978 (amended 1980), ordinarily 400 bighas in the *terai* or 10,000 ropanis (500ha) elsewhere shall be designated as Panchayat protected forest in each Village Panchayat. The Panchayat is obliged to: maintain and protect the forest; prevent poaching of forest produce, fires, destruction and damage to trees and quarrying; and to act in accordance with a specified working plan. It is forbidden to damage, mortgage, sell or alienate, reclaim, or cultivate the forest, or deviate from the agreed working plan.

Religious forest[1] State forest, or part thereof, located at a place of religious importance entrusted to a religious institution for its protection and proper management.

Contract forest[1] State forest, or part thereof, devoid of trees, or has only stray trees, entrusted to any individual or agency for production and consumption of forest products.

Private forest reserve Any person may plant a forest on his land. Such forested land must be registered with the state. Activities may be controlled or prohibited within a private forest reserve by order published in the *Nepal Gazette*. If any order is contravened, management may be transferred to the local forest officer for up to a maximum period of 30 years.

All provisions of this Act relating to state forests are applicable to private forests.

Source: Translation of original legislation

Title : The National Parks and Wildlife Conservation Act, 2029

Date: March 1973; amended 1975 and 1983

Brief description: To provide for national parks and the conservation of wildlife

Administrative authority: Department of National Parks and Wildlife Conservation (Director General)

Designations:

National Park Area Set aside for conservation, management and utilisation of animals, birds, vegetation and landscape together with the natural environment.

Entry is restricted to persons possessing an entry permit or written permission from an authorised officer, except in the case of government officials or persons travelling on an existing right-of-way.

Prohibited activities include: hunting or damaging any animal; building or occupying any form of shelter or house; occupying, clearing or cultivating land; pasturing or watering any domesticated animal; damaging, felling or removing any tree or other plant; mining, quarrying or removing stone, minerals, or earth; carrying or using any weapon, ammunition or poison; carrying any domestic or other animal or trophy, except by a government official on duty or by a person travelling along an existing right-of-way; blocking or diverting any river, stream or other source of water flowing into a national park, or introducing any harmful or poisonous substance therein; and damaging or removing any boundary marks, signposts or notices.

Services or amenities may be provided by HMG or under contract to the government.

Reserve Means controlled natural reserve, wildlife reserve or hunting reserve.

None of the activities prohibited within a national park is permitted without written permission from an authorised officer.

Controlled natural reserve (Strict nature reserve) Area of ecological or other significance set aside for the purpose of scientific study.

Entry is restricted to persons having written permission from an authorised officer.

Wildlife reserve Area set aside for the conservation and management of animals, birds and other resources and their habitats.

Hunting reserve Area set aside for the management of animals, birds and other resources to provide for hunting.

Conservation area (1989 amendment) Area managed in accordance with an integrated plan for the conservation of the natural environment and the sustainable use of natural resources.

Source: Original legislation

Title : Soil and Watershed Conservation Act, 2039

Date: 1982

Administrative authority: Department of Soil Conservation and Watershed Management

Brief description: Not available

Designations:

Protected watershed area Area protected to conserve soil and watersheds, and in which measures for afforestation may be taken.

Official permission is required for cutting trees and other plants or forest products. Land use, including cultivation and planting of trees, may be subject to official controls.

Source: MFSC, 1988

[1]Designations have recently been revised under the Master Plan for the Forestry Sector and are due to be incorporated within new forest legislation. Definitions, as provided by the Department of Forests, are as follows:

National Forest All forests except those designated otherwise.

Community Forest Government forest land entrusted to user groups to encourage sustained use of such resources. It is further subdivided according to the management criteria.

Community plantation forest Any government forest land, devoid of trees or in which only scattered trees or shrubby vegetation is left, which has been notified for forest development through reforestation by the active participation of user groups.

Community forest Any government forest which has been notified for management and conservation by the active involvement of user groups.

Leased forest Forest on land that has been leased by central or local agencies of the government, village development committees or private owners to individuals, cooperatives, institutions or commercial firms for forest production purposes.

Religious forest Forest belonging to religious institutions under the Guthi Act.

SUMMARY OF PROTECTED AREAS

Map ref.	National/international designation Name of area	IUCN management category	Area (ha)	Year notified
	National Parks			
1.	Khaptad	II	22,500	1986
2.	Langtang	II	171,000	1976
3.	Rara	II	10,600	1977
4.	Royal Bardia	II	96,800	1988
5.	Royal Chitwan	II	93,200	1973
6.	Sagarmatha	II	114,800	1976
7.	Shey-Phoksundo	II	355,500	1984
	Wildlife Reserves			
8.	Koshi Tappu	IV	17,500	1976
9.	Parsa	IV	49,900	1984
10.	Royal Sukla Phanta	IV	15,500	1976
11.	Shivapuri WR	IV	11,200	1985
	Hunting Reserve			
12.	Dhorpatan HR	VIII	132,500	1987
	Ramsar Wetland			
	Koshi Tappu Wildlife Reserve	R	17,500	1987
	World Heritage Sites			
	Royal Chitwan National Park	X	93,200	1984
	Sagarmatha National Park	X	114,800	1979

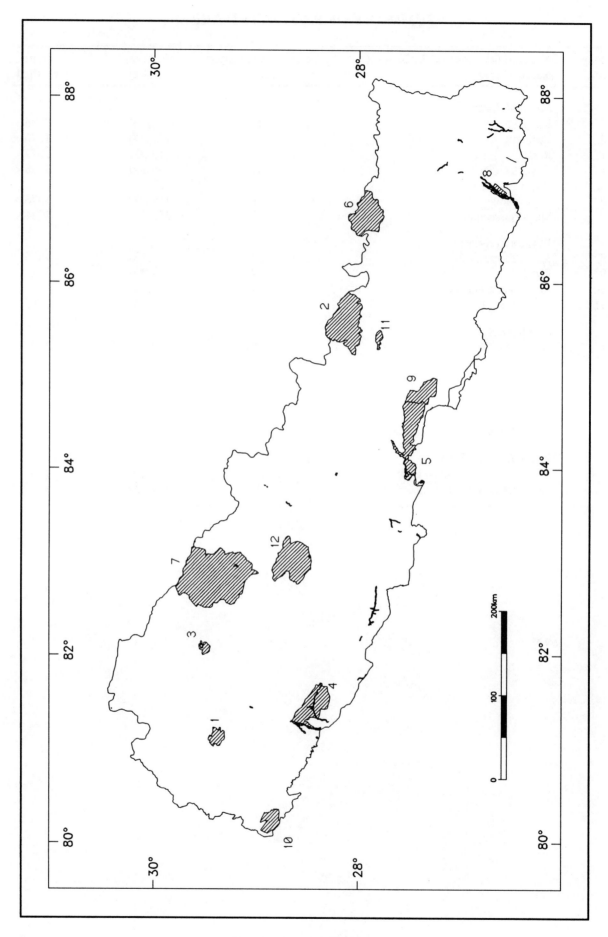

Protected Areas of Nepal

ISLAMIC REPUBLIC OF PAKISTAN

Area 803,940 sq. km

Population 114,600,000 (1990)
Natural increase: 3.0% per annum

Economic Indicators
GNP: US$ 350 per capita (1988)

Policy and Legislation Environmental protection and ecology are included in the concurrent legislative list of Pakistan's 1973 constitution. This initiative, together with the formation of an Environment and Urban Affairs Division in 1973, was largely responsible for enactment of the Environment Protection Ordinance, 1983. The Ordinance is a landmark in Pakistan's legislation and represents official recognition of a holistic approach to environmental issues. It provides for the control of pollution and preservation of a comprehensive national environmental policy, and filing of detailed environmental impact statements by proponents of projects likely to adversely affect the environment. The main drawback of the Ordinance, however, is its much narrower scope – focusing on anti-industrial pollution – than was envisaged in the original draft, which included legal provisions for the protection of Pakistan's natural resource base (Mumtaz, 1989).

A significant step towards meeting the environmental challenge was taken in 1988, with the support of IUCN, by initiating the national conservation strategy (NCS) development process. A secretariat has been set up in the Environment and Urban Affairs Division (Ministry of Housing and Works), which reports to a high-level steering committee comprising representatives of eight ministries directly concerned with natural resources, and five from the private sector. The NCS development process represents an unique policy review of economic issues and their collective impacts on the environment. Public consultations are an integral part of this review and planning exercise (Kabraji, 1986; Mumtaz, 1989). As part of the review process, a national workshop was held in 1986 (IUCN/GOP, 1987). The first phase of the development process, namely the formulation of Pakistan's NCS, was completed in 1990 (JRC, n.d.).

Early Hindu and Muslim rulers, recognising the inadvisability of uncontrolled hunting, were the first to set aside game reserves wherein hunting was restricted during breeding seasons. By the late 16th century, the Mughals had codified regulations pertaining to hunting and these policies were adopted by succeeding Sikh and British administrations (ALIC, 1981). Indiscriminate exploitation of forest resources during the 19th century led to the realisation of the need for a forest policy. Although not of direct relevance to Pakistan, the first forest policy directive issued by the government of India was in the form of a Memorandum (3 August 1855) for the protection and extraction of teak. It restricted the rights of forest dwellers to conserve the forests.

Government of India Circular No. 22-F (19 October 1894) represented a more comprehensive forest policy statement, which emphasised management of forests for timber production, watershed protection and maintenance of productive capacity. It also provided guidelines on basic principles associated with rights of people living adjacent to forest lands (Mumtaz, 1989).

Among the first pieces of legislation that directly benefited wildlife were the rules and regulations formulated in Sind under the Indian Forest Act in 1887 and later incorporated in the Bombay Forest Manual. Under this legislation, forests were protected from grazing by livestock but hunting was not legally controlled. Hunting and other forms of resource exploitation were subsequently controlled within areas declared as reserved or protected forests under the Indian Forest Act, 1927, the title of which was changed to Pakistan Forest Act, 1927 following Pakistan's adoption of the Act after partition in 1947 (Ferguson, 1978; Rao, 1984). The 1927 Act sought to "consolidate the law relating to forests, the transit of forest produce and the duty leviable on timber and other forest produce." It further "empowers the government to set aside forest reserves, appoint officers charged with the management of those territories, enforce rules governing the use of forests, determine the degree to which timber and other products may be exploited, and regulate the movement of cattle upon these lands. Moreover, the Act authorises the government to punish violators of the stipulations contained in it." The 1927 Act has since been amended by the West Pakistan Goats (Restriction) Ordinance of 1959 and the West Pakistan Goats Restriction Rules of 1961, which enable the government to protect rangelands from grazing damage by goats through limiting their numbers and movements.

The 1927 Act is not conservation-oriented, commercial forestry interests being foremost. Subsequent forest policy, under the directives of 1955, 1962 and 1980, has attempted to bring forests under sound scientific management and included provisions for the creation of national parks to conserve major ecosystems, but it has not been successful (Mumtaz, 1989). The need to reassess and redefine policy is being addressed by the Forest Department, following on from a recent evaluation of Pakistan's forest policy at an international seminar organised by the Ministry of Food, Agriculture and Cooperatives in 1989. Existing forest legislation is regulatory in nature. It needs to be revised to meet the requirements of development and extension forestry, with more persuasive rather than punitive provisions (Shekh and Jan, n.d.).

Wildlife conservation legislation inherited from British India was superseded by the now obsolete West Pakistan Wildlife Protection Ordinance, 1959 and the West Pakistan Wildlife Protection Rules, 1960 issued under that ordinance. Apart from prohibiting the killing of certain species of fauna, this legislation made provision for the declaration of game sanctuaries, in which hunting was prohibited, and game reserves, in which hunting was controlled under license, but did not protect the habitat against settlement, cultivation, grazing and other forms of exploitation. Furthermore, both the West Pakistan Wildlife Protection Ordinance and the Pakistan Forest Act applied only to the settled areas of Pakistan (i.e. the flood plains of the Kabul and Indus rivers and all the land to the east of them); neither were applicable to the Special/Tribal Areas, which constituted most of the mountainous half of the country to the west of the Indus and where much of Pakistan's remaining wildlife was to be found (Grimwood, 1969).

A Wildlife Enquiry Committee was set up in 1968 to review *inter alia* the existing conservation legislation, based on recommendations by World Wildlife Fund (Mountfort and Poore, 1967, 1968). Legislation was drafted by this committee (Government of Pakistan, 1971) and, with minor modifications, was subsequently adopted at provincial level through the provision of various acts and an ordinance, namely: Sind Wildlife Protection Ordinance, 1972, Punjab Wildlife (Protection, Preservation, Conservation and Management) Act, 1974, Baluchistan Wildlife Protection Act, 1974, and North-West Frontier Province Wildlife (Protection, Preservation, Conservation and Management) Act, 1975. Separate laws were passed for the Northern Areas, Azad State of Jammu & Kashmir and Federal Capital Territory of Islamabad. These are the Northern Areas Wildlife Preservation Act, 1975, Azad Jammu and Kashmir Wildlife Act, 1975 and the Islamabad Wildlife (Protection, Preservation, Conservation and Management) Ordinance, 1979 (Rau, 1984). This is the first time in the history of Pakistan's wildlife legislation that an attempt has been made to provide for the conservation of habitat (although limited to protected areas) and species other than game species.

All of these statutes provide for the creation and management of national parks, wildlife sanctuaries (synonymous with wildlife reserves in the Northern Areas Act), game reserves (synonymous with controlled hunting areas in the Northern Areas Act) and, in the case of the Punjab, North-West Frontier Province and Islamabad legislation, private game reserves (see Annex). Parts of areas protected under some statutes may be denotified under pressure for agricultural extension or land development (Khan and Hussain, 1985; Rao, 1984). To date, there are no notified private game reserves but a number exist in Baluchistan (e.g. Goth Raisani, Serajabad, Nasirabad area) and Sind (e.g. Khairpur), where there is no legal provision for their establishment, and in Punjab (e.g. Kalabagh). Existing wildlife legislation is reviewed by Rao (1984). Model legislation (Pakistan Wildlife Protection Act) is currently being prepared by the National Council for Conservation of Wildlife (Rao, 1987).

International Activities Pakistan ratified both the Convention concerning the Protection of the World Cultural and Natural Heritage (World Heritage Convention) and the Convention on Wetlands of International Importance especially as Waterfowl Habitat (Ramsar Convention) on 23 July 1976. No natural sites have been inscribed to date under the World Heritage Convention. Nine wetlands were designated at the time of Pakistan's ratification of the Ramsar Convention, of which two (Kandar Dam and Kheshki Reservoir) are no longer considered to be of international importance (Scott, 1989). Pakistan participates in the Unesco Man and the Biosphere Programme, but there does not appear to have been any significant development in recent years. Pakistan also participates in the South Asian Cooperative Environmental Programme.

Administration and Management Originally, the Game Department was responsible for administering the West Pakistan Wildlife Protection Ordinance up until 1967, when it was absorbed into the Forest Department (Grimwood, 1969). Following the recommendations of the Wildlife Enquiry Committee (Government of Pakistan, 1971), a National Council for Conservation of Wildlife was established on 7 July 1974 within the then Federal Ministry of Food and Agriculture. It has an advisory board, and is responsible for coordinating central and provincial government effort in the formulation and implementation of wildlife policies. The Inspector General of Forests is assisted by a Conservator of Wildlife, who acts as an adviser on wildlife, but the actual management of wildlife is handled by the provincial forest (wildlife) departments. Punjab and Sind have separate wildlife administrations, but in Azad State of Jammu & Kashmir, Baluchistan, Northern Areas and North-West Frontier Province wildlife is administered by branches of the respective forest departments. In practice, forest staff look after wildlife in reserved or protected forests, and wildlife staff are responsible for protecting wildlife in other protected areas and elsewhere. In North-West Frontier Province, wildlife staff are responsible for wildlife everywhere. Within the Federal Capital Territory of Islamabad, the Directorate of Environment is responsible for the administration of protected areas. Legal provision has been made for the creation of wildlife management boards to approve wildlife policies and monitor development activities in Punjab, Sind, North-West Frontier Province and Islamabad. Sind has an effective wildlife management board, while those of North-West Frontier Province and Punjab are progressing. Boards exist in Azad State of Jammu & Kashmir, Baluchistan and Northern Areas but only in an advisory capacity. That for Islamabad is not yet active. Provision has also been made for the appointment of honorary officers to help implement wildlife legislation in all political units except Baluchistan and Islamabad. The idea was first introduced in Sind in the 1970s and proved to be very successful in Kirthar National Park, resulting in the recovery of markhor and other large mammal populations. It has since been adopted in Azad State of Jammu & Kashmir and Punjab with the appointment of local dignitaries as honorary game wardens invested with considerable legal power to help enforce the law within protected areas (Ferguson, 1978; Mumtaz, 1990, 1991; NCCW, 1978; Rao, 1984; Roberts, 1983).

Allocation of funds to the forestry sub-sector has increased from 10.2% in the Sixth Five-Year Plan (1983-1988) to 12.5% in the Seventh Five-Year Plan (1988-1993). Of the Rs 2 billion allocated to the sub-sector under the Seventh Plan, Rs 332 million (16.6%) is earmarked for wildlife conservation (Sheikh and Jan, n.d.). Within the wildlife sector, the total budget allocated to the federal units in 1990-1991 is Rs 93.4 million (US$ 4.3 million), of which 52.5% represents recurrent expenditure and 47.5% capital development costs. The total number of staff within the wildlife sector is 3,206: 121 are administrative and executive, 2,375 are protection and operational, and 710 are supporting staff (Malik, 1990).

The Environmental Protection Ordinance is enforced by the Pakistan Environment Protection Council, but this has not yet been formed. The Council is also responsible for establishing a national environmental policy, providing direction to conserve renewable and expendable resources and ensuring that environmental considerations are incorporated within national development plans and policies. Administration of the Ordinance is the responsibility of the Pakistan Environment Protection Agency. Provincial Environment Protection Agencies have been set up, but other implementation procedures have yet to be streamlined (Mumtaz, 1989).

Among the non-governmental organisations involved with conservation is the Pakistan Wildlife Conservation Foundation, a registered charity established in 1979. Its president is appointed by a resolution of the National Council for Conservation of Wildlife. A main objective is to promote wildlife conservation activities through provision of funds in accordance with the policies of the National Council for Conservation of Wildlife. IUCN (The World Conservation Union) has a country office in Karachi. Field programmes concerned with protected areas management issues are focused on Korangi/Phitti Creek in the Indus Delta, juniper forests in Baluchistan and Khunjerab National Park in the Northern Areas. World Wide Fund for Nature-Pakistan (formerly World Wildlife Fund-Pakistan) has offices in Lahore and Karachi. Two bodies are concerned specifically with promoting the conservation of pheasants, namely the World Pheasant Association (Pakistan) and the Pheasant Conservation Forum.

Emphasis on the management of national parks has been given to the development of recreation facilities for tourists rather than nature conservation as, for example, in Lal Suhanra and Margalla Hills national parks. Management categories need to be modified

(Grimwood, 1972; Rao, 1984), perhaps by the introduction of nature reserves and country parks to replace wildlife sanctuaries. Protected and reserved forests continue to be managed under forest working plans after being designated national parks or wildlife sanctuaries, thereby undermining the purpose of their renotification. Hunting in game reserves is not controlled on a sustained yield basis, permits being issued arbitrarily and subject to political influence (Rao, 1984). The Government of Punjab, however, has restricted the number of shoots under an amendment to the Punjab Wildlife Act (Khan and Hussain, 1985). Weak enforcement of the law is an overall constraint, but also safeguards against habitat degradation within protected areas are inadequate (Rao, 1984). This is largely a reflection of inadequate financial and technical resources. In addition, except in Punjab, present administrative arrangements handicap wildlife and protected areas management due to the lack of independence of the wildlife administrations within the federal units (Mumtaz, 1990, 1991).

Systems Reviews Predominantly arid and semi-arid, Pakistan is a land of great contrasts. Nearly 60% of the country consists of mountainous terrain and elevated plateaux; the rest is lowland, generally below 300m. The highlands comprise: the Himalaya and adjacent mountain ranges to the north, rising to 8,611m at the top of K2, the world's second highest peak; the central Sulaiman Range and its southern extensions (Ras Koh, Siahan and Kirthar ranges); and the western Baluchistan Plateau. The lowlands comprise the Indus River plain and a narrow stretch of coastline bordering the Arabian Sea. A profile of the environment has been prepared by the Government of Pakistan (1989).

Pakistan did not inherit a very rich forest resource base, a reflection of its arid climate and the incessant cutting of trees throughout much of the country over the last few centuries. Under extensive reforestation schemes and extension programmes, forest coverage has increased from 1.4 million ha at the time of independence to 4.6 million ha (5.2% of total land area) by 1984. One million ha of forest, for example, was planted in North-West Frontier Province with the cooperation of the people. Forest cover is most extensive in Azad State of Jammu & Kashmir (27.7%), North-West Frontier Province (13.9%) and Northern Areas (13.4%); in the other three states it is below 5%. There are two types of forest in Pakistan: production forest managed for commercial extraction of timber; and protection forest which has no commercial value and is primarily for soil protection. Only 27.6% of forest is commercially used, the bulk (72.4%) of this resource being under protection (Sheikh and Jan, n.d.; JRC, 1989).

Most of Pakistan's remaining wildlife is to be found in the mountainous country west of the Indus, where human pressures have not been as great as in the plains. The two regions of outstanding importance are the Himalayan and Karakoram massifs in the extreme north and the desert in the south-west of the country (Grimwood,

1969). To the east of the Indus, Hazara Division in North-West Frontier Province and several areas in Punjab have a considerable amount of wildlife (M.M. Malik, pers. comm., 1987), as does the Neelum Valley in Azad State of Jammu & Kashmir (G. Duke, pers. comm., 1990). Wildlife resources and their exploitation have been reviewed for Baluchistan (Groombridge, 1988; Mian and O'Gara, 1987; Roberts, 1973) and Sind (Roberts, 1972). Major irrigation systems, built to tap the water resources of the Indus and its tributaries to meet the demands of an increasing human population, have resulted in the disappearance of extensive tracts of the original tropical thorn scrub, riverine swamp and forest in the plains (Roberts, 1977). In a recent review of critical ecosystems in Pakistan, Roberts (1986) identifies the Indus riverine zone, and the Chaghai Desert and juniper forests of Baluchistan as being of unique ecological interest and international conservation importance.

Pakistan possesses a great variety of wetlands distributed throughout much of the country. Inland waters cover 7.8 million ha, over half of which comprises waterlogged areas, seasonally flooded plains and saline wastes. Coastal mangrove swamps cover at least 260,000ha. Pakistan's wetlands are important for waterfowl, particularly those of the Indus Valley – a major wintering ground for a wide variety of central and northern Asian species, as well as being of socio-economic value (Scott, 1989).

Prior to 1966, Pakistan had taken no significant steps towards establishing a protected areas network. That year, at the invitation of the government, the World Wildlife Fund carried out a survey of the country's wildlife resources and recommended measures to arrest their deterioration (Mountfort and Poore, 1967, 1968). These included the establishment of two large national parks and eight wildlife sanctuaries. This initiative was followed by the constitution of a Wildlife Enquiry Committee in 1968, which made further recommendations for the establishment of four national parks, 18 wildlife sanctuaries and 52 game reserves (Government of Pakistan, 1971). These recommendations have been substantially exceeded: four national parks, 44 wildlife sanctuaries and 65 game reserves had been declared by 1978 (ALIC, 1981). During the period 1968-1971, various technical assistance was received from the Food and Agricultural Organisation of the United Nations, which latterly included the appointment of an adviser to the Wildlife Enquiry Committee (Grimwood, 1969, 1972). The network currently comprises 10 national parks, over 80 wildlife sanctuaries and over 80 game reserves, covering 7.2 million ha (9% total land area). Although extensive, given Pakistan's human population, only a fraction of the network is protected. Game reserves, in particular, which are often on private land, receive minimal protection due to lack of legal provisions to control land use. Wildlife sanctuaries enjoy better protection but, in practice, legal restrictions are seldom enforced other than preventing hunting. Most sanctuaries have been designated in reserved forests of

commercial value, where timber and minor forest products are harvested. Enforcement is better in national parks but only Kirthar currently has a management plan. Plans for some of the other national parks are due to be prepared, although that for Khunjerab has met with difficulty due to land ownership disputes between the government and local people (Malik, 1990, 1991).

Protected areas have been created haphazardly, often in the absence of any criteria for their selection, and boundaries drawn with little or no ecological basis. Priorities to develop the existing network of protected areas are identified in the IUCN systems review of the Indomalayan Realm (MacKinnon and MacKinnon, 1986) and further recommendations are made in the Corbett Action Plan (IUCN, 1985). Malik (1990, 1991) recommends a doubling of protected areas coverage. While most major habitats are represented within the existing protected areas system (MacKinnon and MacKinnon, 1986), a comprehensive systems review has never been carried out at the national level. Clearly, this is a priority in order to plan the further development of Pakistan's protected areas network.

Other Relevant Information

Federal Capital Territory The Directorate of Environment, within the Capital Development Authority, is responsible for protected areas management. It is headed by a Director, who is supported by a Deputy Director, two Assistant Directors, a field staff of 68 and 30 other staff. The Directorate is well organised and enjoys good support from other government agencies by virtue of being in the capital (Malik, 1990).

Northern Areas The Northern Areas Forest Department manages protected areas in its jurisdiction. Apart from Khunjerab National Park, which is independently managed under a Park Director, wildlife staff are attached to the territorial forest divisions under Divisional Forest Officers. The total number of wildlife staff is 87, of which 60 are operational/protection personnel. The budget allocated for 1990-1991 is Rs 1.8 million, of which 83.3% is recurrent expenditure and the rest (16.6%) for development costs (Malik, 1990).

North-West Frontier Province Protected areas management has been assigned to an independent Wildlife Wing within the Forest Department, headed by a Conservator of Wildlife. The province is divided into six wildlife divisions, each headed by a Divisional Forest Officer, Wildlife. A wildlife ranger is allocated to each of the 14 districts, as well as to each of the two national parks. The total number of wildlife staff is 502, of whom 357 are operational/protection personnel. Although the Wildlife Wing enjoys considerable independence in its operations, policy and financial constraints are a source of conflict. The budget allocated for 1990-1991 Rs 9.9 million, of which 67.7% is recurrent expenditure and the rest (32.3%) for development costs (Malik, 1990).

Addresses

National Council for Conservation of Wildlife (Conservator, Wildlife), Ministry of Food, Agriculture and Cooperatives, 485 Street 84, G-6/4 ISLAMABAD (Tel. 829756; Tlx 5844 MINFA PK; Cable AGRIDIV)

Ministry of Food, Agriculture and Cooperatives (Inspector-General of Forests), Room 323, Block B, Pakistan Secretariat, ISLAMABAD (Tel. 825289; Tlx 5844 MINFA PK; Cable AGRIDIV)

Forest Department – Wildlife Wing (Wildlife Warden), Azad State of Jammu & Kashmir, MUZAFFARABAD (Tel: 18)

Forestry and Wildlife Department, (Divisional Forest Officer, Wildlife), Government of Baluchistan, Spinny Road, QUETTA (Tel: 71298)

Environment Directorate (Director), Capital Development Authority, Sitara Market, ISLAMABAD (Tel: 826397)

Forest Department (Conservator of Forests), Northern Areas, PO Box 501, GILGIT (Tel: 360)

Forest Department – Wildlife Wing (Conservator, Wildlife), Government of North-West Frontier Province, Shami Road, PESHAWAR (Tel: 73184)

Wildlife Department, (Conservator of Forests, Parks and Wildlife), Government of Punjab, 2 Sanda Road, LAHORE (Tel: 61798/63947)

Sind Wildlife Management Board, (Conservator of Forests, Wildlife), Aiwan-e-Saddar Road, PO Box 3722, KARACHI 1 (Tel: 523176)

IUCN, The World Conservation Union (Country Representative), 1 Bath Island Road, KARACHI 75530 (Tel: 573046/79/82; Tlx: 24154 MARK PK)

Pakistan Wildlife Conservation Foundation, 485 Street 84, G-6/4 ISLAMABAD (Tel: 829756; Tlx: 5844 MINFA PK; Cable AGRIDIV)

Pheasant Conservation Forum (Secretary), c/o National Council for Conservation of Wildlife, Ministry of Food, Agriculture and Cooperatives, 485 Street 84, G-6/4 ISLAMABAD (Tel: 829756; Tlx: 5844 MINFA PK; Cable AGRIDIV)

World Pheasant Association-Pakistan (Chairman), 7 Aziz-Bhatti Road, The Mall, LAHORE

Word Wide Fund For Nature-Pakistan, 1 Bath Island Road, KARACHI 75530 (Tel: 573046/79/82; Tlx: 24154 MARK PK)

World Wide Fund For Nature-Pakistan (Director), PO Box 5180, LAHORE (Tel: 851174, 856177; FAX: 370429; Tlx: 44866 PKGS PK)

References

ALIC (1981). Draft environmental profile: the Islamic Republic of Pakistan. US Agency for International Development/US National Park Service/US Man and the Biosphere Secretariat. Arid Lands Information Centre, Office of Arid Lands Studies, University of Arizona, Tucson, USA. 227 pp.

Ferguson, D.A. (1978). Protection, conservation, and management of threatened and endangered species in Pakistan. US Fish and Wildlife Service, Washington DC. Unpublished report. 62 pp.

Government of Pakistan (1971). *Summary of Wildlife Enquiry Committee Report*. Printing Corporation of Pakistan Press, Islamabad. 44 pp.

Government of Pakistan (1989). *Environmental profile of Pakistan*. Environment & Urban Affairs Division, Government of Pakistan, Islamabad. 248 pp.

Grimwood, I.R. (1969). *Wildlife conservation in Pakistan*. Pakistan National Forestry Research and Training Project. Report No. 17. UNDP/FAO, Rome. 31 pp.

Grimwood, I.R. (1972). *Wildlife conservation and management*. Report No. TA 3077. FAO, Rome. 58 pp.

Groombridge, B. (1988). Baluchistan Province, Pakistan: a preliminary environmental profile. IUCN Conservation Monitoring Centre, Cambridge, UK. Unpublished report. 104 pp.

IUCN (1985). *The Corbett Action Plan for protected areas of the Indomalayan Realm*. IUCN, Gland, Switzerland and Cambridge, UK. 23 pp.

IUCN/GOP (1987). *Towards a national conservation strategy for Pakistan*. Proceedings of the Pakistan Workshop 1986. Asian Art Press, Lahore. 367 pp.

JRC (n.d.) *Towards sustainable development: the Pakistan National Conservation Strategy*. Journalists' Resource Centre for the Environment, IUCN Pakistan, Karachi.

Kabraji, A.M. (1986). A national conservation strategy for Pakistan. In: Carwardine, M. (Ed.), *The nature of Pakistan*. IUCN, Gland, Switzerland. Pp. 69-71.

Khan, A. and Hussain, M. (1985). Development of protected area system in Pakistan in terms of representative coverage of ecotypes. In: Thorsell, J.W. (Ed.), *Conserving Asia's natural heritage*. IUCN, Gland, Switzerland. Pp. 60-68.

MacKinnon, J. and MacKinnon, K. (1986). *Review of the protected areas systems in the Indo-Malayan realm*. IUCN, Gland, Switzerland and Cambridge, UK. 284 pp.

Malik, M.M. (1990). Management status of protected areas in Pakistan. Paper presented at Regional Expert Consultation on Management of protected areas in the Asia-Pacific Region. FAO Regional Office for Asia and the Pacific, Bangkok, 10-14 December 1990. 40 pp.

Malik, M.M. (1991). Management status of protected areas in Pakistan. *Tiger Paper* 18 (1): 21-28.

Mian, A. and O'Gara, B.W. (1987). Baluchistan and wildlife potentials. University of Baluchistan, Quetta, Pakistan and University of Montana, Missoula, USA. Draft. 32 pp.

Mountfort, G. and Poore, D. (1967). The conservation of wildlife in Pakistan. World Wildlife Fund, Morges, Switzerland. Unpublished report. 27 pp.

Mountfort, G. and Poore, D. (1968). Report on the Second World Wildlife Fund Expedition to Pakistan.

World Wildlife Fund, Morges, Switzerland. Unpublished. 25 pp.

Mumtaz, K. (1989). Pakistan's environment: a historical perspective. In: Shirkat Gah-Women's Resource Centre, *Pakistan's environment: a historical perspective and selected bibliography with annotations*. Journalists' Resource Centre for the Environment-IUCN Pakistan, Karachi. Pp. 7-38.

National Council for Conservation of Wildlife (1978). Wildlife conservation strategy: Pakistan. National Council for Conservation of Wildlife, Islamabad, Pakistan. Unpublished report. 73 pp.

Rao, A.L. (1984). A review of wildlife legislation in Pakistan. MSc. thesis, University of Edinburgh, Edinburgh, UK. 66 pp.

Rao, A.L. (1987). Nature conservation in Pakistan. In: *Towards a national conservation strategy*. Pp. 223-250.

Roberts T.J. (1972). A brief examination of ecological changes in the province of Sind and their consequences on the wildlife resources of the region. *Pakistan Journal of Forestry* 22: 89-96.

Roberts, T.J. (1973). Conservation problems in Baluchistan with particular reference to wildlife preservation. *Pakistan Journal of Forestry* 23: 117-127.

Roberts, T.J. (1977). *The mammals of Pakistan*. Ernest Benn, London. 361 pp.

Roberts, T.J. (1983). Problems in developing a national wildlife policy and in creating effective natural parks and sanctuaries in Pakistan. Paper presented at Bombay Natural History Society Centenary Seminar. Powai, Bombay, December 1983. 9 pp.

Scott, D.A. (Ed.) (1989). *A directory of Asian wetlands*. IUCN, Gland, Switzerland and Cambridge, UK. 1,181 pp.

Sheikh, M.I. and Jan, A. (n.d.). Role of forests and forestry in national conservation strategy of Pakistan. Draft for comment. National Conservation Strategy Secretariat, Islamabad. 86 pp.

ANNEX
Definitions of protected area designations, as legislated, together with authorities responsible for their administration

Title: Pakistan Wildlife Ordinance (Draft)

NOTE: This legislation, drafted by the Wildlife Enquiry Committee, was subsequently adopted (with minor modifications) at provincial level through various enactments.

Date: 1971

Brief description: An ordinance to amend and consolidate the laws relating to preservation, conservation and management of wildlife in Pakistan

Administrative authority: Provisional Government

Designations:

National park A comparatively large area of outstanding scenic merit and natural interest, wherein the primary objective is to protect the landscape, flora and fauna in its natural state and to which the public are allowed access for purposes of recreation, education and research.

No hunting or trapping of any wild animal is permitted within one mile of its boundary. Disturbing any wild animal, damaging or destroying any plant, clearing or breaking up land for cultivation, and polluting water flowing though a park are prohibited, unless authorised by the provincial government for scientific or other exceptional purposes. Harvesting of forest produce on a sustained basis is allowed provided park values are not jeopardised. Construction of access roads, accommodation facilities and public amenities must be carefully planned so as not to impair the primary objective of a park's establishment.

Wildlife sanctuary An area set aside as an undisturbed breeding ground, primarily for the protection of wildlife including all natural resources, such as soil, water and vegetation.

Prohibited activities include entry or residence by any person, cultivation of land, damage or destruction of vegetation, hunting, killing or capture of any wild animal within one mile of its boundaries, introduction of any exotic species of animal, entry of any domestic animal, causing any fire, and polluting water flowing through it, unless authorised by the provincial government for scientific or aesthetic purposes.

Game reserve An area declared as such for the protection of wildlife and increase in populations of important species.

The provincial government may permit hunting and shooting of wild animals under a special permit restricting the number of animals killed, and the area and duration of its validity.

Private game reserve An area of private land set aside by its owner for the same purpose as a game reserve, subject to the satisfaction and notification of the provincial government.

Source: Government of Pakistan (1971)

SUMMARY OF PROTECTED AREAS

Map[†] ref.	*National/international designation* Name of area	IUCN management category	Area (ha)	Year notified
	BALUCHISTAN			
	National Parks			
11	Dhrun	II	167,700	1988
12	Hazar Ganji-Chiltan	V	15,555	1980
13	Hingol	II	165,004	1988
	Wildlife Sanctuaries			
14	Buzi Makola	IV	145,101	1972
15	Chorani	IV	19,433	1972
16	Dureji	IV	178,259	1972
17	Gut	IV	165,992	1983
18	Kachau	IV	21,660	1972
19	Khurkhera	IV	18,345	1972
20	Koh-e-Geish	IV	24,356	1969
21	Kolwah Kap	IV	33,198	1972
22	Maslakh	IV	46,559	1968
23	Raghai Rakhshan	IV	125,425	1971
24	Ras Koh	IV	99,498	1962
25	Sasnamana	IV	6,607	1971
26	Shashan	IV	29,555	1972
27	Ziarat Juniper	IV	37,247	1971
	FEDERAL CAPITAL TERRITORY			
	National Park			
42	Margalla Hills	V	17,386	1980
	Wildlife Sanctuary			
43	Islamabad	IV	7,000	1980
	NORTHERN AREAS			
	National Park			
45	Khunjerab	II	226,913	1975
	Wildlife Sanctuaries			
46	Astore	IV	41,472	1975
47	Baltistan	IV	41,457	1975
48	Kargah	IV	44,308	1975
49	Naltar	IV	27,206	1975
50	Satpara	IV	31,093	1975
	NORTH-WEST FRONTIER PROVINCE			
	National Parks			
60	Ayubia	V	1,684	1984
61	Chitral Gol	II	7,750	1984
	Wildlife Sanctuaries			
62	Agram Basti	IV	29,866	1983
63	Borraka	IV	2,025	1976
65	Manshi	IV	2,321	1977
66	Sheikh Buddin	IV	15,540	1977
	PUNJAB			
	National Parks			
94	Chinji	II	6,095	1987
95	Lal Suhanra	V	37,426	1972

Map[†] ref.	National/international designation Name of area	IUCN management category	Area (ha)	Year notified
	Wildlife Sanctuaries			
96	Bajwat	IV	5,464	1964
99	Chashma Lake	IV	33,084	1974
101	Cholistan	IV	660,949	1981
102	Chumbi Surla	IV	55,945	1978
103	Daphar	IV	2,286	1978
115	Rasool Barrage	IV	1,138	1974
116	Sodhi	IV	5,820	1983
117	Taunsa Barrage	IV	6,567	1972
	SIND			
	National Park			
139	Kirthar	II	308,733	1974
	Wildlife Sanctuaries			
142	Dhoung Block	IV	2,098	1977
148	Hab Dam	IV	27,219	1972
149	Hadero Lake	IV	1,321	1977
150	Haleji Lake	IV	1,704	1977
152	Keti Bunder North	IV	8,948	1977
153	Keti Bunder South	IV	23,046	1977
156	Kinjhar (Kalri) Lake	IV	13,468	1977
160	Mahal Kohistan	IV	70,577	1972
166	Nara Desert	IV	223,590	1980
168	Runn of Kutch	IV	320,463	1980
172	Takkar	IV	43,513	1968
	Biosphere Reserve			
	Lal Suhanra National Park	IX	31,355	1977
	Ramsar Wetlands			
	Drigh Lake Wildlife Sanctuary	R	164	1976
	Haleji Lake Wildlife Sanctuary	R	1,704	1976
	Khabbeke Lake Wildlife Sanctuary	R	283	1976
	Kinjhar(Kalri) Lake Wildlife Sanctuary	R	13,468	1976

[†]Locations of most protected areas are shown on the accompanying map.

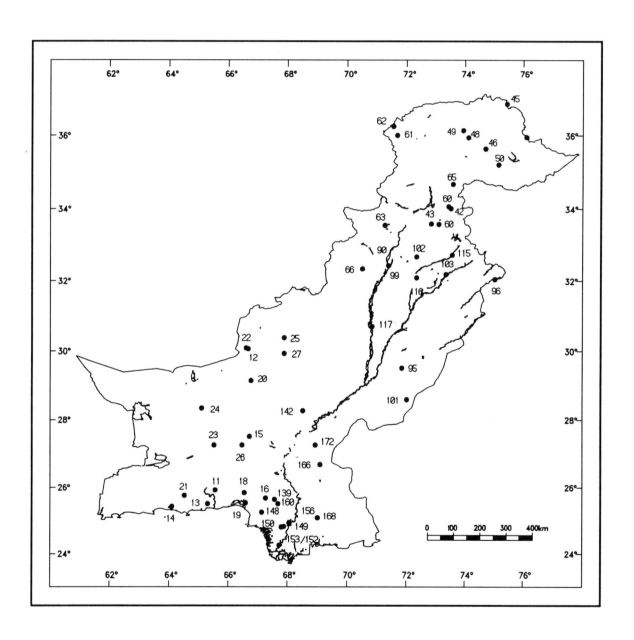

Protected Areas of Pakistan

PHILIPPINES

Area 299,765 sq. km

Population 62,413,000 (1988)
Natural increase: 2.28% per annum

Economic Indicators
GNP: US$ 630 per annum

Policy and Legislation The 1987 Constitution mandates the state ownership of all natural resources, which includes fisheries, forests or timber, wildlife, flora and fauna and other natural resources. Furthermore, it has ordained that the boundaries of forest and national parks must be marked clearly on the ground and there after be conserved and neither enlarged nor diminished except by Act of Congress (PAWB, 1991).

The 1990 Philippine Strategy for Sustainable Development (DENR, 1991) is the outcome of a series of broad consultations between the government, NGOs, the business and academic sectors. The Strategy has been endorsed by the Cabinet, and includes a clear commitment on the behalf of government to establish protected areas as the principal instrument for conservation (WWF, 1991). Ten strategies are promulgated, with the aim of addressing the issues of environment and development in the Philippines. These are: the integration of environmental considerations in decision making; proper pricing of natural resources; property rights reform; establishment of a national integrated protected areas system; rehabilitation of degraded ecosystems; pollution control; integration of population concerns and social welfare in development planning; inducing growth in rural areas; promotion of environmental education; and strengthening citizen's participation. However, the Strategy itself lacks a clear mechanism for funding and implementation (Roque, 1991).

The legal basis for protected areas is complex and since 1900 there have been at least 262 enactments. These have been in a variety of forms, such as Republic acts, proclamations, administrative orders, executive orders, Presidential letters of instruction and others, mostly relating to the establishment or modification of protected areas. In a number of cases enactments have been overlapping or inconsistent, or have not provided sufficient information for a protected area to be accurately delimited. There has been a marked tendency to issue amending enactments which has lead to further confusion. Thus, for example, the legal and administrative status of Quezon Memorial National Park can only be ascertained by reference to at least 12 enactments. Another difficulty arises from overlapping legislation. For example, Presidential Decree No. 389 clearly indicates that the protected areas system comprises national parks, national recreation areas, national historic parks, national seashore parks and national marine parks. Nevertheless, enactments have not used these terms in any clear or consistent manner. Furthermore, this decree was subsequently amended by Presidential decrees nos. 705 and 1559, although without reference to a national system. It is, therefore, unclear whether the entire Presidential Decree No. 389 has been repealed, or merely amended in an undefined manner (NRMC, 1983). An analysis of laws and enactments pertaining to national parks prior to the 1986 revolution is given in NRMC (1983). Forest land use legislation, including that relevant to protected areas, is analysed by Natividad (1987).

Protected areas were first established during the US administration, with the enactment of Executive Order No. 33 on 25 April 1910 establishing the diminutive Rizal (Dapitan) National Park. The first formal protected area legislation was Act No. 3915, "An Act for the Establishment of National Parks Declaring such Parks as Game Refuges and Other Purposes", passed on 1 February 1932. The Act came into force on 1 January 1934 in Forestry Administrative Order No. 7 (National Park Regulations) which provided regulations for the establishment, supervision and special uses of national parks. Act No. 3915 describes national parks as "a portion of the public domain reserved or withdrawn from settlement, occupancy or disposal under the laws of the Philippine Islands which, because of its panoramic, historical, scientific or aesthetic value, are dedicated and set apart for the benefit and enjoyment of the people of the Philippine Islands". A more contemporary definition of national parks is given in the Forestry Reform Code, Presidential Decree No. 705 dated 19 May 1975, as "a forest land reservation essentially of primitive or wilderness character which has been withdrawn from settlement or occupancy and set aside as such exclusively to preserve the scenery, the natural and historic objects and the wild animals or plants therein, and to provide enjoyment of these features in such a manner as will leave them unimpaired for future generations" (NRMC, 1983). Forest Administrative Order No. 7 identifies the purpose of national parks as to: preserve panoramic, scenic or aesthetic interest; provide for recreation; and to preserve flora and fauna, geological features, historic or prehistoric remains and any other feature of scientific or ethnological interest. However, neither a precise definition of national parks, nor specific criteria for selecting areas for national park status is given (Serna, n.d.).

There are numerous confusing and overlapping laws concerning the establishment of marine protected areas. The legislation for, and management of, marine protected areas is discussed in UNEP/IUCN (1988). The first national marine park, Tubbataha Reefs, was established on 11 August 1988 by virtue of Presidential Decree No. 306. The tourist zone, a new category, was devised in the 1970s. These come under the jurisdiction of the Philippines Tourism Authority and within them tourism is given priority over conservation. Puerto Galero Biosphere Reserve falls into this category, although other such zones have been set up under six presidential proclamations (1520, 1522, 1551, 1653, 1667-A and 1801) (NRMC, 1983).

The lack of clear definition and criteria for selecting areas has led to a proliferation of national parks. The total number of national parks created is open to debate, with some authorities stating a total of 62 (NRMC, 1983; PAWB, 1991), 59 (Anon., 1988a), 60 (Basa, 1988) or 72 (Petocz, 1988), the discrepancies probably stemming from the fact that several agencies are responsible for managing protected areas.

A draft 1991 National Integrated Protected Areas System Act has been prepared and is before Congress at present. This will enable the establishment of the Integrated Protected Areas Systems (IPAS) and is being sponsored by the Department of Environment and Natural Resources. Passage of this bill is essential to the implementation of the IPAS because it makes provision for the establishment of the protected areas, funding, administration etc. The Bill has passed the Lower House and is currently before the Senate (C. Roque, pers. comm., 1991). The successful passage is a pre-condition for the release of substantial World Bank funds under the Global Environmental Facility (Roque, 1991).

A review of the current legislative background is being conducted by the Foundation for Sustainable Development as part of the IPAS programme. This includes compiling and reviewing all existing legal instruments, developing guidelines for the proposed IPAS Bill, conducting consultative meetings and submitting the draft legislation to the Philippines legislature (WWF, 1991).

International Activities The Philippines ratified the Unesco Convention Concerning the Protection of the World Cultural and Natural Heritage Convention (World Heritage Convention) on 19 September 1985, although no natural sites have been inscribed, and also participates in the Unesco Programme on Man and the Biosphere. Puerto Galera Biosphere Reserve is internationally recognised as a component of the global biosphere reserve network. The Philippines is a member of ASEAN (Association of South-East Asian Nations) and two parks have been recognised as ASEAN Heritage Sites. However, the country has not yet become a contracting party to the Ramsar Convention on Wetlands of International Importance Especially as a Waterfowl Habitat.

Administration and Management As many as ten different bodies have administrative responsibility for national parks (Lewis, 1988), although the Department of Environment and Natural Resources (DENR) is predominant. Furthermore, the administration has been centralised and subject to several reorganisations. Following the popular revolution of 1986, the Ministry of Natural Resources was reformed into the Department of Environment and Natural Resources, under the provisions of Executive Order 192. Simultaneously, the Bureau of Forest Development was dissolved and a number of new bureaux established within the DENR, including the Protected Areas and Wildlife Bureau and the Forest Management Bureau (Scott, 1989). The Bureau of Forestry in the Department of Agriculture and Natural Resources (DANR) was responsible for the administration of national parks between 1932 and 1952. In August 1952 the Commission on Parks and Wildlife, directly under the Office of the President, was created under provisions in the Republic Act No. 826. Government Re-organization Plan No. 45 of 1956 placed the Parks and Wildlife Office under the control of the DANR. A further reorganisation of the government in 1972 led to the merging of the Parks and Wildlife Office with the Bureau of Forestry and the Reforestation Administration to form the Bureau of Forest Development. When, in 1974, the DANR was divided into the Department of Agriculture and Department of Natural Resources, the Bureau of Forest Development was assigned to the latter Department. Between 1982 and 1987, the development and management of all national parks in the provinces of Leyte, Negros Occidental and Palawan were entrusted to the Natural Conservation Office under the direct supervision of the Minister of Natural Resources, pursuant to Ministry of Natural Resources Administrative Order No. 47.

The Protected Areas and Wildlife Bureau (PAWB) is one of the staff bureaux of the DENR, with a primary responsibility for the establishment and management of protected areas and the conservation of wildlife resources. Its functions are to: formulate and recommend policies, guidelines, rules and regulations for the establishment and management of an IPAS, such as national parks, wildlife sanctuaries and refuges, marine parks and biosphere reserves; formulate and recommend policies, guidelines, rules and regulations for the preservation of biological diversity, genetic resources, the endangered Philippine flora and fauna; prepare an up-to-date listing of endangered Philippine flora and fauna, recommend a programme of conservation and propagation of the same; assist the Secretary (of DENR) in the monitoring and assessment of the management of the IPAS and provide technical assistance to the regional offices in the implementation of programmes for these; and perform any other functions as may be assigned by the Secretary and/or provided by law. The Bureau is headed by a Director assisted by an Assistant Director and complemented by six divisions and two staffs (PAWB, 1991).

PAWB cooperates with the regional offices of the DENR, providing technical assistance, briefing on wildlife policy, resource inventories, education programmes and field operations. the Provincial Environment and Natural Resource Offices and the Community Environment and Natural Resources Offices of the Regional Offices of the DENR perform the actual work of maintaining protected areas. Activities include patrols, habitat restoration, maintenance and interpretation for visitors (PAWB, 1991).

PAWB has a number of major programme areas, but one of the most important is the establishment and management of protected areas. Activities include identification of new sites, boundary marking and periodic assessment, as well as restoration of degraded sites, management of buffer zones and visitor facilities.

Protected areas administration has historically been weak, due to the many institutional reorganisations as well as institutional and juridical struggles. This was particularly true during the period when the Parks and Wildlife Office was subsumed into the Bureau of Forest Development (NRMC, 1983). Further, staffing, funding, training and administrative support have been inadequate (Fernandez, 1988; Penafiel, 1990).

In common with terrestrial areas, marine protected areas also suffer from a confused legislative and administrative background and there is no national governmental mechanism to manage marine areas. In 1977 the Marine Park/Reserve Development Inter-Agency Task Force was set up by Special Order No. 61 to formulate an integrated system of plans and programmes for marine conservation. Under the Presidential Proclamation No. 1801 of 1978 the Philippine Tourism Authority (PTA) assumed responsibility for the promotion and development of aquatic sports. This has led to the establishment of several marine reserves and tourist zones. The Coral Reef Research Team, within the Bureau of Fisheries and Aquatic Resources (Ministry of Natural Resources), is responsible for marine conservation within marine reserves.

The non-governmental organisation movement in the Philippines has been gaining considerable momentum over the last few years. Since the overthrow of the Marcos government, both NGOs and the media have become more vocal against issues of pollution, forest degradation and destructive practices in the marine environment. There are now some 11 conservation and environmental NGOs and 17 university or academic-related organisations in the country. The Haribon Foundation has taken on a leading role for the environmental NGO network and an environmentally sympathetic constituency is forming among the public and in the new senate and congress (Petocz, 1988). The Haribon Foundation has also been directly involved in the development of protected areas (Balete, 1990). University research centres and institutes, such as the Marine Science Institute at University of Philippines, Diliman, the Institute of Terrestrial Ecology and Marine Research Institute at Silliman University, the Institute of Environmental Science and Management at University of Philippines, Los Banos and Ecosystems Research and Development Bureau of the Department of Environment and Natural Resources stationed at University of Philippines, Los Banos have all contributed significantly to developing conservation activities in the country.

Systems Reviews Tropical forests originally covered almost 280,000 sq. km (93%) of the total land area of the Philippines, and still two-thirds of it at the end of World War II (Cox, 1988). There are many published estimates of the current extent of forests, but these are based on different definitions of forest cover and on various information sources covering different periods of time. Consequently, an overall synthesis of the situation is difficult. An estimated 30% (90,000 sq. km) still supports productive or regenerating forest (Cox, 1989). Figures published by Revilla (1986) indicate, however, that forest with at least 40% crown cover may cover only 22% of total land area. The most valuable forests, both economically and biotically, are old-growth dipterocarp formations which cover 9,840 sq. km (3.3%) (Penafiel, 1991). Multispectra/SPOT satellite imagery from 1987-1988 clearly shows the fragmentary nature of the remaining cover, as forests are divided into increasingly isolated vegetation islands (Collins *et al.*, 1991). Furthermore, forest cover is largely restricted to higher land, as human activities preferentially clear lowland vegetation. Major forest blocks, therefore, are restricted to the Sierra Madre and Cordillera Central in Luzon, Central Samar, the highlands around Mt Ragang and other areas to the east in Mindanao. The only extensively forested island is Palawan, principally due to low human population density. However, projected deforestation trends suggest that all but an irreducible minimum of approximately 1,000 sq. km of montane and low quality scrub forest will remain in Palawan by 2010 (HTS, 1985). The annual rate of national deforestation is difficult to define precisely. Estimates range from the highest of 3,790 sq. km between 1972 and 1982 (Ganapin, 1986), 3,235 sq. km between 1970 and 1979 (FAO, 1981), to low estimates of 1,700 sq. km (Cox, 1988) or 910 sq. km (WRI/IIED, 1986). Existing tropical rain forest cover may be lost by the end of the century (Petocz, 1988).

Mangroves, which covered some 4,500 sq. km in 1920, have been depleted both by legal and illegal felling over the last 60 years (Alvarez, 1984). Some 1,461 sq. km of mangroves remained in 1978 (Davies *et al.*, 1990), but only 1,190 sq. km remained in 1990 (Penafiel, 1990). Only 814 sq. km can still be classified as undisturbed (Petocz, 1988) and cover is declining by 50 sq. km each year (Howes, 1987). Other major wetland habitats include estuaries/mudflats, seagrass beds, lakes, freshwater swamps and marshes; a descriptive inventory of wetlands considered to be important for conservation is given by Davies *et al.*, (1990).

The Philippine fauna is exceptionally rich with some 960 terrestrial vertebrates. Many species are forest obligates and are acutely threatened by forest loss. Endemism collectively amounts to 43% of species, with 59% endemic mammals (100 species out of a total of 167) and as many as 85% of non-volant mammals (79 species) (Heaney, 1986; Petocz, 1988). There are 162 endemic bird species out of a total of 388 resident species (Bruce, 1980). Endemism is also unusually high in the herpetofauna at 63% (160 species) whilst 17% (240 species) of the icthyofauna is endemic (Petocz, 1988). Plant endemism is estimated at 44% (3,500 species) but only 5% (75 species) in Palawan (Davies *et al.*, 1986)

The Philippine National Conservation Strategy (Haribon Society, 1983), produced under the guidance of the FAO in 1985, has not yet been formally adopted by the government, but is in the process of being reviewed and revised by the Department of Environment and Natural Resources prior to being presented to the President and legislature for approval. The National Conservation Strategy is intended to reflect the needs and priorities in preserving biological diversity, endorse the new comprehensive national protected areas network (Anon., 1988a) and provide guidance for sustainable development and resource utilisation. The National Conservation Strategy may also be used to promulgate policies on family planning and population stabilisation (Petocz, 1988)..

The Australian Association for Research, Exploration and Aid (AREA), in cooperation with the Philippines Ministry of Natural Resources, founded a Department of Special Projects (1982) to deal with specific conservation problems, e.g. the "Palawan Conservation Programme" set up to coordinate and develop all national parks and conservation activities on Palawan island. A Strategic Environment Plan for Palawan, funded by the EEC, has been completed after eight years' work by national and international consultants. The plan provides a comprehensive framework for sustainable development of Palawan in both terrestrial and marine environments (Petocz, 1988). The extent to which this plan has been implemented is not currently known.

The new government's more enlightened approach to environmental conservation led to an agreement between the Department of Environment and Natural Resources and the Asian Development Bank on 5 April 1988 to provide more than US$ 100 million in soft loans for a national programme of reforestation, social forestry, timber stand improvement, watershed management, development of rattan plantations and forest protection programmes for 1988-1992. The loan is supplemented by a technical assistance grant of US$ 665,000 for studies of forestry development and serves as an excellent model for other Asian governments (Petocz, 1988).

Funding for conservation activities has been made available through "debt swap" agreements. In the first such agreement in Asia, the World Wide Fund for Nature agreed in June 1988 to purchase US$ 2 million of Philippines external debt. The funds are made available in local currency to the Haribon Foundation and the Department of Environment and Natural Resources for a variety of nature conservation activities. These include the development of St Paul's National Park on Palawan, El Nido National Marine Park (Anon., 1988b), Mount Pulog in Benguet (PAWB, 1991), field surveys of several recommended protected areas and other activities (J. Tongson, pers. comm.).

At least 59 national parks have been created since 1900, but they provide little effective protection for the country's terrestrial environment, and in 1986 the Haribon Foundation (Haribon Foundation, 1986) indicated that none would satisfy international protected areas standards established by IUCN. The integrity of virtually all the reserves in the existing Philippines protected areas system, which includes less than 1.3% of the country's total land surface, is poorly maintained. In 1975, it was reported that approximately 72,000 people were permanently settled in park lands and that 54,000ha of the protected area estate were under some form of cultivation (DAP, 1975). A further 4,000ha were being logged by timber companies. Recent information is not available, but it is likely that these figures now drastically underestimate the current situation. Park boundaries are frequently not demarcated (Basa, 1988), law enforcement is lacking and current staffing and financial provisions are such that the PAWB is unable to deploy an effective corps of forest guards and park rangers (WCMC, 1988).

A systematic attempt to identify an integrated protected areas system (IPAS) was initiated in 1986, supported by WWF-US, Department of Environment and Natural Resources and the Haribon Foundation (Anon., 1988a). This identified 27 "priority one" potential protected areas and a further 41 "priority two" areas; only 19 existing national parks were included. The report recommended that the remaining sites be excluded from the IPAS programme. Although the specific recommendations of this study were not accepted by the government, the principle of a planned protected areas system became established. The IPAS was further pursued in the World Bank's FFARM (Forestry, Fisheries and Agricultural Management) Study which included recommendations to the government for the establishment and management of a national protected areas system. The current, World Bank funded IPAS, is effectively a component of the Asian Development Bank-funded Master Plan for Forestry Development (DENR, 1989) and is coordinated by WWF-US, with a number of participatory organisations within the Philippines. It aims, initially, to identify and initiate management for ten priority sites (WWF, 1991). The long term strategy is to place the last adequate, remaining stands of forest, significant marine areas and wetlands under a protected areas regime, with as much as 20% of the country included (Roque, 1991). However, the current funding for the IPAS is short-term, and the full implementation of the system will require both long-term and more substantial funding than is currently available. The proposed IPAS law will recognise an "Indicative List", in which some 330 sites in 15 biogeographical regions are named. In addition, all the remaining old growth forest, some 70,000 sq. km, will be included within the system under various categories according to both size and site (C. Roque, pers. comm., 1991).

Addresses

Protected Areas and Wildlife Bureau, Department of Environment and Natural Resources, Visayas Avenue, Diliman, Quezon City (Tel: 2 978511-15; FAX: 2 981010; Tlx: 7572000 envinar ph)

Ecological Society of the Philippines, 53 Tamarind Road, Forbes Park, Makati, Metro Manila, PO Box 1739, MCC (FAX: 631 7357; Tlx: 29006 JRS PH)

The Haribon Foundation for the Conservation of Natural Resources, Suite 901, Richbelt Towers, 17 Annapolis Street, Greenhills, San Juan, Metro Manila (Tel: 722 7180/722 6357; FAX: 722 7119)

References

Asian Development Bank (1988). Environmental legislation and administration: briefing profiles of selected developing member countries of the Asian Development Bank. *ADB Environment Paper* No. 2. 69 pp.

Anon. (1988a). Development of an integrated protected areas systems (IPAS) for the Philippines. WWF-US/Department of Environment and Natural Resources/Haribon Foundation for the Conservation of Natural Resources. 190 pp.

Anon. (1988b). First debt-for-nature swap in Asia. *CNPPA Newsletter.* No 43.

Alvarez, J.B. (1984). Our vanishing forests. *Greenfields* 14(2): 6-16.

Balete, D.S. (1990). Final report on the faunal and socio-economic surveys of Mt. Isarog National Park, Camarine Sur, Luzon Island. A Debt-for-Nature Swap project of the Haribon Foundation for the Conservation of Natural Resources in collaboration with the Department of the Environment and Natural Resources and World Wildlife Fund-US. Unpublished. 53 pp.

Basa, V.F. (1988). Current report: boundaries of national parks. Integrated protected areas technical workshop. WWF-US/Department of Environment and Natural Resources/Haribon Foundation for the Conservation of Natural Resources. University of the Philippines, Los Banos, 15-17 March. 5 pp.

Bruce, M.D. (1980). A field list of the birds of the Philippines. Traditional Explorations. Sydney, Australia. 8 pp.

Collins, N.M., Sayer, J.A. , and Whitmore, T.C. (Eds) (1991). *The conservation atlas of tropical forests: Asia and the Pacific.* Macmillan Press Ltd, London. 256 pp.

Cox, C.R. (1988). The conservation status of biological resources in the Philippines: a report by the IUCN Conservation Monitoring Centre. Draft. Cambridge. 37 pp.

Cox, C.R. (1989). The Philippines. Draft prepared for IUCN/BP Conservation Atlas of Tropical Forests. World Conservation Monitoring Centre, Cambridge.

David, W.P. (1987). *Soil erosion and soil conservation planning – issues and implications.* College of Engineering and Agro-Industrial Technology, University of the Philippines, Banos.

Davies, J., Magsalay, P.M., Rigor, R., Mapalo, A. and Gonzales, H. (1990). *A Directory of Philippines wetlands.* Two Volumes. Asian Wetland Bureau Philippines Foundation/Haribon Foundation.

Davies, S.D., Droop, S.J.M., Gregerson, P., Henson, L., Leon, C.J., Lamlein Villa-Lobos, J., Synge, H. and Zantovska, J. (1986). *Plants in danger: what do we know?* IUCN, Gland, Switzerland and Cambridge, UK. 461 pp.

DENR (1989). Master plan for forestry development: protected areas management and wildlife conservation. Asian Development Bank TA 993 PHI. Department of Environment and Natural Resources, Quezon City. 82 pp.

DENR (1991). Philippine Strategy for Sustainable Development. Department of Environment and Natural Resources, Quezon City. (Unseen)

Development Academy of the Philippines (1975). *The development plan for the Philippine National Park System.* Volumes I-IX. Development Academy of the Philippines, Quezon City. (Unseen)

FAO (1981). *Tropical forest resources assessment project: the Philippines.* Report prepared by the Food and Agriculture Organisation of the United Nations as cooperating agency with the United Nations Environment Programme. UN32/6. 1301-78-04. Technical Report 3. FAO, Rome. Pp. 391-416.

Fernandez, P.V. (1988). A community care system for national parks (proposal for legislation). Integrated Protected Areas (IPAS) Workshop. University of the Philippines, Los Banos, 15-17 March. Pp. 90-108.

Ganapin, D.J. (1986). Forest resources and timber trade in the Philippines. In: *Proceedings of the Conference on Forest resources Crisis in the Third World.* Sahabat Alam, Kuala Lumpur, Malaysia. Pp. 54-70

Ganapin, D.J. (1987). Forest resources and timber trade in the Philippines. In: *Proceedings of the Conference in Forest Resources Crisis in the Third World,* 6-8 September 1986, Kuala Lumpur, Malaysia. Sahabat Alam, Kuala Lumpur. Pp. 54-70.

Haribon Foundation (1986). Assessment and study of national parks: a proposal. Unpublished manuscript. 16 pp.

Haribon Society (1983). *Philippine national conservation strategy: a strategy for sustainable development.* Haribon Society/IUCN/The Philippine Presidential Committee for the Conservation of the Tamaraw. Manila.

Heaney, L.R. (1986). Biogeography of mammals in SE Asia: estimates of rates of colonisation, extinction and speciation. *Biological Journal of the Linnaean Society* 28: 127-165.

Howes, J. (1987). *Rapid assessment of coastal wetlands in the Philippines.* IPT-Asian Wetland Bureau, Kuala Lumpur, Malaysia. (Unseen)

HTS (1985). Palawan Integrated Area Development Project: environmental monitoring and evaluation system. Annual Report 1985. Hunting Technical Services Limited, Borehamwood, England. 137 pp.

Leong, B.T. and Serna, C.B. (1987). *Status of watersheds in the Philippines*. National Irrigation Administration, Quezon City.

Lewis, R.E. (1988). Mt Apo and the other national parks in the Philippines. *Oryx* 22: 100-109

MacKinnon, J. and MacKinnon, K. (1986). *Review of the protected areas system in the Indo-Malayan Realm*. IUCN, Gland, Switzerland and Cambridge, UK. 284 pp.

MacKenzie D. (1988). Uphill battle to save Filipino trees. *New Scientist* 118 (1619): 42-43.

Myers, N. (1980). *Conversion of moist tropical forests*. National Academy of Sciences, Washington. (Unseen)

Myers, N. (1988). Environmental degradation and some economic consequences in the Philippines. *Environmental Conservation* 15(3): 205-214.

Nepomuceno, P.M. (1977). Status of Philippines parks. *Canopy* 3(5). Forest Research Institute, Philippines.

NRMC (1983). *An analysis of laws and enactments pertaining to national parks*. Volume 1. *Study on national park legislation*. Natural Resources Management Center, Ministry of Natural Resources. Quezon City. 127pp.

Petocz, R. (1988). Philippines: strategy for environmental conservation. A draft plan. Unpublished manuscript. 60 pp.

Philippine National Mangrove Committee (1987). Philippines. In: Umali, R. M. *et al.* (Eds), *Mangroves of Asia and the Pacific: Status and Management*. Natural Resources Management Center and National Mangrove Committee, Ministry of Natural Resources. Manila. Pp. 175-210.

PAWB (1991). The protected areas and biological diversity of the Philippines. Protected Areas and Wildlife Bureau, Department of Environment and Natural Resources, Quezon City. 30 pp.

Rabor, D.S. (1977). *Philippine birds and mammals*. University of the Philippines Press, Quezon City. (Unseen)

Repetto, R. (1988). *The forest for the trees? Government policies and the misuse of forest resources*. World Resources Institute, Washington, DC. 105 pp.

Revilla, A.V. (1986). Fifty-year development program for the Philippines. In: *Proceedings of the Seminar on the Fifty-Year Forestry Development Program of the Philippines*. Philippine Institute for Development Studies and Forestry Development Center, University of the Philippines, Los Banos. Pp. 1-20.

Roque, C. (1991). Prospects for sustainable development in the Philippines. Paper presented to the XVII Pacific Science Congress, 27 May-2 June 1991, Honolulu. 15 pp.

Scott, D.A. (1989). *A directory of Asian wetlands*. IUCN, Gland, Switzerland and Cambridge, UK.

Serna, C.B. (n.d.). The national parks of the Philippines. Bureau of Forest Development. Unpublished report. 4 pp.

UNEP/IUCN (1988). *Coral reefs of the world. Volume 2: Indian Ocean, Red Sea and Gulf*. UNEP Regional Seas Directories and Bibliographies. IUCN, Gland, Switzerland and Cambridge, UK/UNEP, Nairobi, Kenya. 440 pp.

WWF (1991). Inception report: integrated protected areas system of the Philippines. Feasibility studies, preliminary design and other support components. World Wildlife Fund-US, Washington, DC. 31 pp.

WCMC (1988). The conservation status of biological resources in the Philippines. A report by the IUCN Conservation Monitoring Centre prepared for the International Institute for Environment and Development. World Conservation Monitoring Centre, Cambridge, UK. 68 pp.

White, A. (1981) Philippines marine parks: past and current status. *ICLARM Newsletter* 3(14): 17-18

WRI/IIED (1986). *World Resources 1986*. World Resources Institute/International Institute for Environment and Development. Basic Books, New York. 353 pp.

Yaman, L. (1982). An analysis of the National Park System of the Philippines. *Journal of the Natural Resources Management Forum* 3(4).

Zamora, P.M. (1985). Conservation strategies for Philippine mangroves. *Enviroscope* 5(2): 5-9.

Zamora, P.M. (1988). *Diversity of flora in the Philippine mangrove ecosystems*. Paper presented at Technical Workshop in Philippine Biological Diversity, 1-2 March, University of the Philippines, Quezon City. 57 pp.

ANNEX
Definitions of protected area designations, as legislated, together with authorities responsible for their administration

Title: Act No. 3915

Date: 1 February 1932

Brief description: An Act providing for the establishment of national parks, declaring such parks as game refuges and for other purposes

Administrative authority: Director of Forestry/Department of Environment and Natural Resources

Designations:

National parks Areas of the public domain, which because of their panoramic, historical, scientific or aesthetic value, should be dedicated and set apart as a national parks for the benefit and enjoyment of the people of the Philippine Islands.

Game refuges and bird sanctuaries National parks declared as game refuges and bird sanctuaries, and except as provided for in the act, within which it is unlawful for any person to hunt, take, wound or kill, or in any manner disturb or drive away therefrom, any wild bird or wild animal, or take or destroy the nests or eggs of such birds, or take or kill and fish or shellfish.

Title: Forestry Administrative Order No. 7 (National Park Regulations)

Date: 1 January 1934

Brief description: Pursuant to the provisions of section 4 of Act No. 3915, entitled "An Act providing for the establishment of national parks, declaring such parks as game refuges and for other purposes", this Administrative Order setting forth the rules and regulations governing the establishment, supervision and protection, maintenance and use of national parks, is hereby promulgated for the information and guidance of all concerned.

Designations:

National park Such tract of land as shall have been set aside by proclamation of the Governor-General under the provisions of Act No. 3915.

SUMMARY OF PROTECTED AREAS

Map[†] ref.	*National/international designation* Name of area	IUCN management category	Area (ha)	Year notified
	National Parks			
1	Agoo-Damortis National Seashore Park	VI	10,947	1965
2	Aurora Memorial	VIII	5,676	1937
3	Balbalasang-Balbalan	VII	1,338	1974
4	Basilan	II	3,100	1939
5	Bataan	IV	23,688	1945
6	Bicol	VIII	5,201	1934
7	Bulusan Volcano	II	3,673	1935
8	Calauit Island	IV	3,400	1976
9	Central Cebu	VIII	11,894	1937
10	Lake Dapao	IV	1,500	1965
11	Lake Imelda (Dana Lake)	V	2,193	1972
12	Mainit Hot Spring	III	1,381	1957
13	Mayon Volcano	III	5,459	1938
14	Minalungao	III	2,018	1967
15	Mount Apo	II	72,814	1936
16	Mount Arayat	V	3,715	1933
17	Mount Banahaw-San Cristobal	V	11,133	1941
18	Mount Canlaon	II	24,558	1934
19	Mount Data	V	5,512	1940
20	Mount Isarog	II	10,112	1938
21	Mount Malindang	II	53,262	1971
22	Mount Pulog	VII	11,550	1987
24	Mounts Iglit-Baco	IV	75,445	1970
25	Mounts Palay-Palay-Mataas-na-Gulod	VIII	4,000	1976
26	Naujan Lake	IV	21,655	1956
27	Northern Luzon Heroes Hill	VIII	1,316	1963
28	Rajaha Sikatuna	IV	9,023	1987
29	Rizal (Luneta)	III	1,335	1955
30	Roosevelt	VIII	1,335	1933
31	St Paul Subterranean River	II	3,901	1971
32	Taal Volcano	III	4,537	1967
33	Tirad Pass	III	6,320	1938
34	Un-named NP (Proc. No. 1636)	VIII	46,310	
	Wildlife Sanctuaries			
35	F.B. Harrison	IV	140,000	1920
36	Liguasan Marsh GRBS	IV	43,930	1941
37	Magapit	IV	6,002	1932
	Faunal Reserve			
38	Mount Makiling	VII	3,329	1933
	Marine Park			
39	Tubbataha Reefs National Marine Park	II	33,200	1988
	Biosphere Reserves			
	Palawan	IX	1,150,800	1990
	Puerto Galera	IX	23,545	1977

[†]Locations of most protected areas are shown on the accompanying map.

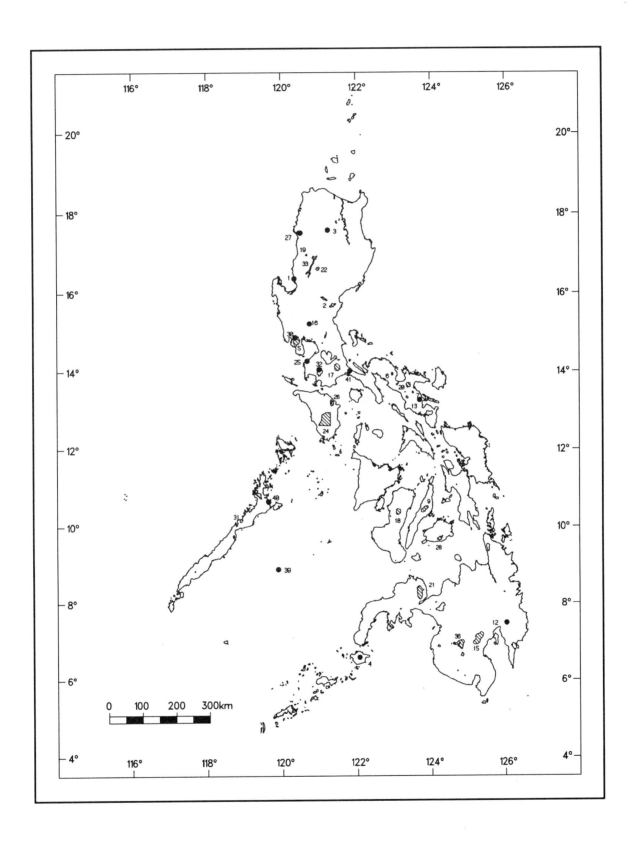

Protected Areas of the Philippines

SINGAPORE

Area 636 sq. km

Population 3,000,000 (1991)
Natural increase: 1.07% per annum

Economic Indicators
GNP: US$ 11,575 per capita (1991)

Policy and Legislation Historically, protection of forests dates from the 1840s, when the Governor "absolutely prohibited the further destruction of forests on the summits of hills" (Logan, 1848). The establishment of forest reserves dates from 1882 when a network was first proposed and a Forest Department established (Cantley, 1884). The reserves were for the supply of timber and firewood, prevention of soil erosion, protection of water supply, and improvement of the climate. Subsequently, in 1936, all existing reserves, except for Bukit Timah, and parts of the mangroves at Pandan and Kranji, were revoked and regazetted in 1939 as forest reserves. The Nature Reserves Act was enacted in 1951 to provide for the establishment of certain lands as nature reserves. The National Parks Act of 1990 repealed the Nature Reserves Act, made provision for national parks and nature reserves and established the National Parks Board (see Annex). The Parks and Trees Act (1985) provides for public parks and for the creation of the office of the Commissioner of Parks and Recreation. The Birds (Sanctuaries) (Amendments) Order (1981) passed pursuant to the Wild Animals and Birds Act, lists a number of areas in the country which are to be sanctuaries for birds.

International Activities Singapore does not participate in any international conventions or programmes concerned with protected areas, such as the Convention Concerning the Protection of the World Cultural and Natural Heritage Convention (World Heritage Convention), the Convention on Wetlands of International Importance Especially as Waterfowl Habitat (Ramsar Convention) or the Unesco Man and the Biosphere (MAB) Programme.

Administration and Management Originally, forest reserves came under the control of the Forest Department; then, in 1939 they became the responsibility of the Director of the Botanic Gardens in his capacity as Conservator of Forests. There are no longer any forest reserves, as such, to administer. Nature reserves, originally the responsibility of the Nature Reserves Board, are now managed, together with national parks, by the National Parks Board. The board is also empowered *inter alia* to acquire and dispose of property and to make rules to give effect to this legislation. Parks are the responsibility of the Parks and Recreation Department, within the Ministry of National Development. The Parks and Recreation department is responsible *inter alia* for constructing new parks and green open spaces. Under Section 25 of the Parks and

Trees Act, the Commissioner for National Development is responsible for making rules for the management and control of public parks. The Parks and Trees Rules (1983) drawn up under this section make strict limitations on activities permitted within public parks. Under Section 21 of the Parks and Trees Act, the Minister for National Development is responsible for making rules for the management and control of public parks.

The principal non-governmental organisation concerned with conservation in Singapore is the Malayan Nature Society (Singapore Branch).

Systems Reviews Singapore consists of one main island (41.8km long and 23km wide) with 58 islets within its territorial waters. The centre of the island has a series of low hills of granite and other igneous rocks. Bukit Timah, the highest hill, attains 163m and only three others exceed 100m in height. In the west and south-west of the island is a series of low ridges aligned north-west to south-east, formed from sedimentary rocks. The coastline is mostly flat and muddy although 5,400ha of this has now been covered by extensive landfill (Hails, 1989).

Prior to the establishment of the British colony in 1819, the great bulk of the island was covered in pristine forest, comprising 82% lowland evergreen dipterocarp rain forest, 13% mangrove and 5% freshwater swamp. As early as 1890, as much as 90% of this vegetation had been cleared. Today, more than half the island is urban in character and natural rain forest is restricted to the 81ha Bukit Timah Nature Reserve and scattered patches, totalling 50ha, in the adjacent Central Catchment Area. There are, in addition, some 1,800ha of 50 to 100 year old secondary forest in the catchment area, on land cleared during the last century, which may have growing value for the conservation of biological diversity (Collins *et al.*, 1991).

The history of the establishment of protected areas and a review of current legislation is given by Corlett (1988). All the primary forest remnants are protected legally in nature reserves. However, there is heavy human impact as the reserves are not fenced, are often surrounded by residential or industrial areas, and, in the case of Bukit Timah, subject to high and largely uncontrolled visitor pressure. The current legislation fails to distinguish between primary and secondary vegetation and areas that should receive strict protection are still used essentially as recreation areas (Collins *et al.*, 1991).

The Malayan Nature Society (Singapore Branch) has formulated a "Masterplan for the Conservation of Nature in Singapore". Twenty-eight sites have been identified, comprising areas which may already have some form of protected status, as well as unprotected areas of the most important sites. Three are within gazetted nature reserves, while the rest includes four wetland areas (Kranji, Khatib Bongsu, Sungei Buloh and Senoko), two islands (Pulau Tekong and Pulau Ubin) and a mangrove area (Mandai), all of which are to the north of Singapore.

Conservation in Singapore must be concentrated on minimising extinction rates in existing reserves and maximising benefits from man-made habitats. Three measures are the most urgent: protection of the margins of existing reserves, restriction of military training in the reserves and diversion of recreational pressure away from the most sensitive areas. A more ambitious idea would be to accelerate succession in the secondary forests of the catchment area by assisting seed dispersal and planting primary forest species. It may also be possible to reintroduce locally extinct forest vertebrates, particularly birds, from Malaysia, now that hunting and trapping have been greatly reduced (Collins *et al.*, 1991).

Major threats to the protected areas are the possibility of degazettement; increasing recreational use of Bukit Timah and the MacRitchie Reservoir area; and the small size of the reserves which exacerbates edge effects, isolation and extinction (Collins *et al.*, 1991). Singapore, however, is the second most densely populated country in the world, and with the great demand for land, conservation has received a low priority (Hails, 1989). Swamp forest felling and reservoir construction have reduced the wetlands to one small area. Similarly, industrial development and exploitation have reduced the mangroves to 2-3% of the original cover (Hails, 1989).

Addresses

National Parks Board (Executive Director), Singapore Botanic Gardens, Cluny Road, SINGAPORE 1025 (Tel: 474 1165; FAX: 475 4295)

Parks and Recreation Department (Commissioner), 5th Storey, National Development Building Annexe B, No. 7 Maxwell Rd, SINGAPORE 0106 (Tel: 322 6410; FAX: 221 3101; Tlx: RS 22603 PRD)

Singapore Institute of Biology, Zoology Department, National University of Singapore, Lower Kent Ridge Road, SINGAPORE 0511 (Tel: 772 2913)

Malayan Nature Society-Singapore Branch (Secretary), Department of Botany, National University of Singapore, Lower Kent Ridge Rd, SINGAPORE 0511 (Tel: 7722858; FAX: 7795671)

References

Cantley, N. (1884). *Report on the Forests of the Straits Settlements.* Singapore Printing Office, Singapore 24 pp. (Unseen)

Collins, N.M., Sayer, J.A. and Whitmore, T.C. (Eds) (1991). *The conservation atlas of tropical forests: Asia and the Pacific.* Macmillan Press Ltd, London. 256 pp.

Corlett, R.T. (1988). Bukit Timah: the history and significance of a small rain-forest reserve. *Environmental Conservation* 15: 37-44.

Hails, C.J. (1989). Singapore. In: Scott, D.A. (Ed.), *A directory of Asian wetlands.* IUCN, Gland, Switzerland and Cambridge, UK. Pp. 899-910.

Logan, J.R. (1848). The probable effects on the climate of Pinang of the continued destruction of the hill jungle. *Journal of the Indian Archipelago and East Asia* 2: 534-536. (Unseen)

MacKinnon, J. and MacKinnon, K. (1986). *Review of the protected areas system in the Indo-Malayan realm.* IUCN, Gland, Switzerland and Cambridge, UK. P. 268.

ANNEX
Definitions of protected area designations, as legislated, together with authorities responsible for their administration

Title: The National Parks Act 1990 (No. 10)

Date: 31 March 1990

Brief description: Makes provision for national parks and nature reserves and establishes a National Parks Board and to repeal the Nature Reserves Act.

Administrative authority: National Parks Board

Designations:

National park and nature reserves An area for the propagation, protection and preservation of the flora and fauna; for the study, research and preservation of objects and places of aesthetic, historical or scientific interest; and also for the study, research and dissemination of knowledge in botany, horticulture, biotechnology and natural and local history.

Unless written authority is previously obtained it is forbidden to:

— remove or damage any plant; prospect, mine, quarry or remove any soil of any sort; disturb or injure any animal or its nest; set traps to harm any animal; erect any building; cultivate any land; follow any activity which damages any plant, animal or property of the board; introduce any animal.

Title: Parks and Trees Act, Chapter 216

Date: 1985

Brief description: Provides *inter alia* for the development, protection and regulation of public parks and for the Commission on Parks and Recreation.

Administrative authority: Parks and Recreation Department, Ministry of National Development

Designations:

Public park Any walk, recreation ground, playground, open space, traffic island, side table or garden maintained by the Commissioner

Unless written permission is previously obtained, activities prohibited include *inter alia*, felling trees with a girth exceeding 1m and damaging trees.

SUMMARY OF PROTECTED AREAS

Map ref.	National/international designation Name of area	IUCN management category	Area (ha)	Year notified
1	*Nature Reserve* Central Catchment	IV	2,715	1951

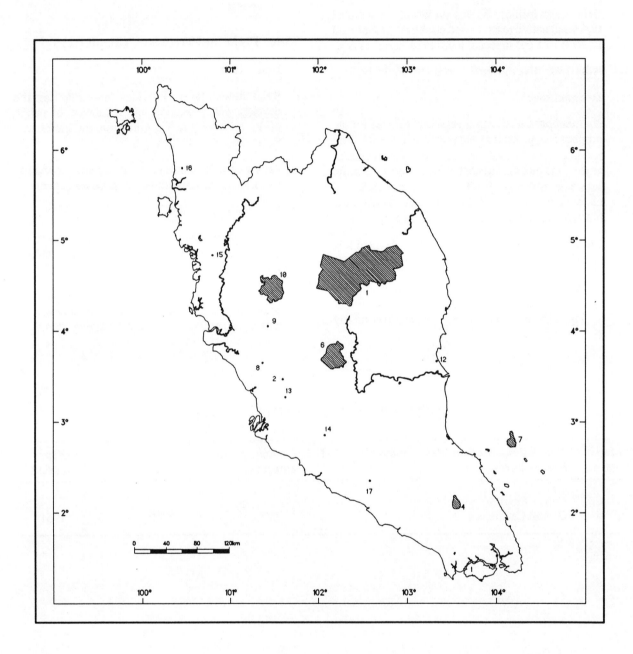

Protected Areas of Singapore (and Peninsular Malaysia)

DEMOCRATIC SOCIALIST REPUBLIC OF SRI LANKA

Area 65,610 sq. km

Population 17.2 million (1990)
Natural increase: 1.5% per annum

Economic Indicators
GNP: US$ 420 per capita (1988)

Policy and Legislation The 1978 Constitution of the Democratic Socialist Republic of Sri Lanka provides the foundation for the government's policies on conserving the environment. According to Chapter 6 Article 27(14) of the Constitution, "The State shall protect, preserve and improve the environment for the benefit of the community", and under Article 28F it is recognised that: "The exercise and enjoyment of rights and freedom is inseparable from the performance of duties and obligations and accordingly, it is the duty of every person in Sri Lanka to protect nature and conserve its riches."

The National Environment Act, 1980, provides a legal and institutional framework for coordinating environmental agencies and policies under the Central Environmental Authority. A serious drawback of the Act, however, is the lack of legal provisions to ensure compliance with environmental requirements. The Central Environmental Authority is also responsible for developing guidelines for environmental impact assessments, with appraisal, approval, and monitoring of development projects delegated to 15 Project Approving Agencies under various ministries. Environmental impact assessments were made mandatory for all development projects in 1984, but only the Coast Conservation Department has made provisions to enforce compliance with such assessments (Jansen, 1989). A synopsis of a national conservation strategy for Sri Lanka has recently been prepared by a task force under the Central Environmental Authority (CEA, 1988). It includes directions for the establishment of a comprehensive system of protected areas and, in the forestry sector, for the identification of forests for protection by statute.

Conservation of nature and culture are ancient traditions in Sri Lanka. King Pandukabaya marked out sanctuaries for aborigines, who always enjoyed royal protection, in the 4th century BC. One of the world's first wildlife sanctuaries was established in the 3rd century BC by King Devanampiyatissa in whose reign Buddhism was introduced to the country. Succeeding kings continued to uphold the Buddhist precept of forbidding the killing of any form of life, as is recorded on rock inscriptions throughout the country. In the 12th century AD, King Kirti-Nissanka-Mala proclaimed that no animals should be killed within a radius of seven gav (35.7km) from the city of Anuradhapura. Large tracts of forest, such as Udawattekele and Sinharaja, were protected as reserves by Sri Lankan rulers, sometimes serving as refuges for themselves, and privileged hunting was enjoyed from the earliest times in preserves specifically created for such purposes. It was not until 1885, however, with the enactment of the Forest Ordinance (No. 10), that legal provision was made to protect wildlife through the establishment of sanctuaries, first Yala (Block II of Ruhuna National Park plus the strict natural reserve) in 1900 and then Wilpattu in 1905 (Crusz, 1973; Gour-Tanguay, 1977; Hoffmann, 1969; Padmalal, 1989).

Game sanctuaries were abolished with the passing of the Fauna and Flora Protection Ordinance (No. 2) in 1937. This Ordinance, last amended in 1970, makes provision for national reserves (embodying only crown land) and sanctuaries (comprising both crown and private land), and their administration by the Department of Wildlife Conservation. National reserves and sanctuaries may be declared, altered or abolished by ministerial order. Full protection is afforded to wildlife in both national reserves and sanctuaries, but in sanctuaries the habitat is totally protected on land belonging to the state, while traditional human activities may continue to be practised on privately-owned land. Three categories of national reserve, *viz.* strict natural reserve, national park and intermediate zone (since deleted from the legislation), were identified in the original legislation and a further two, *viz.* nature reserve and jungle corridor, were introduced in an amendment to the Ordinance in 1964 (see Annex).

The need for a wildlife conservation policy has long been recognised (Hoffmann, 1973). A new National Policy for Wildlife Conservation was approved by the Cabinet in June 1990. Eleven specific policy items are identified, of which the following relate specifically to protected areas: to formulate a manifesto of varied objectives which would suit each area declared for protection; to reassess ways and means of enabling existing and proposed protected areas to meet their objectives; to identify and control sustainable uses of resources within protected areas that are compatible with their objectives; to permit sustained multiple use of resources within protected areas, and zone protected areas accordingly; and to effectively manage protected areas to maintain their resource potential at natural levels. It is proposed that this policy should be incorporated into an effective Conservation Act, which should be sufficiently flexible to permit a variety of conservation practices as advocated in the policy, but should provide strong punitive measures against those who wilfully destroy national wildlife resources through personal greed or prospects of commercial gain (DWC, n.d.).

The national forest policy dates back to 1929, since when it has been modified in the 1970s and later in 1980 (Pushparajah, 1985). A salient feature of the present policy is "To maintain, conserve and create forests for the preservation and amelioration of the environment, soil and water resources and for the protection of the local fauna and flora when they are required for aesthetic, scientific, historical and socio-economic reasons." The Forest Ordinance No. 16, 1907, as amended by Act No. 13 of 1966, makes provision for the establishment of reserved forests and village forests (see Annex), and for the protection of forests and their products, although its primary function has always been to provide for the controlled exploitation of timber. Both reserved and village forests may be declared, altered or abolished by ministerial order. Management of village forests is subject to ministerial regulations. The term reserved forest is interpreted as including plantations and chenas (cultivations established under slash-and-burn practices). Proposed reserves, which are a legacy from the first half of this century, are areas of reserved forest awaiting notification but are nevertheless subject to the provisions of the Forest Ordinance. Under the national Man and Biosphere Programme, arboreta representative of the main bio-climatic zones of the island have been demarcated within reserved forests. Nationally referred to as MAB reserves, timber exploitation and other forms of interference are excluded from these conservation areas (Bharathie, 1979).

To overcome the inherent weaknesses in the Forest Ordinance, the National Heritage Wilderness Areas Act, No. 3 was passed in 1988. Any piece of state land having unique ecosystems, genetic resources or outstanding natural features (including the habitat of threatened species) may be declared a national heritage wilderness area by the minister, subject to the approval of the president and endorsement by parliament (see Annex). Other provisions include the acquisition of land, the preparation of a management plan by the competent authority (Conservator of Forests) in respect of areas declared under the Act, and the over-riding of anything contradictory in the provisions of any other written law other than the Constitution. Entry to a national heritage wilderness area is by permit and restricted for purposes of observing the fauna and flora or conducting scientific research. Both the Forest Ordinance and National Heritage Wilderness Areas Act are administered by the Forest Department.

Other legislation concerned with the conservation of natural resources is reviewed by Crusz (1973), Gour-Tanguay (1977), NPS (1985), Jansen (1989) and, in the case of marine resources, UNEP/IUCN (1988). Special attention has been given to coastal and marine areas under the Coast Conservation Act (No. 57), 1981, which provides for the protection of the coast from erosion or encroachment by the sea and includes the planning and management of development activity within the coastal zone, defined as the area between 300m landwards of the mean high water line and 2km seawards of the mean low water line (Wijewansa, 1985). Under this Act, which is administered by the Coast Conservation Department, sites can be designated as critical habitats for protection (non-development areas).

In general, much of the legislation relating to natural resources has not kept abreast with the changing needs of society and is oriented towards the exploitation rather than sustainable use of such resources. Some laws have suffered from a lack of implementation for social, economic or political reasons, while others have remained fragmentary, with responsibilities spread among a multiplicity of agencies and institutions (Jansen, 1989). In the case of the Fauna and Flora Protection Ordinance, the lack of any criteria to differentiate between the various designations of protected area has resulted in inconsistent application of this system of classification. The national park designation, for example, has been used for some areas which are uninhabited and others, notably in the Mahaweli development area, which have a human population (Bere, 1959; Alwis, 1969; NPS, 1985). The Ordinance is being amended to provide *inter alia* for the creation of new categories of reserve, such as buffer zones, refuges, wetlands and marine reserves, and to enhance the present grossly inadequate penalties for infringements (Hoffmann, 1983; Jansen, 1989).

International Activities At the international level, Sri Lanka has entered a number of obligations and cooperative agreements related to conservation. It is party to the Convention concerning the Protection of the World Cultural and Natural Heritage (World Heritage Convention) which it accepted on 6 June 1980. One natural site has been inscribed to date and initiatives are underway to nominate several more natural sites. Sri Lanka became a contracting party to the Convention on Wetlands of International Importance especially as Waterfowl Habitat (Ramsar Convention) on 15 October 1990, at which time one site was included in the List of Wetlands of International Importance established under the terms of the Convention. Sri Lanka participates in the Unesco Man and the Biosphere Programme (Bharathie, 1979). To date 47 national biosphere reserves have been declared by the Forest Department, of which Sinharaja and Hurulu are designated as part of the international biosphere reserve network.

Sri Lanka is one of seven countries cooperating in the South Asian Cooperative Environmental Programme, which was launched in the 1980s to ameliorate the environment in the region, particularly regarding montane, mangrove, island and marine ecosystems, through legislation, education and training (Rau, 1984).

Administration and Management Under the provisions of the Fauna and Flora Protection Ordinance, the Department of Wildlife Conservation is responsible for wildlife conservation and protected areas management. The Department was set up in 1949 as the Wildlife Conservation Department, prior to which all matters relating to wildlife fell within the remit of the Forest Department. The first warden (director) was appointed in December 1950, and the Department became fully functional as an independent unit in 1957 (Alwis, 1969; Bere, 1959). Originally under the Ministry of Lands and Land Development, and subsequently the

Ministry of Tourism and Shipping (1970-1977) and Ministry of State (1960s and 1980s), the Department of Wildlife Conservation was transferred to the Ministry of Lands, Irrigation and Mahaweli Development in 1989. It is headed by a director, under whom are three deputy directors in charge of administration, technical, and veterinary and research divisions, respectively. Field staff are deployed mostly in the national reserves and marginally outside them. The Department has recently taken on responsibility for the Mahaweli Environment Project, begun in 1982 as part of the Accelerated Mahaweli Development Programme. The Mahaweli Environment Project has been concerned with the development and management of a system of protected areas in the Mahaweli Ganga and adjacent river basins. The Department is presently being restructured in view of the urgent need to upgrade and develop its services. A five-year development plan has been prepared for submission to donor agencies for funding (Kotagama *et al.*, 1990).

The Forest Department, under the provisions of the Forest Ordinance, is *inter alia* responsible for the protection and management of natural and plantation forests, including any declared as biosphere reserves. It is headed by a conservator of forests, supported by seven deputy conservators of forests, 13 assistant conservators of forests, and a number of foresters and forest rangers. The recent Forestry Master Plan, 1985-2020 (Jaakko Poyry International Oy, 1986) received much criticism because of its bias towards the exploitation of timber and fuelwood, with little regard to the ecological, hydrological, social and aesthetic values of forests (Fernando and Samarasinghe, 1988; Jansen, 1989). In response to public agitation, the Plan was reviewed by IUCN (1989) and a conservation review of remaining natural forests has now been incorporated into the Plan. Meanwhile, a moratorium has been imposed on logging in the wet zone, until such time as natural forests are evaluated with regard to their conservation value in maintaining biological diversity and environmental stability.

Conservation of natural resources outside protected areas falls within the mandate of a number of other agencies. The Coast Conservation Department, Ministry of Fisheries, is responsible for planning and managing activities within the coastal zone, including the conservation of marine and coastal resources. Similarly, the National Aquatic Resources and Research Agency, set up under the National Aquatic Resources Research and Development Agency Act (No. 54), 1981, is concerned *inter alia* with the management and conservation of aquatic resources in the inland waters, 200-mile exclusive economic coastal zone and in the deep seas. The Natural Resources, Energy and Science Authority of Sri Lanka was set up to coordinate scientific activity within the country and to advise the President on matters pertaining to the application of science and technology. It has ten working committees, including one for natural resources, and also functions as the focal point for international activities in science and technology. The National MAB Committee, for example, is placed under this authority. The Central Environmental Authority, created in 1981, was intended to be at the apex of governmental agencies concerned with environmental matters. Included within its mandate has been the preparation of a national conservation strategy. To date it has served largely in an advisory capacity, its effectiveness having been limited by its inability to influence the actions of agencies outside its own Ministry of Local Government and Housing (Jansen, 1989).

Non-governmental organisations have recently played an active role in mobilising public opinion and lobbying government on national conservation issues, often successfully as in the case of the Sinharaja rain forest or the national Forestry Master Plan. They have also directly supported various conservation programmes, complementing the work of government agencies. Most non-governmental organisations are registered with the Central Environmental Authority. Some of the more important ones are represented on various national committees, such as the Fauna and Flora Advisory Committee, and the Environmental Council of the Central Environment Authority.

By far the largest and oldest conservation non-governmental organisation is the Wildlife and Nature Protection Society. Originally established on 23 May 1894 to cater for sport-hunting interests, and later incorporated by Act of Parliament in 1968, the Society has become increasingly concerned with the conservation of the island's biological resources through negotiation and cooperation with the state. The Society has about 4,000 members. It produces a biannual wildlife journal, *Loris*. More recently, in 1980, March for Conservation was set up to focus attention, through education, training and research, on the rapid degradation of the island's natural resources. Its particular strength lies in the university basis of its 300 members, providing it with a strong research capability. Members have undertaken research in the Sinharaja rain forest, national parks, coastal zones and the Mahaweli development area. The organisation publishes scientific literature on natural resource issues, and has been instrumental in forming rural-based conservation societies, particularly in the vicinity of the Sinharaja rain forest. The Nation Builders Association is a rural-based organisation, composed largely of village youth and involved in forestry activities mainly in Kandy District. The Environmental Foundation provides the legal teeth to the non-governmental environmental movement. It comprises a group of young lawyers with a mandate to protect the country's natural resources from damaging activities through legal action. Other non-governmental organisations include the Sri Lanka Association for the Advancement of Science, which provides a forum for scientists of all disciplines to promote and advance the sciences, and the Ceylon Bird Club, which has maintained bird records for nearly 50 years and has

129

organised mid-winter waterfowl counts since 1983. Limited financial resources and the lack of time of volunteer members to devote to the conservation effort is a serious constraint. A step in the right direction was taken in 1985 when a number of these voluntary groups joined together under the umbrella of the Sri Lanka Environment Congress in order to pool their resources and generate a more powerful and united front to the environment movement. Further details of these and other voluntary organisations are given elsewhere (Hoffmann, 1969; Jansen, 1989; Wijewansa, 1985).

IUCN (The World Conservation Union) has a country office in Sri Lanka. Projects include implementing management plans in two biologically diverse and critical ecosystems, Knuckles Range and Sinharaja, and executing part of the UNDP/FAO/GoSL Environmental Management in Forestry Development Project.

There are a number of constraints to the effective protection of reserves. Both the Department of Wildlife Conservation and the Forest Department are severely handicapped by a lack of financial resources to administer such areas effectively. Forest reserves, in particular, are often unstaffed with the result that few of the smaller ones exist on the ground, while many of the larger ones are heavily encroached. Enforcement measures are handicapped by the absence of clear-cut boundaries that are easily recognisable on the ground. Lack of local involvement in protected areas is limiting their effectiveness. Political interference is a further obstacle to natural resource conservation (Anon., n.d.; Hoffmann, 1983; Jansen, 1989).

Systems Reviews Sri Lanka became an island probably in the late Miocene. It consists of a south-central massif, rising to some 2,500m, surrounded by a coastal plain which encompasses about 75% of the total geographic area. The massif has a major influence on the climate, with conditions ranging from wet in the south-west to dry in the north and east. Radiating from this massif is a network of major rivers (Fernando, 1984). These features are largely responsible for the wide variety of ecosystems present in Sri Lanka and the associated rich diversity of plants and animals, many of which occur in the wet zone and are endemic (Wijesinghe *et al.*, 1989). The status of the country's natural resources and trends in their use is reviewed by Baldwin (1991).

The pattern of land use has been strongly influenced by the different climatic zones. The first major changes in land-use patterns took place during the colonial era when forests in the wet zone were cleared for cultivation of coconut and rubber in the lowlands and coffee, later to be replaced by tea, in the uplands (Jansen, 1989). The wet zone, which receives in excess of 2540mm of rainfall per year, is now densely populated and intensively used for agriculture, with little potential for further expansion. The dry zone, on the other hand, until about the 13th century AD the centre of a highly advanced civilisation based on irrigated agriculture, receives less than 1905mm of rainfall annually, is sparsely populated and

offers the greatest opportunities for agricultural expansion in conjunction with irrigation schemes. In the face of a lack of food, overpopulation and unemployment in the wet zone, the former colonial government devised a scheme under the Land Development Ordinance, 1935, to translocate peasant farmers and the unemployed from the wet zone to the dry zone (Drijver *et al.*, 1985). After independence, the Sri Lankan government developed the scheme into the much more ambitious Mahaweli Development Programme, with the accelerated phase scheduled for completion in 1986-1987. This is one of the largest development schemes ever undertaken in South Asia, with 360,000ha of land in the basins of the Mahaweli Ganga and the south-eastern rivers irrigated for agricultural purposes for the benefit of nearly one million people to be settled in the area (Wijewansa, 1985). A major limiting factor, however, is the availability of water which will be enough for year-round irrigation of only 200,000ha (Perera, 1984).

The main forest types are the tropical wet evergreen (below 1,000m) and montane (above 1,000m) forests of the wet zone, tropical moist/wet semi-evergreen forests of the intermediate zone, tropical dry mixed evergreen forests of the dry zone and semi-evergreen thorn forest of the arid zone. Large tracts of forest have been cleared during the last hundred years to accommodate the growing human population, which has risen from 3.5 million in 1900 through 5 million in 1950 to nearly 15 million by 1980. Natural forest cover has dwindled rapidly from an estimated 89% in 1889 to 46.5% in 1956 and 28% in 1981-1983 (Nanayakkara, 1987). The annual rate of deforestation of closed forest (3.5%) was among the highest in the world for the period 1981-1985 (Repetto, 1988). It has been estimated from the Forest Cover Map produced by the Survey Department in 1981, which is based on satellite imagery supplemented by ground-truthing, that natural forest cover had shrunk to 24.9% by 1981, of which only 1.5% comprised lowland wet-zone forest, biologically the most valuable forest type. This estimate is lower than the later figure of 28% natural high forest for mid-1983 derived from the FAO inventory (Forest Department, 1986). Extrapolating from the FAO data, natural forests covered an estimated 1.58 million ha (24% total land area) in 1989 (Baldwin, 1991).

The wetlands of Sri Lanka comprise a variety of coastal and inland systems, ranging from estuaries, lagoons and mangroves to rivers, villus and tanks (reservoirs). Many of the tanks date back 1,500 years when they formed part of an intricate irrigation system for rice cultivation. In recent decades, several large reservoirs have been constructed as part of large-scale hydro-power and irrigation projects, notably in the Mahaweli catchment area. There are over 10,000 tanks in Sri Lanka, of which 3,500 are considered to be significant water bodies. Only 60 of these exceed 300ha in size. The wetlands support a variety of wildlife, as well as being of socio-economic importance, particularly for irrigation purposes (Scott, 1989).

Coral reefs are reported to be most extensive off the eastern coast, although the Basses off the south-east coast and those of Dutch Bay off the north-west coast are among the most spectacular. Reef resources are reviewed in UNEP/IUCN (1988).

The present growing awareness of the need to conserve the island's natural resources can be traced back to at least 1873 when the colonial government warned against the replacement of natural forests with plantations in view of its serious impact on the climate (Hooker, 1873 cited in Jansen, 1989). Within the forestry sector, forests were reserved from 1875 onwards but they continued to be maximally exploited for timber until the 1960s. In 1938, a law prohibiting the clearing of forests above 1,500m was enacted. Beginning in 1969 under the national Man and Biosphere Programme, the Forest Department began to establish a national network of MAB reserves, the most important of which (Sinharaja) has since been upgraded to a national heritage wilderness area.

Within the wildlife sector, one-third of the present protected areas network had been set up by 1950, following the enactment of the Fauna and Flora Protection Ordinance. The network grew progressively in subsequent decades, with a major expansion recently following an initiative begun in the early 1980s to mitigate the impact of the Accelerated Mahaweli Development Programme (Jansen, 1985). This initiative, carried out under the Mahaweli Environment Project, followed in the wake of environmental assessments carried out by TAMS (1980,1981). Needs for environmental action were subsequently identified in a study commissioned by the European Community (Driver *et al.*, 1985).

Sri Lanka currently has a fairly extensive system of protected areas, covering nearly 14% of its total land area (12% under the Department of Wildlife Conservation and nearly 2% under the Forest Department), but it is not comprehensive. Priorities to develop the existing network are identified in the IUCN systems review of the Indomalayan Realm (MacKinnon and MacKinnon, 1986), and specific recommendations are made in the Corbett Action Plan (IUCN, 1985). While most habitats are represented within the existing protected areas system, most of the major conservation areas lie within the dry zone. Coverage of tropical wet evergreen and hill forests is still far from adequate (MacKinnon and MacKinnon, 1986). Here lie a number of floristically important regions, such as the forests of Kottawa, Hinidumkanda, Kanneliya and Gilimale, which have not been allocated for conservation (Gunatilleke and Gunatilleke, 1990). In the case of the Knuckles Range, it is the subject of a conservation plan supported by IUCN. Demarcation of protected areas in the wet zone is considered to be a high priority in the new national wildlife conservation policy (DWC, n.d.). Other gaps in the network include coastal and marine protected areas, the only one established to date being Hikkaduwa

Marine Sanctuary (Kotagama *et al.*, 1990). In the dry zone, important areas with inadequate conservation status include the Samanalawewa area in the Walawe Ganga basin and the core of the Minneriya-Giritale Nature Reserve (T.W. Hoffmann, pers. comm., 1990). These and other deficiencies in the existing network are being addressed by a conservation review of remaining natural forests, implemented by IUCN as part of the UNDP/FAO/GoSL Environmental Management in Forestry Development Project.

Addresses

Department of Wildlife Conservation, Director, Ministry of Lands, Irrigation and Mahaweli Development, 82 Rajamalwatte Road, Battaramulla, COLOMBO 6 (Cable: WILDLIFE; Tel: 566601)

Forest Department, Chief Conservator of Forests, Ministry of Lands, Irrigation and Mahaweli Development, 82 Rajamalwatte Road, Battaramulla, COLOMBO 6 (Tel: 566631-3)

Coast Conservation Department, Director, New Secretariat, Maligawatte, COLOMBO 10

Central Environmental Authority (CEA), Director General, Maligawatte New Town, COLOMBO 10 (Tel: 549455/6, 437488/9)

Natural Resources, Energy and Science Authority of Sri Lanka (NARESA), Director General, 47/5 Maitland Place, COLOMBO 7 (Tel: 596771-3)

Ceylon Bird Club, Chairman, PO Box 11, COLOMBO (Cable LANKABAUR; Tlx 21204 BAURCO CE; Tel: 20551)

Environmental Foundation, Executive Director, 29 Siripa Road, Road, COLOMBO 5 (Tel: 588804; FAX: 546518; Tlx 22894 SAGCO CE)

IUCN (The World Conservation Union), Country Representative, No. 7 Vajira Lane, COLOMBO 5 (Tel: 587031; FAX: 580089)

March for Conservation, c/o Department of Zoology, University of Colombo, Minidas Kumaratunga Mawatha, COLOMBO 3 (Tel: 549136)

Wildlife and Nature Conservation Society of Sri Lanka, President, Chaitiya Road, Marine Drive, Fort, COLOMBO 1 (Tel: 25248)

References

Alwis, L. de (1969). *The national parks of Ceylon: a guide*. Department of Wildlife Conservation, Colombo. 89 pp.

Anon. (n.d.). Draft environmental report on Sri Lanka. Paper presented at 25th Working Session of IUCN's Commission on National Parks and Protected Areas. Corbett National Park, India, 4-8 February 1985.

Anon. (1988). Management policies for national parks, nature reserves and sanctuaries in Sri Lanka. Draft. 10 pp.

Baldwin, M.F. (Ed.) (1991). *Natural resources of Sri Lanka: conditions and trends*. Keells Business Systems, Colombo. 280 pp.

Bere, R.M. (1959). Some of the wild life problems of Ceylon. *Oryx* 5: 126-137.

Bharathie, K.P. Sri (1979). Man and Biosphere Reserves in Sri Lanka. *Sri Lanka Forester* 14: 37-40.

CEA (1988). *The national conservation strategy for Sri Lanka: a synopsis.* Central Environmental Authority, Colombo. 28 pp.

Crusz, H. (1973). Nature conservation in Sri Lanka. *Biological Conservation* 5: 199-208.

Drijver, C., Toornstra, F. and Lionel Siriwardena, S.S.A. (1985). *Mahaweli Ganga Project in Sri Lanka: evaluation of environment problems and the role of settler-households in conservation.* Centre for Environmental Studies, State University of Leiden. 77 pp.

DWC (n.d.). *A national policy for wildlife conservation of Sri Lanka.* Department of Wildlife Conservation, Ministry of Lands, Irrigation and Mahaweli Development. 11 pp.

Gour-Tanguay, R. (Ed.) (1977). *Environmental policies in developing countries.* Erich Schmidt Verlag, Berlin. Pp. 818900/00-81900/10.

Guntilleke, I.A.U.N. and Guntilleke, C.V.S. (1990). Threatened woody endemics of the wet lowlands of Sri Lanka and their conservation. *Biological Conservation* 55 (in press).

Fernando, C.H. (Ed.)(1984). *Ecology and biogeography in Sri Lanka.* Dr W. Junk, the Hague. 505 pp.

Fernando, R. and Samarasinghe, S.W.R. de A. (1988). *Forest conservation and the forestry master plan for Sri Lanka - a review.* Wildlife and Nature Protection Society of Sri Lanka, Colombo. 76 pp. (Unseen)

Forest Department (1986). *A national forest inventory of Sri Lanka 1982- 85.* GOSL/UNDP/FAO Project SRL/79/014. Forest Department, Colombo. 238 pp.

Hoffmann, T.W. (1969). The Wildlife Protection Society of Ceylon: some historical reflections on the occasion of its 75th anniversary. *Loris* 11: 1-13.

Hoffmann, T.W. (1973). The need for a policy. *Loris* 11: 10-15.

Hoffmann, T.W. (1983). Wildlife conservation in Sri Lanka: a brief survey of the present status. Bombay Natural History Society Centenary Seminar (1883-1983). Powai, Bombay, India, 6-10 December 1983.

IUCN (1985). *The Corbett Action plan for protected areas of the Indomalayan Realm.* IUCN, Gland, Switzerland and Cambridge, UK. 23 pp.

IUCN (1989). Forest Sector Development Project: Environmental Management Component. IUCN (The World Conservation Union), Gland, Switzerland. Unpublished report. 42 pp.

Jansen, M. (1985). A network of national parks for Sri Lanka. *Tigerpaper* 12(1): 4-7.

Jansen, M. (1989). Sri Lanka – biological diversity and tropical forests. Status and recommended conservation needs. Revised version. USAID/Sri Lanka, Colombo. Unpublished report. 70 pp.

Jaakko Poyry International Oy (1986). Forestry Master Plan for Sri Lanka. Forest Resources Development Project, Ministry of Lands and Land Development. Unpublished. 146 pp.

Kotogama, S.W., Fernando, V.P. and Ishwaran, N. (1990). *A five-year development plan for the wildlife conservation and protected area management sector of Sri Lanka.* Ministry of Lands, Irrigation and Mahaweli Development, Colombo. 50 pp.

MacKinnon, J. and MacKinnon, K. (1986). *Review of the protected areas system in the Indo-Malayan Realm.* IUCN, Gland, Switzerland and Cambridge, UK/UNEP, Nairobi, Kenya. 284 pp.

Nanayakkara, V.R. (1987). Forest history of Sri Lanka. In: Vivekanandan, K. (Ed.), *1887-1987 100 years of forest conservation.* Forest Department, Colombo. (Unseen)

NPS (1985). Policy recommendations for national reserves and sanctuaries of Sri Lanka. Draft. Department of Wildlife Conservation, Colombo. 32 pp.

Padmalal, U.K.G.K. (1989). Some aspects of wildlife management and conservation in Sri Lanka. In: Karim, G.M.M.E., Akonda, A.W. and Sewitz, P. (Eds), *Conservation of wildlife in Bangladesh.* German Cultural Institute/Forest Department/Dhaka University/Wildlife Society of Bangladesh/Unesco, Dacca. Pp. 71-78.

Perera, N.P. (1984). Natural resources, settlements and land use. In C.H. Fernando (Ed.), *Ecology and biogeography in Sri Lanka.* Dr W. Junk, The Hague. Pp. 453-492.

Pushparajah, M. (1985). Forest policy and management of wet zone forests. *Sri Lanka Forester* 17: 6-8.

Rau, A.L. (1984). A review of wildlife legislation in Pakistan. M.Sc. thesis, University of Edinburgh, Edinburgh. 66 pp.

Repetto, R. (1988). *The forest for the trees? Government policies and the misuse of forest resources.* World Resources Institute, Washington. 105 pp.

Scott, D.A. (1989). *A directory of Asian wetlands.* IUCN, Gland, Switzerland and Cambridge, UK. 1,186 pp.

Somasekaram, T. (Ed.) (1988). *The national atlas of Sri Lanka.* Survey Department, Colombo.

TAMS (1980). *Environmental assessment - Accelerated Mahaweli Development Programme.* Volume 2. *Terrestrial Environment.* Tippets, Abbet, McCarthy and Stratton, New York. (Unseen)

TAMS (1981). *Environmental plan of action – Accelerated Mahaweli Development Programme.* Tippets, Abbet, McCarthy and Stratton, New York. (Unseen)

UNEP/IUCN (1988). *Coral reefs of the world. Volume 2: Indian Ocean, Red Sea and Gulf.* IUCN, Gland, Switzerland and Cambridge, UK/UNEP, Nairobi, Kenya. 440 pp.

Wijesinghe, L.C.A. de S., Guntilleke, I.A.U.N., Jayawardana, S.D.G., and Guntilleke, C.V.S. (1989). *Biological conservation in Sri Lanka (a national status report).* Natural Resources, Energy and Science Authority of Sri Lanka, Colombo. 64 pp.

Wijewansa, R.A. (1985). Sri Lanka: natural resources expertise profile. CDC/IUCN Gland, Switzerland. Unpublished report. 130 pp.

ANNEX
Definitions of protected area designations, as legislated, together with authorities responsible for their administration

Title: Fauna and Flora Protection Ordinance (No. 2)

Date: 1937, last amended 1970

Brief description: An ordinance to provide for the protection of fauna and flora of Sri Lanka

Administrative authority: Department of Wildlife Conservation (Director)

Designations:

National reserve

Area of crown land, all or any specified part of which is a strict natural reserve, national park, intermediate zone (this category has since been deleted from the legislation), nature reserve or jungle corridor. Nature reserve and jungle corridor categories were introduced in an amendment to the Ordinance in 1964.

Strict natural reserve No person may enter a strict natural reserve or in any way disturb its fauna and flora, nor may any animal or plant be collected, damaged or destroyed. Entry is prohibited except for the purpose of scientific research, subject to written authorisation by the Warden.

Activities prohibited are: hunting, killing or removal of any wild animal; destroying eggs or nests of birds and reptiles; disturbing wild animals or interfering with their breeding; felling or otherwise damaging any plant; breaking up land for cultivation, mining or any other purpose; kindling or carrying fire; and possessing or using any trap, explosive or poison to damage plant or animal life.

National park No person may enter a national park except to observe the fauna and flora, nor may any animal or plant be collected, damaged or destroyed. Entry is subject to the conditions of a permit issued by the prescribed officer on payment of a prescribed fee.

Activities prohibited are the same as those prohibited in a strict natural reserve.

Intermediate zone Any person, subject to such conditions and restrictions as may be prescribed, may enter an intermediate zone to hunt, kill or collect any wild animal.

Various activities are prohibited.

Originally envisaged as a buffer zone, providing for controlled hunting, this category has since been deleted from the latest revision of the Ordinance.

Nature reserve Similar to a national park, except that human activities traditionally permitted on crown land are allowed to continue.

Jungle corridor Provides for the safe movement of wildlife, particularly migratory elephant, in areas where human activity is permitted.

Sanctuary

Specified area of land (other than land declared as a national reserve) which may include both crown and non-crown land.

Activities prohibited are: hunting, killing or removal of any wild animal; and destroying eggs or nests of birds and reptiles.

Source: Original legislation

Title: Forest Ordinance (No. 16)

Date: 1907, amended by Act No. 13 of 1966

Brief description: An ordinance to consolidate and amend the law relating to forests and the felling and transport of timber

Administrative authority: Forest Department (Conservator of Forests)

Designations:

Reserved forest (i.e. forest reserve) Any crown land declared by ministerial order and published in the Gazette.

Activities prohibited include: trespass by persons or their cattle; felling trees or collecting forest produce; poisoning water; quarrying stone and burning lime or charcoal; making fires or fresh clearings; breaking up land for cultivation or any other purpose; constructing temporary or permanent buildings, sawpits, or roads; and pasturing cattle, hunting, shooting or fishing in contravention of any ministerial regulations.

Village forest Any portion of a forest may be constituted a village forest for the benefit of a village community.

Activities prohibited within village forests include: poisoning of water, and injuring by fire or other means any tree listed in Schedule I of the Ordinance.

Management of village forests, including pasturing of cattle, is subject to ministerial regulations.

Source: Original legislation

Title: National Heritage Wilderness Areas
Act, No. 3

Date: 4 March 1988

Brief description: An act to make provision for the declaration of national heritage wilderness areas; for the protection and preservation of such areas; and for matters concerned therewith or incidental thereto

Administrative authority: Forest Department (Conservator of Forests)

Designations:

National heritage wilderness area Any portion of state land having unique ecosystems, genetic resources or outstanding natural features.

Entry is by permit and restricted only for purposes of observing the fauna and flora or conducting scientific research.

Activities prohibited include: marking, damaging or removing any plant, tree or part thereof or any other forest produce; cutting grass or pasturing cattle; polluting water; disturbing, collecting or destroying any animal (or egg or nest in the case of birds or reptiles); erect (and occupy) any temporary or permanent building (this provision does not apply to the competent authority); make any fresh clearing, or break up any land for cultivation or other purpose; kindle, keep or carry fire; construct any road; and damage, alter or remove any boundary mark.

Source: Original legislation

SUMMARY OF PROTECTED AREAS

Map[†] ref.	National/international designation Name of area	IUCN management category	Area (ha)	Year notified
	National Parks			
1	Flood Plains	II	17,350	1984
2	Gal Oya	II	25,900	1954
3	Horton Plains	II	3,160	1988
4	Lahugala Kitulana	II	1,554	1980
5	Maduru Oya	II	58,850	1983
6	Ruhuna (Yala)	II	97,878	1938
7	Somawathiya Chaitiya	II	37,762	1986
8	Uda Walawe	II	30,821	1972
9	Wasgomuwa	II	37,063	1980
10	Wilpattu	II	131,693	1938
11	Yala East	II	18,149	1969
	Strict Natural Reserves			
12	Hakgala	I	1,142	1938
13	Ritigala	I	1,528	1941
14	Yala	I	28,904	1938
	Nature Reserves			
15	Minneriya-Giritale	IV	7,529	1988
16	Tirikonamadu	IV	25,019	1986
	National Heritage Wilderness Area			
17	Sinharaja	IV	7,648	1988
	Sanctuaries			
19	Anuradhapura	IV	3,501	1938
20	Buddhangala	IV	1,841	1974
21	Bundala	IV	6,216	1969
22	Chundikulam	IV	11,150	1938
23	Gal Oya Valley North-East	IV	12,432	1954
24	Gal Oya Valley South-West	IV	15,281	1954
26	Giant's Tank	IV	3,941	1954
32	Katagamuwa	IV	1,004	1938
36	Kokilai	IV	2,995	1951

Map[†] ref.	National/international designation Name of area	IUCN management category	Area (ha)	Year notified
37	Kudumbigala	IV	4,403	1973
39	Madhu Road	IV	26,677	1968
40	Mahakandarawewa	IV		1966
42	Mihintale	IV	1,000	1938
43	Minneriya-Giritale	IV	6,693	1938
44	Padaviya Tank	IV	6,475	1963
45	Pallekele-Kahalla-Balaluwewa	IV	21,690	1989
47	Parapuduura Nuns Island	IV	190	1988
49	Peak Wilderness	IV	22,380	1940
51	Polonnaruwa	IV	1,523	1938
52	Ravana Ella	IV	1,932	1979
55	Senanayake Samudra	IV	9,324	1954
56	Seruwila-Allai	IV	15,540	1970
57	Sigiriya	IV	5,099	1990
60	Telwatte	IV	1,424	1938
61	Trincomalee Naval Headworks	IV	18,130	1963
63	Vavunikulam	IV	4,856	1963
64	Victoria-Randenigala-Rantambe	IV	42,087	1987
67	Wirawila-Tissa	IV	4,164	1938

Biosphere Reserves				
Hurulu Forest Reservne		IX	512	1977
Sinharaja Forest Reserve		IX	8,864	1978
Ramsar Wetland				
Bundala Sanctuary		R	6,216	1990
World Heritage Site				
Sinharaja National Heritage Wilderness Area		X	8,864	1988

[†]Locations of most protected areas are shown on the accompanying map.

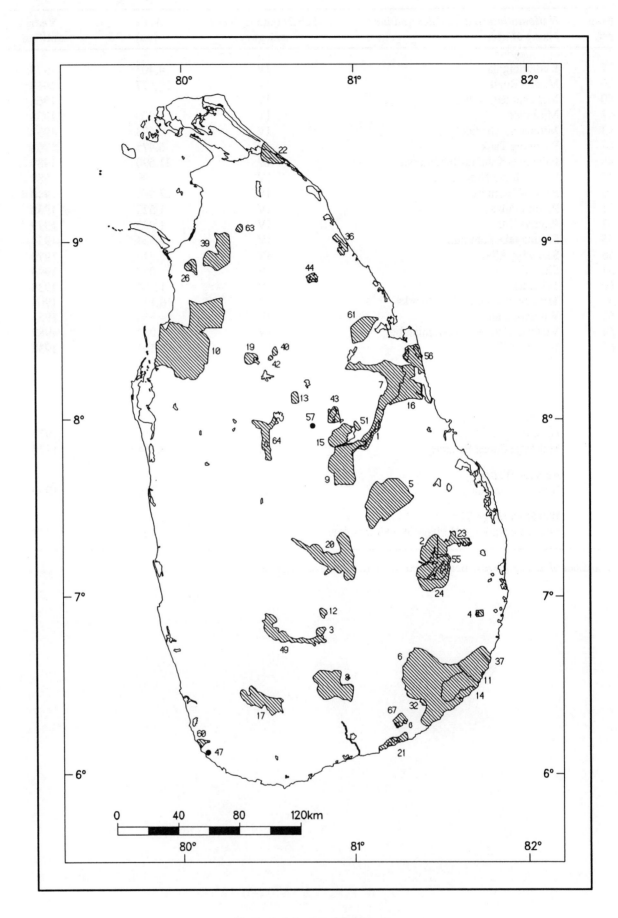

Protected Areas of Sri Lanka

TAIWAN
(REPUBLIC OF CHINA)

Area 36,960 sq. km

Population 20,440,000 (1991)
Natural increase: <2% per annum

Economic Indicators
GNP: US$ 7,997 (1990)

Policy and Legislation The history of protected areas establishment dates back to the Japanese occupation, when three mountainous areas in the centre of the island were designated as national parks in the 1930s, but never placed under management. Following the return to Chinese rule in 1945, the Taiwan Forestry Bureau became the leading nature conservation organisation for many years and established several forest recreation areas, notably at Alishan and Chitou (McHenry and Lin, 1984).

The National Park Law, promulgated on 13 June 1972, is designed to protect some of Taiwan's remaining wilderness and historically important sites, as well as providing for public recreation and scientific research (see Annex). Under this Act the Ministry of Interior is empowered to set up a National Park Planning Commission authorised to designate, alter or abolish areas for national parks and to review national park management plans. All decisions relating to the establishment and abolition of national parks and the declaration and alteration of their boundaries require the approval of the Executive Yuan (Cabinet). A national park may be divided into the following zones for management: existing use areas, recreation areas, cultural/historic areas, scenic areas and ecological protection areas. In general, collection of specimens, use of pesticides and herbicides, and construction of facilities are prohibited in ecological protection areas. Entrance to such areas is by permit only. Building, prospecting, farming, grazing livestock, fishing and other activities may be allowed in existing use areas or recreation areas with approval from the national park authorities (CPA, 1984). It was not until 1977, five years later, that instructions were given by the Executive Yuan to establish a national park in the Kenting area (finally operational in 1984). In April 1979 the Executive Yuan approved the Comprehensive Development Plan for the Taiwan Area in which a total of six national parks was proposed, four of which have since been established (Shueshan and Tapachienshan are outstanding). A National Park Planning Commission was appointed in September of that year to plan these areas (CPA, 1985; McHenry and Lin, 1984).

In addition, seven coastal protection areas have been designated under the Taiwan Coastal Area Natural Environment Protection Plan promulgated by the Executive Yuan (McHenry and Wu, 1985). Coastal protection areas are managed as general conservation or natural preservation zones, the latter being designated as ecological preservation areas or natural preservation areas under the Cultural Heritage Preservation Act 1982. This Act was being reviewed during 1989 in order to clarify this system of classification. Coastal protection areas are protected from any activities detrimental to their environmental quality; environmental impact assessments are required for any proposed developments within them. Under this Act, a Commission on Natural/Cultural Landscapes has been established and nine nature preserves have been declared cooperatively by the Ministry of Economic Affairs and the Council of Agriculture, one of which lies within a coastal protected area, namely Tanshui River Mangrove Forest Nature Preserve. Nature preserves provide for the protection of fauna and threatened species and their habitats, as well as having educational and research roles. The Cultural Heritage Preservation Act also provides the basis for the promotion of the Taiwan Nature Conservation Strategy (CPA, 1985; Lee, 1988).

The Wildlife Conservation Law was passed by the Legislative Yuan in June 1989. It provides for the conservation and protection of wildlife, replacing the earlier Hunting Ban of 1972. Responsibility for its implementation falls on the Council of Agriculture at national level, Taiwan Provincial Government at provincial level, and country and municipal governments at local level. Under this Law, wildlife protection areas may be established for the conservation of any animal listed under its protection (see Annex).

International Activities Participation in international nature conservation activities is one of the objectives of the Taiwan Nature Conservation Strategy (CPA, 1985). Taiwan is not yet a party to any international or regional conventions or programmes concerned with the conservation of natural ecosystems/heritage, such as the World Heritage Convention, the Unesco Man and the Biosphere Programme, Ramsar Convention and ASEAN Environment Programme, principally because the country has been rejected on political grounds.

Administration and Management Under the National Park Law (1972), the Ministry of Interior is the authority ultimately responsible for national parks. Following the reorganisation of the Ministry in March 1981, the Construction and Planning Administration was set up and the National Park Department established under its jurisdiction in December 1981. This Department is responsible for planning and administering national parks, as well as implementing the Taiwan Coastal Area Natural Environmental Protection Plan.

Under the Cultural Heritage Preservation Act, the Council of Agriculture is responsible for the planning and managing nature preserves, as part of its wider responsibilities for wildlife conservation. The Council of Agriculture is also responsible for the establishment and management of wildlife protection areas.

In order to coordinate government agencies with responsibilities for nature conservation, the Council of Agriculture established an inter-agency Commission on Nature/Cultural Landscapes in 1984, with one committee comprising representatives from many different government departments, and another with a membership comprising scientific experts (Patel and Yao-Sung, 1988).

Systems Reviews Taiwan is one of the most densely populated countries in the world. Most of this population is distributed along the western coastal plains. Straddling the Tropic of Cancer, Taiwan exhibits a wide variety of climate/habitat associations from coral reefs and tropical monsoon forests to high mountains snow-capped in winter. Two-thirds of the island is mountainous. Annual rainfall varies from less than 1000mm to more than 5000mm. The distribution of vegetation is influenced by elevation, climate and soils. The original vegetation can be classified as follows: alpine tundra formations, subalpine coniferous forest, cold temperate montane coniferous forest, warm temperate coniferous forest, warm temperate rain forest, tropical rain forest, littoral forest and savanna formation, and these are described by Severinghaus (1989). A brief account of the status, distribution and conservation of tropical rain forest in Taiwan is given in Collins *et al.* (1991). Wetlands include tidal mudflats, mangrove swamps and salt marshes, primarily along the west coast, and a few small lakes and ponds in remote mountain areas. Previously, there were 80,000ha of intertidal mud flats along the west coast, although 11,000ha have been reclaimed and there are plans to reclaim a further 15,000ha (Scott, 1989). Twelve of the most important wetland sites are described in detail in Scott (1989). Corals are found in all the waters around Taiwan, except the sandy area on the west coast. The main reef area is around the southern tip in the vicinity of Kenting where coral communities with small fringing reefs are found. The status and distribution of corals is further discussed in UNEP/IUCN (1988) and number of specific sites are described in detail.

Species diversity and endemism per unit area are among the highest in the world. There are over 4,000 species of vascular plants (1,075 endemic), 61 of mammals (42 endemic), 428 of birds (70 endemic), 135 of freshwater fish, 90 of reptiles, 30 of amphibians, and over 400 species of butterflies (McHenry and Lin, 1984).

The forests of Taiwan cover 1,864,700ha, or 52.1% of the country. The rest of the land area comprises cultivations (32.3%), fish ponds (1.4%), rivers and riverbeds (4.7%) and other lands (9.6%). The distribution of Taiwan's forests coincides approximately with that of the Central Mountain Range, peripheral coastal plains having been largely cleared for agriculture and other uses.

National forest, which occupies a total area of 1,786,500ha, is managed and developed for conservation of water resources, protection of national land, timber production and recreation. Some 411,384ha (23%) of national forest had been reclassified as protection forest by the end of 1983, being land occupied by railways, highways, reservoirs, power supplies, watersheds, etc. (TFB, 1984). There is a moratorium on timber extraction in areas above 2,500m and in the coastal zone. Following severe water shortages experienced several years ago, quotas have been fixed for timber production. A total ban on cutting is anticipated in the next year or so.

Taiwan has developed remarkably quickly, moving from a pre-World War II agricultural society to become one of the world's most prosperous industrial nations. Gross national product per capita has risen from US$ 48 in 1952 to over US$ 5,000 and foreign exchange reserves presently exceed US$ 75 billion, second highest in the world after Japan (Andrews, 1988). Due to this rapid growth in the economy, natural resources have been heavily exploited, species, particularly large mammals (McCullough, 1974), have been lost, and the environment has become seriously polluted. Following in the wake of initiatives taken to conserve the country's biological diversity through the establishment of a protected areas network and protection of wildlife, and recognising the need to redress the balance, the Executive Yuan requested the Ministry of Interior to develop a nature conservation strategy to promote a national nature conservation policy and sound resource management system based on principles of sustainable development (CPA, 1985). Goals identified in the strategy include the establishment of national parks, green belts and urban forests, and the protection of coastal resources. Outstanding, with regard to protected areas, is the proposal to develop a nationwide inventory of natural areas and development of a national park system plan to provide a long-term strategy for protecting nationally important natural (and cultural) resources (Bright, 1987). Natural areas are already in the process of being inventoried by the Taiwan Forestry Research Institute (McHenry and Lin, 1984).

The Society of Wildlife and Nature (SWAN), established on 15 April 1982, is the principal nature conservation society in Taiwan. It currently has 850 individual and group members, including government agencies, corporations and bird and other nature clubs. Other non-governmental conservation organisations include the Animals Protection Society, concerned with the protection of wetlands and waterfowl, and various bird societies. By 1989, there were some 17 non-governmental conservation organisations (L.L. Severinghaus, pers. comm., 1989).

The most serious threats to Taiwan's national parks are logging, surface mining and hydroelectric development. There is pressure on the parks authorities to permit such activities (L.L. Severinghaus, pers. comm., 1989). More subtle is the constant pressure for installation of military and communications facilities and noise pollution from military and civil aircraft. Other threats are major road constructions, adverse land use practices on privately-owned holdings within parks, outdoor advertising, vandalism, junk-yards, litter, and commercial shows and pageants (Bright, 1987). Assessment of visitor carrying capacity and implementation of appropriate controls to maintain the quality of the "wilderness experience" is a priority for the more popular parks (McHenry and Lin, 1984).

Addresses

National Park Department, Construction and Planning Administration, Ministry of Interior, 194 Peihsin Road, Section 3, Hsintien, Taipei Hsien, Taiwan

Council of Agriculture, 37 Nan Hai Road, Taipei, Taiwan 10728 (Tel: 2 314 7213/331 7541/312 4045; Fax: 2 331 0341/312 5857; Tlx: 8515 TAIPEI)

Society for Wildlife and Nature (SWAN), 1F, No. 7, Lane 58, Woh Long Street, Dah Au Chiu, TAIPEI (Tel: 2 741 2827; FAX: 2 741 1512)

References

Andrews, J. (1988). Taiwan: transition on trial. *The Economic*, 5 March.

Collins, N.M., Sayer, J.A. and Whitmore, T.C. (Eds) (1991). *The conservation atlas of tropical forests*: Asia and the Pacific. Macmillan Press Ltd, London. 256 pp.

CPA (1984). *Laws and regulations on construction and planning*. Construction and Planning Administration, Ministry of Interior, Taipei. 225 pp.

CPA (1985). *The Taiwan Nature Conservation Strategy*. Construction and Planning Administration, Ministry of Interior, Taipei. 39 pp.

Lee, San-Wei (1988). The work of nature and ecological conservation in the past three years – retrospective. *Council of Agriculture Forestry Series* 16: 59-65.

McHenry, T.J.P. and Lin, Yau-Yaw (1984). Taiwan's first national parks. *Parks* 9(3/4): 13-16.

McHenry, T.J.P. and Wu, Chuan-An (1985). Coastal zone protection in Taiwan. *Tiger Paper* 12(1): 13-16.

Patel, A.D. and Yao-Sung, L. (1988). History of wildlife conservation in Taiwan. Draft manuscript. Unpublished.

Scott, D.A. (1989). *A directory of Asian wetlands*. IUCN, Gland, Switzerland and Cambridge, UK. 1181 pp.

Severinghaus, L.L. (1989). Natural resources. In: The Steering Committee (Eds), *Taiwan 2000: Balancing Economic Growth and Environmental Protection*. Institute of Ethnology, Academia Sinica, Nankang, Taipei, Taiwan. Pp. 49-127

TFB (1984). *Forestry in Taiwan, Republic of China*. Taiwan Forestry Bureau, Taipei. 35 pp.

UNEP/IUCN (1988). *Coral reefs of the world. Volume 3: Central and Western Pacific*. UNEP Regional Seas Directories and Bibliographies. IUCN, Gland, Switzerland and Cambridge, UK/UNEP, Nairobi, Kenya. 378 pp.

ANNEX
Definitions of protected area designations, as legislated, together with authorities responsible for their administration

Title: Wildlife Conservation Law

Date: 23 June 1989

Brief Description: Enacted to conserve and protect wildlife, and to maintain the balance of natural ecosystems. Whatever is not regulated by this law may be regulated by other compatible legislation. Supersedes 1972 Hunting Ban.

Administrative authority: Council of Agriculture; Taiwan Provincial Government; County and Municipal governments

Designations:

Wildlife protection area Area established for *conservation animals* (defined as those animals protected under this law and including endangered, rare and valuable species, and any other species requiring conservation), and in which restoration activities may be carried out.

Privately-owned land within a wildlife protection area may be purchased or expropriated by the government according to law and managed by the authorities. If not purchased or expropriated, the owner should provide habitat for wildlife according to stipulations made by the authorities.

Creation, cancellation or change in status of a wildlife protection area should be authorised by the Council of Agriculture.

Source: Translation of original legislation

Title: National Park Law

Date: 13 June 1972

Brief description: Enacted for the purpose of preserving the nation's unique natural scenery, wild fauna and flora, and historic sites and providing for public recreation and scientific research.

Administrative authority: Ministry of Interior

Designations:

National park Area which represents the natural heritage of the nation, including unique scenery, landscapes, landforms, fossils, living fossils, and fauna and flora in naturally evolving communities.

Area which has educational significance for the perception of nature and which includes important pre-historic and historic sites and their surroundings which require long-term preservation by the nation.

Area which possesses outdoor recreation resources and unique scenery which are of easy access for public use.

Prohibited activities are: burning vegetation or setting fires to clear land; hunting animals or catching fish; polluting air or water; picking or removing flowers; defacing trees, bark, stones or signs; littering (including fruit skins); driving outside designated areas; and any conduct prohibited by the responsible authority of the national park.

Activities permitted within the different types of management zone are specified in the legislation (see CPA, 1984).

Source: Translation of original legislation (CPA, 1984)

SUMMARY OF PROTECTED AREAS

Map ref.	National/international designation Name of area	IUCN management category	Area (ha)	Year notified
	National Parks			
1	Kenting	V	32,631	1982
2	Taroko	II	92,000	1986
3	Yangmingshan	V	11,456	1985
4	Yushan	II	105,490	1985
	Nature Preserve			
5	Ta-Wu Mountain	IV	47,000	1987

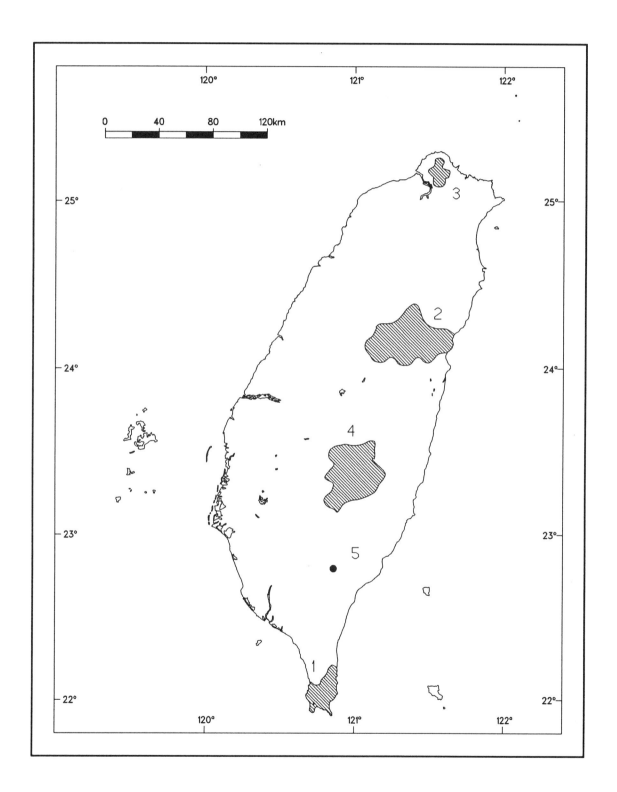

Protected Areas of Taiwan

THAILAND

Area 513,517 sq. km

Population 55,702,000 (1990)
Natural increase: 1.35% per annum

Economic Indicators
GNP: US$ 1,000 per capita (1988)

Policy and Legislation The earliest official conservation measures were taken during the 13th century by King Ram Khamkaeng the Great who established Royal Dong Tan Park. Parks were also established around temples and other religious sites, which, due to the Buddhist prohibition on killing, functioned as wildlife sanctuaries (Kasetsart University, 1987). Thereafter, conservation fell into abeyance until the establishment of the Royal Forest Department in 1896. The Department introduced modern management practices to achieve the maximum, and theoretically sustainable, yield of forest products. The current enabling legislation of the Department is the Forest Act B.E. 2484 (1941).

In the 1940s and 1950s the Royal Forest Department established a limited number of "forest parks" for recreation and habitat protection. Protected areas legislation, and other nature conservation measures, was established in the early 1960s with the passing of two key acts. The Wild Animals Reservation and Protection Act B.E. 2503 (1960) makes provision for the establishment of wildlife sanctuaries and non-hunting areas, the regulation of hunting, control of wildlife trade and provides a limited number of species with total legal protection. The National Parks Act B.E. 2504 (1961) is primarily concerned with the establishment and management of national parks (IUCN, 1979; Kasetsart University, 1987). The National Forest Reserves Act B.E. 2507 (1964) was passed to strengthen the inadequate forest protection afforded by the earlier Forest Acts. This Act also defines two forest types: production and conservation. National parks, wildlife sanctuaries and most non-hunting areas are classified in the latter category. The 1964 act also formalised the legal status of forest parks.

In December 1985, the Thai Cabinet approved a Draft National Forest Policy which reaffirmed the standing policy of maintaining at least 40% forest cover in the country, and further divided the forests into two categories. Protected forests, which should account for 15% of land area, encompasses all those sites set aside for nature conservation, watershed protection, research and recreation, including national parks and wildlife sanctuaries. The second category, accounting for the remainder of all forest land, is productive forest, which was to be utilised for the exploitation of timber and forest products. In May 1989 two Royal decrees were passed to amend and supplement the Forest Acts, making provision for a nationwide ban on logging. Provisions

were also made to automatically rescind logging concessions that remained in force within national parks under Section 30 of the National Parks Act, and in wildlife sanctuaries, following a Thai Judicial Council ruling in 1987 under which logging concessions granted within sanctuaries prior to gazettement remained in force. It is possible that this ban could be lifted, but the logging ban within national parks and wildlife sanctuaries can only be removed with a change in the law. Thus, establishment of these two protected area categories is now seen as the principal means by which remaining forests can be protected from logging. Consequently, the protected areas systems is being enlarged rapidly (Wongpakdee, 1990). However, the logging ban does not cover mangroves, and increased timber prices may have encouraged further illegal logging (Lohmann, 1990).

Nature conservation in all fields was strengthened when the Enhancement and Conservation of National Environmental Quality Act B.E. 2518 (1975) was passed. This provided for the establishment of the National Environment Board, within the Office of the Prime Minister, charged with submitting recommendations on environmental policy to the Cabinet. The National Environment Board operates through its executive arm, the Office of the National Environment Board, which has divisions responsible for environmental information, policy, planning, impact evaluation and standards.

National parks are established solely on government land and by Royal Decree to conserve both living natural resources and landscapes, and to make provision for public education and recreation. Theoretically, complete protection from all activities except research and recreation is provided. Wildlife sanctuaries, termed wild animal preserved areas in the 1960 Act, are established under Royal Decree and, again, exist only on land owned by public bodies. Sanctuaries exist largely for species conservation and protection is rigorous: hunting of all taxa is prohibited and habitat is also protected. Recreation is not encouraged and unauthorised entry and other activities are prohibited. Non-hunting areas are designated by the Ministry of Agriculture and Cooperatives on either private or state land for the sole purpose of conserving named species. Non-hunting areas are frequently defined by discrete features, such as lakes or temple grounds, to protect, for example, seasonal over-wintering or nesting birds. Activities such as agriculture, hunting or timber felling are not prohibited within a non-hunting area and there is no restriction on access or habitation unless proscribed by other legislation, such as the National Forest Reserves Act (FAO, 1981; Kasetsart University, 1987).

Proposals for both parks and sanctuaries originate within the National Parks Division and Wildlife Conservation

Division, respectively, and are considered by either the National Park or the Wild Animals Reservation and Protection Committee as appropriate. The protected area is gazetted when a Royal Decree, with a map showing the boundary lines, is promulgated. The extension or cancellation of a park or sanctuary is also only possible by Royal Decree. Non-hunting areas differ inasmuch that a ministerial order alone is sufficient for their establishment or degazettement.

International Activities Three biosphere reserves have been established since June 1976 under the Unesco Man and the Biosphere programme. Two national parks, Khao Yai and Tarutao, are ASEAN Heritage sites, in recognition of their regional importance. Thailand ratified the Convention Concerning the Protection of the World Cultural and Natural Heritage (World Heritage Convention) in 1987 and three sites were nominated for inscription on the World Heritage List during 1991.

Administration and Management In the early 1960s, the Royal Forest Department devolved its responsibility for protected areas upon the National Parks and the Wildlife Sections of its Silviculture Division. The National Parks Section was upgraded to "Sub-Division" status in 1965 and, with an increasing workload, to divisional level in 1972. For similar reasons, the Wildlife Conservation Division emerged out of the former Section in 1975. Neither the National Park Division nor the Wildlife Conservation Division has regional offices and both operate out of a common headquarters in Bangkok.

The National Park Division is headed by a Director and is controlled by the National Park Committee which includes representatives from government departments, the Office of National Environment Board, Kasetsart University, the Tourist Authority of Thailand and others. The Division comprises seven sub-divisions, viz. administration, technical, national park management, forest park management, extension, planning, and construction and maintenance (Kasetsart University, 1987). A management planning section is to be established for the preparation of management plans (Anon., 1986). The Director has considerable autonomy within the Royal Forest Department and in 1986 the Division had a staff of 128 graduates, 87 officials with technical school certificates and 539 persons without formal qualifications, the majority of whom, however, have received law-enforcement training within the Division. In addition, more than 3,000 daily-paid workers are employed. The number of staff totalled 4,500 during 1990, comprising 250 with professional qualifications, 700 forest guards and the remainder being park assistants and manual workers. A number of training programmes have been organised and are planned for park personnel, covering topics such as management skills, law and administration (Wongpakdee, 1990). The Division's budget nearly doubled from 70 million baht (US$ 2.5 million) in 1982 to 125 million baht (US$ 4.7 million) in 1986, with some 57% being allocated to staff salaries (Kasetsart

University, 1987). The 1990 budget was 165 million baht (US$ 6.6 million), increasing to 286 million baht (US$ 11.44 million) in 1991. Some 20 million baht (US$ 800,000) was generated by entrance fees (Wongpakdee, 1990).

The Wild Animals Reservation and Protection Committee, which includes members from the Ministry of Agriculture and Cooperatives, the Royal Forest Department, Land Department, Office of the National Environment Board, Budget Bureau, Dusit Zoo, government departments of animal husbandry, foreign trade, customs, mineral resources and others, acts as an advisory and decision-making body for the Wildlife Conservation Division. There are nine sub-divisions: extension, technical, wildlife sanctuaries, law enforcement, administration, planning, non-hunting management, propagation and foreign affairs. Professional staff includes 85 graduates, 192 officials with diplomas and 482 guards, of whom all have received in-service law-enforcement training. A substantial but indeterminate number of daily-paid labourers are also employed. The divisional budget in 1986 stood at 116.2 million baht (US$ 4.4 million), of which 42% was allocated to wildlife sanctuaries and 24% to salaries. During 1986 the budget allocation for non-hunting areas, on a per unit area basis, was approximately twice that of wildlife sanctuaries. Recent budget figures for the Wildlife Conservation Division are not available, although they are said to be increasing in line with allocations to the National Parks Division; some 20 million baht (US$ 800,000) was derived from entrance fees during 1990 (Wongpakdee, 1990).

Since 1975, forest reserves have been the responsibility of the Royal Forest Department's National Forest Land Management Division. This has seven sub-divisions, viz. forest land development, forest land use, survey and planning, land use cooperation, construction, forest improvement and administration. The Division has over 500 staff, supported by some 350 regional forestry officials. The 1986 budget allocation for reserved forests was 50 million baht (US$ 1.9 million), which is approximately twice that disbursed annually between 1977 and 1981 (Kasetsart University, 1987).

Protected area management is concentrated on site demarcation, patrols and law enforcement, occasionally supported by local police and regional forestry officers. All existing parks have permanent headquarters, although the provision of guard stations, tourist accommodation and staff housing is generally inadequate. The supply of staff vehicles, weapons, radios, as well as roads, trails, telephones, electricity and water supply tends not to be sufficient to meet demand. These deficiencies hamper the implementation and development of education, interpretation and research activities, although a number of parks are used by visiting scientists and non-governmental nature conservation organisations. The Tourism Authority of Thailand has recently committed some 5 million baht (US$ 190,000) to assist the National Parks Division in developing the recreation and tourism infrastructure of selected parks. Similar, but rather more severe, deficiencies exist in the wildlife sanctuaries. Nearly all sanctuaries have some research facilities, although these tend not to be comprehensive. Despite these failings, the protected area system remains relatively intact and the sites often stand in sharp contrast to their developed surroundings (Kasetsart University, 1987).

Systems Reviews The country divides naturally into six regions. The Northern Highlands extend from the borders with Myanmar and Laos south to about 18° latitude. They mainly comprise ridges oriented north-east to south-east, reaching an elevation of between 1,500m and 2,000m, and separated by wide valleys at between 300m and 500m. Originally, the mountains above 1,000m were clad in evergreen montane rain forest, with mixed deciduous monsoon and dry dipterocarp savanna forests on their flanks. The valleys, however, have long been wholly cultivated. This region suffers from the steady southward push of hill tribes such as the Hmong and Yao, which cultivate upland rice and, at higher elevations, the opium poppy. Undisturbed forest is now restricted to a few scattered patches in remote areas.

The Korat Plateau covers the north-east of Thailand. It forms a shallow depression at 100m and 200m, rimmed by the Petchabun Range in the west, and the Dangrek Range in the south. These reach 500m to 1,400m and meet in the highlands of Khao Yai National Park. The plateau is now largely devoid of forest, but extensive areas still persist on the ranges. Dry monsoon forests on the lower slopes grade into evergreen rain forest on the hills and finally into pine woodlands on the ridge tops.

The Central Plain of the Chao Phraya River is now almost entirely under intensive rice cultivation and its original swamp and monsoon forest has completely disappeared.

The South-East Uplands are an extension of the Cardamom Mountains from across the Cambodian border. Rainfall approaches 5000mm in some areas. Small remnants of the once prevalent tropical rain forest still survive in protected areas.

The Tenasserim Hills extend south from about 18°N in the Northern Highlands, along the Myanmar border to the Kra Isthmus, at about 10°N, rising steeply to about 1,000m. Since the Thai side of the Tenasserim lies in the rain shadow of higher hills on the Burmese side, it is relatively dry, but semi-evergreen rain forest persists of higher elevations along the border. The upper flanks are often precipitous, with bare rock. The slopes, once clothed in deciduous monsoon forest containing some teak and much *Shorea* spp., are now deforested and covered with bamboo and grassland.

The Southern Peninsula extends to the Malaysian border from a line joining Chumphon to Ranong at 10°N. It is an area of heavy rainfall and was originally covered in rain forest. However, most forest in the lowlands has been lost to agriculture. Extensive tracts persist only on the hills, but during the last decade even these have come under assault, principally from rubber plantations, which have often been established with international aid (Collins *et al.*, 1991).

The great variation in topography and climate has led to the development of a complex mosaic of forest types in which the drier, more open deciduous formations in the seasonal areas give way, with increasing rainfall, to variety of semi-evergreen and evergreen facies, including evergreen forest. Mangrove forests also occur in saline, silt-rich coastal waters. In the mid-1940s some 70-80% of Thailand's land mass supported closed forest. Despite state ownership of all forests in the form of national parks, wildlife sanctuaries and forest reserves, more recent estimates indicate a severe decline in cover to 33% in 1978 (FAO, 1981), 30% in 1982 (Round, 1985), and 26% in 1990 (Collins *et al.*, 1990). It is now officially recognised that total open and closed canopy forest cover may have fallen to 127,940 sq. km (25 %) and that much of this remaining forest may have been subject to considerable disturbance (Collins *et al.*, 1990). The proportion of different forest types, according to Round (1988), is 43.6% evergreen, 21.8% mixed deciduous, 31.4% dry dipterocarp, 1.4% coniferous and 1.8% mangrove. Regional variation in forest cover is considerable, ranging from 59.9% forest cover in the northern region to 17.2% cover in the north-eastern region. The 1973-1982 rate of deforestation also shows marked regional variation, with a 10.8% reduction in total forest cover in the peninsular and 48.9% in the less prosperous north-eastern region (Round, 1985). The aggregate figure of forest loss during 1985-1988 is estimated at 2,354 sq. km annually (FAO, 1988). Permanent agricultural encroachment and swidden agriculture are the principal causes of deforestation, in addition to both previously government sanctioned and illegal logging, the widespread practice of annual burning of forest undergrowth, and developments such as hydroelectric projects and highway construction. Despite a cabinet directive in 1981 that 50% of Thailand's land area should be forested, relatively undisturbed protected areas are increasingly isolated, thus possibly contributing to a decline in diversity. A number of protected areas have been adversely affected by development activities, such as hydroelectric projects and highway construction, whilst mineral extraction, resettlement programmes and recreation increase the pressure on wildlife and forests (Jintanugool *et al.*, 1982; Round, 1985). The 1989 logging ban has relieved some pressure from protected areas by nullifying logging concessions, but many of the fundamental causes of deforestation, such as rural poverty and illegal logging, persist. There is also concern that the ban will have an adverse effect on the poorer countries in the region which will be enticed to increase their own rates of felling to satisfy the demands of Thai sawmills (Round, 1989).

Thailand supports an extremely diverse fauna and flora. Situated in the Indo-Chinese peninsula of the

Oriental region, the country has been described as a "zoogeographic cross roads". For example, the avifauna comprises Sino-Himalayan, Indo-burmese, Indo-Chinese and Sundaic elements, to which may be added large numbers of migrant visitors from the Palaearctic region (Round, 1988). Current knowledge of the flora and fauna is not complete, but it estimated that some 20,000 to 25,000 species of vascular plants are present, including 10,000 to 15,000 flowering species. This includes more than 500 tree species and 1,000 orchid species. Approximately 891 bird species have been recorded, of which roughly 638 breed or formerly bred within the country. In a review of the status and conservation of forest birds, Round (1988) estimates that at least 521 species are present in existing national parks and wildlife sanctuaries, although 106 land birds, and water birds associated with forests are considered threatened and six species are considered to be extinct. Mammals number about 265 species (Lekagul and McNeely, 1977), fish approximately 1,450, reptiles at least 300 and amphibians at least 100 (Nuthaphand, 1979; Taylor, 1962, 1963, 1965).

A national conservation plan for Thailand was compiled by IUCN at the request of the National Environment Board and completed in 1979. The guidelines suggested in the plan were included in the fifth Five-Year National Social and Economic Development Plan (1982-1986), in particular the need to expand the protected areas system (IUCN 1978; Kasetsart University, 1987). The national park system was expanded from 16 sites (9,357 sq. km) in 1979 to 45 sites (24,222 sq. km) in 1985. By December 1990 some 63 national parks had been established, including 14 with a marine component, covering 33,687 sq. km (6.6% of total land area). A further 45 are proposed, to cover another 22,617 sq. km. At the same time there were 32 wildlife sanctuaries, covering 24,950 sq. km (4.9% of total land area), with a further six proposed to cover an additional 2,196 sq. km.

The proposed and existing national parks and wildlife sanctuaries will cover 16.3% of the total area of the country (Wongpakdee, 1990). The national forest reserve network covers 232,393 sq. km, but this total includes many national parks and wildlife sanctuaries, as well as 1,218 forest reserves (RFD, 1989).

In 1982 UNDP/FAO presented a report (FAO, 1981) which identified the strengths and weaknesses of the protected area system. In 1984 the first protected area management plan was compiled with assistance from WWF and IUCN. The success of this has prompted the preparation and implementation of management plans for a further 23 protected areas as an integral part of the Sixth Five-Year Plan (1987-1991), funded by US$ 6 million from the government (Anon., 1986). An additional component of the project was the establishment of a programme of rural development for conservation amongst villages surrounding Khao Yai National Park. This channels benefits from the park to local villagers thus helping to stem poaching and encroachment; a similar scheme based on wildlife

farming operates at Phu Khiew Wildlife Sanctuary (Kasetsart University, 1987). In 1986 US-AID funded an assessment of protected areas in Thailand (Kasetsart University, 1987). The report brings together a disparate body of information on protected areas and made recommendations concerning budget, staff numbers, training and services, law enforcement, tourism, recreation and education, the combination of the National Park and Wildlife Conservation divisions and their promotion to departmental level, formulation of explicit policy statements, establishment of regional offices, and amendment of extant legislation to reflect the role of protected areas in socio-economic development.

The leading national non-governmental organisation, since the demise of the Association for the Conservation of Wildlife, is Wildlife Fund Thailand, an affiliate body of WWF and member of IUCN. Amongst the Fund's major projects is a rural development scheme at Baan Saap Tai, ecological surveys in Thung Yai and Huai Kha Khaeng, and the operation of a mobile conservation unit. The Conservation Data Centre at Mahidol University holds information on flora and fauna both within and outside protected areas and manages a computerised conservation database.

The major threat to protected areas is poaching, and habitat loss due to agricultural encroachment. For this reason, parks and sanctuaries tend to be on higher ground that is less favourable for agriculture and lowland forest is not well represented in the protected area system (Round, 1985). Although partly stemming from the lack of a systematic protected area acquisition policy, measures to tackle encroachment were made in 1975. The Cabinet directed that reserved forests that have been heavily degraded by encroachment and settlement be developed into "forest villages". This gives families the right to remain on the land indefinitely which hopefully will prevent further forest destruction. Approximately 474,000 families, occupying some 770,000ha, have officially taken part in the programme (Kasetsart University, 1987). Similarly, a number of projects, some with Royal patronage, are intended to encourage hill tribes in the north to abandon swidden agriculture which is a major cause of deforestation and a serious threat to parks such as Doi Inthanon.

Most national parks and wildlife sanctuaries were originally designated as national forest reserves, with an emphasis on production, and with a tendency for villages to be established in and around them. Logging concessions were frequently granted. Subsequently, encroachment has continued and considerable areas are lost from protected areas each year. An additional problem has been the continuing validity of logging concessions in areas that were designated as national parks and wildlife sanctuaries. Until the 1989 national logging ban it had been legally possible for logging companies to continue their operations within protected areas, under Section 30 of the National Park Act (Wongpakdee, 1990).

A systematic problem across the country is the poor or even absent marking of protected area boundaries, making patrolling and law enforcement difficult. Budget allocations are generally inadequate, due to the rapid expansion of the protected areas systems; a delay of three to five years may pass from the date of gazettement to the provision of a management budget. There is also conflict with other land uses, promoted by government agencies, such as road construction, military facilities etc., with protected areas being opened to further encroachment along newly-constructed access roads. Only a limited number of sites have management plans, and there is a lack of both skilled personnel and funding to address this omission. Similarly, there are insufficient staff and funding to carry out research and resource inventories within protected areas which hampers good management (Wongpakdee, 1990).

An economic analysis of the costs and benefits of protected areas has been conducted, with case studies focusing on Khao Yai National Park, Khao Soi Dao Wildlife Sanctuary and Thale Noi Non-hunting Area (Dixon and Sherman, 1990). Nature conservation in Thailand, with an emphasis on the protected areas systems, is discussed by Arbhabhirama *et al.* (1988).

Addresses

Royal Forest Department, (Director-General), Phaholyothin Road, Jatujak, Bangkok 10900

National Parks Division, (Director), Royal Forest Department, Phaholyothin Road, Jatujak, Bangkok 10900 (Tel: 2 579 4842/0529; FAX: 2 579 8532/2791)

Wildlife Conservation Division, (Director), Royal Forest Department, Phaholyothin Road, Jatujak, Bangkok 10900

Conservation Data Centre, Mahidol University, Rama 6 Road, Bangkok 10400 (Tel: 246 1358-74 Ext. 431)

Wildlife Fund Thailand, 251/88-90 Phaholyothin Road, Bang Khen, Bangkok 10220 (Tel: 521 3435, 552 2111/2790; FAX: 552 6083)

References

Arbhabhirama, A., Phantumvanit, D., Elkington, J. and Ingkasuwan, P. (1988). *Thailand Natural Resources Profile*. Oxford University Press, Singapore. Pp. 179-236.

Anon. (1983). *ASEAN heritage parks and reserves*. The ASEAN Experts Group On The Environment and The United Nations Environment Programme. Bangkok. 94 pp.

Anon. (1986). Thailand : park planning unit. *CNPPA Members Newsletter*. IUCN, Gland, Switzerland. Pp. 35.

Collins, N.M., Sayer, J.A. and Whitmore, T.C. (Eds) (1991). *The conservation atlas of tropical forests: Asia and the Pacific*. Macmillan Press Ltd, London. 256 pp.

Dixon, J.A. and Sherman, P.B. (1990). *Economics of protected areas: a new look at benefits and costs*. Island Press, Washington, DC. 234 pp.

Kasetsart University (1987). *Assessment of national parks, wildlife sanctuaries and other preserves in Thailand. Final Report*. Kasetsart University, Royal Forest Department, Office of the National Environment Board and US Agency for International Development. 130 pp.

FAO (1981). *National parks and wildlife management: Thailand. A review of the nature conservation programmes and policies of the Royal Forest Department*. THA 77/003. Bangkok. 104 pp.

FAO (1988). *An interim report on the state of forest resources in the developing countries*. FAO, Rome, Italy. 18 pp.

IUCN (1978). *National Conservation Plan for Thailand 1980-1984. IUCN/UNEP/FAO, Morges, Switzerland. 131 pp.*

IUCN (1979). *Conservation for Thailand – policy guidelines*. Vol. 2. Appendices. IUCN/UNEP, Morges, Switzerland. 139 pp.

Jintanugool, J., Eudey, A.A., Brockelman, W.A. (1982). Species conservation priorities in Thailand. Species conservation priorities in the tropical forests of southeast Asia. *Occasional paper* No. 1. IUCN Species Survival Commission (SSC). 58 pp.

Lekagul, B. and McNeely, J.A. (1977). *Mammals of Thailand*. Association for the Conservation of Nature, Bangkok.

Nutaphand, W. (1979). The turtles of Thailand. Siamfram Zoological Garden, Bangkok. (Unseen)

Round, P.D. (1985). *Status and conservation of resident forest birds in Thailand*. Association for the Conservation of Wildlife. 143 pp.

Round, P.D. (1989). Implications of the logging ban for the conservation of Thai wildlife. *WWF Reports* October/November. 4 pp.

Royal Forest Department (1989). *Forestry statistics of Thailand 1989*. Forest Statistics Sub-Division, Planning Division, Bangkok. 79 pp.

Taylor, E.H. (1962). The amphibian fauna of Thailand. *University of Kansas Scientific Bulletin*: 45(9). (Unseen)

Taylor, E.H. (1963). The lizard fauna of Thailand. *University of Kansas Scientific Bulletin*: 44(19).

Taylor, E.H. (1965). The serpents of Thailand and adjacent waters. *University of Kansas Scientific Bulletin* 45(9). (Unseen)

Vejaboosakorn, S. (1985). The development of a protected area system for Thailand in terms of representative coverage of ecotypes. In: Thorsell, J.W. (Ed.), *Conserving Asia's natural heritage*. IUCN, Gland, Switzerland and Cambridge, UK. 237 pp.

Wongpakdee, S. (1990). Thailand national parks and wildlife sanctuaries. Paper presented at the Regional Expert Consultation on Management of Protected Areas in the Asia-Pacific Region. FAO Regional Office for Asia and the Pacific, Bangkok, 10-14 December. 15 pp.

ANNEX
Definition of protected area designations, as legislated, together with authorities responsible for their administration

Title: National Parks Act

Date: 3 October 1961 (B.E. 2504)

Brief description: Act making provision for the establishment and management of national parks

Administrative authority: Ministry of Agriculture and Cooperatives (Director, National Park Division, Royal Forest Department)

Designations:

National park Any land or natural feature which is of interest to be maintained with a view to preserving it for the benefit of public education and pleasure, with the provision that such land in not owned or legally possessed by any person other than a public body.

Prohibited activities are defined in Chapter 3 of the Act to provide comprehensive habitat and wildlife protection, but with provision for recreation.

Source: Translated original legislation

Title: Wild Animals Reservation and Protection Act

Date: 26 December 1960 (B.E. 2503)

Brief description: A law for the reservation and protection of wild animals

Administrative authority: Ministry of Agriculture and Cooperatives (Director, Wildlife Conservation Division, Royal Forest Department)

Designations:

Wild animal preserved area (Wildlife sanctuary)[1] Any area deemed appropriate for the preservation of the breed of animal and on land not owned or legally possessed by any person other than a public body.

Non-hunting area Any place used for official service or public interest or place for common use of the public in which hunting of any wild animal of any kind or category is prohibited.

Monastic precincts Within the precinct of a monastery or place provided for religious observance of the public, hunting, collecting or endangering any wild animal is prohibited.

Source: Translated original legislation

[1] Wildlife sanctuary is the commonly used terminology

Title: National Forest Reserves Act

Date: 16 April 1964 (amended 1989)

Administrative authority: Minister of Agriculture and Cooperatives (Director General, Royal Forest Department)

Brief description: An act to amend the law on the protection and reservation of forest

Designation:

National reserved forest A forest designated as such under this or, preceding but repealed, acts on protection and reservation of forest (1938, 1953, 1954) for the purpose of preserving forest, timber, forest products or other natural resources, by recourse to a Ministerial Regulation Within such a forest it is forbidden to hold or possess land, clear land, burn forest, work timber, gather forest products or do any act detrimental to the nature of the national reserved forest with the exception of:

(1) working timber or gathering forest products under Section 15, utilising or dwelling under Section 16, acting under Section 17, putting to use under Section 18 or acting under Section 19 or 20;

(2) working prohibited timber or gathering prohibited forest products under the laws on forest.

NB: Amended in 1989 to remove the right of exploitation.

SUMMARY OF PROTECTED AREAS

Map[†] ref.	*National/international designation* Name of area	IUCN management category	Area (ha)	Year notified
	National Parks			
1	Ao Phangnga	II	40,000	1981
2	Chae Son	II	59,200	1988
3	Chaloem Rattanakosin (Tham Than Lot)	II	5,900	1980
4	Chat Trakan	II	54,300	1987
5	Doi Inthanon	II	48,240	1972
6	Doi Khuntan	V	25,529	1975
7	Doi Suthep-Pui	II	26,106	1981
8	Erawan	II	55,000	1975
9	Hat Chao Mai	II	23,088	1981
10	Hat Nai Yang (+ Ko Phuket reefs)	II	9,000	1981
11	Hat Nopharat Thara - Mu Ko Phi Phi	II	38,996	1983
12	Huai Huat	II	82,856	1988
13	Kaeng Krachan	IV	291,000	1981
14	Kaeng Tana	II	8,000	1981
15	Khao Chamao-Khao Wong	II	8,368	1975
16	Khao Khitchakut	II	5,870	1977
17	Khao Laem Ya - Mu Ko Samet	V	13,100	1981
18	Khao Lam Pi - Hat Thai Muang	II	7,200	1986
19	Khao Luang	II	57,000	1974
20	Khao Pu - Khao Ya	II	69,400	1982
21	Khao Sam Lan	V	4,457	1981
22	Khao Sam Roi Yot	II	9,808	1966
23	Khao Sok	II	64,552	1980
24	Khao Yai	II	216,863	1962
25	Khlong Lan	II	30,000	1982
26	Laem Son	II	31,500	1983
27	Lansang	II	10,400	1979
28	Mae Ping	II	100,300	1981
29	Mae Wong	II	89,400	1987
30	Mae Yom	II	45,475	1986
31	Mu Ko Chang	II	65,000	1982
32	Mu Ko Phetra	II	49,438	1984
33	Mu Ko Similan	II	12,800	1982
34	Mu Ko Surin	II	13,500	1981
35	Mukdahan	II	4,550	1988
36	Nam Nao	II	96,600	1972
37	Namtok Mae Surin	II	39,660	1981
38	Namtok Phlui (Khao Sabup)	II	13,450	1975
39	Pang Sida	II	84,400	1982
40	Phu Chong - Na Yoi	II	68,600	1987
41	Phu Hin Rong Kla	II	30,700	1984
42	Phu Kao - Phu Phan Kham	II	32,200	1985
43	Phu Kradung	II	34,812	1962
44	Phu Phan	II	66,470	1972
45	Phu Rua	II	12,084	1979
46	Ramkamhaeng	II	34,100	1980
47	Sai Yok	II	50,000	1980
48	Si Laana	II	140,600	1989
49	Si Nakarin	II	153,200	1981
50	Si Phangnga	II	24,608	1988
51	Si Satchanalai	II	21,320	1981
52	Tarutao	II	149,000	1974
53	Tat Ton	II	21,718	1980
54	Thaleban	II	10,168	1980

Map[†] ref.	*National/international designation* Name of area	IUCN management category	Area (ha)	Year notified
55	Thap Lan	II	224,000	1981
56	Thung Salaeng Luang	II	126,240	1972
57	Ton Krabak Yai	II	14,900	1981
58	Wiang Kosai	II	41,000	1981
	Wildlife Sanctuaries			
59	Doi Chiang Dao	IV	52,100	1978
60	Doi Luang	IV	9,705	1984
61	Doi Pha Chang	IV	57,108	1980
62	Doi Pha Muang	IV	58,311	1980
63	Huai Kha Khaeng	IV	257,464	1972
64	Huai-Sa-la	IV	38,000	1990
65	Khao Ang Ru Nai	IV	10,810	1977
66	Khao Banthat	IV	126,695	1977
67	Khao Phanom Dong Rak	IV	31,600	1978
68	Khao Pra Bang Kram	IV	18,640	1987
69	Khao Sanam Phriang	II	10,001	1985
70	Khao Soi Dao	IV	74,502	1972
71	Khlong Nakha	IV	48,000	1972
72	Khlong Phraya	IV	9,500	1980
73	Khlong Saeng	IV	115,530	1974
74	Mae Tuen	IV	117,300	1978
75	Mae Yuam Fang Khwa	IV	29,200	1986
76	Maenam Phachi	IV	48,931	1978
77	Omgoy	IV	122,400	1983
78	Phu Khieo	IV	156,000	1972
79	Phu Luang	IV	84,799	1974
80	Phu Miang-Phu Thong	IV	54,500	1977
81	Phu Wua	IV	18,650	1975
82	Phu-si-tan	IV	25,000	1990
83	Prince Chumphon Park	IV	45,400	1988
84	Salawin	IV	87,500	1978
85	Sub-langka	IV	15,500	1986
86	Thung Yai Naresuan	IV	320,000	1974
87	Ton Nga Chang	IV	18,195	1978
88	Umphang	IV	251,564	1989
89	Wang-pong	IV	14,800	1987
90	Yod Dom	IV	20,255	1977
	Non-Hunting Area			
91	Mu Ko Libong	VIII	44,749	1979
	Biosphere Reserves			
	Hauy Tak Teak Reserve	IX	4,700	1977
	Mae Sa-Kog Ma Reserve	IX	14,200	1977
	Sakaerat Environmental Research Station	IX	7,200	1976

[†]Locations of most protected areas are shown on the accompanying map.

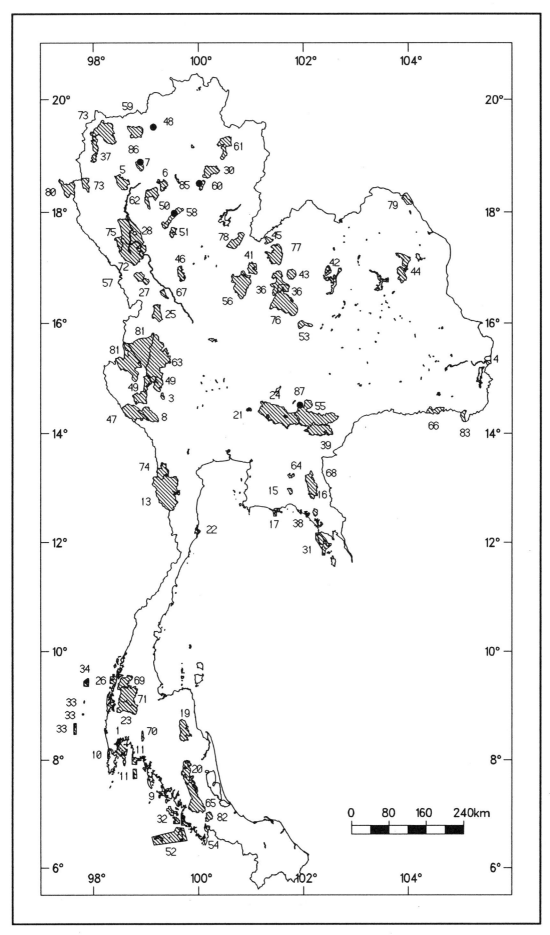

Protected Areas of Thailand

THE SOCIALIST REPUBLIC OF VIET NAM

Area 331,690 sq. km

Population 66,693,000 (1990)
Natural increase: 2.1% per annum

Economic Indicators
GNP: US$ 198 per capita
GDP: Not available

Policy and Legislation Article 5 of the Law on the Protection of Forests, promulgated on 5 September 1972, makes provision for the establishment of protected areas. The Article states "the government delimits forest preserves aimed at protecting flora and fauna, historical and cultural relics and public health, conducting scientific research or other special interests". Within forest preserves it is not permitted to fell trees (other than for management purposes) or shoot birds and other wildlife (MoF, 1985). Some activities are allowed, with the permission of either the Minister of Forests or the Council of Ministers, for example erecting buildings and collecting fuelwood or scientific specimens (CoM, 1986). In general, implementation of this law has been weak, due to socio-economic difficulties, low literacy levels and weak provisions within the law itself (Hoang and Vo, 1990).

The current protected areas system is largely based on Council of Ministers Decision No. 194/CT, promulgated on 9 August 1986. This was made on the basis of Article 5, and the later decisions No. 41/TTg of 24 January 1977, No. 360/TTg of 7 July 1978 (which established Nam Cat Tien National Park), No. 65/HDBT of 7 April 1982, No. 85/CT of 1 March 1984 and No. 79/CT of 31 March 1986 (which established Cat Ba National Park). The Ministry of Forestry has set a target of 900,000-1,000,000ha of forest to be set aside for the purpose of protecting gene pools, historic relics and for tourism (Anon., 1984) and Decision No. 194/CT lists 73 "forbidden forests" which are categorised as national parks, nature reserves and historical and cultural reserves.

An Environment Law has been prepared with the IUCN, although no details are available at present. A Forest Law has been formulated and was due to be submitted to the National assembly by the end of 1990.

The forest estate is divided into special-use forests, protection forests and production forests. Special-use forests have been further classified into 87 protected areas, comprising 7 national parks, 49 nature reserves and 31 historical, cultural and environmental sites, covering some 9,000 sq. km. The Ministry of Forestry is reviewing boundaries with a view to extending some protected areas. Each area has a management board for protection and administration. Protection forests cover 57,000 sq. km. Management boards are to be established for controlling these areas, and in general they are to be transferred to cooperatives, households and individuals who will be obliged to implement management plans applicable to each forest type. Protection forests are further classified into forest for coastal protection (1,500 sq. km) and watershed protection (55,000 sq. km). Production forests cover 122,000 sq. km and have been allocated to state enterprises (60,000 sq. km), cooperatives and individuals (62,000 sq. km), on the basis of long term leases (Hoang and Vo, 1990).

A preliminary revised classification of protected areas has been proposed (Le Trong, 1991), as follows: national park, provincial park, natural reserve, cultural protected area, no hunting area, protected forest and biosphere reserve.

International Activities On 20 September 1987 the Government of Viet Nam became a Contracting Party to the Convention on Wetlands of International Importance Especially as Waterfowl Habitat (Ramsar Convention), and designated part of the Red River Delta for inclusion in the List of Wetlands of International Importance. Viet Nam ratified the Convention concerning the World Cultural and Natural Heritage (World Heritage Convention) on 6 October 1987, although to date no sites have been inscribed on the World Heritage list. Nam Ba Cat Tien National Park (35,000ha) has been proposed as a biosphere reserve under the Unesco Man and the Biosphere Programme (Thai van Trung, 1985).

An international agreement has been signed with Laos and Cambodia, under the auspices of the tripartite Commission for Economic and Cultural Cooperation, in order to protect kouprey *Bos sauveli* (E), sarus crane *Grus antigone* and migratory waterfowl; mutual measures may include the establishment of transfrontier reserves (Kemf, 1986c; MacKinnon, 1986; MacKinnon and Stuart, 1989). International assistance for the management of protected areas has been made available under FAO's Technical Cooperation Programme, including support for the development of Cuc Phuong National Park (G. Child, pers. comm., 1989). However, poor economic performance and a lack of aid from much of the international community considerably limits conservation and development (Agarwal, 1984; Westing and Westing, 1981).

There is hope that bilateral and multilateral development assistance may be renewed through the mechanism of the Tropical Forestry Action Plan. With technical assistance to national foresters, a strategic plan that will embrace industrial forestry, fuelwood production, ecosystem protection and institution building is being developed (Collins *et al.*, 1991). Training programmes implemented at the Centre for Natural Resources Management and Environmental Studies have been supported by IUCN and WWF. The World Bank is formulating a project profile to support the Ministry of Forestry to train conservationists, with an estimated input of US$ 2.7 million.

The state research programme 52D, implemented by the Natural Resources Management and Environmental Studies Centre, has received assistance from IUCN, WWF, International Waterfowl and Wetlands Research Bureau, International Council for Bird Preservation, International Crane Foundation and the Swedish International Development Agency, with a focus on elaborating the National Conservation Strategy, studying and formulating the Environment Law and implementing wildlife surveys.

Administration and Management The main administrative body is the Department of Basic Inventory in the State Committee for Science and Technology. This Department is responsible for submitting plans for the establishment of protected areas to the government and implementing effective coordination between different research institutes. The Ministry of Forestry is responsible for the development and management of protected areas. The Ministry submits its outline master plans for protected areas both to the government and to the Department of Basic Inventory for their approval. The Department of Forest Management and Protection in the Ministry of Forests is responsible for the management of protected forests, while national parks, nature reserves and historic and cultural reserves, including protected waterbird colonies, are managed by the forestry officers of the approximately 360 local People's Committees (Scott, 1989).

The Forestry Inventory and Planning Institute has a particular responsibility for surveying protected areas and preparing management plans for them. By 1989 management plans had been prepared for Cuc Phuong and Cat Ba national parks and a plan was being prepared for Nam Bai Cat Tien (Collins *et al.*, 1991) as well several other sites (Hoang and Vo, 1991). However, even basic resource inventories are generally lacking for most protected areas (Le Trong, 1991).

The Ministry of Water Resources and the Ministry of Agriculture are involved in the establishment of protected wetland areas. The Division of Agricultural Water Supply in the Ministry of Water Resources plays an important role in the decision-making process, while the Ministry of Agriculture is responsibility for resolving conflicts between agricultural development and wetland conservation (Scott, 1989).

A definition of the responsibilities of the Ministry of Forests and the People's Committees is given in Decision No. 194/CT. In particular, the Ministry of Forestry is responsible for coordinating State Committees, Ministries and People's Committees in establishing, planning and protecting forbidden forests, as well as defining boundaries both on maps and on the ground. People's Committees that have forbidden forests

within their area are responsible for providing information on the regulations to local residents, as well as organising law enforcement patrols. The Decision also sets out a five-year work plan (1986-1990) for the Ministry of Forests, to include natural resource surveys, provision of management plans, establishing rules for management of forbidden forests and investigating the potential of exploiting some sites for commercial or tourist uses.

Management of reserves, undertaken by forestry personnel, appears to be generally inadequate, with insufficient staff and irregular budget allocations (Trung, 1985). Despite some 10,000 forestry police, even priority areas are not adequately protected (IUCN, 1985), although there are a number of notable exceptions such as Cuc Phuong National Park (MacKinnon and MacKinnon, 1986). A number of army units have been actively and successfully involved in forest protection, as well as contributing to establishment of plantations.

Training has been implemented although to a limited degree. The Centre for Natural Resources Management and Environmental Studies has organised two six-month duration post-graduate training course for 80 officers who have since been deployed to national parks and nature reserves.

Systems Reviews Viet Nam extends from 8°30'N to 23°30'N along the south-eastern margin of Mainland Southeast Asia. Three-quarters of the country is hilly, with peaks rising to more than 3,000m in the north-west, but grading into rolling dissected plateaux in the south. The Annamite Mountain chain forms the natural boundary between Viet Nam, Laos and Cambodia. Land suitable for agriculture covers approximately 100,000 sq. km and is mostly situated in the larger fertile plains of the Nam Bo and Bac Bo, which include the Mekong and Red River deltas respectively (Vu Tu Lap, 1979). The climate varies from humid tropical in the southern lowlands to temperate conditions in the northern highlands. Mean annual sea temperatures vary accordingly from 27°C in the south to 21°C in the extreme north. The approximate mean annual rainfall is 2000mm but this increases on the narrow central mountainous region to 3000mm, sufficient to support tropical rain forest. There are three monsoon seasons, namely the north-east winter monsoon, and the south-east and western summer monsoons. Destructive typhoons sometimes develop over the East Sea during hot weather (Scott, 1989).

The original vegetation of Viet Nam almost entirely comprised tropical forest, two-thirds of which was dry evergreen and semi-evergreen forest (IUCN, 1985; MacKinnon and MacKinnon, 1986). By 1943 forest cover had declined to around 43% of total land area, with extensive areas cleared in coastal regions and in the agriculturally valuable flood plains of the Mekong and Red rivers.

Forest cover has declined even more rapidly since hostilities ceased (Kemf, 1986b) principally due to agricultural clearance, forest fires, fuelwood and timber gathering and urban expansion (IUCN, 1985; Kemf, 1988; Vo Quy, 1985), migration into the uplands and poorly managed forest exploitation (Le Trong, 1991). Deforestation is particularly severe in the midlands where some 130,000 sq. km has been cleared leading to siltation, floods, drought and heavy losses of topsoil.

An assessment of forest cover was made between 1973 and 1976 with the aid of Landsat satellite imagery (FAO/UNEP, 1981). Estimated total broad-leaved closed forest cover in 1980 was 74,000 sq. km (22% of total land area), of which only 15,000 sq. km (4%) were in an undisturbed, natural state. Of the total cover, 36,700 sq. km (11%) were considered unsuitable for logging but potentially at risk from agricultural encroachment. Legally protected forest covered only 5,600 sq. km (2%). Figures published by FAO (1987) suggest that earlier estimates may have been exaggerated and that closed forest cover in 1980 was only 61,650 sq. km (19%). The same report estimated that closed forest covered 48,620 sq. km (15%) in 1985. Projections for 1990 suggest closed forest may have fallen to 34,060 sq. km (10%), with just 3,000 sq. km (1%) in an undisturbed state (FAO, 1987). Unpublished information from the Ministry of Forestry in 1989 provides another set of statistics. Based on interpretation of 1987 Landsat data, it indicates that 87,254 sq. km (26%) of natural forest remain, 79,054 sq.km (24%) of which are closed broad-leaved forest. Some 189,000 sq. km (57%) are classified as forest land and 24,200 sq. km (73%) of natural forest has been allocated as protection forest. However, these latter statistics are seen as unduly optimistic (Collins *et al.*, 1991).

Extensive mangrove forest and associated brackish water forests of *Melaleuca* occur in the Mekong Delta. Small areas are also found in the Red River Delta and along the northern coast close to the Chinese border. The Red River mangroves, once extensive, have now been almost entirely cleared for agriculture, fisheries and forestry (Collins *et al.*, 1991).

The period from 1945 to 1975 witnessed almost uninterrupted warfare, with severe damage to natural resources. An estimated 22,000 sq. km of forest and farmland were destroyed, mainly in the south of the country, by intensive bombing, tactical spraying of 72 million litres of herbicides (including 44 million litres of "Agent Orange"), mechanical forest clearance, napalming of flammable *Melaleuca* forests and direct attacks on potentially useful wildlife such as elephants (Agarwal, 1984; Kemf, 1986a; Vo Quy, 1985). The most severely damaged ecosystem was the mangrove forests of the Mekong Delta, of which an estimated 40% (12,400 sq. km) were destroyed (Kemf, 1985). A massive mangrove replanting programme was started after the war, which, after initial setbacks, has now led to the successful re-establishment of thousands of hectares, providing fuelwood, fish and prawns, as well

as providing wildlife habitat. However, many areas continue to remain barren (Collins *et al.*, 1991).

The legacy of war continues to hamper development and conservation efforts, with unexploded munitions and discarded military equipment found throughout the country. Large areas of secondary grass *Imperata cylindrica* and *Pennisetum polystachyum* have replaced forests and some 20 to 25 million bomb craters hamper agriculture and irrigation (IUCN, 1985). Persistent dioxin impurities in "Agent Orange" herbicide has contaminated soil, entered food chains and is a suspected carcinogen (Agarwal, 1984; Kemf, 1986b). Furthermore, this contamination appears to have blunted reafforestation efforts with some areas remaining barren.

Between 100,000ha to 150,000ha of forest are replanted annually but due to fires, pests and illegal felling only about one-third survives (IUCN, 1985; Kemf, 1985b). In subsequent years the survival rate has steadily increased, thanks to intensified efforts by forestry officials, and in some areas between 85% and 100% of saplings survive (E. Kemf. pers. comm., 1987). During 1987 the area replanted totalled 160,000ha, and annual targets of up to 300,000ha have been identified (Kemf, 1988). The Ministry of Education has made tree planting part of the school curriculum and in 1985 and 1986 students planted some 52 million trees. It is estimated that full recovery of mangroves may take more than 50 years (Kemf, 1986a), and large areas, for example 20% of the mangroves on the Camua Peninsula, remain barren. Rear mangroves of *Melaleuca* have also been slow to recover from defoliation and napalming because invasive grasses rapidly colonised the cleared area. However, areas of the Mekong Delta are already in an advanced state of recovery with a sustainable supply of fuelwood being produced at Rung Sat, south-east of Ho Chi Minh City (Kemf, 1988). A popular summary of the rehabilitation of the environment in Viet Nam is given in Kemf (1988).

In recognition of the severe environmental problems facing the country, the Programme for the Rational Utilisation of Natural Resources and Environmental Protection (Committee 52 02) was established in 1981 to assemble researchers from leading education and research institutions to review the state of the environment, identify issues and find appropriate solutions. In 1984, in collaboration with IUCN, a National Conservation Strategy was drafted by the Committee (IUCN, 1985). The major environmental threats were identified and it was recommended that priority be given to: reducing the population growth rate to zero; increasing the rate of reafforestation to 1,000,000ha per annum, with a final target of covering 50% of the land mass; and establishing a National Board of Environmental Coordination at ministerial level, with wide cross-sectoral powers, to formulate and enforce new environmental legislation and regulations. Also recommended is the establishment of a parks and protected forests system, a model national park at Cuc Phuong and protected marine areas (IUCN, 1985). A

National Board for the Conservation of Nature was established in 1986, thus endorsing the strategy at the highest level (Kemf, 1986a), although the Council of Ministers has not formally adopted the Strategy (Collins *et al.*, 1991). Several important decisions have been made on the basis of the Strategy, for example, the establishment of the Centre for Natural Resources Management and Environmental Studies, participation in the Unesco MAB Programme and the establishment of 87 protected areas. A revised National Conservation Strategy is due to be prepared in 1995 (Hoang and Vo, 1991). As a continuation and expansion of the programme for research, in 1986 the old Committee (52 02) was replaced by a new Committee (52D) to administer the national programme for the period 1986-1990, with an enlarged scope and increased budget. The programme for this period paid much attention to research for establishing nature reserves and national parks to protect viable examples of all main natural ecosystems, supplemented where necessary with *ex-situ* conservation (Le Trong, 1991).

The government started to establish nature reserves in 1962, and the system grew rapidly during the 1980s. However, a lack of staff, resources and management experience prevents the system fulfilling its maximum conservation potential. Some protected areas cover substantial tracts of land, but many are small due to the fragmented condition of the remaining forest. Nevertheless, forests retained for watershed protection are important for wildlife and as corridors between reserves and isolated forest patches. It is highly desirable that a forest corridor be created down the length of the Annamite Mountains, since this is an important habitat for elephant and other large mammals. The extensive wetlands in the Mekong Delta are also of critical importance. Forest cover in this area is vital to ensure the proper flow of water through the Delta's many channels, for protecting banks, fish and prawn nurseries and providing refuges for water birds. The conservation of wetlands in the Mekong Delta and in other parts of the country is discussed in detail in Scott (1989).

One of the major difficulties in the management of protected areas is the presence of settlements within reserves. These populations are expected to increase, with a concomitant increase in the level of shifting agriculture, hunting and forest exploitation. Some resettlement programmes have been carried out, although these are hampered by a lack of resources and suitable land (Hoang and Vo, 1990).

Addresses

Centre for Natural Resources Management and Environmental Studies (CRES), University of Hanoi, 19 Le Thanh Tong, Hanoi

Forestry Inventory and Planning Institute; Ministry of Agriculture; Ministry of Forestry; Ministry of Water Resources (addresses not known)

References

Agarwal, A. (1984). Vietnam after the storm. *New Scientist* 1409: 10-14.

Anon. (1984). Vietnam. *Tigerpaper* 11(2): 14.

Collins, N.M., Sayer, J.A. and Whitmore, T.C. (Eds) (1991). *The conservation atlas of tropical forests: Asia and the Pacific*. Macmillan Press Ltd, London. 256 pp.

CoM (1986). Decision by the Chairman of the Council of Ministers on stipulations of forbidden forests. Council of Ministers, Ministry of Forestry, Ha Noi. 2 pp.

Le Trong, C. (1989). The current issues of natural conservation in Vietnam. *Tigerpaper*. July-September. Pp. 7-10.

Le Trong, C. (1991). Biodiversity conservation in relation to development in Vietnam. Paper presented at the XVII Pacific Science Congress, Honolulu, 27 May to 2 June. 33 pp.

FAO (1987). *Special Study on Forest and Utilization of Forest Resources in the Developing Region. Asia-Pacific Region*. Assessment of Forest Resources in Six Countries. FO:MISC/88/7. FAO, Rome. 18 pp.

FAO/UNEP (1981). *Tropical Forest Resources Assessment Project* Volume 3. FAO, Rome. 475 pp.

Hoang, H. and Vo, Q. (1990). Nature conservation in Vietnam: an overview. Paper presented at the Regional Expert Consultation on Management of Protected Areas in the Asia-Pacific Region. FAO Regional Office for Asia and the Pacific, 10-14 December, Bangkok. 40 pp.

IUCN (1985). *Vietnam: national conservation strategy*. Draft. Prepared by the Programme for Rational Utilisation of Natural Resources and Environmental Protection (Programme 52-02) with assistance from IUCN. Environmental Services Group, World Wildlife Fund-India, New Delhi, India. 71 pp.

Kemf, E. (1985). "Ecocide" in Vietnam. *WWF News* 34. 8 pp.

Kemf, E. (1986a). The re-greening of Vietnam. *WWF Monthly Report*: 85-89.

Kemf, E. (1986b). The re-greening of Vietnam. *WWF News* 41: 4-5.

Kemf, E. (1986c). Indochina unites to save the kouprey. *WWF News* 41. 8 pp.

Kemf. E. (1988). The re-greening of Vietnam. *New Scientist* 1618: 53-57.

MacKinnon, J. (1983). Report on a visit to Hanoi. Programme of Natural Resources and Environmental Research and Protection. Bogor. 8 pp.

MacKinnon, J. (1986). Bid to save the kouprey. *WWF Monthly Report*: 91-97.

MacKinnon, J. and MacKinnon, K. (1986). *Review of the protected areas system in the Indomalayan Realm*. IUCN, Cambridge, UK.

MacKinnon, J.R. and Stuart, S.N. (Eds) (1989). *The kouprey: an action plan for its conservation*. Prepared by the Species Survival Commission of

IUCN and WWF. IUCN, Gland, Switzerland and Cambridge, UK. 20 pp.

MoF (1985). Rung cam Viet Nam (Forest Preserves in Vietnam). Ministry of Forestry. 40 pp.

Pfeiffer, E.W. (1984). The conservation of nature in Viet Nam. *Environmental Conservation* 11: 217-221.

Scott, D. A. (1989). *A directory of Asian wetlands.* IUCN, Gland, Switzerland and Cambridge, UK. 1181 pp.

Trung, Thai van. (1985). The development of a protected area system in Vietnam (condensed from an original paper presented in French). In: Thorsell, J.W. (Ed.),

Conserving Asia's natural heritage. IUCN, Gland, Switzerland and Cambridge, UK. 251 pp.

Westing, A.H. and Westing, C.E. (1981). Endangered species and habitats of Viet Nam. *Environmental Conservation* 8: 59-61.

Vo Quy (1985). Rare species and protection measures proposed for Vietnam. In: Thorsell, J.W. (Ed.), *Conserving Asia's natural heritage.* IUCN, Gland, Switzerland and Cambridge, UK. 251 pp.

Vu Tu Lap (1979). *Viet Nam Geographical Data.* Foreign Language Publishing House, Hanoi. (Unseen)

ANNEX
Definition of protected area designations, as legislated, together with authorities responsible for their administration

Title: Law on the Protection of Forests

Date: 5 September 1972

Brief Description: Article 5 makes provision for the establishment of protected areas.

Administrative Authority:

Designations:

Forest preserves Delimited by government and intended to protect flora and fauna, historical and cultural relics and public health, and for conducting scientific research or other special interests.

It is not permitted to fell trees (other than for management purposes) or shoot birds and other wildlife (MoF, 1985). Some activities are allowed, with the permission of either the Minister of Forests or the Council of Ministers, for example erecting buildings and collecting fuelwood or scientific specimens (CoM, 1986).

Title: Council of Ministers Decision No. 194/CT

Date: 9 August 1986

Brief Description: Makes provision for the present system of protected areas

Administrative authority: Ministry of Forestry

Designations:

Special-use forest
National park
Nature reserve
Historical, cultural and environmental reserve
Production forest
Protection forest
Coastal protection
Watershed protection

SUMMARY OF PROTECTED AREAS

Map[†] ref.	National/international designation Name of area	IUCN management category	Area (ha)	Year notified
	National Parks			
1	Ba Be	II	5,000	1977
2	Ba Vi	II	2,144	1977
3	Bach Ma Hai Van	II	40,000	1986
4	Cat Ba	II	27,700	1986
5	Con Dao	II	6,043	1982
6	Cuc Phuong	II	25,000	1962
7	Nam Bai Cat Tien	II	36,500	1978
	Nature Reserves			
8	Anh Son	IV	1,500	1986
9	Ba Mun	IV	1,800	1977
10	Bana-Nui Chua	IV	5,217	1986
11	Ben En	IV	12,000	1986
12	Binh Chan Phuoc Buu	IV	5,474	1986
13	Bu Gia Map	IV	16,000	1986
14	Bu Huong	IV	5,000	1986
15	Chiem Hoa Nahang	IV	20,000	1986
16	Chu Yang Sinh	IV	20,000	1986
17	Cu Lao Cham	IV	1,535	1986
18	Dao Ngoan Muc	IV	2,000	1986
19	Dao Phu Quoc	IV	5,000	1986
20	Duoc Nam Can	IV	4,000	1986
21	Huu Lien	IV	3,000	1986
22	Kalon Song Mao	IV	2,000	1986
23	Khu Dao Thac Ba	IV	5,000	1986
24	Kon Kai Kinh	IV	28,000	1986
25	Kong Cha Rang	IV	16,000	1986
26	Langbian Plateau	IV	4,000	1977
27	Lo Go Sa Mat	IV	10,000	1986
28	Mom Ray	IV	45,000	
29	Muong Cha	IV	182,000	1986
30	Nam Dun	IV	18,000	1986
31	Nam Lung	IV	20,000	1986
32	Ngoc Linh	IV	20,000	1986
33	Nui Ba	IV	6,000	1986
34	Nui Cam	IV	1,500	1986
35	Nui Dai Binh	IV	5,000	1986
36	Nui Hoang Lien	IV	5,000	1986
37	Nui Pia Hoac	IV	10,000	1986
38	Nui Yen Tu	IV	5,000	1986
39	Pa Co Nang kia	IV	1,000	
40	Quang Xuyen	IV	20,000	1986
41	Rung Kho Phan Rang	IV	1,000	1986
42	Sop Cop	IV	5,000	1986
43	Suoi Trai	IV	19,000	
44	Tanh Linh	IV	2,000	1986
45	Tay Bai Cat Tien	IV	10,000	
46	Thanh Thuy	IV	7,000	1986
47	Thuong Da Nhim	IV	7,000	1986
48	Thuong Tien	IV	1,500	1986
49	Tieu Tao-Easup	IV	20,000	1986
50	Trung Khanh	IV	3,000	1986
51	U Minh	IV	2,000	1986
52	Vu Quang	IV	16,000	1986

Map[†] ref.	National/international designation Name of area	IUCN management category	Area (ha)	Year notified
53	Xuan Nha	IV	60,000	1986
54	Xuan Son	IV	4,585	1986
55	Yok Don R	IV	57,500	1988
	Reserve			
56	Tram Chin Sarus Crane Reserve	IV	5,500	1986
	Historic/Cultural Sites			
57	Ban dao Son Tra	IV	4,000	1977
58	Duong Minh Chau	IV	5,000	1986
59	Nui Tam Dao	IV	19,000	1977

[†]Locations of most protected areas are shown on the accompanying map.

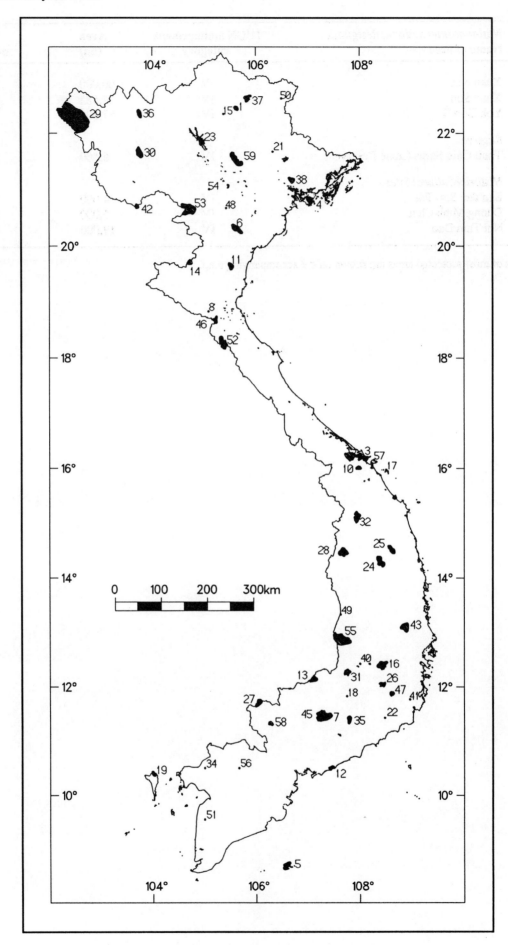

Protected Areas of Viet Nam

Oceania

COOK ISLANDS

Area The territorial seas and Exclusive Economic Zone cover nearly 2 million sq. km, whilst total land area is only about 240 sq. km (Utanga, 1989).

Population 17,463 (SREP, 1989)
Natural increase: No information

Economic Indicators
GNP: No information

Policy and Legislation The Cook Islands Act (1915) constitutes the earliest conservation legislation. Sections 356 and 357 cover the acquisition and reservation of land for public purposes, such as recreation. Section 487 provides for the establishment of native reserves to protect sites of historic or scenic interest and sources of water supply. Local Island Council Ordinances also make provision for the reservation of land for public purposes. The 1915 Act was effectively superseded by the Conservation Act (1975) which became the principal legislative instrument for the conservation of nature and natural resources, protection of historic sites and the environment, and the establishment of national parks and other protected areas (SPREP, 1985b).

The 1975 Act was largely unused and was repealed and replaced in April 1987 by the 1986-87 Conservation Act. The 1986-87 Act is essentially similar to the 1975 Act, but is equally binding on both government and the public. The principal difference is that the Conservation Service is established as an independent corporation, whereas previously it was within the Ministry of Internal Affairs and Conservation. The Act, which applies in full only to Rarotonga and Aitutaki, provides for the post of Director of Conservation with wide-ranging powers to protect, conserve, manage and control parks, wildlife, forests, water catchments and resources. Under the Act (Sections 27 and 28), any land, lagoon, reef or island, or portion of the seabed with its superjacent waters, can be declared a national park or reserve. The Director is obliged to prepare a management plan for any national park or reserve declared under the Act and, after approval, to implement it. National parks, as defined in Section 30, are intended for the protection, conservation and management of wildlife and natural features, the encouragement and regulation of the appropriate use, appreciation and enjoyment of the park by the public and the protection of special features, e.g. archaeological sites, water catchments and soil resources. The Act does not, however, define the purpose of reserves. There are also specific provisions for the protection of the coastal zone and Cook Island waters from unauthorised activities. Finally, the Act has provisions for the control of soil erosion, siltation, aggregate extraction, pollution and agricultural encroachment.

As the Conservation Act (1986/87) does not provide adequate protection for the Outer Islands, the Conservation Service, in liaison with each Island Council, has started preparing separate conservation plans for these islands. The Service has proposed developing legal mechanisms under which parks and reserves could be established on native freehold land. The outline concepts include land-leasing and shared-management regimes (McCormack, 1989).

The traditional system of resources management was based on the subdivision of land on high islands with boundaries running along dividing ridges to include entire valleys, alluvial plains, storm ridges on the coastal plain, the beach and lagoons fronting the valley and out to the outer edge of the reef. Such subdivisions, known as ra'hui, and invested with supernatural powers (tapu or taboo), were held by a Sub-Chief (Mataiapo or Rangatira) under a Chief, for members of the tribe. Tribal members were allocated different areas for planting, gathering and fishing. In addition, areas within the sub-division could be set aside for exclusive use of the Chief, and comparable prohibitions on removing specific natural resources, such as brilliant plumage required in ceremonies, fruits, trees, ferns or medicinal plants, animals and others. This system was curtailed when all land below mean high water mark was defined as Crown property, and in the Southern Group is largely used today for the control of trespass, and conservation of fruit trees, fruit bats, wildfowl and other game. In contrast, the traditional systems of resource conservation are more intact in the northern atolls, for example Pukapuka. Although there has been some codification of ra'hui as Island Council by-laws, a system of reserved areas for the protection of food plants, fishing grounds, sea birds, coconut crab, and turtles controlled by the community, is essentially traditional. The management of natural resources allows an usually high population to inhabit a relatively small island (Utanga, 1989).

Other legislation affecting protected areas includes the 1966 Local Government Act, which provides for the creation of Island Councils and enables them to regulate the use of any reserve or park under their control. Under the Trochus Act (1975), three fishing reserves have been established, at Aitutaki, Palmerston and Manuae, respectively, in which unlicensed diving and fishing for trochus shell is prohibited. The Territorial Sea and Exclusive Economic Zone Act, 1979, controls the management, conservation, exploitation and exploration of marine resources within the territorial sea.

International Activities The Cook Islands is not yet party to any of the international conventions or programmes that directly promote the conservation of natural areas, namely the Convention concerning the Protection of the World Cultural and Natural Heritage (World Heritage Convention), Unesco Man and the Biosphere Programme and the Convention on Wetlands of International Importance especially as Waterfowl Habitat (Ramsar Convention).

At a regional level, the Cook Islands ratified the 1976 Convention on the Conservation of Nature in the South Pacific on 24 June 1987. Known as the Apia Convention, it entered into force during 1990. The Convention is coordinated by the South Pacific Commission and represents the first attempt within the region to cooperate on environmental matters. Among other measures, it encourages the creation of protected areas to preserve indigenous flora and fauna.

The Cook Islands is also party to the South Pacific Regional Environment Programme (SPREP) and the Convention for the Protection of the Natural Resources and Environment of the South Pacific Region, 1986 (SPREP Convention) was signed on 25 November 1986 and ratified on 9 September 1987. The Convention entered into force during August 1990. Article 14 calls upon the parties to take all appropriate measures to protect rare or fragile ecosystems and threatened or endangered flora and fauna through the establishment of protected areas and the regulation of activities likely to have an adverse effect on the species, ecosystems and biological processes being protected. However, as this provision only applies to the Convention area, which by definition is open ocean, it is most likely to assist with the establishment of marine reserves and the conservation of marine species.

Other international and regional conventions concerning environmental protection to which the Cook Islands is party are reviewed by Venkatesh *et al.* (1983).

Administration and Management The 1986-87 Conservation Act is administered by the Conservation Service. The Conservation Service is run by a Council appointed by the Minister of Conservation. The Director, responsible for administration of the Service, is the Chairman of the Council (G. McCormack, pers. comm., 1989). The function of the Conservation Service is to promote the conservation of the environment for the use and enjoyment of present and future generations. Within the Conservation Act there are specific regulations concerning some aspects of the environment, but there are also provisions under which the Conservation Service can establish additional regulations (McCormack, 1989).

Systems Reviews The Cook Islands are extremely remote oceanic islands, lying approximately 3,000km north-east from the nearest major land mass, namely New Zealand, and are defined by statute as all islands lying between 08°S and 23°S and 156°W and 167°W. It is divided geographically into a Northern Group, comprising atolls, and a Southern Group, mainly comprising high volcanic islands, but also Manuae, an atoll and Takutea, which is a sand cay. The depths of the surrounding ocean, reaching 1,300-1,500m, precluded the formation of land bridges during glacial periods. The islands are also located remotely on a biological diversity gradient which diminishes westward from continental land masses, and both northward and southward away from the equator. These biogeographical dimensions have contributed to distinctive ecosystems,

which are also shaped by annual periodic climatic effects, and episodic events such as drought, inundation and cyclones, but the principal influence on biological diversity is the physical structure of the islands. The Cook Islands, however, have a long history of human habitation and consequently much of the natural environment has been modified, especially on coastal flats (Dahl, 1980a).

Vegetation varies from montane rain forest on Rarotonga (Merlin, 1985), through lowland limestone rain forest on Mauke, beach forest on atolls and reef islets, to scrub and grassland formations. Freshwater marsh is found on Mangaia, Rarotonga, Mauke, Mitiaro and Atiu, whilst tidal salt marsh is restricted to Ngatangiia Harbour, Rarotonga (Dahl, 1980a). Natural vegetation on coastal flats has been largely modified by man (Stoddart, 1975c) and lowland forest has been almost totally destroyed (Davis *et al.*, 1986). However, upland forest above 250m remains largely intact (Sykes, 1983) and the forest cover of deep valley heads and sharp ridges and peaks appears to be pristine (Philipson, 1971). Coastal vegetation on Rarotonga has been heavily modified and burning has spread to such a degree that valleys are covered by introduced grasses and weeds. Hills near the sea mostly support *Gleichenia* thickets or forest (G. McCormack, pers. comm., 1989). The mainland and reef island vegetation of Aitutaki is described by Stoddart (1975b and 1975c).

All Cook Islands feature coral formations, which are frequently fringing and lagoon reefs. Within the southern Cooks, windward and leeward atoll reefs are restricted to Manuae and Palmerston, whilst barrier reefs are found only at Aitutaki (Dahl, 1980b). UNEP/IUCN (1988) provides a brief summary of each island's reefs, and more detailed accounts for Aitutaki, Manihiki, Ngatangiia Harbour and Muri Lagoon, Pukapuka and Suwarrow Atoll National Park. Crossland (1928) and Dana (1898) gave early descriptions of the fringing reefs around Rarotonga and more recent work has been carried out by, for example, Dahl (1980b), Gauss (1982), Lewis *et al.* (1980) and Stoddart (1972; 1975a; 1975b; 1975c).

Proposals to designate protected areas in the central uplands of Rarotonga date back to the 19th century, but Suwarrow Atoll National Park is the only major gazetted protected area at present. Previously recorded as having an area of 13,468ha, a recent legal clarification has reduced this to only 160ha, comprising the land area above mean high water mark. This represents 0.07% of the terrestrial Cook Islands and an even less significant proportion of the marine area, and whereas previously algal ridge, reef flat, patch reef and open lagoon were protected, these are now excluded from the protected areas system.

One of the most recent proposals, originating in the Department of Internal Affairs and Conservation, envisaged a 1,000ha kakerori reserve specifically for the conservation of Rarotongan flycatcher *Pomarea dimidiata* (Anon., 1985; SPREP, 1985a). Difficulties in establishing agreement between customary land owners and the government have led to this proposal being dropped in favour of two smaller sites in the same area (McCormack, 1988). Proposals to gazette Takutea Island, and Rarotonga water catchment and wildlife reserves have been identified as priority actions (SPREP, 1985a).

An Action Strategy for Protected Areas in the South Pacific (SPREP, 1985a) provides a work programme to implement conservation and protected areas objectives. The principal goals of the strategy cover conservation education, conservation policy development, establishment of protected areas, protected area management and regional and international cooperation. Priority recommendations for the Cook Islands are as follows: conduct baseline survey, prepare management plan and recruit and train personnel for Suwarrow Atoll National Park; establish Kakerori Reserve on Rarotonga Island, including completion of baseline survey, preparation of management plans, and conducting a public education and awareness campaign; develop and implement a public education and awareness campaign for the establishment of Rarotonga Water Catchment and Wildlife Reserve; develop a national conservation strategy; designate Takutea Island as a wildlife sanctuary through negotiation with customary land owners and a public awareness campaign followed by preparation of management plans (SPREP, 1985a).

Despite poor cooperation between private land owners and the government (G. McCormack, pers. comm., 1989), considerable progress was made in implementing these priorities during the period 1985-89. This included the establishment of the Conservation Service; development of Suwarrow National Park; control of rats threatening Rarotongan flycatcher (kakerori) and the preparation of a concept document detailing proposals for a kakerori reserve; preparation of proposals for Takutea and Te Manga nature reserves; and preparation and implementation of species management plans and public education and awareness programmes (SPREP, 1989).

Dahl (1986) recommends as priority the establishment of a major protected area in central Rarotonga. Further recommendations include protection of endemic species in natural areas in Mangaia, Mitiaro and possibly other islands, and the establishment of coastal and marine reserves, as contributions to enhanced environmental management (Dahl, 1986). These are based largely on early, more extensive and specific recommendations which include proposals to protect higher areas, swamps, marshes and other terrestrial vegetation, endemic bird habitats, elevated limestone (makatea) regions, barrier and fringing reefs, lagoons, motus (islets), banks of major streams and historical features (Dahl, 1980a).

Rarotonga, as the capital and most densely populated island, exhibits the most pronounced development and the severest environmental problems. The traditional subsistence economy has long since been replaced with a commercial plantation export economy. The deeply incised, mountainous interior, rising to 652m, restricts agriculture to a circumferential belt some 1km wide (Johnston, 1959). Reefs, reef flats and lagoons play a leading role in protein supply in the Cook Islands (Hambuechen, 1973a). Islanders may depend on marine resources for 90% of their protein in the northern group and 60% in the southern group. In addition to fish, turtle meat, but not eggs, is consumed on Palmerston and in the Northern Cook Islands (G. McCormack, pers. comm., 1989).

Airport and hotel construction, port improvement, pollution and soil erosion have contributed to coastal degradation in Rarotonga. The reefs are considered to be in an advanced stage of degradation (Dahl, 1980b). Chemical run-off may be a significant problem (Hambuechen, 1973a; 1973b), and problems created by the increased use of pesticides are described by Hambuechen (1973b). Fish poisoning and dynamiting has been reported (Dahl, 1980b; Hambuechen, 1973a). Phosphate mining, a potentially serious environmental threat, may be investigated on Rakahanga and Manihiki (UNEP/IUCN, 1988).

A major crown-of-thorns starfish *Acanthaster planci* plague occurred on the north-west reefs of Rarotonga during 1972-73; there was a minor infestation on Aitutaki in 1973. There were no known previous infestations and there have been none since (G. McCormack, pers. comm., 1989). Other marine threats actually or potentially include offshore activities such as aggregate extraction, eutrophication and inappropriate recreational use. Fish poisoning and dynamiting has occurred and pearl oyster stocks have been depleted. Principal threats on land include agricultural encroachment into remaining forest, indiscriminate burning, deforestation and propagation of exotic pests such as rhinoceros beetle, rats and feral animals. Coastal resource management is weak due to a division of administrative responsibilities and a lack of legislated authority (CSC, 1985a and 1985b; Dahl, 1980b; SPREP, 1985b; UNEP/IUCN, 1988). There is a risk of severe economic and social disruption in the event of increased sea level rises caused by global warming of the atmosphere (Pernetta, 1988). The agricultural belt of Rarotonga could be reduced in size and altered in profile by enhanced erosion (Nunn, 1988). More catastrophically, low-lying islands, such as Suwarrow or Takutea, could be inundated and completely destroyed (Pernetta, 1988).

Protected habitats are restricted to beach and wet atoll forest (Dahl, 1980a), in addition to important seabird breeding areas on Suwarrow (SPREP, 1985b). Omissions from the protected area system include lowland rain forest on Aitutaki and Mauke, limestone forest, the montane and ridge rain forest of central Rarotonga, freshwater marsh, tidal salt marsh, permanent lake, mountain stream, closed lagoon and fringing reef (Dahl, 1980a; 1986).

Addresses

The Director, Conservation Service, PO Box 371, Tupapa, Rarotonga

References

Anon. (1985). Endangered species management needs in the Cook Islands. In: Thomas, P.E.J. (Ed.), *Report of the Third South Pacific National Parks and Reserves Conference.* Volume II. *Collected key issue and case study papers.* South Pacific Commission, Noumea, New Caledonia. Pp. 271-275.

CSC (1985a). Environmental planning programme: coastal zone management of tropical islands. *Proceedings of the workshop/planning meeting on coastal zone management of the South Pacific region, Tahiti.* Commonwealth Science Council Technical Publications Series, London. No. 180.

CSC (1985b). *Environmental planning programme: coastal zone management of tropical islands. SOPACOAST: the South Pacific coastal zone management programme.* CSC Technical Publications Series, London. No. 204.

Crossland, C. (1928). Coral reefs of Tahiti, Moorea and Rarotonga. *Journal of the Linnean Society* 36: 577-620. (Unseen)

Dahl, A.L. (1980a). Regional ecosystem survey of the South Pacific Area. *SPC/IUCN Technical Paper* 179. South Pacific Commission, Noumea, New Caledonia. 99 pp.

Dahl, A.L. (1980b). Report on marine surveys of Rarotonga and Aitutaki (November 1976). South Pacific Commission, Noumea, New Caledonia.

Dahl, A.L. (1986). *Review of the protected areas system in Oceania.* IUCN, Gland, Switzerland and Cambridge, UK. 328 pp.

Dana, J.D. (1898). *Corals and Coral Islands.* Dodd, Mead and Co., New York.

Davis, S.D., Droop, S.J.M., Gregerson, P., Henson, L., Leon, C.J., Lamlein Villa-Lobos, J., Synge, H. and Zantovska, J. (1986). *Plants in danger: what do we know?* IUCN, Gland, Switzerland and Cambridge, UK. 488 pp.

Gauss, G.A. (1982). Sea bed studies in nearshore areas of the Cook Islands. *South Pacific Marine Geology Notes, CCOP, ESCA* 2: 131-154. (Unseen)

Hambuechen, W.H. (1973a). Cook Islands. *Proceedings and Papers, Regional Symposium on Conservation of Nature – Reef and Lagoons, 1971.* South Pacific Commission, Noumea, New Caledonia. (Unseen)

Hambuechen, W.H. (1973b). Pesticides in the Cook Islands. *Proceedings and Papers, Regional Symposium on Conservation of Nature – Reef and Lagoons, 1971.* South Pacific Commission, Noumea, New Caledonia. (Unseen)

IUCN (1988). *From strategy to action: how to implement the report of the Commission on Environment and Development.* IUCN, Gland, Switzerland. 116 pp.

Johnston, W.B. (1959). The Cook Islands: land use in an island group in the south-west Pacific. *Journal of Tropical Geography* 13: 38-57.

Lewis, K.B., Utanga, A.T., Hill, P.J. and Kingan, S.G. (1980). The origin of channel-filled sands and gravels on an algal-dominated reef terrace, Rarotonga, Cook Islands. *South Pacific Geology Notes* 2: 1-23

McCormack, G. (1988). Kakerori Nature Reserve. A concept document. Cook Islands Conservation Service. Unpublished. 4 pp.

McCormack, G. (1989). Cook Islands Conservation Service: two years under the 1986/87 Act. Fourth South Pacific Conference on Nature Conservation and Protected Areas, Port Vila, Vanuatu, 4-12 September. South Pacific Commission, Noumea, New Caledonia. 5 pp.

Merlin, M.D. (1985). Woody vegetation in the upland region of Rarotonga, Cook Islands. *Pacific Science* 39: 81-99. (Unseen)

Nunn, P. (1988). Potential impacts of projected sea level rise on Pacific Island States (Cook Islands, Fiji, Kiribati, Tonga, and Western Samoa): a preliminary report. In: MEDU joint meeting of the task team on the implications of climatic change in the Mediterranean. Split, Yugoslavia, 3-7 October. Pp. 53-81.

Pernetta, J.C. (1988). Projected climate change and sea level rise: a relative impact rating for the countries of the Pacific Basin. In: MEDU joint meeting of the task team on the implications of climatic change in the Mediterranean. Split, Yugoslavia, 3-7 October. Pp. 1-11.

Philipson, W.R. (1971). Floristics of Rarotonga. *Bulletin of the Royal Society of New Zealand* 8: 49-54. (Unseen)

Sloth, B. (1988). *Nature legislation and nature conservation as part of tourism development in the island Pacific.* Pacific Regional Tourism Development Programme. Tourism Council of the South Pacific, Suva, Fiji. 82 pp.

SPREP (1985a). *Action strategy for protected areas in the South Pacific Region.* South Pacific Commission, Noumea, New Caledonia. 24 pp.

SPREP (1985b). Cook Islands. In: Thomas, P.E.J. (Ed.), *Report of the Third South Pacific National Parks and Reserves Conference.* Volume III. *Country reviews.* South Pacific Commission, Noumea, New Caledonia.

SPREP (1989). Cook Islands. Paper presented at the Fourth South Pacific Conference on Nature Conservation and Protected Areas, Port Vila, Vanuatu, 4-12 September. 5 pp.

Stoddart, D.R. (1972). Reef islands of Rarotonga, with list of vascular flora by F.R. Fosberg. *Atoll Research Bulletin* 160: 1-14.

Stoddart, D.R. (1975a). Scientific studies in the Southern Cook Islands: background and bibliography. In: Stoddart, D.R. and Gibbs, P.E. (Eds), Almost-atoll of Aitutaki: reef studies in the Cook Islands, South Pacific. *Atoll Research Bulletin* 190: 1-30.

Stoddart, D.R. (1975b). Mainland vegetation of Aitutaki. In: Stoddart, D.R. and Gibbs, P.E. (Eds), Almost-atoll of Aitutaki: reef studies in the Cook Islands, South Pacific. *Atoll Research Bulletin* 190: 117-122.

Stoddart, D.R. (1975c). Vegetation and floristics of the Aitutaki motus. In: Stoddart, D.R. and Gibbs, P.E. (Eds), Almost-atoll of Aitutaki: reef studies in the Cook Islands, South Pacific. *Atoll Research Bulletin* 190: 87-116.

Sykes, W.R. (1983). Conservation on South Pacific islands. In: Given, D. R. (Ed.), *Conservation of Plant Species and Habitats*. Nature Conservation Council, Wellington, New Zealand. Pp 37-42. (Unseen)

UNEP/IUCN (1988). *Coral reefs of the world. Volume 3: Central and Western Pacific.* UNEP Regional Seas Directories and Bibliographies. IUCN, Gland, Switzerland and Cambridge, UK/UNEP, Nairobi, Kenya. 378 pp.

Utanga, A. (1989). Customary tenure and traditional resource management in the Cook Islands. In: *SPREP Report of the workshop on customary tenure, traditional resource management and nature conservation.* South Pacific Commission, Noumea, New Caledonia. Pp. 101-105.

Venkatesh, S, Va'ai, S. and Pulea, M. (1983). An overview of environmental protection legislation in the South Pacific countries. SPREP *Topic Review* No. 13. South Pacific Commission, Noumea, New Caledonia. 63 pp.

Wilder, G.P. (1931). Flora of Rarotonga. *Bulletin of the Bernice P. Bishop Museum* 86. 113 pp. (Unseen)

ANNEX
Definitions of protected area designations, as legislated, together with authorities responsible for their administration

Title: Conservation Act 1986-87

Date: 15 April 1987

Brief description: An Act to establish a Conservation Service as a corporation and to make provision for the conservation and protection of the environment and national resources, and the establishment of national parks and reserves.

Administrative authority: Conservation Service (Director)

Designations:

National park Any land, lagoon, reef, or island, or any Cook Islands waters, or portion of the sea-bed of those waters may be proclaimed as a national park subject to the Act.

National parks, as defined in Section 30, are intended for the protection, conservation and management of wildlife and natural features, the encouragement and regulation of the appropriate use, appreciation and enjoyment of the park by the public and the protection of special features, e.g. archaeological sites, water catchments and soil resources.

Reserve Any land, lagoon, reef, or island, or any Cook Islands waters, or portion of the sea-bed of those waters may be proclaimed as a reserve subject to the Act.

Title: The Cook Islands Act

Date: 1915

Brief description: Constitutes the earliest conservation legislation. Numerous subsequent amendments.

Administrative authority: Island Councils

Designations:

Recreation reserve Established under Sections 356 and 357

Native reserve Established under Section 487 for protection of sites of historic or scenic interest and sources of water supply.

Title: Trochus Act

Date: 1975

Brief description: Provides for the establishment of fishing reserves.

Administrative authority: No information

Designations:

Fishing reserve Unlicensed diving and fishing for trochus shell is prohibited.

SUMMARY OF PROTECTED AREAS

Map ref.	*National/international designation* Name of area	IUCN management category	Area (ha)	Year notified
1	*National Park* Suwarrow Atoll	IV	160	1978

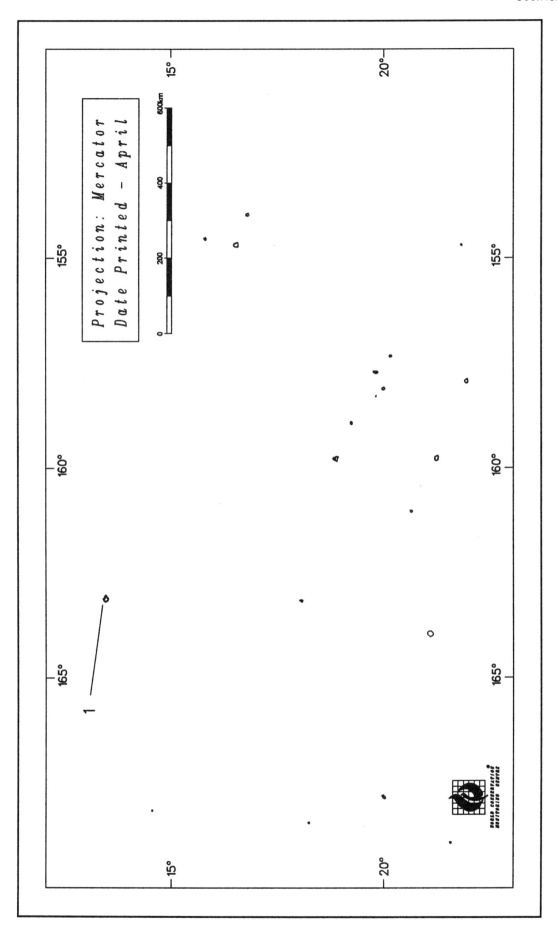

Protected Areas of the Cook Islands

FEDERATED STATES OF MICRONESIA

Area 702 sq. km

Population 86,094 (1988)
Natural increase: No information

Economic Indicators
GNP: No information

Policy and Legislation The Federated States of Micronesia consists of the states of Yap, Truk, Pohnpei and Kosrae and became an independent nation in free association with the USA on 3 November 1986, having previously been part of the United Nations Trust Territory of the Pacific Islands. Legally a sovereign nation, the Federation is loose, with each state having an elected governor and a unicameral assembly (Paxton, 1989), and in general the national government is relatively weak compared to the USA (C. Dahl, pers. comm., 1990).

Some United States Federal legislation and Trust Territory legislation applied while this country was a part of the Trust Territory. This included the Trust Territory Environment Act enacted by the former Congress of Micronesia in 1972. This provided for the establishment of a Trust Territory Environmental Protection Board. However, neither the Act, nor the work of the Board related specifically to protected areas. The Trust Territory Endangered Species Act (TTPI Public Law 6-55 of 1975) allowed for acquisition of land or water for the purpose of conserving threatened species. Other relevant Trust Territory Acts included the Fishing Law and the Land Use Planning Act, both 1972. US legislation relevant to the Trust Territories included various pollution laws and housing acts, as well as the Fish and Wildlife Coordination Act and the National Environmental Policy Act. With the termination of the Trust Territory, US Federal provisions only apply to the now very limited US actions within the Federation and do not apply to private individual or national government actions.

Neither the legislation, nor the policy on protected areas was clear while the country was a part of the Trust Territory, and no protected areas were actually established. This is largely because the United States, while administering the Territory, did not own property and hence had no direct jurisdiction to cover designation of protected areas. Virtually all land and reefs and coastal areas are in private or traditional ownership and this has continued to inhibit the establishment of protected areas (S.L. Anefal, pers. comm., 1990).

Legislation to support a coastal resource management programme in Kosrae has been drafted and submitted to the legislature. However, it will be some time before the bill is acted upon, and even if passed, there is likely to be a considerable delay before any protected areas are established. In Pohnpei, a watershed management act was passed in 1987 which gave the state government substantial authority to protect much of the island's interior and mangrove areas. However, implementation is slow and there is little awareness among the populace that the uplands are off-limits for traditional agricultural use (C. Dahl, pers. comm., 1990).

International Activities The Federated States of Micronesia is not yet party to any of the international conventions or programmes that directly promote the conservation of natural areas, namely the Convention concerning the Protection of the World Cultural and Natural Heritage (World Heritage Convention), Unesco Man and the Biosphere Programme and the Convention on Wetlands of International Importance especially as Waterfowl Habitat (Ramsar Convention).

The Convention on the Conservation of Nature in the South Pacific (1976) has been neither signed nor ratified. Known as the Apia Convention, it entered into force during 1990. The Convention is coordinated by the South Pacific Commission and represents the first attempt within the region to cooperate on environmental matters. Among other measures, it encourages the creation of protected areas to preserve indigenous flora and fauna.

The Federated States of Micronesia is party to the South Pacific Regional Environment Programme (SPREP) and the 1986 Convention for the Protection of the Natural Resources and Environment of the South Pacific Region (SPREP Convention) has been signed (9 April 1987) and ratified (29 November 1988). The Convention entered into force during August 1990. Article 14 calls upon the parties to take all appropriate measures to protect rare or fragile ecosystems and threatened or endangered flora and fauna through the establishment of protected areas and the regulation of activities likely to have an adverse effect on the species, ecosystems and biological processes being protected. However, as this provision only applies to the convention areas, which by definition is open ocean, it is most likely to assist with the establishment of marine reserves and the conservation of marine species.

Administration and Management Pohnpei State has a Division of Parks and Recreation within its Department of Lands.

Systems Reviews The Federated States of Micronesia includes most of the Caroline Islands, running west-north-west from Kosrae to Yap, and consists of volcanic and metamorphic islands and atolls. "Almost atolls" are also present, making the group one of the more typical island chains. The development and origin of the Caroline Islands is described by Scott and Rotondo (1983); general information is given in SPREP (1980). The islands' traditional economies have been undermined by a century of foreign influence, trade,

169

control and warfare. The Japanese in particular disrupted land-use controls and ownership customs, permitted greater public access to resources and undermined traditional conservation (Maragos, 1986).

The islands support a rich flora and fauna with many endemic forms. Major ecosystems include lowland, montane, cloud, riverine, swamp, mangrove and atoll forest, savanna and grassland, seagrass beds, lagoons, and extensive coral reefs (Dahl, 1980). Much of the natural vegetation has been cleared for coconut plantations (e.g. on Yap and Puluwat) or disturbed by phosphate mining (e.g. on the raised coral island of Fais). Few areas of native vegetation remain on the Truk Islands, except on the high volcanic islands of Moen, Dublon, Uman, Fefan, Udot and Tol. Although the lowland forests on the Pohnpei Islands have been much disturbed, both Kusaie and the island of Pohnpei retain upland forests (IUCN, 1986). A description of forests and conservation problems is given by Fosberg (1973). An account of the reefs and reef resources is given in UNEP/IUCN (1988).

Environmental issues, reviewed by Maragos (1986), include poorly-planned coastal development leading to degraded or destroyed mangoves, reefs and sea grass beds and consequently reduced fisheries output, and the disposal of wastewater discharges and solid waste. Crowding and land shortages in Pohnpei has spurred immigration from the outer atolls and led to additional landfilling along the shorelines and mangrove areas in the Kolonia-Sokehs area for residences. This has not yet had serious environmental consequences but the rapidly expanding population may have escalating effects in the future. A major emerging problem is the increased fishing pressure from rising populations and over-fishing of preferred species. Truk is economically severely depressed and overcrowded conditions on some islands in the lagoons are affecting health and welfare. In particular, water supply and quality are unreliable. Truk has also experienced considerable reef degradation through dynamiting.

Some degree of protection has been recommended for Gaferut (Dahl, 1980), although this has not been implemented, and turtle and bird populations are still traditionally exploited by the owners (S.L. Anefal, pers. comm., 1990). Turtle reserves have been recommended for Elato, Pikelot, West Fayu, all in Yap State, and Orulok. At present no turtle sanctuaries have been established in Yap State (S.L. Anefal, pers. comm., 1990). Dahl (1980; 1986) stresses that there is an urgent need to inventory the biomes of the Federated States of Micronesia in view of the great richness of the area and the likelihood of increasing pressure in the near future.

Yap State's Marine Resources Management Division is working on a Marine Resources Coastal Management Plan, incorporating traditional customs and laws (S.L. Anefal, pers. comm., 1990). This is likely to include recommendations for the protection of specific areas. However, a strong system of reef ownership still exists in Yap so the establishment of protected areas in Yap is likely to be on a significantly different basis than elsewhere (C. Dahl, pers. comm., 1990).

The Pohnpei Coastal Resources Management Plan (1987) recommended numerous marine areas for protection, although these recommendations have not been acted upon. There has also been some interest at community level in the establishment of a mangrove reserve in Pohnpei, with the intention of generating revenue from tourists. Protected status has been conferred on inland areas of Pohnpei Island which contains a range of tropical high island forest, important both as wildlife habitat and watershed protection (Anon, 1989). Surveys of Oroluk Atoll are planned for 1991, with the intention of recommending some form of protection. The traditional management of resources, and the pressures placed upon on it by economic and social pressures, is discussed by Yinug *et al.* (1989) with the conclusion that some form of legislated resource management is becoming necessary. A coastal resource management plan for Kosrae is being developed (Dahl, 1989) which includes proposals for the establishment of a contiguous marine reserve and terrestrial park and a resource reserve on the south coast of the island in which subsistence use would be allowed.

Addresses

Department of Human Resources, PO Box 490, Kolonia, Pohnpei, Eastern Caroline Islands 96941, FSM

Yap Institute of Natural Resources, PO Box 215, Yap, West Caroline Islands 96943, FSM

References

Anon. (1989). Progress with the action strategy for protected areas in the South Pacific. *Information Paper* 3. Fourth South Pacific Conference on Nature Conservation and Protected Areas. Port Vila, Vanuatu, 4-12 September. 19 pp.

Dahl, A.L. (1980). Regional ecosystems of the South Pacific area. SPC/IUCN *Technical Paper* 179. South Pacific Commission, Noumea, New Caledonia. 99 pp.

Dahl, A.L. (1986). *Review of the protected areas system in Oceania*. IUCN, Gland, Switzerland and Cambridge, UK/UNEP, Nairobi, Kenya. 328 pp.

Dahl, C. (1989). Developing a coastal resource management plan for Kosrae State, Federated States of Micronesia. *Case Study* No. 23. Fourth South Pacific Conference on Nature Conservation and Protected Areas, Port Vila, Vanuatu, 4-12 September. 7 pp.

Fosberg, F.R. (1973). On present condition and conservation of forests in Micronesia. In: Pacific Science Association, *Planned Utilisation of the Lowland Tropical Forests*. Proceedings of the Pacific Science Standing Committee Symposium, August 1971, Bogor, Indonesia. Pp. 165-171.

Maragos, J.E. (1986). Coastal resource development and management in the US Pacific Islands: 1.

Island-by-island analysis. Office of Technology Assessment, US Congress. Draft.

Paxton, J. (1989). *The Statesman's Yearbook 1989-1990*. The Macmillan Press Ltd, London. 1691 pp.

Scott, G.A.J. and Rotondo, G.M. (1983). A model for the development of types of atoll and volcanic islands on the Pacific lithospheric plate. *Atoll Research Bulletin* 260. 33 pp.

SPREP (1980). Trust Territory of the Pacific Islands. Country Report 14. South Pacific Commission, Noumea, New Caledonia.

UNEP/IUCN (1988). *Coral reefs of the world. Volume 3. Central and Western Pacific*. UNEP Regional Seas Directories and Bibliographies. IUCN, Gland, Switzerland and Cambridge, UK/UNEP, Nairobi, Kenya. 378 pp.

Yinug, M., Falanruw, M., Manmaw, C. (1989). Traditional and current resource management in Yap. In: SPREP, *Report on the workshop on customary tenure, traditional resource management and nature conservation*, Noumea, 28 March – 1 April. South Pacific Commission, Noumea. Pp. 113-116.

ANNEX
Definitions of protected area designations, as legislated, together with authorities responsible for their administration

Title: Pohnpei Watershed Forest Reserve and Mangrove Protection Act

Date: 1987

Brief description: Provides for the protection of soil, water and mangrove ecosystems in Pohnpei State

Administrative authority: No information

Designation:

Watershed forest reserve Designated in areas of soil instability, permitted uses include limited agriculture, research, dispersed recreation, gathering of wild plants, and some timber harvesting, provided environmental protection requirements are met. Forbidden uses include permanent occupancy, pesticide use, unauthorised tree-cutting, land clearing with fire, and grazing.

Important watershed areas Areas which have already been settled; restrictions include no additional building of roads or structures, no rebuilding or improvement of existing structures, and strict enforcement of other regulations.

Unauthorised cutting of trees in watershed forest reserves and in mangrove forests carry mandatory fines of US$ 1,000 per tree cut.

FIJI

Area 18,330 sq. km

Population 732,000 in 1988 (World Bank, 1990)
Natural increase: 2% per annum

Economic Indicators
GNP: US$ 1,520 per capita (1988) (World Bank, 1990)

Policy and Legislation Government policy on establishment of protected areas, as expressed in Development Plan DP8 and DP9, is to conserve and protect important and unique aspects of the country's natural heritage through the establishment of a comprehensive system of parks and reserves. It is intended that a National Parks and Reserves Act will be enacted which will "effectively provide for the preservation and protection of the natural environment including unspoilt landscape, reefs and waters, indigenous flora and fauna, habitats and ecological systems, features of scenic, historic or archaeological interest or other scientific interest". In addition, the bill would allow the use of the parks by the public for enjoyment, recreation and education purposes (Rabuka and Cabaniuk, 1989).

Legislative Council Paper No. 5 (1950) describes the Government of Fiji's forest policy, which includes the following aims: to protect and develop natural vegetation where its retention is necessary for climatic reasons and for the conservation of resources of soil and water for agriculture and to ensure adequate and continuous supplies of forest produce; to maintain and improve the fertility of the soil by preserving and, where necessary, extending the forest cover; to check soil erosion and to recover areas already eroded; and to provide and preserve amenities. The Ministry of Forestry has recognised the need to set aside forest from production and the currently proposed ten-year development plan for the Extension Division for Conservation, Environment and Parks Service outlines plans to set aside 15% of the total natural forest for permanent protection, with a suggestion of 7,000ha to be set aside annually for 20 years.

Eighty-three per cent of land is under constitutionally-inalienable native communal ownership, ten per cent is freehold and seven per cent is Crown (government) land. Native land owned by indigenous people is reserved for use only by native Fijians, although areas may be leased through the Native Land Trust Board. Much of Fiji's native forest is "privately" owned in communal tribal tenure which severely constrains the Forestry Department's ability to manage forest resources. In view of this, the Native Land Trust Board prepared and approved a Forestry Policy in 1985. The main objectives include: ensuring that the forest is managed in accordance with sound forest management and land conservation practices; protecting the environment; and establishing a sound forest management policy to provide the basis for a national integrated land use policy. Both the Native Land Trust Board's Forestry Policy and Tourism Policy lend support to protected areas establishment, and the establishment of forest parks and reserves in particular (Rabuka and Cabaniuk, 1989).

Relevant legislation, described in Annex 1, comprises the 1923 Birds and Game Protection Act, the earliest nature conservation law, the Native Land Trust Act (1940), the Forest Act (1953) and the National Trust for Fiji Act (1970). The Native Land Trust Act provides for the establishment of the Native Land Trust Board, the principal agency through which Fijian land claims are administered. The Forest Act provides for the constitution of nature reserves, and the protection of any unalienated Crown land, or land leased to the Crown, as a reserved forest. Furthermore, nature reserves may be declared within reserved forests. Protected forests may also be established on native land, with the consent of the Native Land Trust Board (Sloth, 1988). The Forestry Act, however, has a number of serious weaknesses. For example, legal loopholes permit clearfelling of forests over which the Forestry Department has no control and all protected areas established under the provisions of the Forestry Act are subjected to dereservation at ministerial level; reserve forests have frequently been dereserved. This indicates an inadequate level of protection for sites of national or international importance (Watling, 1988a). The National Trust for Fiji Act provides for the preservation, protection and management of natural heritage in general.

In addition, the 1953 Land Conservation and Improvement Ordinance (Cap. 120) provides for the establishment of a Land Conservation Board with powers to issue conservation orders prohibiting or controlling land clearance for any purpose, and also makes provision for the appointment of conservation officers. The 1946 Town Planning Ordinance (Cap. 109) provides for the conservation of the "natural beauties of the area" in the preparation of town planning schemes.

International Activities Fiji is not yet party to either the Unesco Man and the Biosphere Programme or the Convention on Wetlands of International Importance especially as Waterfowl Habitat (Ramsar Convention), international conventions that directly promote the conservation of natural areas. However, a Cabinet decision to accede to the Convention concerning the Protection of the World Cultural and Natural Heritage (World Heritage Convention) has been taken.

At a regional level, Fiji signed the 1976 Convention on the Conservation of Nature in the South Pacific on 18 September 1989. Known as the Apia Convention, it entered into force during 1990. The Convention is coordinated by the South Pacific Commission and represents the first attempt within the region to cooperate on environmental matters. Among other measures, it encourages the creation of protected areas to preserve indigenous flora and fauna.

Fiji is party to the South Pacific Regional Environment Programme (SPREP) and has ratified (18 September 1989) the Convention for the Protection of the Natural Resources and Environment of the South Pacific Region, 1986 (SPREP Convention), which entered into force during August 1990. Article 14 calls upon the parties to take all appropriate measures to protect rare or fragile ecosystems and threatened or endangered flora and fauna through the establishment of protected areas and the regulation of activities likely to have an adverse effect on the species, ecosystems and biological processes being protected. However, as this provision only applies to the Convention area, which by definition is open ocean, it is most likely to assist with the establishment of marine reserves and the conservation of marine species.

Other international and regional conventions concerning environmental protection to which Fiji is party are reviewed by Venkatesh *et al.* (1983).

Administration and Management The National Trust for Fiji has broad legal responsibility for nature conservation and comprises a council of ten members, including a chairman and vice-chairman, appointed by the Minister for a period not exceeding two years. The National Trust is currently responsible for three reserves. The Forestry Department, under the Ministry of Forests, is primarily concerned with the management of indigenous forests and plantations for commercial production purposes, but is also responsible for administering nature reserves, as well as recreation and amenity areas (Dunlap and Singh, 1980; SPREP, 1985b). The Department currently administers seven nature reserves, established within forest reserves, under the 1953 Act (SPREP, 1985b), 18 forest reserves and a number of forest parks and amenity reserves.

Although the Native Land Trust Board (NLTB) has no specific nature conservation policy, its stated role in protected areas establishment is to lend support to and facilitate the efforts of the National Trust and the Forestry Department in securing native lands for parks and reserves *via* the native lands leasing procedure (Rabuka and Cabaniuk, 1989). Protected area establishment on native reserves must proceed through established leasing procedures, to ensure that the landowners are fully aware from the outset of the purpose of such areas and of the benefits (socio-economic, environmental and otherwise) which will accrue to them through their lands being given park/reserve status; the landowners are fully aware of the implications of protected area status as it affects customary land use rights; landowner benefits are maximised, not only in terms of lease premiums, income, and visitor receipts, but also with regard to employment and, if appropriate, training opportunities; the lessee

receives long-term security over the lands in question; and should any dispute arise between the landowners and lessee, both sides have recourse to arbitration through the NLTB. It should be noted that Namenalala Island Nature Reserve (43ha, established in 1984 through lease agreement between the Native Land Trust Board, the landowners and Namenalala Island Resort Ltd) is the only protected area to have been established in Fiji to date through the Native Land Trust Leasing Procedure (Lenoa *et al.*, 1989).

There are also lease agreements between the Forestry Department, the Fiji Pine Commission, and the Native Land Trust Board on the lease of land for 50 years for afforesting grasslands and reafforesting logged areas with hardwoods.

System Reviews Fiji comprises approximately 300 islands and islets scattered across an Exclusive Economic Zone located between 10-25°S and 173°E-176°W. The islands form several distinct groups: Rotuma; Vanua Levu and associated islands, including Taveuni and Ringgold Isles; the Lau Group; the Lomaiviti Group; the Yasawas; Viti Levu and associated islands; and Kadavu and associated islands.

Natural forest cover in 1986 was estimated at 8,049 sq. km, 44% of total land area (S. Siwatibau, pers. comm., 1989). Rain forest (veikauloa) occurs in the south and east of larger islands and on small limestone islands. There is virtually no forest cover remaining on small volcanic islands (D. Watling, pers. comm., 1989). Although the islands are not high enough to exhibit altitudinally stratified forest, upland montane formations with depleted tree, and enhanced epiphyte, species component, are found at the highest levels. The highest and wettest ridges support a distinctive moss forest, although the total area of this habitat is very small. Dry zone vegetation (talasiga), including dry forests, savanna woodlands and grasslands, predominates on northern and western slopes of large islands, and inland up to an altitude of 450m. Dry forest has largely been replaced by scrub, grassland and sugar cane. A zone of intermediate vegetation, comprising a mosaic of light forest, bamboo stands and grassland, lies immediately leeward of wet forests. Beach forest has been almost entirely lost to agriculture throughout the country and exists only as isolated pockets on some of the smaller islands. Extensive mangrove forests, covering some 45,000ha, occur as deltaic formations of larger rivers and on sheltered coasts (Davis *et al.*, 1986; Watling, 1988b). Highly localised multiple landslides and cyclones are regularly occurring natural disasters. These have moulded forest such that secondary associations are a widespread and integral part of the Fijian forest ecosystem. This may have favourably selected fauna adapted to disturbed environments (Watling, 1988b). Schmid (1978) provides an account of the vegetation and forest distribution maps (after Parham, 1972) for Viti Levu and Vanua Levu. Freshwater wetland vegetation of Viti Levu is described by Ash and Ash (1984).

Coral reefs occur around all island groups in barrier, fringing and platform formations. Shelf atolls are restricted to Wailagilala, in the Lau Group, and Qelelevu, east of Vanua Levu. Barrier reefs include the Great Sea Reef which extends for 200km and is globally significant. Fringing reefs are found from the southern end of the Mamanuca Group almost to Beqa, south of Viti Levu. Platform reefs are restricted to shelfwaters and are common inside the Great Sea Reef, including the Yasawa-Mamanuca arc, and within the Mabualau-Ovalau series of barrier reefs of eastern Viti Levu (UNEP/IUCN, 1988). Detailed accounts for a number of specific sites are given in UNEP/IUCN (1988).

In 1986 the Fiji Forestry Department classified 2,373 sq. km of forest as production forest, of which 1,413 sq. km (60%) was under logging concessions and licences. The gross area under concessions/licences, including non-commercial forest, was 2,780 sq. km, which is 35% of the total forest area (S. Siwatibau, pers. comm., 1989). However, on Vanua Levu nearly 50% of forests have been selectively logged and 90% is destined to be legitimately logged by Fiji Forest Industries (CMC, 1987). By December 1986, plantations covered 80,000ha, comprising 45,160ha of Fiji Pine Commission pine forests established on former grassland, and much of the remainder comprising exotic hardwoods in reafforestation plantations on cut-over native land (S. Siwatibau, pers. comm., 1989). Figures from Drysdale (1988) indicate 1986 totals of 53,449ha of pine plantation and 25,669ha of hardwood plantations; such plantations have a low animal diversity, especially with respect to endemic fauna (CMC, 1987; Watling, 1988b).

Forest cover is being depleted at an overall rate of approximately 1% per annum. Between 1969 and 1988 there has been a possible 30% reduction in the area of non-commercial forest, a 5.4% loss of production forests and an 8% reduction in protection forests, although these are intended to remain under forest cover in perpetuity. There is no systematic monitoring of forest loss, which varies greatly by district and is probably much worse on Viti Levu than elsewhere (Drysdale, 1988). The loss of forest cover is particularly threatening to the native flora which, with frequently restricted distributions in the forest, is vulnerable to extinction (Watling, 1988a).

The existing protected areas network covers a total of 6,592.5ha, or 0.36% of Fiji's terrestrial area, a clearly inadequate sample to provide for long-term conservation. Habitats that are included within protected areas include montane rain forest, cloud forest, atoll/beach forest, limestone island scrub and mountain stream. Habitats that at present receive no protection include lowland forest, swamp and mangrove forests, savanna and grasslands, sea turtle nesting areas, reefs and other more restricted ecosystems (Dahl, 1980). The present system is inadequate for several reasons. No ecological or heritage considerations were involved in the selection of protected areas. Protection forests have no long-term conservation value, given their present inadequate legal status and management. Forest and

nature reserves are under the management of departmental rather than national institutions. Coupled with inadequate legislative and institutional support, this suggests a likely failure to meet increasing political and social pressures in the future (Watling, 1988a).

An Action Strategy for protected areas in the South Pacific region (SPREP, 1985a) was developed by field managers at the technical session and adopted by the ministerial meeting of the 1985 Third South Pacific National Parks and Reserves Conference. The strategy provides a work programme to implement conservation and protected areas objectives. The principal goals of the strategy cover conservation education, conservation policy development, establishment of protected areas, protected area management and regional and international cooperation. Priority recommendations for Fiji are as follows: development of a national conservation strategy; final drafting of the National Parks and Reserves Act; development of Garrick Reserve as a national park pilot project; development of Tai-Elevuka Reef environs as a marine national park pilot project; designation of Sigatoka Sand Dunes as a project area; and the provision of staff training, etc.

A number of objectives have been achieved, including government agreement to prepare a national environment strategy, in a joint project with IUCN and the Asian Development Bank; establishment of Sigatoka Sand Dunes National Park and Reserve; development of the concept of "landowner parks" by NLTB, matching nature conservation with local tourism ventures, as expressed in the Waikakata Archaeological Park and Forest Reserve Plan (1988) and the Bouma Forest Park and Reserve Plan (1989); and an ecological survey has been carried out, culminating in "A representative national parks and reserves system for Fiji's tropical forests" (Maruia Society, 1989). The latter includes proposals for protected areas that cover 6.8% of the unlogged indigenous production-zoned forest resource, and comprises a number of national parks, nature reserves, forest parks and "special feature" forest reserves. Progress has also been made in the development of a mangrove management plan, protected areas legislation, and management plans for Garrick Memorial Reserve and Sigatoka Sand Dunes National Park (Anon., 1989).

Two previous reports (Dahl, 1980; Dunlap and Singh, 1980) have already recommended the establishment of reserves and although these were not implemented they have provided valuable input to the Maruia Society (1989) report. In turn, the recently initiated National Environment Strategy will incorporate the recommendations embodied in the 1989 report. Dahl (1980) recommends the establishment of preferably contiguous mountain and lowland forest reserves in east and west Viti Levu and Vanua Levu; forest reserves on a number of larger islands and mangrove reserves at the mouth of Samambula River; and, in total, 44 potential conservation sites are identified. In the IUCN/UNEP *Review of the Protected Areas Systems in Oceania,* Dahl

(1986) makes a number of less specific recommendations, including increasing the number of protected areas on Viti Levu, Vanua Levu and smaller islands with special features, in addition to including representative examples of each island type and a selection of marine areas within the protected areas network.

A series of national conservation congresses has been initiated and the most recent, in June 1988, called for more efficient and sustainable management of forest resources (NTF, 1988a, 1988b). Support was given to the National Trust to develop a plan of action to remove the final bureaucratic impediments to the creation of Dakua Reserve at Waisali, Vanua Levu, and to foster the development and designation of other areas including Waikatakata and Vatura as national pilot projects and Garrick Memorial Park as a functional conservation unit. The National Trust is currently seeking support for a biological survey of a number of islets in the Yasawa-Mamanuca chain, with the aim of establishing a protected area for crested iguana, to be managed for compatible nature tourism (Juvik and Singh, 1989).

A major review of the forestry sector has been completed and translated into a draft Fiji Tropical Forestry Action Plan which was due to be finalised during 1990 (Forestry Department, 1990).

Addresses

Conservator of Forests, Ministry of Forests, PO Box 2218, Government Buildings, Suva

National Trust for Fiji, PO Box 2089, Government Buildings, Suva

Native Land Trust Board, PO Box 116, Suva

References

Ash, J. and Ash, W. (1984). Freshwater wetland vegetation of Viti Levu, Fiji. *New Zealand Journal of Botany* 22: 337-391.

CMC (1987). Conservation issues in Fiji. A briefing document prepared for Deutsche Gesellschaft für Technische Zusammenarbeit (GTZ) GmbH. IUCN Conservation Monitoring Centre, Cambridge, UK. Unpublished report. 33 pp.

Dahl, A.L. (1980). Regional ecosystem survey of the South Pacific Area. *SPC/IUCN Technical Paper* 179. South Pacific Commission, Noumea, New Caledonia. 99 pp.

Dahl, A.L. (1986). *Review of the protected areas system in Oceania*. IUCN, Gland, Switzerland and Cambridge, UK/UNEP, Nairobi, Kenya. 328 pp.

Davis, S.D., Droop, S.J.M., Gregerson, P., Henson, L., Leon, C.J., Lamlein Villa-Lobos, J., Synge, H. and Zantovska, J. (1986). *Plants in danger: what do we know?* IUCN, Gland, Switzerland and Cambridge, UK. 488 pp.

Drysdale, P.J. (1988). Rainforest management and conservation in Fiji: a prescription for action. In: *Proceedings of the Second National Conservation Congress 9-10 June. Volume 2*. National Trust for Fiji, Suva, Fiji. 264 pp.

Dunlap, R.C. and Singh, B.B. (1980). *A national parks and reserves systems for Fiji*. A report to the National Trust for Fiji. Three volumes. 117 pp.

Forestry Department (1989). *Annual report for the year 1989*. Paper No. 21 of 1990. Ministry of Forests, Suva. 22 pp.

Juvik, J.O. and Singh, B. (1989). Conservation of the Fijian crested iguana: a progress report. *Case Study* 22. Fourth South Pacific Conference on Nature Conservation and Protected Areas, Port Vila, Vanuatu, 4-12 September. South Pacific Commission, Noumea, New Caledonia. 7 pp.

Lenoa, L., Waqaisavou, T. and Lees, A. (1989). A representative national parks and reserves system for Fiji. *Case Study* 7. Fourth South Pacific Conference on Nature Conservation and Protected Areas, Port Vila, Vanuatu, 4-12 September. South Pacific Commission, Noumea, New Caledonia. 15 pp.

Maruia Society (1989). A representative national parks and reserves system for Fiji's tropical forests. *Maruia Society Policy Reports Series* No. 9. 110 pp.

National Trust for Fiji (1988a). Fiji's rainforests: our heritage and future. *Proceedings of the Second National Conservation Congress, Suva, Fiji*, 9-10 June. Volume 1. National Trust for Fiji. 189 pp.

National Trust for Fiji (1988b). Fiji's rainforests: our heritage and future. *Proceedings of the Second National Conservation Congress, Suva, Fiji*, 9-10 June. Volume 2. National Trust for Fiji. 264 pp.

Parham, J.W. (1972). *Plants of the Fiji Islands*. Suva, Government Press. 490 pp. (Unseen)

Rabuka, M. and Cabaniuk, S. (1989). The role of the Native Land Trust Board in the establishment of parks and reserves in Fiji. In: Thomas, P.E.J. (Ed.), *Report of the workshop on customary tenure, traditional resource management and nature conservation*. SPREP, Noumea, New Caledonia. Pp. 71-86.

Schmid, M. (1978). The Melanesian forest ecosystem (New Caledonia, New Hebrides, Fiji Islands and Solomon Islands). In: Unesco/UNEP/FAO, *Tropical forest ecosystems*. Unesco, Paris. Pp. 654-683.

Sloth, B. (1988). *Nature legislation and nature conservation as part of tourism development in the island Pacific*. Pacific Regional Tourism Development Programme. Tourism Council of the South Pacific, Suva. 82 pp.

SPREP (1985a). *Action strategy for protected areas in the South Pacific Region*. South Pacific Commission, Noumea, New Caledonia. 24 pp.

SPREP (1985b). Fiji. In: Thomas, P.E.J. (Ed.), *Report of the Third South Pacific National Parks and Reserves Conference*. Volume III. *Country reviews*. South Pacific Commission, Noumea, New Caledonia.

SPREP (1989). Progress with the action strategy for protected areas in the South Pacific. Information Paper presented at the Fourth South Pacific Conference on Nature Conservation and Protected Areas. Port Vila, Vanuatu, 4-12 September. 19 pp.

UNEP/IUCN (1988). *Coral reefs of the world. Volume 3: Central and Western Pacific.* UNEP Regional Seas Directories and Bibliographies. IUCN, Gland, Switzerland and Cambridge, UK/UNEP, Nairobi, Kenya. 378 pp.

Venkatesh, S, Va'ai, S. and Pulea, M. (1983). An overview of environmental protection legislation in the South Pacific countries. SPREP *Topic Review* No. 13. South Pacific Commission, Noumea, New Caledonia. 63 pp.

Watling, D. (1988a). The forestry sector development study. FIJ/86/004. Report of the environmental scientist. Unpublished. Suva. 44 pp.

Watling, D. (1988b). The effects of logging on Fijian wildlife. In: *Proceedings of the Second National Conservation Congress* 9-10 June. *Volume 2.* National Trust for Fiji, Suva. 25 pp.

World Bank (1990). *World Tables* 1989-1990 Edition. John Hopkins University Press, Baltimore. 646 pp.

ANNEX
Definitions of protected area designations, as legislated, together with authorities responsible for their administration

Title: Birds and Game Protection Act

Date: 1923

Brief description: Proscribes hunting of specified bird species, establishes closed seasons and quotas for others, and enables the Minister to establish reserved areas. Original legislation is not available.

Administrative authority: No information

Designation:

Reserved areas Details not available

Title: Native Land Trust Act

Date: 1940/1943

Administrative authority: Native Land Trust Board

Description: Establishes and requires the Native Land Trust Board to declare and administer native reserves for the use, maintenance and support of Fijian owners. Original legislation is not available.

Designation: Native reserves may be declared on any native land.

Title: An Act Relating to Forest and Forest Produce (The Forest Act)

Date: 1 October 1953, last amended 1978

Administrative authority: Conservator of Forests, Ministry of Forests

Description: Provides for the protection of any unalienated Crown land, or land leased to the Crown, as a reserved forest, which may also be sub-classified into nature reserve, protected forest and silvicultural area.

Designation: Reserved forest

Nature reserve Part of a reserved forest not being a silvicultural area. Excludes people except between 6am and 9pm. Prohibits logging and other damage and the Conservator may approve operations only if they protect flora and fauna. The Native Land Trust Board's right of dereservation is not specified.

Protected forest Established with Native Land Trust Board consent on any native land not being part of a reserved forest or alienated and not being part of a silvicultural area. Once protected, the Native Land Trust Board may not alienate without consent of the Conservator. Licences may be issued but illegal logging and grazing are prohibited.

Silvicultural area Part of a protected forest so declared. Excludes people except between 6am and 9pm. Objectives are not clearly stated, but may suggest production forestry or conversion to plantation.

Section 12 defines offences against the Act. In addition to the restricted times of entry, the following general activities are forbidden: grazing; cutting or damaging any trees or forest produce; erecting buildings; setting fires; clearing land; constructing or obstructing roads; setting or possessing traps etc.; entering areas closed under forest regulations; damaging forest property; hunting and fishing.

NB: All areas protected under the Act are subject to dereservation by the Minister of Forestry.

Title: The National Trust for Fiji Act

Date 1970

Brief description Makes provision for the establishment of the National Trust for Fiji, a statutory body under the Ministry of Housing and Urban Development.

Administrative authority National Trust for Fiji (Chairman)

SUMMARY OF PROTECTED AREAS

Map[†] ref.	*National/International designation* Name of area	IUCN management category	Area (ha)	Year notified
	Nature Reserves			
2	Draunibota and Labiko Islands	I	2	1959
3	Nadarivatu	I	93	1956
4	Namenalala Island	III	43	1984
5	Naqarabuluti	I	279	1958
6	Ravilevu	I	4,020	1959
7	Tomaniivi	I	1,322	1958
8	Vunimoli	I	19	1966
9	Vuo Island	I	1	1960
10	Yadua Taba Island Crested Iguana Reserve	IV	70	1981
	Forest Reserve			
11	J H Garrick Memorial	IV	428	1983

[†]Locations of most protected areas are shown on the accompanying map.

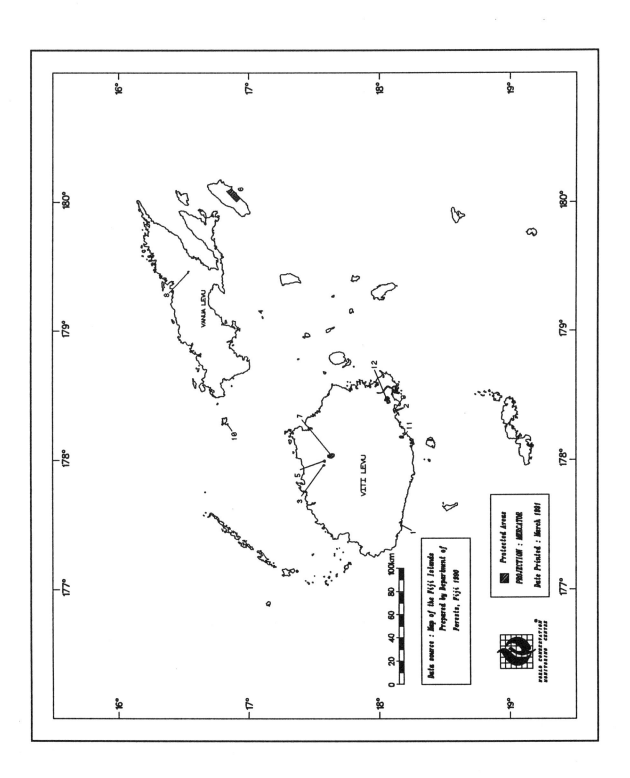

Protected Areas of Fiji

FRANCE – FRENCH POLYNESIA

Area 3,521 sq. km (spread over approximately 4,198,000 sq. km of ocean)

Population 188,814 (1988)
Natural increase: 1.9% per annum (Dahl, 1986)

Economic Indicators
GNP: Not available

Policy and Legislation The status of French Polynesia, a French overseas territory, under Law No. 77-772 of 12 July 1972 gives full power to the territory in environmental matters. The Governor of French Polynesia also has authority for Clipperton Island, some 5,000km north-west of Tahiti, although this is not part of French Polynesian territory.

There are no laws specifically providing for the creation and management of territorial parks and other protected areas. There is a general planning code, the Délibération de l'Assemblée Territoriale No. 61-44 of 8 April 1961, which was modified in 1982 to take account of economic restraints. Book 1, Title 3 of this code makes provision for the creation of protected areas, by defining the procedure leading to classification of an area. A regulation specifically providing for the creation and management of protected areas is planned (Siu and Vernaudon, 1989). Law No. 56-836 of 26 July 1956 gives the Territorial Assembly the power to regulate nature protection. This is reinforced by the new statute in Law No. 84-260 of 6 September 1984 which made provision for the nomination of a Minister of Environment and a Délégation à l'Environnement. Legal power rests with the Territorial Assembly and executive power with the President of the Territorial Government and the Council of Ministers. Book 1, Chapter VIII, Title V, Articles D.151 1-14 of the 1984 Code provide for a list of sites and monuments for conservation. The Article states that a list of buildings and a list of natural sites and monuments, of which the conservation or preservation is of historic, artistic, scientific, legendary or folkloric interest, are established in the territory and published in the Journal Officiel. Inscription on these lists is announced by decree of the head of the territory in government council following proposal by the Commission on Natural Sites and Monuments (CNSM) under Article D.100-2 of this Code. The regional administrative authority will notify the owners of properties, sites and monuments of inscription. Inscription obliges the owner not to modify the aspect of any natural property, site or monument, nor to undertake any activity other than normal exploitation, running repairs and maintenance without advising the interested head of regional administration at least two months before the proposed date for the start of work.

The procedure for classifying natural protected areas is undertaken by the CNSM, which, since 1985, has been presided over by the Minister of the Environment. On the advice of the CNSM, and with the approval of the Council of Ministers, the classification of a protected area may be pronounced after consultation with the Territorial Assembly in the case of public property, or after informing land owners in the case of private property. In the latter case, a delay of three months is mandatory to allow representations to be made. An intermediary procedure involves inscribing proposed sites on a list, after which the site must be maintained whilst legal formalities are prepared. However, a delay of more than one year will render this null and void (Siu and Vernaudon, 1989).

The Forestry Regulations, resulting from nine legal instruments dating from 1942 to 1978, contain provisions for the protection of soil, vegetation and wildlife, and the establishment of strict nature reserves (réserves intégrales). However, there is no provision for supervision. Other regulations cover hunting and fishing, but with no provision for protected areas.

Recommendations and outlines of possible protected area legislation has been put to the government as part of reports prepared and submitted by SPREP (Anon., 1989).

International Activities At the international level, France has entered into a number of obligations and cooperative agreements related to conservation. It is party to the Convention concerning the Protection of the World Cultural and Natural Heritage (World Heritage Convention) which it accepted on 27 June 1975. France became a contracting party to the Convention on Wetlands of International Importance especially as Waterfowl Habitat (Ramsar Convention) on 1 October 1986. At present there are no World Heritage nor Ramsar sites within French Polynesia. France participates in the Unesco Man and the Biosphere Programme and one site in French Polynesia has been inscribed as a biosphere reserve.

France signed the Convention on the Conservation of Nature in the South Pacific on 12 June 1976 with subsequent ratification on 20 January 1989. Known as the Apia Convention, it entered into force during 1990. The Convention is coordinated by the South Pacific Commission and represents the first attempt within the region to cooperate on environmental matters. Among other measures, it encourages the creation of protected areas to preserve indigenous flora and fauna.

The Convention for the Protection of the Natural Resources and Environment of the South Pacific Region (SPREP Convention) was signed on 25 November 1986 and ratified on 17 July 1990. The Convention entered into force during August 1990. Article 14 calls upon the parties to take all appropriate measures to protect rare or fragile ecosystems and threatened or endangered flora and fauna through the establishment of protected areas and the regulation of activities likely to have an adverse effect on the species, ecosystems and biological processes being protected. However, as this provision only applies to the Convention area, which by definition is open ocean, it is most likely to assist with the establishment of marine reserves and the conservation of marine species. However, the Instrument of Approval is accompanied by the following reservation: "The Government of the French Republic, in signing the present Convention, declares that, insofar as it is concerned, the prescription of the aforesaid Convention will not cover wastes and other matter entailing a level of pollution caused by radioactivity to a degree less than that prescribed by the recommendations of the International Atomic Energy Agency".

Administration and Management The CNMS was established in 1962 to propose sites for classification or protection, and to provide advice on environmental issues. The Fisheries Department manages the marine environment, while the Forest Branch of the Agriculture Department is responsible for implementation of the Forestry Act. The Délégation à l'Environnement was created on 30 May 1985 (Délibération 85-1040/AT), responsible to the Minister of the Environment, with a staff complement that has increased from two in 1985 to twelve in 1989. Objectives are to: monitor environmental protection and rehabilitation; monitor environmental risks arising from development; develop or coordinate studies and direct research into environmental issues; develop training programmes and information services to enhance environmental awareness; and to identify and classify natural and cultural sites. To achieve these objectives, the Délégation acts in concert with the relevant ministries and services and coordinates environmental policies (Siu and Vernaudon, 1989).

The Délégation has identified certain priorities in applying its statutory powers. It has established new regulations for environmental protection classification (Délibération 87-80/AT of 12 June 1987), modifying Book IV of the Code de l'aménagement du territoire. The aim is to increase the efficiency of pollution prevention with means better adapted to current economic constraints and local values, while at the same time allowing the development of economic activity in harmony with the environment. It has also established an environmental charter, of which two sections are at an advanced stage of preparation: environmental impact assessment and protection of natural areas (Anon., 1990).

Systems Reviews French Polynesia extends from 134°28'W (Temoe) to 145°40'W (Manuae or Scilly) and from 7°50'S (Motu One) to 27°36'W (Rapa). The emergent land area (3,521 sq. km) is augmented by about 7,000 sq. km of lagoon. The islands are situated on a general north-west to south-east orientation, their age decreasing from north-west to south-east, and they form five distinctive archipelagos: Society, Tuamotu, Gambier, Marquesas and Tubai or Austral Islands. There are around 130 islands, of which 84 are atolls; most of the remainder are high volcanic islands, many being very mountainous with inaccessible interiors.

Habitats range from atoll and lowland rain forest, through riverine and bamboo forest to cloud and montane formations. Most coastal forest has been destroyed, but that which remains is of low economic value and relatively unthreatened. Lying at the easternmost extremity on the Indo-Pacific Province, French Polynesia is at the limit of the axis of decreasing species richness and has a comparatively poor coral fauna. The main reef formations are found around the high islands and atolls, although there are several oceanic banks of variable forms (e.g. Ebrill Reef in the Gambiers and Moses Reef in the Australs). A detailed summary account of coral reefs in French Polynesia is given in UNEP/IUCN (1988).

The islands with the highest conservation interest have very few protected areas. On the basis of biological diversity, endemism and other factors, Nuku Hiva is rated as the most important island for conservation within the country and one of the most important volcanic islands in the Oceanian realm. Tahiti, Rapa and Moorea are all at risk from human activities and only Tahiti, with the recently established Vallée de Faaiti, includes a protected area. All have significant species endemism, and the establishment of appropriate protected areas should be a priority. Some habitat restoration may be needed on Moorea and other islands. In addition, protected area establishment should be considered at least on Raiatea and possibly Huahine and Tahaa. In the Austral Islands, Rimatara and Raivavai should be given priority for their endemism, as should Mangareva in the Gambier Islands. The Tuamotu atolls are simpler island ecosystems, with one reserve at Taiaro Atoll. Additional reserves are needed at least on Matureivavao, Niau, Napuka and the raised coral island of Makatea, which, despite former mining damage, retains some conservation value. The Marquesas are quite distinctive biologically and the four present island reserves are inadequate: protected areas should be established on each island. In addition to Nuku Hiva, priority should be given to protected areas on Hiva Oa and Ua Pou, and to the general control of the feral animals which are causing great destruction. A representative system of marine reserves across the great expanse of French Polynesia should also be developed (Dahl, 1986). Currently proposed protected areas comprise a territorial park at Atimaono, Pari Historic and Archaeological Site, Atoll de Scilly, Atoll de Tetiaroa, and Hemeni and Keokeo Ilets in the Marquesas for the protection of sea birds, particularly *Sterna fuscata*. The establishment of marine reserves and other terrestrial protected areas is being studied (Siu and Vernaudon, 1989).

The Délégation à l'Environnement has become actively involved *inter alia* in the organisation of environmental courses ranging from primary to university level, providing material for television broadcast, posters, information signs and the organisation of an annual environment day (SPREP, 1989).

No overall environmental policy has existed for many years, although protection has been applied *ad hoc* for a certain number of problems. Principal environmental problems include introduced mammals (particularly rats) and birds, and soil erosion and consequent sedimentation of lagoons caused mainly by terracing undertaken during urbanisation. The recent expansion of *Miconia calvescens* (Melastomataceae), introduced in 1937 for its ornamental value, now threatens the indigenous flora of the islands of Tahita, Moorea and Raiatea. Arrêté territorial 290 of 14 March 1990 taken by the Council of Ministers forbids its propagation on the other Polynesian islands and a research programme by ORSTOM over the last two years continues to seek a method of counteracting the plant.

Addresses

Délégation à l'Environnement, Ministère de l'Environnement, BP 4562, Papeete, Tahiti

References

Anon. (1989). Progress with the action strategy for protected areas in the South Pacific. *Information Paper* 3. Fourth South Pacific Conference on Nature Conservation and Protected Areas. Port Vila, Vanuatu, 4-12 September. 19 pp.

Anon. (1990). Législation et administration du milieu naturel. Unpublished. 6 pp.

Dahl, A.L. (1986). *Review of the protected areas system in Oceania.* IUCN, Gland, Switzerland and Cambridge, UK/UNEP, Nairobi, Kenya. 328 pp.

Paxton, J. (1989). *The Statesman's yearbook: 1989-90.* The Macmillan Press Ltd, London. 1691 pp.

SPREP (1980). French Polynesia. *Country Report* 5. South Pacific Commission, Noumea, New Caledonia. 11 pp.

SPREP (1989). Polynésie Française. Paper presented at the Fourth South Pacific Conference on Nature Conservation and Protected Areas, Port Vila, Vanuatu, 4-12 September 1989. 5 pp.

Siu, P and Vernaudon, Y. (1989). Programme de mise en place d'un système de périmètres protégés en Polynésie Française. *Case study* No. 18. Fourth South Pacific Conference on Nature Conservation and Protected Areas. Port Vila, Vanuatu, 4-12 September. 6 pp.

UNEP/IUCN (1988). *Coral reefs of the world. Volume 3. Central and Western Pacific.* UNEP Regional Seas Directories and Bibliographies. IUCN, Gland, Switzerland and Cambridge, UK/UNEP, Nairobi, Kenya. 378 pp.

SUMMARY OF PROTECTED AREAS

Map ref.	National/international designation Name of area	IUCN management category	Area (ha)	Year notified
	Strict Nature Reserve			
1	Atoll de Taiaro (W.A. Robinson)	IV	2,000	1977
	Nature Reserves			
2	Eiao Island	IV	5,180	1971
3	Hatutu Island	IV	1,813	1971
	Park			
4	Vallee de Faaiti	II	750	1989
	Unclassified			
5	Mohotani	IV	1,554	1971
6	Sable Island (Motu One)	IV		1971
7	Scilly Atoll (Manuae)	IV	200	1971
	Biosphere Reserve			
	Atoll de Taiaro (W.A. Robinson)	IX	2,000	1977

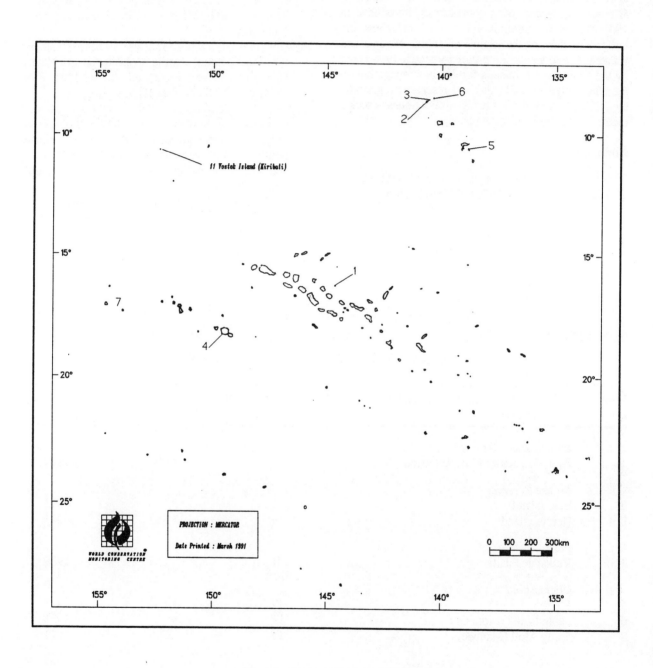

11 Vostok Island (Kiribati)

PROJECTION : MERCATOR

Date Printed : March 1991

WORLD CONSERVATION
MONITORING CENTRE

0   100   200   300km

Protected Areas of French Polynesia

FRANCE – NEW CALEDONIA

Area 19,000 sq. km

Population 164,173 (April 1989) (Institut Territorial de la Statistique et des Etudes Economiques, Noumea)
Natural increase: No information

Economic Indicators
GNP: No information

Policy and Legislation New Caledonia was annexed by France in 1853 and became an Overseas Territory in 1958. Following constitutional changes in 1985 and 1988, the Territory is divided into three autonomous provinces, administered by a High Commissioner assisted by a four-member Consultative Committee. Law No. 88.1028 of 9 November 1988 gives the three provinces full powers in environmental matters.

Protection of natural heritage is one of the objectives of the "long-term economic and social plan for New Caledonia" adopted by the Territorial Assembly on 21 February 1979. Revised legislation pertaining to parks and reserves came before the Territorial Assembly in 1980 (Délibération No. 108 of 9 May 1980 enforced by Decree No. 1504 of 21 May 1980). Previously, existing laws covered establishment of complete reserves, botanical reserves, forest reserves and marine reserves. The revised legislation introduced the terms réserve naturelle intégrale (strict nature reserve), parc territorial (territorial park) and réserve spéciale (special reserve), and provided more complete definitions of these terms, which in principal relate to IUCN management categories I (strict nature reserve), II (national park) and IV (managed nature reserve), respectively (see Annex).

Protected areas can only be established on land belonging to the Territory, to the State, or on public land. The establishment of such protected areas on private property may be allowed on demand of private individuals. Reserves are not protected against mining activity unless located within mining reserves established under the mining laws (Decree No. 54-1110 of 13 November 1954).

Protection zones can also be established under the Water Resources and Pollution Law (Délibération No. 105 of 26 August 1968), whereby activities likely to endanger water quality can be prohibited or controlled. Decree No. 51-100 of 26 January 1951 lays down conditions for forest classification. Areas of archaeological, historic or tourist interest can also be declared as classified sites under Law No. 56-1106 of 3 November 1956 and Délibération No. 225 of 17 June 1965.

There are a few small zones where activities are restricted for the purpose of "good management of public land". These are afforestation zones and lands allocated to the Forestry Department for the production of timber in which certain measures for the protection of wildlife may also be applied (SPREP, 1985a).

International Activities At the international level, France has entered into a number of obligations and cooperative agreements related to conservation. It is party to the Convention concerning the Protection of the World Cultural and Natural Heritage (World Heritage Convention) which it accepted on 27 June 1975. France became a contracting party to the Convention on Wetlands of International Importance especially as Waterfowl Habitat (Ramsar Convention) on 1 October 1986.

France signed the Convention on the Conservation of Nature in the South Pacific on 12 June 1976, and subsequently ratified it on 20 January 1989. Known as the Apia Convention, it entered into force during 1990. The Convention is coordinated by the South Pacific Commission and represents the first attempt within the region to cooperate on environmental matters. Among other measures, it encourages the creation of protected areas to preserve indigenous flora and fauna.

France also signed the Convention for the Protection of the Natural Resources and Environment of the South Pacific Region (SPREP Convention) on 25 November 1987, with ratification on 17 July 1990. The Convention entered into force during August 1989. Article 14 calls upon the parties to take all appropriate measures to protect rare or fragile ecosystems and threatened or endangered flora and fauna through the establishment of protected areas and the regulation of activities likely to have an adverse effect on the species, ecosystems and biological processes being protected. However, as this provision only applies to the Convention area, which by definition is open ocean, it is most likely to assist with the establishment of marine reserves and the conservation of marine species. It should be noted that the instrument of Approval is accompanied by the following reservation: "The Government of the French Republic, in signing the present Convention declares that, insofar as it is concerned, the prescription of the aforesaid Convention will not cover wastes and other matter entailing a level of pollution caused by radioactivity to a degree less than that prescribed by the recommendations of the International Atomic Energy Agency".

Administration and Management The Service de l'Environnement et Gestion des Parcs et Réserves (Direction du Développement Rural) is responsible for the administration of parks and reserves which are controlled on a day-to-day basis by individual forestry districts, divided between three administrative centres. Délibération No. 38-90/APS of 28 March 1990 makes provision for the creation of a Comité pour la Protection de l'Environnement dans la Province Sud. This body comprises a number of members from government departments, ORSTOM and, at present, the President of the Association pour la Sauvegarde de la Nature Néo-Calédonienne. The Committee's rôle is largely to act in an advisory capacity to the provincial authorities.

Systems Reviews The natural vegetation comprises tropical evergreen rain forest up to 1,000m and tropical montane rain forest above 1,000m. A variant of evergreen rain forest, sometimes with *Araucaria columnaris*, is dominant near the coast on raised coral, especially in the Loyalty Islands and Ile des Pins. Dry sclerophyllous forest is characteristic of drier western slopes. Various types of scrub on acidic and ultrabasic rocks cover about 30% of the land area. Mangroves occur along western coasts. According to a 1974/75 forest inventory, the major vegetation types are dense evergreen forest (22.8% of total land area), Niaouli savanna woodland (13.8%), low scrub in mining areas (25.1%), savanna grassland (21.7%) and scrub (8.3%) (SPREP, 1989). Hunter Island has some grassland with occasional trees; Walpole Island is covered by dense scrub; and Matthew Island has almost no vegetation (Davis *et al.*, 1986). A more detailed vegetation account and vegetation maps are given by Schmid (1978).

A summary account of coral reefs is given in UNEP/IUCN (1988), with specific descriptions of Chesterfield Islands, D'Entrecasteaux Reefs, Great Reef Marine Reserve (Réserves Tournantes sur le Grand Recif), Maître and Amédée Islets Nature Reserves and Yves Merlet Marine Reserve. Grande Terre is surrounded by a barrier reef more than 1,600 km in length, which borders a lagoon of clear and rather shallow water covering 16,000 sq. km, second in size only to the Australian Great Barrier Reef (UNEP/IUCN, 1988).

The protected areas system has relatively good coverage, although most sites were established to protect areas of botanical interest. However, enforcement is weak and it is not certain whether the existing system is adequate to continue to protect such a rich and localised flora.

The two principal threats to the natural flora and fauna are mining and bushfires. Since mining is the basis of the country's economy, measures taken to conserve natural areas do not necessarily include the exclusion of mining activities. In these cases, no disturbance is permitted without prior review and approval. Moreover, before any mine can be developed an environmental impact assessment is carried out by the Commission de Prévention des Dégâts miniers. The environmental impact of mining in New Caledonia is discussed by Dupon (1986). Large areas are burnt annually, frequently deliberately, to attract rain, encourage fresh grass growth or to clear undergrowth. This has led to habitat destruction, denudation of soils and progressive shrinking of forests. There is now pressure for logging in many areas and, as there is still no complete forest classification, destruction can occur without the Department being empowered to take action. Forest

exploitation has been uncontrolled and has contributed to the destruction of forests, or depletion of the more valuable species. Today, the Direction du Développement Rural exercises control to prevent indiscriminate or excessive felling (SPREP, 1989).

An Action Strategy for Protected Areas in the South Pacific Region (SPREP, 1985b) has been prepared. The principal goals of the strategy cover conservation education, conservation policy development, establishment of protected areas, protected areas management and regional and international cooperation. Priority recommendations for New Caledonia are intensification of education and public information activities relating to nature conservation, with special attention being paid to school text-books for use at all levels; adoption of a Territorial "charter" on the protection of the natural and cultural heritage; establishment of new protected areas as follows: a bird reserve on Chesterfield Islands, rotating marine reserves in reef areas, extension of Rivière Bleue Territorial Park and preservation of the lake ecosystems of the Plaine des Lacs; legislative measures to ensure permanency of parks and reserves, redrafting of forestry legislation, taking into account customary ownership; protection of endemic endangered species; establishment of a central structure for the management and planning of protected areas; and support of SPREP's activities and promotion of cooperation among international scientists and experts with a view to furthering knowledge in the field of ecology. The extent to which these objectives have been achieved is not known. It has also been recommended that a protected area be established on Huon Island, to the north of the Territory for the protection of marine turtles (J-L. d'Auzon, pers, comm., 1991).

Addresses

Service de l'Environnement et Gestion des Parcs et Réserves, Direction du Développement Rural, BP 256, Noumea

References

Davis, S.D., Droop, S.J.M., Gregerson, P., Henson, L., Leon, C.J., Lamlein Villa-Lobos, J., Synge, H. and Zantovska, J. (1986). *Plants in danger: what do we know?* IUCN, Gland, Switzerland and Cambridge, UK. 488 pp.

Dupon, J.F. (1986). The effects of mining on the environment of high islands: case study of nickel mining in New Caledonia. *Environmental Case Study* No. 1. SPREP, South Pacific Commission, Noumea, New Caledonia. 5 pp.

Paxton, J. 1989. *The Statesman's yearbook 1989-90*. The Macmillan Press Ltd, London. 1691 pp.

Schmid, M. (1978). The Melanesian forest ecosystem (New Caledonia, New Hebrides, Fiji Islands and Solomon Islands). In: Unesco/UNEP/FAO, *Tropical forest ecosystems*. Unesco, Paris. Pp. 654-683.

SPREP (1980). New Caledonia. Country Report 8. South Pacific Commission, Noumea, New Caledonia.

SPREP (1985a). New Caledonia. In: Thomas, P.E.J. (Ed.), *Report of the Third South Pacific National Parks and Reserves Conference.* Volume III. *Country reviews.* South Pacific Commission, Noumea, New Caledonia. Pp. 125-133.

SPREP (1985b). *Action strategy for protected areas in the South Pacific Region.* South Pacific Commission, Noumea, New Caledonia. 24 pp.

SPREP (1989). New Caledonia. Paper presented at the Fourth South Pacific Conference on Nature Conservation and Protected Areas, Port Vila, Vanuatu, 4-12 September 1989. 11 pp.

UNEP/IUCN (1988). *Coral reefs of the world.* *Volume 3. Central and Western Pacific.* UNEP Regional Seas Directories and Bibliographies. IUCN, Gland, Switzerland and Cambridge, UK/UNEP, Nairobi, Kenya. 378 pp.

ANNEX
Definitions of protected area designations, as legislated, together with authorities responsible for their administration

Title: Délibération No. 108

Date: 9 May 1980

Brief description: Introduced and defined the terms réserve naturelle intégrale (strict nature reserve), parc territorial (territorial park) and réserve spéciale (special reserve), and provided more complete definitions of these terms, which in principal relate to IUCN management categories I (strict nature reserve), II (national park) and IV (managed nature reserve), respectively. Enforced by Decree No. 1504 of 21 May 1980.

Administrative authority: Service de l'Environnement et Gestion des Parcs et Réserves (Direction du Développement Rural)

Designations:

Réserve naturelle intégrale (Strict nature reserve) Areas in which all hunting, fishing, forestry, agriculture or mining, all activities likely to modify the landscape or vegetation, disturbance to flora and fauna, all floral and faunal introductions, whether indigenous, wild, exotic or domestic, and collection of botanical and geological samples are strictly forbidden. It is also forbidden to enter, move about, camp, or carry out scientific research in such reserves without written permission of the Service de l'Environnement et Gestion des Parcs et Réserves, and flying over the reserves is regulated. Réserves naturelles intégrales may not be established except where all prospecting, mineral research or exploitation have been prohibited under Article 2210-1 of the present code.

Parc territorial (Territorial park) Set aside for the conservation of flora and fauna but also to cater for public education and recreation. Flora and fauna are fully protected and gathering of botanical or geological samples is forbidden without express permission of the Service de l'Environnement et Gestion des Parcs et Réserves and only for scientific purposes. There is provision for permission to be given by the service for facilities such as roads, paths, restaurants, hotels and associated infrastructure to be built for educational and recreational purposes. NB Subsequent to the recently introduced constitutional changes, parc territorial might be more properly termed parc provincial, although it is not known if this is the case.

Réserve spéciale (Special reserve) Areas where certain activities may be prohibited or regulated for specific environmental protection objectives. Réserves spéciales are divided into:

Réserve spéciale botanique (Special botanical reserve) Created for the restoration and conservation of habitats, and the conservation of rare, noteworthy or threatened plant species. It is expressly forbidden to modify the landscape or carry out any acts detrimental to the vegetation except with authorisation given by the Head of the Service de l'Environnement et Gestion des Parcs et Réserves.

Réserve spéciale de faune (Special faunal reserve) In which particular measures for the protection of one or more animal species may be taken. Réserves spéciales de faune include a number of marine areas and islands that are totally or partially protected both for the purpose of species conservation and for stock management and scientific studies.

Réserve spéciale marine (Special marine reserve)

SUMMARY OF PROTECTED AREAS

Map[†] ref.	*National/international designation* Name of area	IUCN management category	Area (ha)	Year notified
	Strict Nature Reserve			
1	Montagne des Sources	I	5,870	1950
	Special Botanical Reserves			
2	Cap Ndua (Southern Botanical Reserve #5)	IV	830	1972
3	Chutes de la Madeleine	IV	400	1990
4	Fausse Yaté (Southern Botanical Reserve #2)	IV	386	1972
5	Forêt Cachée (Southern Botanical Reserve #7)	IV	635	1972
6	Forêt Nord (Southern Botanical Reserve #4)	IV	280	1972
7	Forêt de Sailles	IV	1,100	1983
8	Mont Humboldt	IV	3,200	1950
9	Mont Mou	IV	675	1950
10	Mont Oungoué (Southern Botanical Reserve #3)	IV	307	1972
11	Mont Panié	IV	5,000	1950
12	Pic Ningua	IV	350	
13	Pic du Pin (Southern Botanical Reserve #6)	IV	1,482	1972
14	Yaté Barrage (Southern Botanical Reserve #1)	IV	546	1972
	Special Fauna and Flora Reserves			
15	L'Ilot Maître	V	154	1981
16	Mont Do	IV	300	1981
	Special Faunal Reserves			
17	Aoupinie	IV	5,400	1975
18	Haute Yaté (Includes Rivière Bleue Territorial Park)	IV	15,900	1960
19	L'Etang de Koumac	VIII	53	1989
20	L'Ile Pam	IV	460	1966
21	L'Ilot Leprédour	IV	760	1941
	Special Marine Reserves			
22	La Dieppoise	V	1	1990
23	Réserve Spéciale Tournante de Marine Faune	VIII	35,570	1981
24	Yves Merlet	I	16,700	1970
	Territorial Parks			
25	Ouen-Toro	II	44	1989
26	Parc Territorial du Lagon Sud: Amédée Islet	V	154	1981
27	Parc Territorial du Lagon Sud: Bailly Island	V	314	1989
28	Parc Territorial du Lagon Sud: Canard Island	V	50	1989
29	Parc Territorial du Lagon Sud: Laregnare Island	V	362	1989
30	Parc Territorial du Lagon Sud: Signal Island	V	181	1989
31	Rivière Bleue	IV	9,054	1980
32	Thy	II	1,133	1980
	Forest Reserves			
33	"South" of New Caledonia	VIII	4,665	1980
34	Col d'Amieu	VIII	12,368	1970
35	Kuebini	VIII	58	1980
36	Mont Mou	IV	4,363	
37	Ouenarou	VIII	1,171	1959
38	Povilla	VIII	600	1971
39	Tangadiou	VIII	1,016	1970
40	Tango	VIII	29,089	1979
41	Tiponite	VIII	1,085	1970

Map[†] ref.	Name of area	IUCN management category	Area (ha)	Year notified
	Unclassified			
42	"Michel Corbasson"	II	35	
45	Branch Nord Dumbea et Couvelee	VI		

[†]Locations of some protected areas are shown on the accompanying map.

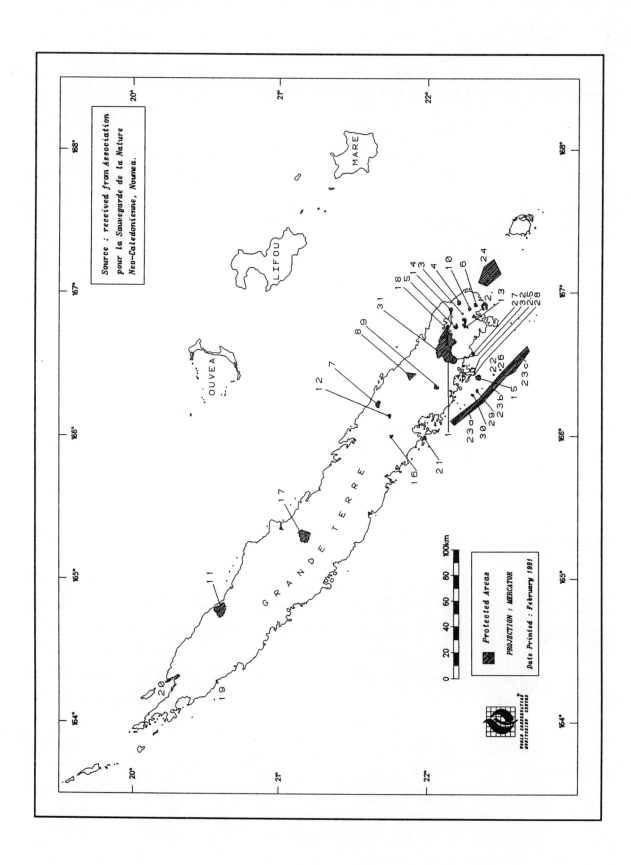

Protected Areas of New Caledonia

FRANCE – WALLIS AND FUTUNA

Area 210 sq. km

Population 13,000 (1986) (SPREP, 1986)
Natural increase: 4.0% per annum (Dahl, 1986)

Economic Indicators
GNP: No information

Policy and Legislation Wallis and Futuna is an Overseas Territory of France but the legal status, as provided for by Law No. 61-814 of 29 July 1961, gives the Territory jurisdiction in environmental affairs. As all matters relating to land tenure lie exclusively within the jurisdiction of the traditional leaders, the administration's scope of action is extremely limited. Traditional law is applied to the day-to-day settlement of local affairs. There is no written protected area legislation. As a result of population increases and the Territory's exposure to outside influences, traditional laws no longer provide a sufficient means of control to preserve the existing land and sea resources. Supplementary legislation is desirable in respect of town planning and the protection of wooded areas, the water table, water catchments and coastal zones etc. (SPREP, 1982).

The Long-term Economic and Social Development Plan, passed by the Territorial Assembly on 24 July 1979, includes a chapter on the protection of the natural heritage of the islands (Chapter 9). Two of the priorities listed are protection of the coastal zone, and protection of natural sites (Alofi Forest). Restrictions and regulations are applied to specific areas and are decreed orally by the King and the Council of Ministers and relayed by word of mouth by the district or village chiefs to their subjects. Today, radio broadcasts are also used to disseminate such instructions (SPREP, 1985).

International Activities At the international level, France has entered into a number of obligations and cooperative agreements related to conservation. It is party to the Convention concerning the Protection of the World Cultural and Natural Heritage (World Heritage Convention) which it accepted on 27 June 1975. France became a contracting party to the Convention on Wetlands of International Importance especially as Waterfowl Habitat (Ramsar Convention) on 1 October 1986.

France signed the Convention on the Conservation of Nature in the South Pacific on 12 June 1976, and subsequently ratified it on 20 January 1989. Known as the Apia Convention, it entered into force during 1990. The Convention is coordinated by the South Pacific Commission and represents the first attempt within the region to cooperate on environmental matters. Among other measures, it encourages the creation of protected areas to preserve indigenous flora and fauna.

France also signed the Convention for the Protection of the Natural Resources and Environment of the South Pacific Region (SPREP Convention) on 25 November 1987, with ratification on 17 July 1990. The Convention entered into force during August 1990. Article 14 calls upon the parties to take all appropriate measures to protect rare or fragile ecosystems and threatened or endangered flora and fauna through the establishment of protected areas and the regulation of activities likely to have an adverse effect on the species, ecosystems and biological processes being protected. However, as this provision only applies to the Convention area, which by definition is open ocean, it is most likely to assist with the establishment of marine reserves and the conservation of marine species. It should be noted that the instrument of Approval is accompanied by the following reservation: "The Government of the French Republic, in signing the present Convention, declares that, insofar as it is concerned, the prescription of the aforesaid Convention will not cover wastes and other matter entailing a level of pollution caused by radioactivity to a degree less than that prescribed by the recommendations of the International Atomic Energy Agency".

Administration and Management No department has overall responsibility for coordination of activities relating to national environmental policy.

Systems Reviews Wallis and Futuna comprises three main islands: Uvea (Iles Wallis), Futuna and the uninhabited Alofi (Iles de Horne).

Uvea (96 sq. km) is a low volcanic island, with a maximum altitude of 145m, with volcanic lakes and seasonal springs. It is surrounded by a barrier reef, some 3-4km offshore, with about 22 reef islets. By contrast, Futuna (80 sq. km) and Alofi (36 sq. km), 250km to the south-west, are mountainous, with steep slopes interrupted by a series of uplifted coral tiers, reaching 524m and 416m, respectively.

When the islands were first settled the natural vegetation appears to have consisted largely of forest, today in great jeopardy. Still used for supplying timber for building, it has also been cleared for agricultural land. The harmful effects of systematic and localised clearing of the vegetation have been further aggravated by bushfires occurring during the dry season. In the entire low-altitude zone that is permanently settled, the primary vegetation has been radically modified by pre- and post-European plant introductions and the large-scale extensions of coconut groves.

Evergreen closed forest covered about 15% of Uvea in 1983, the remainder including secondary forest and toafa (a poor, scrubby formation) with *Pandanus*, grasses and ferns (Morat *et al.*, 1983). Much of the original vegetation has been cleared for coconut, breadfruit and bananas. The establishment of two airfields and a network of roads during World War Two increased the damage described by Burrows (1937). Now only patches of forest remain in the southern central and western region. Everywhere else, the forest has been replaced by gardens, fallow areas carrying a secondary bush, and, especially in the northern half of the island, by very poor lands on which only toafa grows.

In Futuna, which is believed to have been settled earlier than Uvea, the primary forest has receded to the upper parts of the hills and into almost inaccessible gorges. About 30% of Futuna still has dense forest cover, particularly in the montane regions above 400m (Davis *et al.*, 1986), but an extensive area of the central plateau bears only the very poor toafa vegetation. The steep outer slopes running down to the coast are cleared and given over to dryland crops grown under often quite difficult conditions (SPREP, 1986).

Alofi, which is uninhabited due to the lack of reliable water supplies, continues to support primary rain forest (Dahl, 1986) on as much of 70% of its surface (SPREP, 1986). Despite this predominance, toafa vegetation exists in a few places in central areas.

Analysis of aerial photographs suggests that the limits of toafa on Uvea and Futuna have not changed greatly during the last 20-40 years, possibly due to the unsuitability of soils for further agricultural expansion. However, aerial photography also confirms the diminishing primary cover on Uvea, being replaced by perennial plantations or secondary fallow cover. Although shortening of the fallow period to less than ten years is likely to lead to the appearance of toafa vegetation, reafforestation with pines accompanied by the development of a thick undergrowth, coconut plantations and trials with fodder crops have been successful in the toafa areas of both Uvea and Futuna (SPREP, 1986).

An account of the coral reefs, which occur extensively around Uvea and Futuna and the north-west of Alofi, is given in UNEP/IUCN (1988). The local population obtains the greater part of its protein from marine products, particularly fish, and reefs and lagoons have been heavily fished. Fishing is carried out mainly by line, net, speargun or harpoon. Deep water bottom fishing is increasing. *Trochus* spp. from Uvea lagoon are exported in large numbers to New Caledonia (SPREP, 1982).

Although there are no designated protected areas, the "Vao Tapu" forests are protected by customary taboos. In the Uvea districts, utilisation of the primary forest is subject to control of the customary authority, but damage appears to have increased since the mid-1960s, particularly as a result of the extension of individual houses along roads, which in turn is directly related to the increase in motor vehicles and extension of water and electricity supplies.

Despite as much as half of the population living overseas, there is a serious question over the long-term availability of land. Consequently, there is an awareness that all land requires some form of protection. Customary leaders attempt to dissuade the sale of land to foreigners, in order to better meet the future needs of the indigenous population (SPREP, 1985). A discussion of environmental issues is given in SPREP (1982). Major problems include damage caused by semi-feral pigs, soil erosion, loss of forest on Wallis and Futuna, and overfishing around Futuna. In general, the modest level of development and inhospitable terrain, for example on Alofi, has led to few severe environmental problems. However, with increasing pressure for land, resources and development, the current situation may change rapidly.

Addresses

Chef de la Mission des Affaires intérnationales et de la Coopération, Direction de l'Espace rural/Forêt, Ministère de l'Agriculture, 1 ter Ave. de Lowendal, 75700 Paris, France

References

Burrows, E.C. (1937). Ethnology of Uvea (Wallis Island). *Bulletin of the Bernice P. Bishop Museum.* No. 138. 239 pp.

Dahl, A.L. (1986). *Review of the protected areas system in Oceania.* IUCN, Gland, Switzerland and Cambridge, UK/UNEP, Nairobi, Kenya. 328 pp.

Davis, S.D., Droop, S.J.M., Gregerson, P., Henson, L., Leon, C.J., Lamlein Villa-Lobos, J., Synge, H. and Zantovska, J. (1986). *Plants in danger: what do we know?* IUCN, Gland, Switzerland and Cambridge, UK. 488 pp.

Morat, P. Veillon, J.M. and Hoff, M. (1983). Introduction à la végétation et à la flore du territoire de Wallis et Futuna. ORSTOM, Nouméa, New Caledonia. 24 pp. (Unseen)

SPREP (1982). Wallis and Fortuna Islands. *Country Report* 19. South Pacific Commission, Noumea, New Caledonia. 14 pp.

SPREP (1985). Wallis and Futuna. In: Thomas, P.E.J. (Ed.), *Report of the Third South Pacific National Parks and Reserves Conference.* Volume III. *Country reviews.* South Pacific Commission, Noumea, New Caledonia. Pp. 229-231.

SPREP (1986). Wallis and Futuna: man against the forest. *South Pacific Study* No. 2. South Pacific Commission, Noumea, New Caledonia. 6 pp.

UNEP/IUCN (1988). *Coral reefs of the world. Volume 3. Central and Western Pacific.* UNEP Regional Seas Directories and Bibliographies. IUCN, Gland, Switzerland and Cambridge, UK/UNEP, Nairobi, Kenya. 378 pp.

SUMMARY OF PROTECTED AREAS

National/international designation Name of area	IUCN management category	Area (ha)	Year notified
Unclassified Lalolalo Vao Tupu (Forbidden Forest)	I	30	

REPUBLIC OF KIRIBATI

Area The total land area is approximately 684 sq. km, scattered over more than 5 million sq. km of ocean.

Population 66,250 (1987 estimate) (Paxton, 1990)
Natural increase: 2.1% per annum (Dahl, 1986)

Economic Indicators
GNP: US$ 649 per capita (1988)

Policy and Legislation In 1980 the Government of Kiribati published a statement of its policy concerning nature conservation in the Line and Phoenix Islands. This recognised the need to integrate conservation and development with respect to the islands' natural resources. The role of conservation was defined in terms of providing for the present and future social and economic needs of the country (Garnett, 1983).

The legal basis for nature conservation is the Wildlife Conservation Ordinance (1975), amended in 1979. Under the new Ordinance, the 1938 Gilbert and Ellice Islands Colony Wild Birds Protection Ordinance was repealed and the status of bird sanctuaries was changed to wildlife sanctuaries. The 1975 Ordinance makes, *inter alia*, the following provisions. First, all 31 regularly occurring bird species, and their nests and eggs, are fully protected throughout Kiribati. All turtles are fully protected on land, although it is not clear from the legislation whether protection extends to marine areas. Secondly, under Section 8(1), wildlife sanctuaries may be declared and closed areas may be declared within such wildlife sanctuaries. Thirdly, Section 11 provides wildlife wardens with powers of search and arrest.

This current legislation is weak, because measures for the protection of vegetation, prohibiting the introduction of plants and animals, preventing fire, removal of soil, and dumping of refuse, and the control of vehicles are lacking. In particular, there is a lack of effective protection for wildlife sanctuaries, within which it is possible to clear vegetation without contravening the law. It would seem that under current legislation the only areas adequately protected are those additionally designated as closed areas.

It should be noted that, although Article 14(1) of the Constitution guarantees freedom of movement, restrictions required in the interests of environmental conservation are deemed not to be in contravention of the Article.

Prior to contact with Europeans, land owners in Kiribati held tenure of reefs and lagoons adjacent to their lands and had exclusive rights to fisheries and passage. Most land, particularly in the South Gilberts, was owned by groups of extended families (utu) who lived in small, scattered hamlets (kaainga), although in the northern atolls the ruling king had control of a large area of land, reefs and lagoons, and dispensed fishing rights to the various clans in the domain. In the late 19th century this system began to break down: under British colonial rule, sea tenure *per se* was not recognised, although there was an attempt to modify traditional fishing rights and the government did recognise tenure of fish weirs, reclaimed areas, fish ponds and other accretions (Zann, 1985). However, there are still a number of laws and customs regulating different aspects of fishing activities on many of the atolls. These are frequently formulated and applied by individual Island Councils.

International Activities Kiribati is not yet party to any of the international conventions or programmes that directly promote the conservation of natural areas, namely the Convention concerning the Protection of the World Cultural and Natural Heritage (World Heritage Convention), Unesco Man and the Biosphere Programme and the Convention on Wetlands of International Importance especially as Waterfowl Habitat (Ramsar Convention).

At a regional level, Kiribati has not ratified the Convention on the Conservation of Nature in the South Pacific, 1976. Known as the Apia Convention, it entered into force during 1990. The Convention is coordinated by the South Pacific Commission and represents the first attempt within the region to cooperate on environmental matters. Among other measures, it encourages the creation of protected areas to preserve indigenous flora and fauna.

Kiribati is party to the South Pacific Regional Environment Programme (SPREP) but has not ratified the Convention for the Protection of the Natural Resources and Environment of the South Pacific Region, 1986 (SPREP Convention). The Convention entered into force during August 1990. Article 14 calls upon the parties to take all appropriate measures to protect rare or fragile ecosystems and threatened or endangered flora and fauna through the establishment of protected areas and the regulation of activities likely to have an adverse effect on the species, ecosystems and biological processes being protected. However, as this provision only applies to the Convention area, which by definition is open ocean, it is most likely to assist with the establishment of marine reserves and the conservation of marine species.

Other international and regional conventions concerning environmental protection to which Kiribati is party are reviewed by Venkatesh *et al.* (1983).

Administration and Management The Wildlife Conservation Unit, established in 1977 under the Ministry of the Line and Phoenix Islands and based on Kiritimati, is the only government division responsible for conservation management in the Line and Phoenix Islands (Garnett, 1983). In 1985 the Unit comprised one wildlife warden and two assistants. The bulk of the

Unit's work is on Kiritimati; the other sanctuaries being protected by their remoteness (SPREP, 1985a). The Unit has as its principal responsibilities enforcement of the Wildlife Conservation Ordinance, education/public awareness, survey and research, advice to governments, control of introduced species, and tourism, though it is not the only agency concerned in each of these activities. There is no administrative body responsible for conservation in the Gilbert Islands.

Systems Reviews The Republic of Kiribati comprises all the Gilbert and Phoenix Islands, eight of the eleven Line Islands (the other three, Jarvis, Palmyra and Kingman Reef being dependencies of the USA) and Banaba (Ocean Island), lying approximately between longitudes 169°E and 147°W and latitudes 5°N and 12°S. The islands extend nearly 5,000km from east to west and straddle both the Equator and the International Date Line. All the islands are low-lying coral atolls (UNEP/IUCN, 1988), rising to no more than 3-4m above mean sea level, with the exception of Banaba which is a raised reef.

Most of the natural vegetation of the larger islands has been replaced by coconut plantations, breadfruit and *Pandanus* spp. (Davis *et al.*, 1986). However, considerable ecological diversity remains on the different islands, largely because they fall within a zone of steep rainfall gradients. Thus, Kiritimati has tracts of open scrub and grassland, whilst Teraina, 380km to the north-west, has a freshwater lake surrounded by peat bogs and woodland rich in epiphytes and fern undergrowth (Perry, 1980). There is relatively little information on the marine environment, although reefs at Tabuaeran, Teraina, Tarawa and Onotoa have been described (UNEP/IUCN, 1988).

The current protected areas network comprises wildlife sanctuaries covering the whole of three of the Phoenix Islands and four of the Line Islands. In addition, closed areas have been established within Kiritimati Wildlife Sanctuary, and covering all of Malden and Starbuck islands. There are no protected areas in the Gilbert Islands.

Most habitats found in Kiribati are represented in the protected areas system, although five are omitted: *Guettarda* forest; *Pemphis acidula* scrub; freshwater marsh; freshwater lake; and brackish lagoon. The existing network includes some of the most important sea bird colonies in the Pacific (Hay, 1986).

The most profound threat, both to protected areas and the country as a whole, lies in the putative rise in sea levels, caused by global climatic warming. This could lead to salinisation of freshwater aquifers, increased erosion and possibly inundation rendering the country uninhabitable (Pernetta, 1988). Alien species introduced by man affect all islands in the Line Group, with the exception of Vostok Island. Feral cats, which may number up to 2,000 on Kiritimati, have driven 10 or 11 bird species (out of 18) to nest only on isolated islets (Perry, 1980). Cats

appear to be increasing in numbers, and bird populations are also threatened by poaching, especially red-tailed tropic bird *Phaethon rubricauda*, red-footed booby *Sula sula* and masked booby *Sula dactylatra*. Many new vehicles have been brought to the island, facilitating much greater human access to the bird colonies (E.A. Schreiber, pers. comm., 1989). Exotic plant species have been extensively introduced. Although the majority may have no effect on sea birds, a few may be deleterious. Other natural threats include periodic droughts (Dahl, 1986) and climatic perturbations such as the El Niño Southern Oscillation (ENSO). The latter is believed responsible for the precipitous decline in sea bird populations on Kiritimati in 1982-83 (Schreiber and Schreiber, 1984). The 1986-1987 ENSO set back the recovery of bird populations from the 1982-1983 ENSO (E.A. Schreiber, pers. comm., 1989)

An Action Strategy for Protected Areas in the South Pacific Region (SPREP, 1985b) has been developed. The principal goals of the strategy cover conservation education, conservation policy development, establishment of protected areas, protected area management and regional and international cooperation. Priority recommendations for Kiribati are as follows: formulate a national conservation strategy; and implement the Feral Animal Eradication Programme on Kiritimati. Proposals for a cat eradication programme were prepared by the Wildlife Conservation Unit in 1983 in conjunction with the New Zealand Wildlife Service (SPREP, 1985a). However, this has not yet been implemented (UNEP/IUCN, 1988).

Dahl (1980) recommends both the upgrading of the existing network and the establishment of new reserves, to include appropriate samples of atoll, forest, marine and lagoon environments. This would entail a communication link and surveillance centre on Kanton, retaining the extant reserves on Birnie, McKean and Rawaki (Phoenix) and creating new sites on Enderbury, Orona and possibly Manra. Most of the Line Islands, in particular Caroline and Malden, are considered candidates for reserve status, especially if existing predators can be controlled. Flint, Caroline, Kanton and Enderbury require protection as turtle breeding areas. The bogs, and possibly the lake on Tabuaeran, including adequate areas of Polynesian warbler habitat, should be protected, in addition to seabird breeding areas (Dahl, 1980).

Garnett (1983) makes a comprehensive set of recommendations, principally covering Kiritimati, which addresses aspects of legislation, development and land use planning, and conservation management. Specific recommendations include: amending Section 8 of the 1975 Wildlife Conservation Ordinance to give adequate protection to important ecosystems; gazetting Enderbury Island as a wildlife sanctuary and closed area; upgrading Birnie, McKean, Rawaki and Vostok islands to closed area status; and designating specific areas on Kiritimati as wildlife sanctuaries and closed areas, as opposed to the whole island as is currently the case (Garnett, 1983).

Dahl (1986) notes that there are no protected areas in the Gilbert Islands and suggests that small forested islands with seabird rookeries on Butaritari and Nonouti might be considered for reserves under local management. Marine reserves may be required for fisheries management. The principal omissions from the protected areas systems are bogs in the Line and Phoenix Islands and other natural habitats on Teraina (Dahl, 1986).

Addresses

The Director, Wildlife Conservation Unit, Kiritimati

References

Dahl, A.L. (1980). Regional ecosystems survey of the South Pacific area. *SPC/IUCN Technical Paper 179*. South Pacific Commission, Noumea, New Caledonia. 99 pp.

Dahl, A.L. (1986). *Review of the protected areas system in Oceania.* IUCN, Gland, Switzerland and Cambridge, UK/UNEP, Nairobi, Kenya. 328 pp.

Davis, S.D., Droop, S.J.M., Gregerson, P., Henson, L., Leon, C.J., Lamlein Villa-Lobos, J., Synge, H. and Zantovska, J. (1986). *Plants in danger: what do we know?* IUCN, Gland, Switzerland and Cambridge, UK. 488 pp.

Douglas, G. (1969). Checklist of Pacific Oceanic Islands. *Micronesica* 5(2): 327-463.

Fosberg, F.R. (1953). Vegetation of central Pacific atolls. *Atoll Research Bulletin* 23: 1-26. (Unseen)

Garnett, M.C. (1983). A management plan for nature conservation in the Line and Phoenix Islands. Two volumes. Unpublished. 436 pp.

Hay, R. (1986). Bird conservation in the Pacific Islands. *ICBP Study Report* No. 7. International Council for Bird Preservation, Cambridge, UK. 102 pp.

Pernetta, J.C. (1988). Projected climate change and sea level rise: a relative impact rating for countries of the South Pacific Basin. In: MEDU joint meeting of the task team on the implications of climatic change in the Mediterranean. Split, Yugoslavia, 3-7 October. Pp. 1-11.

Paxton, J. (Ed.). (1990). *The Statesman's yearbook.* 127th Edition. The Macmillan Press Ltd, London. 1690 pp.

Perry, R. (1980). Wildlife Conservation in the Line Islands, Republic of Kiribati (formerly Gilbert Islands). *Environmental Conservation* 7: 311-318.

Schreiber, R.W. and Schreiber, E.A. (1984). Central Pacific seabirds and the El Niño Southern Oscillation: 1982-1983 perspectives. *Science* 225: 713-716.

SPREP (1985a). Kiribati. In: Thomas, P.E.J. (Ed.), *Report of the Third South Pacific National Parks and Reserves Conference.* Volume III. *Country reviews.* South Pacific Commission, Noumea, New Caledonia. Pp. 115-124.

SPREP (1985b). *Action strategy for protected areas in the South Pacific Region.* South Pacific Commission, Noumea, New Caledonia. 24 pp.

UNEP/IUCN (1988). *Coral reefs of the world. Volume 3. Central and Western Pacific.* UNEP Regional Seas Directories and Bibliographies. IUCN, Gland, Switzerland and Cambridge, UK/UNEP, Nairobi, Kenya. 378 pp.

Venkatesh, S., Va'ai, S. and Pulea, M. (1983). An overview of environmental legislation in the South Pacific countries. *SPREP Topic Review* No. 13. South Pacific Commission, Noumea, New Caledonia. 63 pp.

Zann, L.P. (1985). Traditional management and conservation of fisheries in Kiribati and Tuvalu atolls. In: Ruddle, K. and Johannes, R.E. (Eds), *The traditional knowledge and management of coastal systems in Asia and the Pacific.* Unesco/ROSTEA, Jakarta. Pp. 53-77.

ANNEX
Definitions of protected area designations, as legislated, together with authorities responsible for their administration

Title : Wildlife Conservation Ordinance

Date: 1975, amended in 1979

Brief description: Under the Ordinance the 1938 Gilbert and Ellice Islands Colony Wild Birds Protection Ordinance was repealed and the status of bird sanctuaries was changed to wildlife sanctuaries.

Administrative authority: The Director, Wildlife Conservation Unit, Kiritimati

Designation:

Wildlife sanctuary Under Section 8(1), the Minister of the Line and Phoenix Islands, after consultation with the Council of Ministers, may declare any area to be a wildlife sanctuary. Under Section 8(2), it is stipulated that "no person shall in a wildlife sanctuary hunt, kill or capture any bird or other animal (other than a fish) or search for, take or wilfully destroy, break or damage the eggs or nest of any bird or other animal, except under and in accordance with the terms of a valid written licence granted to that person by the Minister under this section".

Closed area The Ordinance makes further provision by allowing closed areas to be declared within wildlife sanctuaries. With the exception of appropriate officials and licensed individuals, entry into a closed area is prohibited under Section 8(6).

SUMMARY OF PROTECTED AREAS

Map[†] ref.	*National/international designation* Name of area	IUCN management category	Area (ha)	Year notified
	Closed Areas			
1	Cook Islet Closed Area (Kiritimati WS)	I	3	1975
2	Motu Tabu Islet Closed Area (Kirimati WS)	I	1	1975
3	Motu Upua Closed Area (Kiritimati WS)	I	4	1975
4	Ngaontetaake Islet Closed Area (Kiritimati WS)	I	2	1979
	Wildlife Sanctuaries			
5	Birnie Island	IV	20	1975
6	Kiritimati	Unassigned	32,100	1960
7	Malden Island (Closed Area)	I	3,930	1975
8	McKean Island	IV	57	1975
9	Phoenix Island (Rawaki)	IV	6,500	1975
10	Starbuck (Closed Area)	I	16,200	1975
11	Vostok Island	IV	24	1979

[†]Locations of most protected areas are shown on the accompanying map.

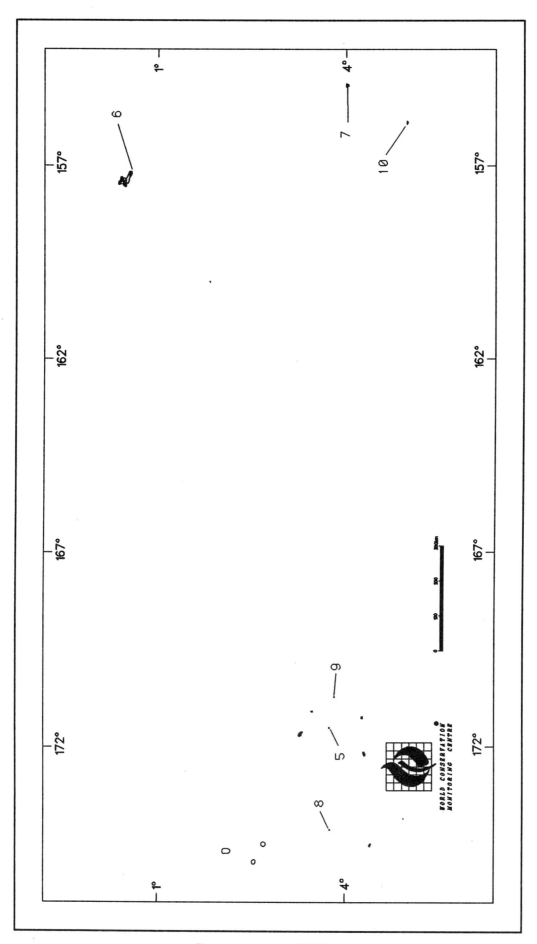

Protected Areas of Kiribati

REPUBLIC OF THE MARSHALL ISLANDS

Area 181 sq. km

Population 40,609 (July 1988)
Natural increase: No information

Economic Indicators
GNP: No information

Policy and Legislation Since October 1986 the Marshall Islands have been a sovereign state in free association with the United States of America which remains responsible for defence and provides financial support. A number of Trust Territory regulations remain in force, but will be revised in due course and cover topics such as water supply, pesticides and sewage disposal. The Trust Territory Endangered Species Act (TTPI Public Law No. 6.55 of 1975), which allowed for acquisition of land or water for the purpose of conserving threatened species, is no longer applicable.

US regulations still apply at the Kwajalein Missile Range and other areas, and Kwajalein is likely to experience considerable development pressure due to activities relating to the Strategic Defence Initiative. Other US legislation relevant to the Trust Territories includes the National Environmental Protection Act, enacted by the Nitijela (Parliament) on 19 December 1984. This Act includes many Trust Territory regulations which are under review such that the Act will address the needs of the Marshall Islands in a more specific manner (SPREP, 1989). The National Environmental Protection Act, along with the Coastal Conservation Act (1988), charges the Marshall Islands Environmental Protection Authority with responsibility, *inter alia*, to "preserve and improve the quality of the environment". Amongst other measures, the Act makes provision for the preservation of important historical, cultural and natural aspects of the nation's heritage.

There is currently no protected areas legislation, although most other environmental issues are quite well covered by existing legislation. Protected areas established prior to independence are no longer recognised. Thus, two reserves, Bokaak (Taongi) and Bikar, set up by Order of the District Administrator during the 1950s, are not recognised by the government nor by the Irooj Laplap (paramount chief of the Ratak Chain). At present (1990) it is not clear whether protected areas legislation will take the form of additional regulations within the National Environmental Protection Act, or be part of an entirely new Conservation Act. Thomas *et al.* (1989) present a suggested outline of legislation which would enable the establishment of a legal and administrative framework for the coordinated development and implementation of policies and programmes for nature conservation and protected areas in the Marshall Islands. Virtually all land is held in traditional ownership and any legislation must make provision for strong landowner involvement in protected area management.

International Activities The Marshall Islands are not party to the three major international conventions concerned with the protection of nature, namely the Convention concerning the Protection of the World Cultural and Natural Heritage (World Heritage Convention), Unesco Man and the Biosphere Programme and the Convention on Wetlands of International Importance Especially as Waterfowl Habitat (Ramsar Convention).

The Convention on the Conservation of Nature in the South Pacific (1976) has been neither signed nor ratified. Known as the Apia Convention, it entered into force during 1990. The Convention is coordinated by the South Pacific Commission and represents the first attempt within the region to cooperate on environmental matters. Among other measures, it encourages the creation of protected areas to preserve indigenous flora and fauna.

The Convention for the Protection of the Natural Resources and Environment of the South Pacific Region (SPREP Convention) has been signed and was ratified on 4 May 1987. The Convention entered into force during August 1990. Article 14 calls upon the parties to take all appropriate measures to protect rare or fragile ecosystems and threatened or endangered flora and fauna through the establishment of protected areas and the regulation of activities likely to have an adverse effect on the species, ecosystems and biological processes being protected. However, as this provision only applies to the Convention area, which by definition is open ocean, it is most likely to assist with the establishment of marine reserves and the conservation of marine species.

Administration and Management Until independence, this role would have been the responsibility of the Chief Conservationist for the Trust Territory. The Director of Coast Conservation, who may be the General Manager of the Environmental Protection Authority, is responsible for submitting a comprehensive Coastal Zone Management Plan, and permits for development activities consistent with that plan. The Environmental Protection Authority is an autonomous body established under the provisions of the 1984 National Environmental Protection Act, with a broad remit in environmental protection and management, including, potentially, the protection of natural sites. The General Manager heads a staff of ten, including a full-time Public Education Officer, two Environmental Specialists, a Consultant to the General Manager and two US Department of the Interior assisted positions (Director of Water Quality Monitoring and Legal Counsel) (SPREP, 1989). Thomas *et al.* (1989) propose a three-fold administrative structure comprising an *ad hoc* National Conservation Service, local Atoll Conservation Authorities and a Conservation Service within the Ministry of Resources and Development or the Environmental Protection Authority.

Systems Reviews The Marshall Islands are the easternmost island group of Micronesia, comprising two island chains, Ralik (18 atolls) and Ratak (15 atolls), between 8°-12°N and 162°-172°E. Small remnants of atoll/beach forest (mostly comprising pan-Pacific species) occur on some of the northern atolls, for example Wotho, Ujae, and some of the islets of Kwajalein. Small areas of mangrove forest are found on Jaluit, Ailinglaplap and Mejit (Dahl, 1980). Vegetation on almost all the Marshall Islands has been extensively modified; most atolls have coconut and breadfruit plantations. Fosberg (1973) provides an account of the condition and status of the forests and brief summaries for each of the islands are given by Dahl (1986) and Douglas (1969). There have been few descriptions of the coral reefs, but in general windward ocean reef slopes have submarine terraces and often descend gradually, in contrast to leeward ocean reef slopes which descend nearly vertically. Detailed accounts for Bikini, Kawajalein, Enewetak, Majuro and Arno are given in UNEP/IUCN (1988).

The Marshall Islands have abundant marine resources and the potential for commercial fisheries and mariculture is high. Copra production could be developed further. There may be some important mineral resources, including deep sea cobalt-rich manganese crust deposits and possibly some phosphate rock deposits in the lagoon. However, the potential for developing mineral resources is either unknown or not economically feasible at present. Consequently, the major options for economic development appear to lie in increased use of living marine resources, coastal tourism, mariculture and enhanced low-input agriculture. Most of the atolls and individual islands are inhabited, and potable water is available on the southern islands but scarce in the northern islands. Only six atolls and one island are currently unoccupied, namely: Ailinginae, Bikar, Bikini, Erikub, Rongerik, Taka and Taongi Atolls, and Jemo Island (L.S. Hamilton, pers. comm., 1989).

Bikini and Enewetak were used as atomic weapon test sites by the USA from 1946 to the 1960s. The full extent of the disruption to the atoll ecosystems has not been fully documented or evaluated. The 1954 Bravo test, the first and largest thermonuclear explosion by the US, spread fallout to several of the northern Marshall atolls and caused inestimable damage and social disruption to many reef and island communities (UNEP/IUCN, 1988). The potentially most threatening environmental hazard is that posed by increasing sea-levels due to global warming. Threats to freshwater supplies, land loss and episodic destruction through hurricanes may make the country uninhabitable if worst case scenarios are realised (Pernetta, 1988).

The Marshall Islands were included in a review of the protected areas systems of Oceania (Dahl, 1986), which indicated a need to reinstate the protected status of Taongi and Bikar as well as gazetting protected areas for the conservation of birds, remaining natural vegetation and at least one major coral reef. A more recent and more detailed review of the biological diversity of the northern atolls and their potential as protected areas (Thomas *et al.*, 1989) also confirms the significance of Taongi and Bikar and identifies sites of special conservation significance, especially for seabird protection. The survey, a cooperative venture, indicates a willingness on the behalf of the government to contemplate the establishment of protected areas.

Following independence there has been a strong desire to establish the economic foundation for the future prosperity of the nation, indicating a need for careful natural resource management. Neither the legislation nor the policy on protected areas was clear while the country was a part of the Trust Territory, and this is one reason why so few protected areas were actually established. This is largely because the United States, while administering the Territory, did not own property, and hence had no direct jurisdiction over designation of protected areas. However, there is a strong possibility that a protected areas system will be implemented in the near future (L.S. Hamilton, pers. comm., 1989).

Addresses

Ministry of Resources and Development, Majuro, Republic of the Marshall Islands, RMI 96960

Environmental Protection Authority, PO Box 1322, Majuro, RMI 96960

References

Douglas, G. (1969). Check list of Pacific Oceanic islands. *Micronesica* 5: 327-463.

Dahl, A.L. (1980). Regional ecosystems survey of the South Pacific area. SPC/IUCN *Technical Paper* No. 179. South Pacific Commission, Noumea, New Caledonia. 99 pp.

Dahl, A.L. (1986). *Review of the protected areas system in Oceania.* IUCN, Gland, Switzerland and Cambridge, UK/UNEP, Nairobi, Kenya. 328 pp.

Fosberg, F.R. (1973) On present condition and conservation of forests in Micronesia. In: Pacific Science Association, Planned Utilisation of Lowland Tropical Forests. *Proceedings of the Pacific Science Standing Committee Symposium, Bogor, Indonesia,* August 1971. Pp. 165-171.

Pernetta, J.C. (1988). Projected climate change and sea level rise: a relative impact rating for countries of the South Pacific Basin. In: MEDU joint meeting of the task team on the implications of climatic change in the Mediterranean. Split, Yugoslavia, 3-7 October. Pp. 1-11.

SPREP (1989). Republic of the Marshall Islands. Paper presented at the Fourth South Pacific

Conference on Nature Conservation and Protected Areas, Port Vila, Vanuatu, 4-12 September. 7 pp.

Thomas, P.E.J., Fosberg, F.R., Hamilton, L.S., Herbst, D.R., Juvik, J.O., Maragos, J.E., Naughton, J.J. and Streck, C.J. (1989). *Report on the Northern Marshall Islands natural diversity and protected areas survey: 7-24 September 1988.* South Pacific Regional Environment Programme, Noumea, New Caledonia and East-West Center, Honolulu, Hawaii. 133 pp.

UNEP/IUCN (1988). *Coral reefs of the world. Volume 3. Central and Western Pacific.* UNEP Regional Seas Directories and Bibliographies. IUCN, Gland, Switzerland and Cambridge, UK/UNEP, Nairobi, Kenya. 378 pp.

REPUBLIC OF NAURU

Area 20.7 sq. km

Population 8,100 (1983) (Paxton, 1989)
Natural increase: No information

Economic Indicators
GNP: US$ 9,091 per capita

Policy and Legislation Information on policies is not available. There is no protected areas legislation.

International Activities Nauru is not yet party to any of the international conventions or programmes that directly promote the conservation of natural areas, namely the Convention concerning the Protection of the World Cultural and Natural Heritage (World Heritage Convention), Unesco Man and the Biosphere Programme and the Convention on Wetlands of International Importance especially as Waterfowl Habitat (Ramsar Convention).

The Convention on the Conservation of Nature in the South Pacific (1976) has been neither signed nor ratified. Known as the Apia Convention, it entered into force during 1990. The Convention is coordinated by the South Pacific Commission and represents the first attempt within the region to cooperate on environmental matters. Among other measures, it encourages the creation of protected areas to preserve indigenous flora and fauna.

Nauru is party to the South Pacific Regional Enviroenment Programme (SPREP), and the 1986 Convention for the Protection of the Natural Resources and Environment of the South Pacific Region (SPREP Convention) has been signed (15 April 1987) but not yet ratified. The Convention entered into force during August 1990. Article 14 calls upon the parties to take all appropriate measures to protect rare or fragile ecosystems and threatened or endangered flora and fauna through the establishment of protected areas and the regulation of activities likely to have an adverse effect on the species, ecosystems and biological processes being protected. However, as this provision only applies to the Convention area, which by definition is open ocean, it is most likely to assist with the establishment of marine reserves and the conservation of marine species.

Administration and Management Not applicable

Systems Reviews Nauru is a single raised coral island in the west-central Pacific Ocean located at 0°32'S, 166°56'E. The highest point is 71m, surrounded by a terrace and fringing reef. The plateau is largely composed of phosphate rock and is encircled by cliffs which give way to a flat, fertile coastal belt, 90-270m wide. The soil is an a mixture of sand and fine corals,

which with an irregular rainfall, restricts cultivation of the coastal belt (UNEP/IUCN, 1988).

Vegetation comprises mixed plateau forest, dominated by *Calophyllum*, a few remaining areas of atoll forest, with *Pandanus* and *Cocos* (Douglas, 1969) and just two hectares of mangroves (Dahl, 1986). The vegetation in the interior has been greatly modified by phosphate mining, although the extent to which vegetation has re-established on the worked-out phosphate pits has been examined by Manner *et al.* (1984). Land with little or no top soil has been denuded to allow mining, but there has generally been no run-off as the bedrock is very porous (Manner *et al.*, 1984). Continuing denudation may cause long-term environmental micro-climate changes but there has been no monitoring of the impact of phosphate mining on the island (UNEP/IUCN, 1988).

There is no true reef and no lagoon; the island is surrounded by an almost consistent 150-200m wide intertidal platform, cut into the original limestone of the island and typified by the presence of numerous emergent coral pinnacles. The platform is dominated by large yellow-brown algae and little or no coral growth occurs on the reef flat. However, a rich fauna is evident in deeper water, although species diversity has not been documented. The benthic fauna is quite well represented with many common Indo-Pacific species. Reef crabs and other crustaceans are found as well as molluscs, urchins, sea cucumbers and other invertebrates. There around 80 commonly caught fish species (Petit-Skinner, 1981; UNEP/IUCN, 1989).

Most of the population, which enjoys one of the highest per capita incomes in the world, lives near the main coastal road and/or shoreline. There is no evidence of over-exploitation of marine resources. Coastal waters are relatively unpolluted although there may have been one or two instances of silt accumulating on some parts of the reef flat. To date there has been no recorded damage to reef flora and fauna (UNEP/IUCN, 1988). Phosphate is the only significant resource on Nauru, and therefore commercial development of phosphate mining takes priority over the conservation of the natural environment. It was stated at the Second South Pacific Conference on National Parks and Reserves that it was not possible to consider conservation of the remaining phosphate-bearing areas on the island. About 60% of the island is phosphate bearing, and of this about two-thirds has already been mined.

Restoration of mined land is a key environmental problem. A Commission of Inquiry into the Rehabilitation of the Worked-Out Phosphate Lands in Nauru has been established to look at the issue of rehabilitation of the island and its cost and feasibility (Anon., 1987). It was charged with examining all forms of alternative land usage, including agriculture, and the impact of phosphate mining on fisheries and marine resources. Financial compensation is being sought from Australia, the United Kingdom and New Zealand, the main consumers of phosphate. An account of the British Phosphate Commissioners, responsible for phosphate extraction prior to independence, with reference to the extensive mining on Nauru and other Pacific islands, is given by Williams and Macdonald (1985). A study should be started immediately to provide baseline data for future environmental impact monitoring. Efforts should be made to establish protected areas, including reefs and important cultural sites (Dahl, 1980) and any forested areas should be protected from any further mining (Dahl, 1986).

Addresses

No information

References

Anon. (1987). Commission of Inquiry into the Rehabilitation of Worked-out Phosphate Lands in Nauru. Republic of Nauru.

Anon. (1989). *Proceedings of the Second South Pacific Conference on National Parks and Reserves*, 24-27 April 1979. Situation Report Nauru. National Parks and Wildlife Service, Sydney, New South Wales, Australia.

Dahl, A.L. (1980). Regional ecosystem surveys of the South Pacific Area. *SPC/IUCN Technical Paper 179*. South Pacific Commission, Noumea, New Caledonia. 99 pp.

Dahl, A.L. (1986). *Review of the protected areas system in Oceania*. IUCN, Gland, Switzerland and Cambridge UK/UNEP, Nairobi, Kenya. 328 pp.

Davis, S.D., Droop, S.J.M., Gregerson, P., Henson, L., Leon, C.J., Lamlein Villa-Lobos, J., Synge, H. and Zantovska, J. (1986). *Plants in danger: what do we know?* IUCN, Gland, Switzerland and Cambridge, UK. 488 pp.

Manner, H.I, Thaman, R.R and Hassall, D.C. (1984). Phosphate mining induced changes on Nauru Island. *Ecology* 65(5): 1454-1465.

Petit-Skinner, S. (1981). *The Nauruans*. MacDuff Press, San Francisco. (Unseen)

Williams, M and Macdonald, B. (1985). *The Phosphateers*. Melbourne University Press. Carlton, Australia. 586 pp. (Unseen)

NIUE

Area 260 sq. km

Population 2,270 (1989) (G.S.T. Talagi, pers comm., 1990)
Natural increase: – 3.7% per annum

Economic Indicators
GNP: No information

Policy and Legislation Niue is a self-governing nation in free association with New Zealand. The 1980-1985 Niue National Development Plan (NNDP) had a number of objectives relating to the environment "as government is aware of the dangers from unwise practices" (Sloth, 1988). The 1985-1990 NNDP emphasises socio-economic development and social services and does not specify protection of the biological environment, although it is recognised in one chapter.

The role of traditional law in the management of island affairs is very strong. Traditional conservation measures are in effect for different times of the year and for different species (UNEP/IUCN, 1988). Customary restrictions, or "fono", are applied from time to time in certain temporary "reserves" which allow the recovery of exploited resources (Yaldwyn, 1973). There is no protected areas legislation and no formally protected areas, although there is a "tapu" forest area. Despite a lack of government legislation under which reserves may be created, the existing tapu forest is probably secure. Niueans and the Government own all the land; no alienation is permitted, although the government may lease areas for up to 60 years (Hay, 1986; SPREP, 1980; UNEP/IUCN, 1988).

International Activities Niue is not yet party to any of the three major international conventions concerned with with the conservation of nature, namely the Unesco Man and the Biosphere Programme, the Convention on Wetlands of International Importance especially as Waterfowl Habitat (Ramsar Convention) and the Convention concerning the Protection of the World Cultural and Natural Heritage (World Heritage Convention).

At a regional level, Niue has neither signed nor ratified the Convention on the Conservation of Nature in the South Pacific, 1976. Known as the Apia Convention, it entered into force during 1990. The Convention is coordinated by the South Pacific Commission and represents the first attempt within the region to cooperate on environmental matters. Among other measures, it encourages the creation of protected areas to preserve indigenous flora and fauna.

Niue is party to the South Pacific Regional Environment Programme (SPREP) but has neither signed nor ratified the Convention for the Protection of the Natural Resources and Environment of the South Pacific Region, 1986 (SPREP Convention). The Convention entered into force during August 1990. Article 14 calls upon the parties to take all appropriate measures to protect rare or fragile ecosystems and threatened or endangered flora and fauna through the establishment of protected areas and the regulation of activities likely to have an adverse effect on the species, ecosystems and biological processes being protected. However, as this provision only applies to the Convention area, which by definition is open ocean, it is most likely to assist with the establishment of marine reserves and the conservation of marine species.

Systems Reviews Niue is a single, isolated island located at 169°53'W, 19°03'S, 480km north-east of Tonga and 560km south-east of Western Samoa. The island is roughly circular and comprises a raised atoll of coralline limestone about 62m high with coastal terraces, the most prominent being 20-28m above sea level. A number of submerged terraces also occur. The island has a slightly depressed upper surface representing the "lagoon" of the original atoll (Yaldwyn, 1973).

The original tropical high rain forest which once covered the island has now been reduced to fragments, generally in the east, and totalling 3,200ha (12.3% of total land area) in 1981. There is also a narrow (200m-800m width) strip of coastal forest encircling the island's lower terrace. This forest is still largely intact and covered some 2,500ha (9.6%) in 1981. Second-growth or regenerating forest is more widespread, totalling 12,000ha (46.2%) in 1981.

Unproductive "fernland", principally comprising *Nephrolepis hirsutula*, covering approximately 3,200ha (12.4%), is found largely in the south in the old lagoon basin. This is generally thought to have arisen as a result of prolonged burning, overcropping and subsequent soil impoverishment (Wodzicki, 1971; Yaldwyn, 1973), and early attempts at mechanised cultivation using bulldozers and discing equipment (G.S.T. Talagi, pers. comm., 1990).

There is no true reef or lagoon. The island is partly surrounded by a platform reef, varying from a few metres to several hundred metres in width and cut in the limestone of the island. Large parts of this are subtidal, the remainder being intertidal. Much of the south and east sides of the islands are entirely devoid of reef flats; some parts have 1m-8m wide pools about 1.5m-2.5m above sea level. The flat has a thin discontinuous veneer of living corals on its upper (intertidal) surface and rich coral growth over the edge in sub-tidal waters. At least 43 coral genera occur and there is a rich, though largely undocumented, invertebrate fauna (UNEP/IUCN, 1988; Yaldwyn, 1973).

Principal environmental issues include forest destruction by clearing and milling; soil loss due to inappropriate cultivation; and decline in bird and flying fox populations through habitat loss and over-hunting. Reef blasting has been strictly limited in extent (van Westendorp, 1961; Yaldwyn, 1973).

The existing protected areas network is restricted to Huvalu Tapu Forest, and temporary "fono" marine reserves. Dahl (1986) identifies five ecosystems, viz. limestone rain forest, coastal forest on terraces, secondary formations, scrub and fern, and fringing reefs. Only the first of these ecosystems is protected within Huvalu Tapu Forest. Nevertheless, Huvalu probably provides adequate protection for Niue's original terrestrial habitat (Hay, 1986). Dahl (1980) recommends a number of specific sites for designation as protected areas. These include: caves, reefs, historic sites, coastal features, chasms and freshwater springs. Huvalu is one of the few remaining traditional taboo protected areas in Oceania. Dahl (1986) recommends that it should be maintained, and reinforced with legislation if necessary. Other sites may also need protection if they are being degraded. Forest reserves additional to Huvalu would help conserve the Polynesian triller sub-species (Hay, 1986).

Other Relevant Information   Huvalu Tapu Forest comprises approximately 150ha (0.6% of total land area) of Huvalu Forest, which covers 800-1,200ha (3.1%-4.6%), in eastern-central Niue (Wodzicki, 1973). The forest was set aside as a tapu area in pre-European times, and is believed to house the remains of pre-Christian gods. The tapu still exists, and the area represents a fragment of primeval Niuean forest which has survived since before the arrival of the Polynesians (Wodzicki, 1971). The site is maintained by the Village Council of Hakupu which lies some 3km to the south. Human entry and any use of the forest is forbidden, except very occasionally in the case of privileged individuals who have been permitted to enter. The forest is pristine and contains very large trees and an abundant fauna including pigeon and flying fox as well as diurnal, as opposed to the more usual nocturnal, land crabs (Yaldwyn, 1973).

Addresses

The Department of Agriculture, Forestry and Fisheries, PO Box 74, Alofi

Secretary to Government, Administrative Department, Alofi

References

Dahl, A.L. (1980). Regional ecosystem surveys of the South Pacific Area. *SPC/IUCN Technical Paper* 179. South Pacific Commission, Noumea, New Caledonia. 99 pp.

Dahl, A.L. (1986). *Review of the protected areas system in Oceania*. IUCN, Gland, Switzerland and Cambridge, UK/UNEP, Nairobi, Kenya. 328 pp.

Hay, R. (1986). Bird conservation in the Pacific Islands. *Study Report* No. 7. International Council for Bird Preservation, Cambridge, UK. 102 pp.

Schofield, J.C. (1959). The geology and hydrology of Niue Island, South Pacific. *New Zealand Geological Survey Bulletin* 62.

Sloth, B. (1988). *Nature legislation and nature conservation as part of tourism development in the island Pacific*. Tourism Council of the South Pacific, Suva, Fiji. 82 pp.

SPREP (1980). *Niue*. Country Report No. 4. South Pacific Commission, Noumea, New Caledonia. 7 pp.

UNEP/IUCN (1988). *Coral reefs of the world. Volume 3. Central and Western Pacific*. UNEP Regional Seas Directories and Bibliographies. IUCN, Gland, Switzerland and Cambridge, UK/UNEP, Nairobi, Kenya. 378 pp.

van Westendorp, F.J. (1961). Agricultural development on Niue. *South Pacific Bulletin* 11: 67-69. (Unseen)

Yaldwyn, J.C. (1973). The environment, natural history and special conservation problems of Niue Island. In: SPC, *Regional symposium on conservation of nature: reefs and lagoons*. Part 2. South Pacific Commission, Noumea, New Caledonia. 49-55.

Wodzicki, K. (1971). The birds of Niue Island, South Pacific: an annotated checklist. *Notornis* 18(4): 291-304.

COMMONWEALTH OF THE
NORTHERN MARIANA ISLANDS

Area 477 sq. km

Population 20,350 (1985 estimate)
Natural increase: No information

Economic Indicators
GNP: No information

Policy and Legislation The Commonwealth of the Northern Mariana Islands (CNMI) was officially granted the status of a Commonwealth of the USA in October 1986, having previously been part of the Trust Territory of the Pacific Islands; it had, however, been functioning as a Commonwealth since 9 January 1978 (UNEP/IUCN, 1988).

Prior to October 1986, the islands were formally under United States federal legislation and Trust Territory legislation. This included the Trust Territory Environment Enabling Act, enacted by the former Congress of Micronesia in 1972, which provided for the establishment of a Trust Territory Environmental Protection Board. However, neither the Act, nor the work of the Board related specifically to protected areas. The Trust Territory Endangered Species Act (TTPI Public Law No. 6-55 of 1975) allowed for acquisition of land or water for the purpose of conserving threatened species. Other relevant Trust Territory Acts included the Fishing Law and the Land Use Planning Act, both 1972. US legislation relevant to the Trust Territories included various pollution laws and housing acts, as well as the Fish and Wildlife Coordination Act and the National Environmental Policy Act. Neither the legislation nor the policy on protected areas was clear while the country was a part of the Trust Territory, and this is why so few protected areas were actually established over the whole area. Furthermore, the United States, while administering the Territory, did not own property, and hence had no direct jurisdiction over designation of protected areas. New national legislation is under development.

Under the original constitution, passed on 5 December 1976, the islands of Managaha, Sariguan (sic) and Maug were protected, with the provision that Sariguan might be substituted by another island by the legislature in due course. This substitution did not occur until the constitutional amendment of 3 November 1985 which removed Sariguan's protected status. The island was considered a poor candidate for protection due to the depredations of past human habitation, large populations of rats and goats and the destruction of many native species and habitats. In place of Sariguan, which was opened for possible development, Uracas (Farallon de Parajos), Asuncion and Guguan, as well as Maug, were given strict protection under the Constitution (Anon., 1985).

Thus, Section 2 of Article XIV of the Commonwealth Code, as amended in 1985, states that "the island of Managaha shall be maintained as an uninhabited place and used only for cultural and recreational purposes. The islands of Maug, Uracas, Asuncion, Guguan and other islands specified by law shall be maintained as uninhabited places and used only for the preservation and protection of natural resources, including but not limited to bird, wildlife and plant species".

Public Law No. 1-8, Chapter 13, Sections (a), (b), (c), (e) and (f) empowers the Department of Natural Resources to protect and enhance natural resources. Public Law No. 2-51 establishes the Division of Fish and Wildlife within the Department of Natural Resources and mandates this Division to provide for the conservation of fish and wildlife and to acquire areas for the protection of fish and wildlife resources (SPREP, 1985).

International Activities The Commonwealth of the Northern Mariana Islands is not yet directly party to any of the international conventions concerned with protected areas, namely, the Unesco Man and the Biosphere Programme, the Convention on Wetlands of International Importance especially as Waterfowl Habitat (Ramsar Convention) and the Convention concerning the Protection of the World Cultural and Natural Heritage (World Heritage Convention).

The compact between the USA and CNMI leaves international pacts and conventions in the control of the former which itself is a party to all three conventions. However, it is not clear whether the conventions are applicable to CNMI (J.D. Reichel, pers. comm., 1989). At a regional level, CNMI has neither signed nor ratified the Convention on the Conservation of Nature in the South Pacific, 1976. Known as the Apia Convention, it entered into force during 1990. The Convention is coordinated by the South Pacific Commission and represents the first attempt within the region to cooperate on environmental matters. Among other measures, it encourages the creation of protected areas to preserve indigenous flora and fauna.

CNMI is party to the South Pacific Regional Environment Programme (SPREP) but has neither signed nor ratified the Convention for the Protection of the Natural Resources and Environment of the South Pacific Region, 1986 (SPREP Convention). The Convention entered into force during August 1990. Article 14 calls upon the parties to take all appropriate measures to protect rare or fragile ecosystems and threatened or endangered flora and fauna through the establishment of protected areas and the regulation of activities likely to have an adverse effect on the species, ecosystems and biological processes being protected.

However, as this provision only applies to the Convention area, which by definition is open ocean, it is most likely to assist with the establishment of marine reserves and the conservation of marine species.

Administration and Management Administration of official parks and reserves lies in the CNMI Department of Natural Resources, which includes Divisions of Fish and Wildlife, Parks and Recreation and Plant Industry (Forestry). All lands not held with deeds by other agencies are controlled by the Marianas Public Land Corporation (J.D. Reichel, pers. comm., 1989).

Systems Reviews Vegetation comprises pioneer stands of *Casuarina* sp., broad-leaved evergreen thickets, mixed scrub forest, with some *Miscanthus* sp. and *Nephrolepsis* sp. herbaceous communities on the northern islands. Broad-leaved evergreen forest on old lava flows and *Miscanthus* sp. and tree ferns on ash slopes of those northern islands with dormant volcanoes are found (Douglas, 1969). Tinian has mostly secondary forest while Rota has some closed evergreen and limestone forests (Fosberg, 1973). Small areas of cloud forest occur on the volcanic islands of Saipan, Agrihan, Alamagan and Anatahan (Dahl, 1980). The lower slopes of many islands have been cleared for cultivation (Davis *et al.*, 1986) and Saipan, Rata, Tinian and Aguijuan are much disturbed. Remaining areas of natural vegetation, mostly on cliffs, contain rare native and endemic species of plants and birds. The islands to the north, from Farallon de Medinilla to Uracas, are less disturbed and are of prime interest for the study of biotic colonisation under natural conditions (Dahl, 1980). Coral reefs around Rota, Tinian, Saipan and elsewhere are described in UNEP/IUCN (1988).

There are permanent human populations on only the three largest islands to the south, namely Saipan, Tinian and Rota, but two other islands (Almagan and Agrihan) are regularly inhabited. Pagan was continuously inhabited by the largest of the northern islands' populations until an eruption in May 1985 forced evacuation of the island. Over 87% of the population lives on Saipan, where tourism is the main source of income; 163,000 tourists visited Saipan in 1986, the majority being Japanese (IUCN/UNEP, 1988). There has subsequently been a considerable increase, with 233,291 tourists visiting during 1988 and an estimated 300,000 during 1989 (L. Eldredge, pers. comm., 1989).

Moves to protect some of the northerly islands commenced in the 1960s, when Farallon de Medinilla, Guguan, Maug and Uracas were designated as "islands for science" through the efforts of the International Biological Programme and IUCN. Subsequently, Farallon de Medinilla lost its conservation importance due to military training exercises.

Addresses

Division of Parks and Recreation, Department of Natural Resources, Saipan, MP 96950, USA

References

Anon. (1985). CNMI northern islands win preservation. *Coastal Views* 8(3): 1, 8-9.

Dahl, A.L. (1980). Regional ecosystems survey of the South Pacific Area. SPC/IUCN *Technical Paper* 179. South Pacific Commission. Noumea, New Caledonia. 99 pp.

Davis, S.D., Droop, S.J.M., Gregerson, P., Henson, L., Leon, C.J., Lamlein Villa-Lobos, J., Synge, H. and Zantovska, J. (1986). *Plants in danger: what do we know?* IUCN, Gland, Switzerland and Cambridge, UK. 488 pp.

Douglas, G. (1969). Draft checklist of Pacific Islands. *Micronesica* 5: 327-463.

Fosberg. F.R. (1973). On present condition and conservation of forests in Micronesia. In: Pacific Science Association Standing Committee on Pacific Botany Symposium: planned utilization of the lowland tropical forests, August 1971. Bogor, Indonesia.

Owen, R.P. (1973). A conservation program for the Trust Territory. Country Report. Regional Symposium on Conservation of Nature - Reefs and Lagoons. South Pacific Commission, Noumea, New Caledonia.

SPREP (1980). Trust Territory of the Pacific Islands. *Country Report* 14. South Pacific Commission, Noumea, New Caledonia.

SPREP (1985). Northern Mariana Islands. In: Thomas, P.E.J. (Ed.), *Report of the Third South Pacific National Parks and Reserves Conference.* Volume III. *Country reviews.* South Pacific Commission, Noumea, New Caledonia. Pp. 159-161.

UNEP/IUCN (1988). *Coral reefs of the world. Volume 3. Central and Western Pacific.* UNEP Regional Seas Directories and Bibliographies. IUCN, Gland, Switzerland and Cambridge, UK/UNEP, Nairobi, Kenya. 378 pp.

ANNEX
Definitions of protected area designations, as legislated, together with authorities responsible for their administration

Title: Commonwealth Code

Date: 5 December 1976; Amendment No. 37 passed 3 November 1985

Brief description: Section 2 of Article XIV states that "the island of Managaha shall be maintained as an uninhabited place and used only for cultural and recreational purposes. The islands of Maug, Uracas, Asuncion, Guguan and other islands specified by law shall be maintained as uninhabited places and used only for the preservation and protection of natural resources, including but not limited to bird, wildlife and plant species".

Administrative authority: Not specified

Designation: Not specifically stated

SUMMARY OF PROTECTED AREAS

Map ref.	*National/international designation* Name of area	IUCN management category	Area (ha)	Year notified
	Preserves			
1	Asuncion Island	I	722	1985
2	Guguan Island	I	412	1985
3	Maug Island	I	205	1976
4	Uracas Island (Farallon de Pajaros)	I	202	1985

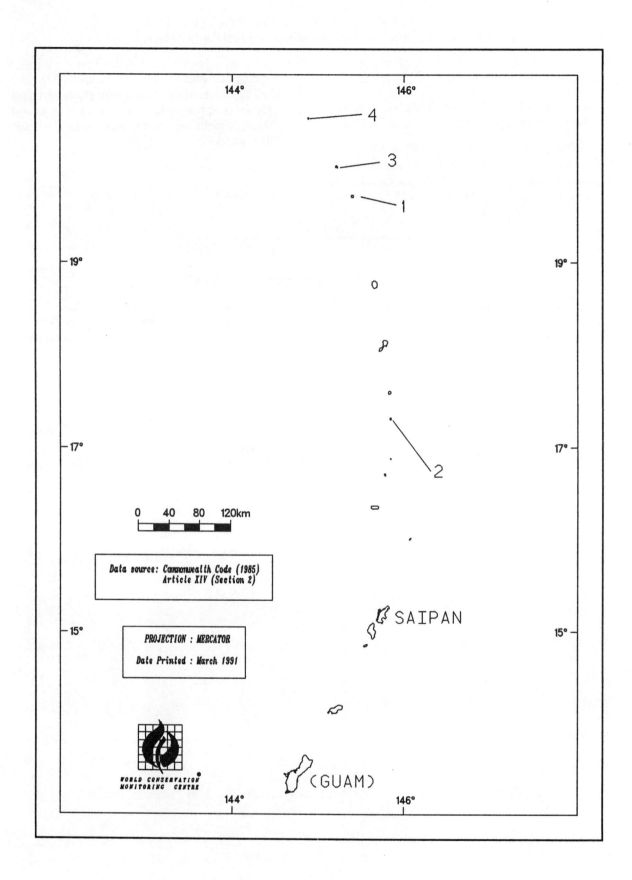

Protected Areas of the Northern Mariana Islands

REPUBLIC OF PALAU

Area 492 sq. km

Population 13,873 (1986); 14,106 (1988 estimate)
Natural increase: – 0.3% per annum

Economic Indicators
GNP: No information

Policy and Legislation Prior to 1981, some United States federal legislation and Trust Territory legislation was applicable, including the Trust Territory Environment Enabling Act enacted by the former Congress of Micronesia in 1972. This provided for the establishment of a Trust Territory Environmental Protection Board. However, neither the Act, nor the work of the board related specifically to protected areas. The Trust Territory Endangered Species Act (TTPI Public Law No. 6-55 of 1975) allowed for acquisition of land or water for the purpose of conserving threatened species. Other relevant Trust Territory Acts included the Fishing Law and the Land Use Planning Act, both 1972. US legislation relevant to the Trust Territory included various pollution laws and housing acts, as well as the Fish and Wildlife Coordination Act and the National Environmental Policy Act. Prior to 1981, neither the legislation, nor the policy on protected areas was clear, with the result that few areas were actually established. This is largely because the United States, while administering the Trust Territory, did not own property, and hence had no direct jurisdiction over designations of protected areas.

The Republic of Palau was established as a constitutional democratic government on 1 January 1981. Article VI of the Constitution places responsibility upon the national government to take positive action to conserve a "beautiful, healthful and resourceful natural environment". There are 16 states within the Republic, each with an individual constitution; 14 states are within the Palauan archipelago and two are in the South-east Islands group (BRD, 1989). On 21 February 1986 Palau and US Government representatives formally signed a Compact of Free Association. To date, six plebiscites have been held on the Compact, but the constitutionally mandated 75% approval vote has not been achieved. A subsequent simple majority vote has been overturned in the courts as unconstitutional. Consequently, Palau has continued to be a remnant of the former Trust Territory of the Pacific.

National protected areas legislation includes Palau National Code, Title 24, Division 3 (Reserves and Protected Areas), Chapter 30 (Ngerukewid Islands Wildlife Preserve), Sections 3001 to 3004, which provides for the creation, prohibitions, penalties and regulations relating to Ngerukewid Islands Wildlife Preserve. As of 1988, Ngerukewid Islands Wildlife Preserve was the only legally established perennially protected natural area in the Republic of Palau (Thomas

et al., 1989). Palau National Code, Sections 3101 to 3103 provide for the seasonal protection of the Ngerumekaol grouper spawning area. There are also a number of State Ordinances, including Ordinance of Koror No. 150-69 (48-69) which provides for the establishment of *Trochus* sanctuaries in Koror. Legal instruments which provide for the protection of cultural and sunken resources include, for example, the Palau Lagoon Monuments Law (Palau National Code, Section 301). A fuller listing of environmental legislation is given in BRD (1989).

International Activities Palau is not yet party to the Unesco Man and the Biosphere Programme, the Convention on Wetlands of International Importance especially as Waterfowl Habitat (Ramsar Convention) nor the Convention concerning the Protection of the World Cultural and Natural Heritage (World Heritage Convention). It has been recommended that Palau become a State Party to the World Heritage Convention in order that Ngerukewid Islands Wildlife Preserve can be inscribed on the World Heritage List (Thomas *et al.*, 1989).

The Convention on the Conservation of Nature in the South Pacific (1976) has been neither signed nor ratified. Known as the Apia Convention, it entered into force during 1990. The Convention is coordinated by the South Pacific Commission and represents the first attempt within the region to cooperate on environmental matters. Among other measures, it encourages the creation of protected areas to preserve indigenous flora and fauna.

Palau signed the Convention for the Protection of the Natural Resources and Environment of the South Pacific (SPREP Convention) on 25 November 1986, although this has not been ratified. Article 14 calls upon the parties to take all appropriate measures to protect rare or fragile ecosystems and threatened or endangered flora and fauna through the establishment of protected areas and the regulation of activities likely to have an adverse effect on the species, ecosystems and biological processes being protected. However, as this provision only applies to the Convention area, which by definition is open ocean, it is most likely to assist with the establishment of marine reserves and the conservation of marine species.

Administration and Management Prior to independence a Conservation Officer for Palau, and the rest of the Trust Territory, was hired to work under the Chief of Agriculture for the Trust Territory. Management of parks, recreation areas and historical sites fell under the Bureau of Community Services in the Ministry of Social Services, while responsibility for conservation areas was placed upon the Bureau of Resources and Development in the Ministry of Natural Resources. Enforcement of the laws pertaining to these areas was the responsibility of the Bureau of Public Safety in the Ministry of Justice (SPREP, 1985).

Since 1981, Ngerukewid Islands and Ngerumekaol have been administered by the Division of Marine Resources, Bureau of Resources and Development, Ministry of National Resources. Parks and recreation areas are administered by the Chief of the Division of Parks and Recreation in the Bureau of Community Services, Ministry of Social Services. The Division of Conservation and Entomology, also within the Bureau of Resources and Development, was created under Executive Order No. 70, with a broad remit to prepare conservation programmes and pest control and entomological activities (Otobed, 1989). *Trochus* breeding sanctuaries are under the protection of the governors of each State (BRD, 1989). In general, surveillance and patrolling activities have been hindered by a lack of staff and resources (D. Otobed, pers. comm., 1989).

Systems Reviews Palau comprises an archipelago of eight large and 18 small high volcanic islands, a number of low limestone islands and about 350 islets surrounded by a complex of fringing and patch reefs. Babeldaob is the largest island, covering 397 sq. km. The climate is maritime, tropical and wet, with only small seasonal and diurnal temperature fluctuations. Rainfall is high throughout the year but especially intense during the June to October monsoon (Smith, 1977).

Undisturbed forest is predominant, varying little between the high volcanic islands and low platform islands. Principal formations are species-rich broad-leaved lowland rain forest on weathered basalt, limestone forest on coralline limestone with very little soil, dense riverine forest along rivers on all high islands and mixed broad-leaved atoll and beach forest on the central portion of atolls and islets and on level areas behind the sand beaches of high islands (Dahl, 1980). Some 75% of Babeldaob is covered by mature forest in upland, mangrove and swampy areas (Merlin and Keene, n.d.), and mangroves occur along as much as 80% of the coast of Babeldaob although they are not prevalent around the shorelines of the limestone islands (Smith, 1977). Palau is considered to have the richest coral reefs in the Pacific with the highest diversity (Faulkner, 1974; Smith, 1977); 300 species have been recorded. A summary account is given in UNEP/IUCN (1988), including detailed accounts of the Chelbacheb and Helen Island reefs.

In general, Palau's natural environment remains in good condition, partly due to a low population density. Abundant and diverse terrestrial and marine ecosystems occur, particularly the very widespread reefs and lagoons. This natural beauty, and the numerous islands and beaches, give the country considerable tourism potential. There have been a number of major capital projects and there is now a risk that unplanned development could have serious environmental consequences. The improvement of living standards on

Koror, the capital island where some 40% of the population dwells, is a major priority. The provision of sewage and waste disposal facilities, and the disposal of hazardous waste and unexploded munitions, such as on Peleliu, constitute major environmental issues (Maragos, 1986). The compact of Free Association between Palau and the United States entitles, *inter alia*, the latter to use large areas of the archipelago for military purposes (Caufield, 1986) which may lead to environmental degradation. Smith (1977) provides an overview of the major natural resources and some of the threats to them.

Palau has been included in a recent review of the protected areas systems in Oceania (Dahl, 1986), and the establishment of terrestrial reserves on Babeldaob and in the Chelbacheb Islands, including marine lakes, is recommended. The richness of the marine environment warrants protection in some significant marine reserves at sites such as Ngemlis. Helen should probably also be protected because of the problem with poaching.

Addresses

Division of Conservation and Entomology, PO Box 100, Biology Laboratory, Koror 96940, USA

References

Bureau of Resources and Development (1989). Palau. *Country Review* No. 10. Fourth South Pacific Conference on Nature Conservation and Protected Areas. Port Vila, Vanuatu, 4-12 September. South Pacific Commission, Noumea. 12 pp.

Caufield, C. (1986). Peace makes waves in the Pacific. *New Scientist* 3 April.

Dahl, A.L. (1980). Regional ecosystem survey of the South Pacific Area. *SPC/IUCN Technical Paper* 179. South Pacific Commission, Noumea, New Caledonia. 99 pp.

Dahl, A.L. (1986). *Review of the protected areas system in Oceania.* IUCN, Gland, Switzerland and Cambridge, UK/UNEP, Nairobi, Kenya. 328 pp.

Faulkner, D. (1974). *This living reef.* Quadrangle/New York Books, New York. 179 pp.

Merlin, M. and Keene, T. (n.d.). Dellomel er a Belau: plants of the Belauan Islands. East-West Centre, Honolulu, Hawaii and Office of the Chief Conservationist and Bureau of Education, Oreor, Belau. 48 pp.

Smith, S.V. (1977). Palau environmental study: a planning document. Contribution to the IUCN Marine Project 3.7.70. 102 pp.

SPREP (1980). Trust Territory of the Pacific Islands. *Country Report* 14. South Pacific Commission, Noumea, New Caledonia.

SPREP (1985). Republic of Palau. In: Thomas, P.E.J. (Ed.), *Report of the Third South Pacific National Parks and Reserves Conference.* Volume III. *Country reviews.* South Pacific Commission, Noumea, New Caledonia. Pp. 162-174.

Thomas, P.E.J., Holthus, P.F. and Idechong, N. (1989). Ngerukewid Islands Wildlife Preserve: proposed management plan. IUCN/World Wide Fund for Nature/South Pacific Regional Environment Programme. South Pacific Commission, Noumea, New Caledonia. 43 pp.

UNEP/IUCN (1988). *Coral reefs of the world. Volume 3. Central and Western Pacific.* UNEP Regional Seas Directories and Bibliographies.

IUCN, Gland, Switzerland and Cambridge, UK/UNEP, Nairobi, Kenya. 378 pp.

Otobed, D. (1989). Conservation in Palau – rebuilding the Conservation Service. *Case Study* No. 34. Fourth South Pacific Conference on Nature Conservation and Protected Areas. Port Vila, Vanuatu, 4-12 September. South Pacific Commission, Noumea, New Caledonia. 3 pp.

ANNEX
Definitions of protected area designations, as legislated, together with authorities responsible for their administration

Title: Palau National Code

Date: Not known

Brief description: Division 3, Title 24, Chapter 30 provides for the creation (S.3001) and protection (S.3002-3004) of Ngerukewid Islands Wildlife Preserve

Administrative authority: Division of Marine Resources, Ministry of National Resources

Designations: The Ngerukewid Islands Preserve is hereby established to include all land, water, reef and underwater areas of the island group known as Ngerukewid (Orukuisu) Islands, bounded by the grid coordinates 91 and 94 and 18 and 22 on sheets 1043-I SW and 1043 II NW on Army Map Series W 856. The preserve is henceforth to be retained in its present primitive condition where the natural plant and animal life shall be permitted to develop undisturbed.

SUMMARY OF PROTECTED AREAS

Map ref.	*National/international designation* Name of area	IUCN management category	Area (ha)	Year notified
1	*Wildlife Reserve* Ngerukewid Islands	III	1,200	1956

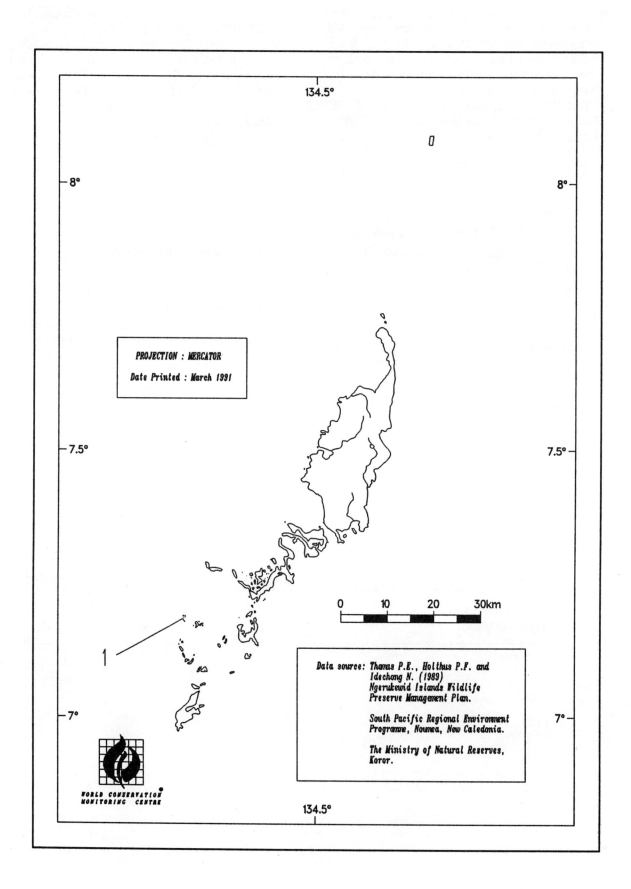

PROJECTION : MERCATOR

Date Printed : March 1991

1

Data source: Thomas P.E., Holthus P.F. and
Idechong N. (1989)
Ngerukewid Islands Wildlife
Preserve Management Plan.

South Pacific Regional Environment
Programme, Noumea, New Caledonia.

The Ministry of Natural Reserves,
Koror.

WORLD CONSERVATION
MONITORING CENTRE

Protected Areas of Palau

PAPUA NEW GUINEA

Area 462,842 sq. km

Population 3,900,000 (Population Reference Bureau, Washington DC, 1989)
Natural increase: 2.7% per annum

Economic Indicators
GNP: US$ 790 per capita (1988)

Policy and Legislation

An Environment and Conservation Policy was adopted by the National Parliament in 1977, in recognition that development must be ecologically, socially and culturally suitable for Papua New Guinea. The Policy was drawn up in response to the Fourth Goal of the National Constitution:

"4. Natural Resources and Environment. We declare our fourth goal to be for Papua New Guinea's natural resources and environment to be conserved and used for the collective benefit of us all, and be replenished for the benefit of future generations."

The Fourth Goal provides for: (1) wise use of natural resources, (2) conservation and replenishment of the environment and (3) protection of flora and fauna for the benefit of present and future generations (SPREP, 1985b).

In order to implement the constitutionally-based policies, various laws have been introduced. Of particular relevance to the establishment of protected areas are the National Parks Act, Conservation Areas Act and the Fauna (Protection and Control) Act (Venkatesh *et al.*, 1983) (see Annex).

The National Parks Act (1982) replaced the amended 1971 Act, which in turn superseded the original National Parks and Gardens Act (1966). It provides for:

"the preservation of the environment and of the national cultural inheritance by – (1) the conservation of sites and areas having particular biological, topographical, geological, historical, scientific or social importance"

and thereby upholds the Fourth National Goal and Directive Principle of the Constitution.

The Act contains provisions for reserving government land and for leasing and accepting gifts of land. Powers to make regulations to control hunting, fishing, sports, vehicles and domestic animals, and law enforcement provisions are contained in the Act. Although comprehensive in its coverage of different types of protected area, the Act does not define or even list the various categories nor is there any statutory requirement for the provision of park management plans (Eaton,

1985; SPREP, 1985b). The procedure for establishment of protected areas under this Act involves three stages: proposal, approval and declaration (Kwapena, 1984).

The Conservation Areas Act (1978) has similar objectives to the National Parks Act but is more comprehensive and, to some extent, remedies deficiencies in the other legislation. For example, provisions include the establishment of a National Conservation Council to advise on the identification and management of protected areas, and the formation of management committees for each area to be responsible for *inter alia* the production of management plans. Conservation areas may be established on land under public, private or customary ownership. The Act awaits implementation due to financial constraints (Eaton, 1985; SPREP, 1985b).

Although concerned primarily with the protection of endangered species, the Fauna (Protection and Control) Act (1966) provides for the establishment of wildlife management areas (WMAs) on land held under customary ownership, of which there are three categories, defined in the Annex. The Act provides for the setting up of wildlife management committees to administer them, thereby involving customary land-owners in the control of wildlife resource exploitation. The committees advise on the provision of specific rules for each area for "the protection, propagation, encouragement, management, control, harvesting and destruction of fauna" (Eaton, 1985; SPREP, 1985b). The procedure for the establishment of WMAs is fully described by Kisokau and Lindgren (1984) and Asigau (1989), and also outlined by Eaton (1986). The WMA concept recognises customary land ownership, and places landowners in direct control. Further, the establishment of WMAs is invariably initiated by the local landowners. The major failings in the WMA system are the generally inadequate size of each area, lack of local resource management expertise, delays in responding to requests for WMA establishment, leading to apathy and weak enforcement of regulations (Asigau, 1989).

The following are the principal classifications and general criteria for protected areas (SPREP, 1985b):

National parks are extensive areas of outstanding scenic and scientific interest which are of national significance. They should be of at least 1,000ha and preferably in excess of 2,000ha. Ideally, the whole range of land-forms and environments found in Papua New Guinea should be represented. National parks have two main functions; first for public use and education and second, for the conservation of nature through protection of undisturbed habitat.

Provincial parks are less extensive natural areas than national parks; frequently less than 2,000ha and often less than 1,000ha. Not necessarily of national significance, they are of scenic and recreational importance at provincial level. Their main role is to provide for outdoor recreation in a natural setting close to urban centres.

Historical sites are areas of historic significance, covering prehistory and recent history. They may be of any size and, in many cases, adjacent areas will be developed for recreational purposes. They should provide for the preservation of areas of historic and prehistoric significance and their interpretation to the public.

Nature reserves can be areas of any size in which samples of ecosystems and habitats are preserved, either for their intrinsic value or for the protection of wildlife. Scientific research is permitted, but access by members of the public is very limited.

National walking tracks are physically challenging and scenic primitive routes through natural landscape that provide for walking in natural surroundings over long distances. Wherever possible, there should be a minimum easement of 10m of natural vegetation on either side of tracks. Advantage may be taken of existing national parks or other large areas of reserved natural landscape.

Sanctuaries are areas set aside primarily for breeding and research on indigenous wildlife and its display to the public for education and recreation purposes. They can be of any size but should contain some natural habitat in addition to the display area.

Wildlife management areas are areas reserved at the request of the land-owners for the conservation and controlled utilisation of the wildlife and its habitat. Declaration of a wildlife management area does not in any way affect ownership of the land, only the way in which resources are used. Thus, wildlife management areas represent an attempt to develop conservation on a customary basis, using traditional methods of resource management (Eaton, 1986).

The customary land tenure system and associated subsistence economy traditionally contains many forms of resource management and conservation. The shifting cultivation system, for example, with its long periods (10-35 years) of fallow helps to maintain soil fertility. Customary rules may prevent the felling of trees along river banks. There are also prohibitions against cutting down trees near villages, while other trees of special economic value or of particular importance for certain types of wildlife may be protected from indiscriminate felling. In addition to controls consciously imposed by village societies, there are a great many associated traditional beliefs and practices that have often proved extremely effective in protecting certain habitats and species. In many Papua New Guinean societies there are prohibitions or *tambu* against entering certain areas or hunting or felling trees within them. These may be sites of old settlements, burial grounds or physical features,

such as mountain tops, caves, ponds and forests. Some areas may be protected permanently, in others the restriction may be for a limited period as may happen after a death in the group (Eaton, 1985).

While traditional beliefs and customs have helped to protect the environment in the past and are often still operative, the integrity of the environment is under increasing threat from pressures associated with population growth, increased mobility and growth of the cash economy. The establishment of a protected areas system has proved to be extremely difficult on account of the traditional land tenure system. New legislation and novel approaches to environmental management have proved necessary.

Other environmental legislation is reviewed by Eaton (1985). Some of this legislation is relevant to protected areas. The Forestry Act (Amalgamated) (1973) is the main legislation responsible for the conservation and management of forest resources. Under this Act the government purchases timber rights from customary landowners for a certain period and then grants a licence to commercial companies to extract the timber. Royalties are paid to the government and a proportion of these is passed on to the provincial government and landowners. Environmental safeguards are provided for in the agreements between the government and logging companies. For example, logging is not allowed within 20m of permanent watercourses, or 50m in the case of major rivers, nor on gradients above 25-30. The interests of customary land-owners are also protected. They retain rights of access for gardening, hunting and collection of wood for fuel and construction purposes. Reforestation is not provided for in the forestry legislation but depends on arrangements between the landowners and permit-holders. The Forestry Act is seen to be inadequate to cope with the modern system of provincial government, and is readily circumvented by recourse to the provisions of the Forestry (Private Dealings) Act which enables landowners to enter into private agreements with logging companies with few if any statutory controls over the scale and manner of the operations (TFAP, 1989). The Act is also discussed further by Sargent (1989). An important statute is the Environmental Planning Act (1978) which calls for an assessment of the impact of a development project on the environment. Both the Environmental Planning Act and permits issued under the Forestry Act may also require logging companies to leave certain areas undisturbed as reserves for wildlife (Venkatesh *et al.*, 1983). There is evidence, however, that much of this legislation is not effectively enforced, with some 70% of logging companies continuing to operate despite the mandatory provision of environmental impact assessments (Anon., 1990).

International Activities Papua New Guinea is not yet party to any of the international conventions or programmes that directly promote the conservation of natural areas, namely the Convention concerning the Protection of the World Cultural and Natural Heritage

(World Heritage Convention), Unesco Man and the Biosphere Programme and the Convention on Wetlands of International Importance especially as Waterfowl Habitat (Ramsar Convention).

At a regional level, Papua New Guinea signed the 1976 Convention on the Conservation of Nature in the South Pacific on 12 June 1976. Known as the Apia Convention, it entered into force during 1990. The Convention is coordinated by the South Pacific Commission and represents the first attempt within the region to cooperate on environmental matters. Among other measures, it encourages the creation of protected areas to preserve indigenous flora and fauna.

Papua New Guinea is also party to the South Pacific Regional Environment Programme (SPREP) and has signed (3 November 1987) and ratified (15 September 1989) the 1986 Convention for the Protection of the Natural Resources and Environment of the South Pacific Region (SPREP Convention). The Convention entered into force during August 1990. Article 14 calls upon the parties to take all appropriate measures to protect rare or fragile ecosystems and threatened or endangered flora and fauna through the establishment of protected areas and the regulation of activities likely to have an adverse effect on the species, ecosystems and biological processes being protected. However, as this provision only applies to the Convention area, which by definition is open ocean, it is most likely to assist with the establishment of marine reserves and the conservation of marine species.

Other international and regional conventions concerning environmental protection to which Papua New Guinea is party are reviewed by Venkatesh *et al.* (1983).

Administration and Management The Department of Environment and Conservation, headed by a Secretary and with its own Ministry, was originally established in 1974 as the Office of Environment and Conservation, Department of Lands, Surveys and Mines. It successively passed through the departments of Natural Resources, Lands, National Mapping & Environment, Lands, Surveys & Environment, and Physical Planning & Environment before being upgraded to departmental status in 1985. The constitutional basis for the existence and operations of the Department of Environment and Conservation is the Fourth Goal of the National Constitution. The Department is divided into four main divisions: Environment, Nature Conservation, Water Resources and Management Services (DEC, 1988).

Administration of acts directly concerned with protected areas, i.e. Fauna (Protection and Control), Conservation Areas, National Parks, is the responsibility of the Nature Conservation Division, which comprises three branches: Conservation Surveys, National Parks, and Wildlife Conservation (DEC, 1988). Under the National Parks Act, 1982 the First Assistant Secretary of the National Parks Service is responsible for the administration and management of national parks, marine national parks, provincial parks, historical sites, nature reserves, national walking tracks and other protected areas. This differs from the previous Act whereby powers were assigned to a National Parks Board (Eaton, 1985; SPREP, 1985b). The National Parks Service internal revenue, collected from park entrance fees, totals some K 30,000 (approximately US$ 35,300). Wildlife management areas, which are declared by the Minister for Environment and Conservation under the Fauna (Protection and Control) Act, are managed by the landowners themselves, who are also responsible for making the rules (SPREP, 1985b; Eaton, 1986).

Systems Reviews Papua New Guinea lies between the Equator and latitude 12°S and between longitudes 141°E and 164°E. It comprises the eastern half of New Guinea and includes the Bismarck Archipelago (principally New Britain, New Ireland, New Hanover and Manus), d'Entrecasteaux Islands, the Louisiade Archipelago and the North Solomon islands of Bougainville and Buka. The western half of New Guinea forms the Indonesian province of Irian Jaya.

Forest of some sort, including successional forest, covers 71% (328,617 sq. km) of mainland Papua New Guinea. Undisturbed rain forest constitutes 65% (300,847 sq. km) of the total area and man-disturbed lands (grassland, gardens, degraded forest, plantation) some 20% (92,568 sq. km) (Beehler, 1985). Summary estimates for 1990 indicate natural forest cover of 78% (361,250 sq. km), including 420 sq. km of forest plantation (FAO, 1987). Discrepancies between the two sets of data can probably be ascribed to different definitions of forests, and different survey and analytical techniques. Approximately 20% of the total land area of Papua New Guinea is currently used for agriculture and 10%, or 46,000 sq. km, is under intensive cultivation (Freyne and McAlpine, 1985). An account of the forests, the threats to them and maps depicting current distribution is given in Collins *et al.* (1991).

The following description of the vegetation, based on Johns (1982) and an unpublished account by M.D.F. Udvardy reflects marked altitudinal zonation. A fringe of mangrove occurs along much of the coastline. Inland, swamps are extensive and covered by high forest with screw "palm" *Pandanus* and sago palm *Metroxylon sagu* forming a lower canopy. On drier land, mixed lowland rain forest is widespread and comprises complex communities, with epiphytes, orchids, tree and ground ferns. In contrast to rain forests elsewhere in Malesia, dipterocarp species are poorly represented. Throughout the lowlands, rain forests have been extensively destroyed or modified by shifting agriculture. Few areas of rain forest have escaped some form of cataclysmic destruction over the past 200-300 years (Johns, 1982). Areas having a markedly seasonal climate support monsoon forest which is characterised by the presence of a number of species that remain leafless for prolonged periods. Savanna vegetation, a degraded form of monsoon forest, occurs in areas receiving an annual rainfall of less than 1000-1300mm.

Dominated by *Eucalyptus* spp., it is quite distinct from lowland alluvial plains vegetation elsewhere in Papua New Guinea and resembles that of northern Australia.

Above 700m coniferous trees appear in the rain forest. Various altitudinally overlapping forest types can be distinguished within the montane zone, which usually extends from 700m to 2,700-3,000m. The upper montane forest, which may extend to 3,300m, is a cloud forest, with 10-25m tall moss-covered trees and a dense understorey. In the subalpine zone, the "high mountain forest" has a closed canopy at about 10m, with moss carpeting the forest floor. Ericaceous (heather family) shrubs supplant the forest near its upper limit at 3,800-4,100m, and are in turn replaced by grasslands, tarns and bogs (Smith, 1982). These are supplanted by tundra, which extends from about 4,400m to 4,700m. With the possible exception of the montane grasslands around Henganofi, all grasslands below 3,000m probably originate from a combination of agriculture and firing. The vegetation of the various island groups is mainly lowland rain forest, and at higher elevation, montane rain forest. The enclaves of grasslands and savannas are likely to be anthropogenous.

The various types of wetlands are described by Paijmans (1976) and Scott (1989), the latter providing detailed accounts of 33 wetland sites. Among the most extensive are mangrove swamps which occupy large parts of the coastal areas of Papua New Guinea, predominantly along protected bays and near the mouths of rivers. The largest expanses are in the south, notably in the Gulf of Papua with 162,000-200,000ha of mangroves. The north coast is not as rich in mangroves as the south coast.

The coral reefs of Papua New Guinea are virtually pristine compared to those of many countries, although they are coming under increasing threat from higher siltation and effluent loads in coastal areas and from commercial exploitation (UNEP/IUCN, 1988). The total area of reefs and associated shallow water to depths of 30m or less is estimated to be 40,000 sq. km (Wright and Kurtama, 1987; Wright and Richards, 1985), with the greatest concentration (12,870 sq. km) lying off Milne Bay Province (Dalzell and Wright, 1986).

The present protected areas system is very inadequate, particularly for a country of such biological importance as Papua New Guinea. Together with the rest of New Guinea, it ranks third in importance to Lord Howe and New Caledonia among 226 Oceanic islands of particular conservation interest (Dahl, 1986). The other islands within Papua New Guinea that fall within the top 12 most important Oceanic islands for conservation are New Britain, Goodenough, and Bougainville in descending order.

Although Dahl (1986) assesses the conservation importance of the different islands within Papua New Guinea and the extent of protected areas coverage, gaps in the protected areas system are not highlighted at national level. Previously (Dahl, 1980), an attempt was made to identify whether the various habitat types within Papua New Guinea are conserved within protected areas, but this review no longer reflects the present situation because many protected areas (notably wildlife management areas) have since been established.

It is instructive to compare the existing network of protected areas with that proposed by Diamond (1976). This proposed system, although more extensive, is largely analogous to a scheme earlier outlined by Specht *et al*. (1974) in which areas of habitat that might be expected to incorporate an almost complete range of biogeographical and ecological patterns are defined. The majority of existing protected areas lie outside the 22 areas of conservation importance identified by Diamond (1976); moreover, most of these conservation areas are not even represented in the protected areas network.

The protected areas network proposed by Diamond (1976) is based largely on bird distributions because these have been studied in most detail. (Available information suggests that fairly similar patterns hold for other animals and for plants.) Less ambitious and focused principally on conserving birds of paradise and their rain forest habitat throughout New Guinea is a 4,882 sq. km system of eight reserves proposed by Beehler (1985). Similarly, Parsons (1983) has proposed the establishment of a network of 20 reserves to meet the conservation requirements of birdwing butterflies. Many of these proposed sites coincide or overlap with those recommended under the schemes already discussed.

An action strategy for protected areas in the South Pacific Region has already been launched (SPREP, 1985a). Principal goals of the strategy cover conservation education, conservation policies, establishment of protected areas, effective protected areas management, and regional and international cooperation. Priority recommendations for Papua New Guinea are as follows: review conservation legislation; develop public awareness programmes in environmental education; review administrative structures to effect efficient implementation of environmental and conservation policies; review "protected areas register"; draw up a list of endangered species of plants and animals; review the effectiveness of the current system of protected areas; undertake a comprehensive survey of terrestrial and marine ecosystems and design a representative system of protected areas; secure assistance and support for the preparation of management plans for Mt Wilhelm National Park, McAdam National Park and Mt Gahavisuka Provincial Park; and develop a national conservation strategy. The outline of an environmental management programme for sustainable developed has been compiled (Kula, 1989), including a timetable for its implementation by 1992. However, the degree of progress with this programme is not known.

Of paramount importance is the need to develop a national conservation plan and identify priorities for the establishment of a comprehensive protected areas network. Not only are more protected areas required to conserve the great diversity of life on the mainland, but attention should also be directed towards developing the network on other large islands such as New Britain, New Ireland, Manus, Goodenough, Fergusson and Bougainville. Smaller islands with significant levels of endemism, such as Ninigo Islands and Luf (Hermit) Islands, may require priority action, however, because they may be under greater relative human threat (Dahl, 1986). Similarly, marine sites need to be identified and incorporated within the protected areas network (Dahl, 1986; Genolagani, 1984). Preliminary recommendations from the 1989 Tropical Forestry Action Plan donor coordination mission suggest that improvements in the existing protected areas network should receive a higher priority than the establishment of new areas. Twenty protected areas are identified as suitable for rehabilitation under any TFAP operations (Srivastava and Bützler, 1989).

The most significant natural resource problems facing Papua New Guinea are forest depletion, soil loss and soil fertility in the mid-montane valley systems, degradation from large-scale mining and agricultural activities, and exploitation of reef fisheries which are among the richest in the world (ADB, 1987; UNEP, 1987; Viner, 1984). Forests are being destroyed at an estimated rate of 80,000ha per year, commercial logging accounting for some 60,000ha yearly, and shifting cultivation 10,000-20,000ha yearly (ADB, 1987; WEI, 1988). This rate is increasing: in 1981-1985, it is estimated to have been 22,000ha per year (Repetto, 1988). These estimates contrast with the more conservative FAO figure of some 12,000ha deforestation annually, with a further 60,000ha disturbed in some way by logging (FAO, 1987). About 1,000,000ha of former forest have now been converted to grassland as a result of over-intensive shifting agriculture (Collins *et al.*, 1991). Papua New Guinea is relatively free from industrial pollution, except in coastal areas where much of the industry is sited. Considerable environmental damage has also been caused by mining activities, notably those of New Guinea Goldfields near Wau, Bougainville Copper in the Jaba catchment area and Ok Tedi in the Fly River region (ADB, 1987; Hughes, 1989; Viner, 1984; WEI, 1988).

Addresses

Department of Environment and Conservation, PO Box 5749, Boroko

References

ADB (1987). Papua New Guinea. Environmental natural resources briefing profile. Asian Development Bank, Manila. 6 pp.

Anon. (1990). PNG bans logging permits. *Christian Science Monitor*: 19.

Asigau, W. (1989). The wildlife management area system in Papua New Guinea. *Case Study* 15. Fourth South Pacific Conference on Nature Conservation and Protected Areas, Port Vila, Vanuatu, 4-12 September. 17 pp.

Beehler, B.M. (1985). Conservation of New Guinea forest birds. *ICBP Technical Publication* No. 4. International Council for Bird Preservation, Cambridge, UK. Pp. 223-246.

Collins, N.M., Sayer, J.A. and Whitmore, T.C. (Eds) (1991). *The conservation atlas of tropical forests: Asia and the Pacific*. Prepared by the International Union for Conservation of Nature, Switzerland and the World Conservation Monitoring Centre, Cambridge, UK. Macmillan Press Ltd, London. 256 pp.

Cragg, S.M. (1987). Papua New Guinea. In: Umali, R.M. *et al.* (Eds), *Mangroves of Asia and the Pacific: status and management*. Natural Resources Management Center and National Mangrove Committee, Ministry of Natural Resources, Manila. Pp. 299-309.

Dahl, A.L. (1980). Regional ecosystems survey of the South Pacific. *South Pacific Commission Technical Paper* No. 179. 99 pp.

Dahl, A.L. (1986). *Review of the protected areas system in Oceania*. IUCN, Gland, Switzerland and Cambridge, UK/UNEP, Nairobi, Kenya. 328 pp.

Dalzell, P. and Wright, A. (1986). An assessment of the exploitation of coral reef fishery resources in Papua New Guinea. In: Maclean, J.L., Dizon, L.B. and Hosillos, L.V. (Eds), *The first Asian fisheries forum*. Asian Fisheries Society, Manila, Philippines.

DEC (1988). Department of Environment and Conservation Handbook. Department of Environment and Conservation, Boroko. 64 pp.

Diamond, J.M. (1976). A proposed natural reserve system for Papua New Guinea. Unpublished report. 16 pp.

Eaton, P. (1985). Land tenure and conservation: protected areas in the South Pacific. *SPREP Topic Review* No. 17. South Pacific Commission, Noumea, New Caledonia. 103 pp.

Eaton, P. (1986). Grass roots conservation. Wildlife management areas in Papua New Guinea. *Land Studies Centre Report* 86/1. University of Papua New Guinea. 101 pp.

FAO (1987). Assessment of forest resources in six countries. Special study on forest management, afforestation and utilization of forest resources in the developing regions. *Field Document* 17. GCP/RAS/106/JPN. FAO, Bangkok. 104 pp.

Freyne, D.F. and McAlpine, J.R. (1985). Land clearing and development in Papua New Guinea. In: *Tropical land clearing for sustainable agriculture*. IBSRAM, Jakarta.

Genolagani, J.M.G. (1984). An assessment of the development of marine parks and reserves in Papua New Guinea. In: McNeely, J.A. and Miller, K.R. (Eds), *National parks, conservation, and development: The role of protected areas in sustaining society*. Smithsonian Institution Press, Washington, DC. Pp. 322-329.

Hughes, P.J. (1989). The effects of mining on the environment of high islands: a case study of gold mining on Misima Island, Papua New Guinea. *Environmental Case Studies* 5. South Pacific Regional Environment Programme, SPC, Noumea, New Caledonia. 6 pp.

Johns, R.J. (1982). Plant zonation. In: Gressitt, J.L. (Ed.), Biogeography and ecology of New Guinea. *Monographiae Biologicae* 42: 309-330.

Kisokau, K. and Lindgren, E. (1984). Ndrolowa Wildlife Management Area. A report on proposals to establish a wildlife management area for a variety of wildlife resources in Manus Province. Office of Environment and Conservation, Department of Physical Planning and Environment. 12 pp.

Kwapena, N. (1984). Wildlife management by the people. In: McNeely, J.A. and Miller, K.R. (Eds) (1984), *National parks, conservation, and development. The role of protected areas in sustaining society.* Smithsonian Institution Press, Washington, DC. Pp. 315-321.

Kula, G.R. (1989). Environmental management for sustainable development programme. *Case Study* No. 20. Fourth South Pacific Conference on Nature Conservation and Protected Areas, Port Vila, Vanuatu, 4-12 September. 8 pp.

Paijmans, K. (1975). Vegetation map of Papua New Guinea (1: 1,000,000) and explanatory notes to the vegetation map of Papua New Guinea. *CSIRO Land Research Series* 35: 1-25.

Parsons, M.J. (1983). A conservation study of the birdwing butterflies *Ornithoptera* and *Troides* (Lepidoptera: Papilionidae) in Papua New Guinea. Final Report to Department of Primary Industry, Papua New Guinea. 111 pp.

Repetto, R. (1988). *The forest for the trees? Government policies and the misuse of forest resources.* World Resources Institute, Washington, DC. 105 pp.

Sargent, C. (1989). Papua New Guinea Tropical Forest Action Plan: land use issues. Draft. International Institute for Environment and Development/United Nations Development Programme. Unpublished. 78 pp.

Scott, D.A. (Ed.). (1989). *A directory of Asian wetlands.* IUCN, Gland, Switzerland and Cambridge, UK. Pp. 1111-1155.

Seddon, G. (1984). Logging in the Gogol Valley, Papua New Guinea. *Ambio* 13: 345-350.

Sloth, B. (1988). *Nature legislation and nature conservation as a part of tourism development in the island Pacific.* Tourism Council of the South Pacific, Suva, Fiji. 82 pp.

Smith, J.M.B. (1982). Origin of the tropicoalpine flora. In: Gressitt, J.L. (Ed.), Biogeography and ecology of New Guinea. *Monographiae Biologicae* 42: 287-308.

Specht, R.L., Roe, E.M. and Boughton, V.H. (1974). Conservation of major plant communities in Australia and Papua New Guinea. *Australian Journal of Botany* Supplementary Series 7: 591-605.

SPREP (1985a). *Action strategy for protected areas in the South Pacific region.* South Pacific Commission, Noumea, New Caledonia. 21 pp.

SPREP (1985b). Papua New Guinea. In: Thomas, P.E.J. (Ed.), *Report of the Third South Pacific National Parks and Reserves Conference.* Volume III. *Country reviews.* South Pacific Commission, Noumea, New Caledonia. Pp. 175-194.

SPREP (1989). Papua New Guinea. Paper presented at the Fourth South Pacific Conference on Nature Conservation and Protected Areas, Port Vila, Vanuatu, 4-12 September. 13 pp.

Srivastava, P. and Bützler, W. (1989). Protective development and conservation of the forest environment in Papua New Guinea: priority needs and measures proposed under the Tropical Forest Action Plan. Draft. Unpublished. 40 pp.

TFAP (1989). Tropical Forestry Action Plan. Papua New Guinea: forestry sector review. Draft. Unpublished. 165 pp.

UNEP (1987). *Environmental management in Papua New Guinea.* Volume 2. *Review of background information.* A programme document submitted by the Government of Papua New Guinea to the United Nations Environment Programme for funding through the UNEP clearing house programme. 108 pp.

UNEP/IUCN (1988). *Coral reefs of the world. Volume 3: Central and Western Pacific.* UNEP Regional Seas Directories and Bibliographies. IUCN, Gland, Switzerland and Cambridge, UK/UNEP, Nairobi, Kenya. 378 pp.

Venkatesh, S., Va'ai, S. and Pulea, M. (1983). An overview of environmental protection legislation in the South Pacific countries. *SPREP Topic Review* 13. South Pacific Commission, Noumea, New Caledonia. 63 pp.

Viner, A.B. (1984). Environmental protection in Papua New Guinea. *Ambio* 13: 342-344.

WEI (1988). *Protecting the environment. A call for support.* Wau Ecology Institute, Wau, Papua New Guinea. 19 pp.

Wright, A. and Richards, A.H. (1985). A multispecies fishery associated with coral reefs in the Tigak Islands, Papua New Guinea. *Asian Marine Biology* 2: 69-84.

Wright, A. and Kurtama, Y.Y. (1987). Man in Papua New Guinea's coastal zone. *Resource Management and Optimization* 4: 261-296.

ANNEX
Definitions of protected area designations, as legislated, together with authorities responsible for their administration

Title: The National Parks Act

Date: 1982

Brief description: Replaced the amended 1971 Act, which in turn superceded the original National Parks and Gardens Act, 1966.

Administrative authority: Nature Conservation Division, Department of Environment and Conservation

Designation: Not defined

Title: The Conservation Areas Act

Date: 12 September 1978

Brief description: Provides (a) for the preservation of the environment and of the national cultural inheritance by (i) the conservation of sites and areas having particular biological, topographical, geological, historic, scientific or social importance; and (ii) the management of those sites and areas, in accordance with the fourth goals of the National Goals and Directive Principles; and (b) to give effect to those goals and Principles under Section 25 of the Constitution, and (c) to establish a National Conservation Council and (d) for other purposes.

Administrative authority: No information

Designation: No information

NB: The Act awaits implementation due to financial constraints.

Title: Fauna (Protection and Control) Act

Date: 1966

Brief description: Although concerned primarily with the protection of endangered species, the Act provides for the establishment of wildlife management areas (WMAs) on land held under customary ownership, of which there are three categories.

Administrative authority: The Act provides for the setting up of wildlife management committees, thereby involving customary land-owners in the control of wildlife resource exploitation. The committees advise on the provision of specific rules for each area for "the protection, propagation, encouragement, management, control, harvesting and destruction of fauna".

Designation:

Wildlife management area

— Category I WMAs are either terrestrial or marine areas reserved at the landowner's request for the conservation and controlled utilisation of all wildlife and habitat.
— Category II WMAs (sometimes referred to as "protected areas") are areas where only specific named species are protected.
— Category III WMAs (or "sanctuaries") are areas where most resources, excluding specific named animal species, are fully protected.

SUMMARY OF PROTECTED AREAS

Map[†] ref.	National/international designation Name of area	IUCN management category	Area (ha)	Year notified
	National Parks			
1	Jimi Valley	II	4,180	1986
3	McAdam	II	2,080	1970
5	Varirata	II	1,063	1969
	Nature Reserve			
7	Talele Islands (Bismarck Archipelago)	IV	40	1973
	Wildlife Management Areas (Categories I to III)			
13	Bagiai (I)	VIII	13,760	1977
14	Balek (III)	IV	470	1977
15	Baniara Island (II)	VIII	15	1975
16	Crown Island (III)	IV	5,969	1977
17	Garu (I)	VIII	8,700	1976
18	Iomare (I)	VIII	3,837	1987
19	Lake Lavu (I)	VIII	2,640	1981
20	Long Island (III)	IV	15,724	1977
21	Maza (I)	VIII	184,230	1978
22	Mojirau (I)	VIII	5,079	1978
23	Ndrolowa (I)	VIII	5,850	1985
24	Neiru (I)	VIII	3,984	1987
25	Nuserang (I)	VIII	22	1986
26	Oia-Mada Wa'a (I)	VIII	22,840	1981
27	Pirung (I)	VIII	44,240	1989
28	Pokili (I)	VIII	9,840	1975
29	Ranba (I)	VIII	41,922	1977
30	Sawataetae (I)	VIII	700	1977
31	Siwi Utame (I)	VIII	12,540	1977
32	Tonda (I)	VIII	590,000	1975
33	Zo-Oimaga (I)	VIII	1,488	1981
	Provincial Parks			
35	Nanuk Island	IV	12	1973
36	Talele Islands	IV	40	1973

[†]Locations of some protected areas are shown on the accompanying map.

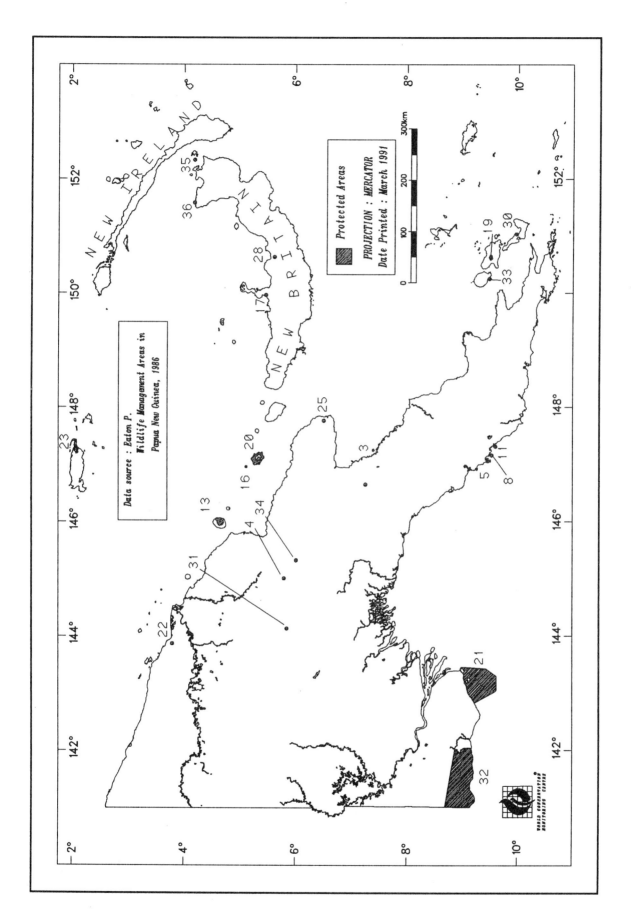

Protected Areas of Papua New Guinea

225

PITCAIRN ISLANDS
(UK DEPENDENT TERRITORY)

Area 43 sq. km (Pitcairn 7.5 sq. km; Henderson 37 sq. km; Oeno 5 sq. km; Ducie 2.5 sq. km)

Population Approximately 48 on Pitcairn Island; all other islands uninhabited

Economic indicators
GNP: Not available
GDP: Not available

Policy and Legislation The Pitcairn Islands, being a UK Dependent Territory, are nationally governed by the Island's Governor, from the British Consulate in Auckland, New Zealand. "Local government" is *via* an elected Island Magistrate and Island Council.

Until recently, there was little consideration of conservation in the Pitcairn Islands. In recent years, however, the Pitcairners have become increasingly aware of the need for conservation on their own island, and the publicity generated by the inclusion of Henderson Island in the World Heritage Convention has helped bring this about. Being such a small community there is little need for legislation on conservation: it is an attitude brought about by caring rather than being told to care.

International Activities The United Kingdom ratified the Convention Concerning the Protection of the World Cultural and Natural Heritage (World Heritage Convention) on 29 May 1984 and Henderson Island was inscribed on the World Heritage List in 1988. The UK Government is considering the possibility of listing Ducie Atoll under the Convention Concerning the Protection of Wetlands Especially as Waterfowl Habitat (Ramsar Convention).

Between January 1991 and April 1992 some 20 scientists from universities and institutions around the world have been spending three-month stretches on Henderson, but also visiting the other islands in the group. The Sir Peter Scott Commemorative Expedition to The Pitcairn Islands, also known as the Pitcairn Islands Scientific Expedition, was organised to explore thoroughly Henderson Island in all its aspects (biology, archaeology, geology). Results from the expedition will be published in the scientific literature.

Administration and Management As part of the Pitcairn Islands Scientific Expedition, a management plan is being compiled for Henderson Island, and the conservation requirements for the Pitcairn Islands as a whole are being investigated. The proposals will be forwarded to the UK government in March 1992.

Systems Reviews The Pitcairn Islands are on the edge of the tropical south Pacific (Pitcairn: 25°04'S, 130°06'W), and are extremely isolated, being over 4,800km from the nearest continent. To the west is French Polynesia (the Gambier Islands being some 560km west of Pitcairn). To the east is Easter Island, 1,760km away. The only inhabited island is Pitcairn, the others being Oeno Atoll (some 160km north-north-east), Henderson Island (176km east-north-east) and Ducie Atoll (464km east). Available information is somewhat sketchy, but will be vastly improved when all the information collected by the Pitcairn Islands Scientific Expedition is drawn together in 1992.

Pitcairn is a small, high island, reaching 333m. There is one settlement, Adamstown, below which is the only sheltered landing on the island, Bounty Bay. There is much evidence of extensive Polynesian occupation on Pitcairn, where the local pitchstone was quarried and exported throughout Polynesia. However, this large population had disappeared by the time of the Island's colonisation by the Bounty mutineers in 1790. Until relatively recently the small population of islanders lived a subsistence existence. As a result, much of the cultivatable land on the island has been used to grow a wide variety of crops. In addition, many non-food plants have been introduced to the island (e.g. Norfolk Island pine and roseapple). The feral goat population, allowed to run wild, has also seriously affected the local habitat. Much of the local woods, used for fuel, building and carving for export, have also been overexploited. As a result, the island has been extensively changed by the islander's occupation: invasive plants have overrun native ones, goats have damaged, tree-felling and the use of the bulldozer have carved up the hill sides and contributed to a considerable erosional damage (caused by some 1600mm of annual tropical rain). There is one strip of native vegetation left on the island, and this runs along a ridge at the Highest Point. Here there are the remnants of "cloud forest" with a small number (c.30) of surviving tree ferns.

Henderson Island is a raised coral island, uplifted by the lithospheric flexure caused by the crustal loading of the volcanic island of Pitcairn. The old lagoon floor of Henderson is now raised to an elevation of about 30m. As a result of the elevation, the island plateau is protected from the periodic inundation of the sea during cyclones which is so typical of lower atolls. This has allowed the continued existence on the island by colonisers, and, as a result, a considerable fauna and flora has built up, much of which is now endemic.

The island typically has steep sides. There is a fringing reef around much of the island, and three beaches (North-west, North, East) on all of which it is possible to get across the reef in an inflatable boat. There are no sheltered anchorages. Henderson is characterised by difficult terrain and very dense vegetation. As a result of the porosity of the limestone, there has been little build-up of soil, so in the island interior the underlying fossil coral is bare. There are gross structures, such as old reef units and erosion features (coral pinnacles, reaching up to 5m high), as well as the smaller fragments of individual coral colonies, fossil clams (often still in growth positions) and other gastropods. Fossil urchins have also been found in the interior. Growing on top of this terrain is a vegetation made up of some 60 species (11 endemic taxa) (Brooke *et al.*, 1991; Fosberg *et al.*, 1983; Paulay and Spencer, 1989).

Insect life is common, but not abundant, with approximately 180 species on the island (Brooke *et al.*, 1991). In addition to the arthropods, there is an abundant terrestrial snail fauna, with at least 18 species (Fosberg *et al.*, 1983). There are four endemic land birds and three species of land reptiles. The only mammal is the introduced Polynesian rat, but seabirds are common around and on the island. Crabs are very abundant. The fringing reefs are in excellent condition. Being at the geographical limit of reef growth, their diversity is not great; nor is that of the reef inhabitants. However, the abundance is large and fish grow to huge sizes, having never been harvested.

There is considerable evidence of an extensive period of Polynesian occupation on Henderson. The population may have reached as many as 100. However, with the exception of the Polynesian rat and some clumps of the sterile Corclyline palm (and possibly *Pandanus* palm), there are few obvious changes to the native fauna and flora. The Polynesians are likely to have lived solely on the coastal fringes of the island, and the plants they cultivated became extinct after the Polynesians disappeared (Brooke *et al.*, 1991). Henderson remains perhaps the only example of a Pacific Island where the present vegetation and fauna even approximates its native condition.

Henderson is threatened from two main sources. First, the Pitcairners harvest Miro (now seriously depleted) and visit the island annually. It is possible that they will one day introduce, as attempted in the past, a destructively invasive plant. However, with their growing awareness of the importance of Henderson this is unlikely. Secondly, there is a small, but significant number of boats that anchor off Henderson and from which people come ashore. These are mainly yachts on circumnavigations of the globe, but also include occasional cruise liners and commercial fishing vessels. It is possible that invasive plants and/or animals may be introduced. It is not possible to "police" the landings onto Henderson, as Pitcairn is over 160km away, and is often unaware of boats in the waters around Henderson.

Ducie Atoll is the most remote of the Pitcairn Islands. It is 469km east of Pitcairn itself, and never actually visited by the islanders. It is an atoll, about 1.6km in diameter, consisting of four islands surrounding some 60% of the lagoon. The largest is Acadia, about 2.4km miles long and up to 250m wide, with a maximum elevation of about 3m.

Essentially, there is only one plant species on the island: *Tournefortia argentea* (one specimen of one other species has been found). There is, therefore, a paucity of arthropods (25 species in total) and terrestrial animals in general. However, Ducie Atoll is an excellent seabird colony. It is perhaps the world's main breeding station of Murphey's petrel (Brooke *et al.*, 1991), with several thousand breeding pairs; in addition, many other seabirds found in the locality breed here in large numbers (Rehder and Randall, 1975; T. Benton, pers. comm., 1991).

Other Relevant Information During 1983 issues relating to conservation of the uninhabited islands, and in particular Henderson, came to the attention of conservation organisations with the proposal by an American millionaire to live on Henderson. Response was quick, resulting in the preparation of Serpell *et al.* (1983) and various other submissions. Henderson, which is a forested atoll, largely unaltered with ten endemic plant taxa, four endemic land birds and various other endemics, is in fact of great scientific and conservation value. In December 1983 the government made it clear that the land would not be leased for this purpose. It has been suggested that the Pitcairn Group, or at least the uninhabited islands, be nominated as World Heritage sites now that the United Kingdom has ratified the World Heritage Convention.

Addresses

Office of the Governor of the Pitcairn Islands, The British Consulate-General, Private Bag, Auckland, New Zealand (FAX: 64 9 3031836)
The Island Magistrate, Pitcairn Island, South Pacific Ocean, *via* New Zealand
Pitcairn Islands Scientific Expedition, Dr M de L Brooke (Expedition Field Leader and Ornithologist) Department of Zoology, University of Cambridge, Downing Street, Cambridge CB2 3EJ, UK

References

Brooke, M. de L, Spencer, S. and Benton, T. (1991). Pitcairn Islands Scientific Expedition, Interim Report. Unpublished.
Fosberg, F.R., Sachet, M.H. and Stoddart, D.R. (1983). Henderson Island (Southeastern Polynesia): Summary of current knowledge. *Atoll Research Bulletin* 272. Smithsonian Institution.
Holloway, J.D. (1990). The Lepidoptera of Easter, Pitcairn and Henderson Islands. *Journal of Natural History* 24: 719-29
Mathis, W.N. (1989). Diptera (Insecta) or true flies of the Pitcairn Group. *Atoll Research Bulletin*, Henderson Island Issue.
Paulay, G. and Spencer T. (1989). Vegetation of Henderson Island. *Atoll Research Bulletin*, Henderson Island Issue.

Rehder, H.A. and Randall, J.E. (1975). Ducie Atoll: Its history, physiography and biota. *Atoll Research Bulletin* 183. Smithsonian Institution.

Schubel, S.E. and Steadman, D.W. (1989). More bird bones from Polynesian archaeological sites on Henderson Island, Pitcairn Group, South Pacific. *Atoll Research Bulletin*, Henderson Island Issue.

Serpell, J., Collar, N., Davis, S. and Wells, S. (1983). Submission to the Foreign and Commonwealth Office on the Future Convention of Henderson Island in the Pitcairn Group. Unpublished report prepared for WWF-UK, IUCN and ICBP.

SPREP (1980). Pitcairn. *Country Report* 11. South Pacific Regional Environment Programme. South Pacific Commission, Noumea.

Steadman, D.W. and Olson, S.L. (1985). Bird remains from an archaeological site on Henderson Island, South Pacific: Man-caused extinctions on an "uninhabited" island. *PNAS* 82: 6191-6195

SOLOMON ISLANDS

Area 28,450 sq. km of land distributed over 1,340,000 sq. km of sea

Population 304,000 (1988) (World Bank, 1990)
Natural increase: 3.6% per annum

Economic Indicators
GNP: US$ 630 per capita (1988)

Policy and Legislation The Solomon Islands Constitution protects the right of land owners to utilise their land and forests as they wish. Consequently, any restriction resulting from designation of a reserve or forest park would impose upon this basic right. Given that some 87% of all land is held in customary or tribal tenure (C. Turnbull, pers. comm., 1990), it is the attitudes and activities of local people, not government, that determine the extent and nature of land reservation for nature conservation.

Formal protected areas legislation principally comprises two acts, both dating from the colonial era. The Wild Birds Protection Act (1914) enabled the Minister, under Section 14, to declare any island or islands, or any part or parts of any island or any district, as a bird sanctuary. The National Parks Act (1954) makes provision for the establishment of strictly protected natural areas as national parks. This legislation is considered to be inconsistent with current concepts of resource use, by making inadequate allowance for genuine customary needs and by placing no obligation upon the administering agency to manage a national park in accordance with stated park objectives (SPREP, 1985b). Some provincial governments have passed ordinances with provisions for the establishment of protected areas, including Temotu and Isabel provinces (Leary, 1990). However, provisional legislation is equally subject to the restrictions of the constitution and traditional practices.

The forest policy of the Solomon Islands was reviewed in 1983, and a National Forest and Timber Policy was approved by Government in 1984. The Policy called for the maximum desirable log processing, minimum wastage and increased investment in forests, aims that were restated in the 1985-89 National Development Plan. Failure to achieve these and more detailed aims is attributed to shortcomings in the forest legislation, institutional weakness and a lack of public participation and awareness. Consequently, a revised forest policy has been promulgated. This identifies six imperatives, viz. protection, sustainable use, basic needs, development, participation and distribution. Six objectives are established including "(VI) set aside areas for environmental, ecological, scientific and heritage reserves taking into account landowner needs and customary values; protect sensitive areas".

The Forestry and Timber Act (1969, amended 1977) has a narrow perspective, but makes some provision for controlled forest areas for forest water catchment protection. The Forest Resources and Timber Utilisation (Amendment) Act 1987 was approved in March 1987, with a retrospective commencement date of 16 June 1978. The North New Georgia Timber Corporation Act (1979) provides for the establishment of a corporation for the promotion of timber utilisation in New Georgia. The Corporation can impose conditions on licensed felling and is charged with the duty of encouraging replanting in felled areas (Sloth, 1988). There was a new Forestry Bill 1989 that was due to go before Parliament in November 1989, which included provisions for the establishment of conservation areas. The Town and Country Planning Act (1979) provides for tree preservation orders for "any tree, groups of trees or woodlands ... in the interests of amenity". The Lands and Titles Act (1968, amended 1970) makes provision for preservation orders to be applied to land of "historic or religious" value, and permits the establishment of nature reserves.

Regulations prescribing measures for the protection and preservation of the marine environment can be promulgated under the Delimitation of Marine Waters Act (1978). A new forestry bill was due to be presented to parliament in November 1989. This includes provisions for environmental protection and the establishment and management of conservation areas (Isa, 1989).

International Activities The Solomon Islands is not yet party to any of the international conventions or programmes that directly promote the conservation of natural areas, namely the Convention concerning the Protection of the World Cultural and Natural Heritage (World Heritage Convention), the Unesco Man and the Biosphere Programme and the Convention on Wetlands of International Importance especially as Waterfowl Habitat (Ramsar Convention).

The Convention on the Conservation of Nature in the South Pacific (1976) has been neither signed nor ratified. Known as the Apia Convention, it awaits sufficient countries to deposit instruments of ratification, acceptance, approval or accession to enter force. The Convention is coordinated by the South Pacific Commission and represents the first attempt within the region to cooperate on environmental matters. Among other measures, it encourages the creation of protected areas to preserve indigenous flora and fauna.

The Solomon Islands is party to the South Pacific Regional Environment Programme (SPREP) and has ratified (10 August 1989) the 1986 Convention for the Protection of the Natural Resources and Environment of the South Pacific Region (SPREP Convention). The convention entered into force during August 1990. Article 14 calls upon the parties to take all appropriate measures to protect rare or fragile ecosystems and threatened or endangered flora and fauna through the establishment of protected areas and the regulation of activities likely to have an adverse effect on the species, ecosystems and biological processes being protected. However, as this provision only applies to the Convention area, which by definition is open ocean, it is most likely to assist with the establishment of marine reserves and the conservation of marine species.

Other international and regional conventions concerning environmental protection to which the Solomon Islands is party are reviewed by Venkatesh *et al.* (1983).

Administration and Management General responsibility for environmental matters is part of the portfolio of the Minister of Lands, Energy and Natural Resources. The Environment and Conservation Division of the Ministry of Natural Resources has responsibility for environmental protection and the conservation of natural resources and in 1989 had five members of staff, with the expectation that this would increase to nine during 1990 (Isa, 1989).

The Forestry Division comprises management, plantation, research and development, timber control (logging and utilisation) and herbarium and extension sections, with, in 1987, 54 professional staff and a labour force of 595. Training of foresters takes place in Papua New Guinea, at the Solomon Islands College of Higher Education, Australia and elsewhere (MNR, 1987). The Conservator of Forests is responsible for implementation of the Forestry and Timber Act.

The Solomon Islands Development Trust is a rural development organisation which places its operations, with a particular focus on community education, in an environmental context and is a significant influence in the rain forest logging debate (Baines, 1990). Community-based associations have been formed to counter the growing threat of over-exploitation, including the Fauro Peoples Association and The Vella 2000. Their principal objectives are to: promote community unity; promote resource management and utilisation; derive economic returns for their members; promote resource-management awareness amongst their membership; and promote respect for traditional values of their communities. In addition, these groups have approached government for advice on the establishment and management of protected areas, with three sites covering 6,000ha already earmarked (Isa, 1988).

Nominal protected areas are few in number and currently cover some 0.2% of the terrestrial territory of the Solomons, and an even less significant percentage of the total national area. This can be largely attributed to the failure to consider traditional attitudes toward land and natural resources and the unwillingness amongst local people to be permanently alienated from their lands (Isa, 1988). For example, Arnavon Wildlife Sanctuary was gazetted without consultation with parties claiming rights under customary law, leading to serious management implementation problems and a failure to protect the site. A number of other reserves exist on paper but none is functional. Customary land owners are frequently willing to grant temporary access rights for timber felling, but reluctant to be alienated permanently from land following the establishment of a protected area (Isa, 1989). The best conservation practice at present is that carried out by some of the customary land-owning groups, although these efforts have thus far escaped recognition and lack the support of formal legislation (G.B.K. Baines, pers. comm., 1989). However, this is likely to be rectified with formal government recognition of these community based protected areas (Isa, 1988).

As an additional hindrance to the establishment of formal protected areas, state-owned land covers only 257 sq. km (9% of total land area), of which 240 sq. km is committed to forestry plantations or operations (C. Turnbull, pers. comm., 1990). Therefore, at present, there is little opportunity to develop conservation areas on uncommitted government land (Leary, 1989). The existing protected areas system is thus very weak, none of the formally gazetted protected areas is managed nor has field staff, and they have very little conservation significance or official recognition. Indeed, as there are no legally gazetted forest reserves, the Solomon Islands effectively has no formal protected areas system.

Systems Reviews The Solomon Islands occupies a central position in the Melanesian Arc that extends from Papua New Guinea in the west to Fiji in the east. They consist of two roughly parallel island chains, forming the western continental margin of the Pacific Basin. The central archipelago lies between latitudes 5°S and 12°S and longitudes 152°E and 163°E. A brief summary of the physical features and coral reefs is given in UNEP/IUCN, 1988.

Prior to recent, extensive logging, the Solomons supported a rain forest mantle, unbroken except for areas of, probably anthropogenic, grassland and heath. Grasslands are probably a fire-climax, with invasion by tree species checked. Lowland formations are largely uniform throughout the archipelago. There is, however, considerable local variation in the floristics and structure, due largely to tropical cyclones. There are about a dozen very common tree species which constitute the lowland forest canopy at 30-45m height. Alluvial valleys sometimes support high forest, to which a number of large tree species are restricted. In many places, however, valley floors carry a dense low scrub, presumably the result of former cultivation, or destructive floods. Forests have also been shaped by natural events. Natural catastrophes, such as earthquakes, landslips, lightning and cyclone-force winds, are not infrequent. High forest may take decades to recover from such events (Whitmore, 1969). The mangrove forest has never been heavily exploited and the canopy is typically 24m tall (Whitmore, 1969). Mangroves around Marovo Lagoon have been felled due to the mistaken belief that they inhibit coconut plantation production (Baines, 1985) and there is some exploitation for firewood and construction timber in rural areas (Collins *et al.*, 1991).

Some 24,200 sq. km, 80% of the total land area, were forested in 1976 (Hansell and Wall, 1976), of which 20,000 sq. km were accounted for by various communities at low and medium altitudes, on basalts or andesites. Revised data from December 1988 indicate 25,500 sq. km of forest cover (C. Turnbull, pers. comm., 1990), of which 23,000 sq. km is unproductive for physical reasons. This discrepancy does not reflect a substantive increase in forest cover but is most probably the result of different definitions of forests, and different survey and analytical techniques. By way of illustration, some 95% of logging takes place on land in customary tenure, whilst reforestation on the same class of land is restricted to just 98ha on Malaita (Wenzel, 1989), indicating a serious imbalance between reforestation and logging operations.

Prior to the 1970s, forest cover was largely pristine, but has since been heavily exploited by foreign interests. During the last 20 years, a timber industry has developed. Widespread logging has resulted in the loss of primary forest on several islands, especially Ghizo and Kolombangara, with consequent loss of wildlife habitat (Dahl, 1986), soil compaction and erosion, leading to reef sedimentation (Anon., 1986b). Historically, forestry practice has not given due consideration to the ecological effects of logging, and small-scale operations are underway or planned in several small areas, rather than utilising larger blocks on larger islands (Hoyle, 1978). However, since 1986 logging companies have been required to fell only trees with a diameter in excess of 60cm, thus preventing the removal of supersmalls (trees with diameters as little as 35cm) (MNR, 1988). This regulation is likely to reduce damage substantially to forests during logging and enhance natural regeneration.

Plantations are being established at the rate of approximately 1,500ha annually (Collins *et al.*, 1991), with 24,000ha established by December 1989 (C. Turnbull, pers. comm., 1990). It is hoped that plantations of fast-growing and commercially-valuable introduced species, such as teak, mahogany and gmelina, will replace the volume felled from natural forests, as a source of export revenue (Anon., 1988). Extensive plantations are also planned for Kolombangara, in a joint venture between the government and the Commonwealth Development Corporation (G.B.K. Baines, pers. comm., 1989).

Very little scientific work has been carried out on coral reefs. The 1965 Royal Society Expedition, which produced the only substantial account, found that in general the Solomon Islands lacked the luxuriant reefs of many parts of the Pacific, due to unfavourable environmental conditions such as steep and exposed shores with little suitable substrate for coral growth. Coral reefs are further discussed in UNEP/IUCN (1988), which also gives a detailed account of Marovo Lagoon.

Of the 31 terrestrial and marine habitats identified by Dahl (1980), only three are included within even nominal protected areas, namely some lowland rain forest types, mixed species montane rain forest and grassland. The much wider range of habitats which at present receive no protection broadly include: various lowland rain forest formations; *Neonauclea/Sloanea* montane rain forest; cloud, riverine, swamp, mangrove and atoll/beach forests; *Casuarina* woodland and scrub; serpentine vegetation and dwarf-shrub heath; various freshwater habitats; sea turtle nesting areas; algal and seagrass beds with associated fauna; various reef formations, rocky coast, lagoons, estuaries and offshore environments (Dahl, 1980). In terms of conservation importance, Rennell Island is ranked ninth in the entire Oceanian Realm, and is the second most important raised coral island, after Guam (Dahl, 1986). According to Collins *et al.* (1991), critically important sites, none of which is protected at present, include Rennell, the Arnavon islands and the kauri *Agathis macrophylla* stands in Santa Cruz. Coastal, mangrove, lowland, freshwater swamp and montane forests are all critically unprotected.

On the basis of the distribution of birds, Diamond (1976) proposed a system of three large high-priority, three smaller high-priority and eight medium-priority protected areas. Mammal, insect and plant species would also benefit as their distribution patterns are probably similar (Diamond, 1976). Dahl (1980) reiterated these recommendations, but also identifies a number of and adds reef, lagoon, mangrove and small botanical reserves. Less specific recommendations in Dahl (1986) include the establishment of protected areas on Rennell, San Cristobal, Guadalcanal, Malaita and Vanikolo, with smaller areas to protect interesting sites and species on other islands. Reforestation with native species might be considered alongside Kolombangara Reserve, to reinforce it and to buffer it from encroachment (Dahl, 1986).

An Action Strategy for Protected Areas in the South Pacific Region (SPREP, 1985a) has been prepared. The principal goals of the strategy cover conservation education, conservation policy development, establishment of protected areas, protected area management and regional and international cooperation. Priority recommendations for the Solomon Islands are as follows: develop a national conservation strategy, and review of all environmental aspects of development; develop, expand and implement a national environmental awareness programme, including radio broadcasts, publications and the establishment of a mobile unit; review and update all existing legislation relating to protected areas and conservation, and, where necessary, draw up new legislation; undertake staff training in the fields of biology, botany, ecology and park management; develop the following as protected areas: Oema Atoll and the island of Oema as a wildlife sanctuary; Mount Austin,

Mount Gallego; Island of Arnavon; Lake Te Nggano and its flora and fauna.

A number of recommendations, constituting a putative conservation strategy aimed at removing obstacles to conservation-based development, are made in SPREP (1985b). These broadly comprise environmental education, environmental surveys, conservation-based rural development, a zoned forest reserve system and an effective environmental law and policing system.

A comprehensive survey of conservation area needs and priorities is required prior to implementing any protected areas network (G.B.K. Baines, pers. comm., 1989). In response to this need, a SPREP-funded project, in cooperation with the Overseas Service Bureau, Australian Volunteer Abroad Programme and the National Environment and Conservation Division, has examined wildlife management issues, with emphasis on species trade, changes in traditional practice and habitat destruction. The project aims to provide guidelines for the sustainable use and management of Solomon Islands fauna, possibly stimulating the formulation of a National Policy on the Environment and Conservation (Leary, 1989) and draws the following broad conclusions. In the long term it is a priority that a comprehensive Conservation Act is developed for the Solomon Islands. Such an act should encompass the establishment of conservation areas, the protection of flora and fauna, regulation of wildlife trade, and the conservation of traditional resources management knowledge. A system of natural areas protection for the Solomon Islands urgently needs to be developed. In addition to large scale reserves, consideration needs to be given to the protection, in terms of good management, of small areas. One such means is the concept of wildlife management areas which has proven successful in Papua New Guinea. It is recommended that wildlife management areas legislation be enacted at the provincial or national level as soon as possible (Leary, 1990). This work has been further augmented by Lees (1991) with the design of a system of nationally representative protected forest areas. In addition to biological factors, social factors were considered in an attempt to bypass some of the problems that have dogged previous designs, and a total of 20 reserves is proposed.

Rennell has been the subject of logging proposals, which once again were being actively considered in 1989 (Collins *et al.*, 1991). At one time bauxite mining posed a considerable threat, particularly to Rennell. However, these proposals have now been dropped. Conversely, there is a current nationwide threat from gold prospecting and there has been localised environmental damage from prospecting activities, a notable example being Fauro Island (G.B.K. Baines, pers. comm., 1989).

Other Relevant Information Tourism is currently at a very low level, with some 11,000 visitors annually, although proposals have been made to develop this sector (Anon., 1986a; Glaser, 1987). Attention is currently being paid to developing "conservation tourism", centred on Rennell's freshwater Te Nggano Lake and emphasising the natural attractions of the Solomon Islands (G.B.K. Baines, pers. comm., 1989). Proposals for tourism expansion are being made on the basis of assistance from the European Development Fund of the European Economic Community, within the framework of the Pacific Regional Tourism Development Programme (Sloth, 1988).

Addresses

Ministry of Natural Resources, PO Box G24, Honiara

References

Anon. (1986b). Deforestation problems in the Solomon Islands. *Forest* 15: 11.

Anon. (1987). SPREP makes considerable progress with implementation of action strategy for protected areas. *Environment Newsletter, South Pacific Regional Environment Programme* No. 10. South Pacific Commission, Noumea, New Caledonia.

Anon. (1988). Projects in the Pacific. *British Overseas Development: Forestry Supplement.* Overseas Development Administration, London.

Baines, G.B.K. (1981). Environmental management for sustainable development in the Solomons: a report on environment and resources. Prepared for the Government of the Solomon Islands. Unpublished. 57 pp.

Baines, G.B.K. (1985). Study Area One: Marovo Lagoon, Solomon Islands. Working paper on pilot project for Commonwealth Science Council. 15 pp.

Baines, G.B.K. (1990). *South Pacific Conservation Programme.* World Wide Fund for Nature, Gland, Switzerland and Australia. 50 pp.

Collins, N.M., Sayer, J.A. and Whitmore, T.C. (Eds) (1991). *The conservation atlas of tropical forests: Asia and the Pacific.* Prepared by IUCN – the World Conservation Union, Switzerland and the World Conservation Monitoring Centre, Cambridge, UK. Macmillan Press Ltd, London. 256 pp.

Dahl, A.L. (1980). Regional ecosystem survey of the South Pacific Area. *SPC/IUCN Technical Paper* 179. South Pacific Commission, Noumea, New Caledonia. 99 pp.

Dahl, A.L. (1986). *Review of the protected areas systems in Oceania.* IUCN, Gland, Switzerland and Cambridge, UK. 239 pp.

Diamond, J.M. (1976). A proposed forest reserve system and conservation strategy for the Solomon Islands. Summary. Unpublished report. 19 pp.

Glaser, T. (1987). Solomon Islands: paradise lost and found. *The Courier* 102: 44-51.

Hoyle, M.A. (1978). Forestry and conservation in the Solomon Islands and the New Hebrides. *Tigerpaper* 5(2): 21-24.

Isa, H. (1988). The current trend in protected area development in Solomon Islands. In: Thomas, P.E.J. (Ed.), *Report on the workshop on customary tenure, traditional resource management and nature conservation*. South Pacific Regional Environment Programme, Noumea, New Caledonia. Pp. 97-99.

Isa, H. (1989). Solomon Islands. *Country Review* 6. Fourth South Pacific Conference on Nature Conservation and Protected Areas, Port Vila, Vanuatu, 4-12 September. South Pacific Commission, Noumea, New Caledonia. 9 pp.

Leary, T. (1989). Wildlife management in the Solomon Islands. *Case Study* 10. Fourth South Pacific Conference on Nature Conservation and Protected Areas, Port Vila, Vanuatu, 4-12 September. South Pacific Commission, Noumea, New Caledonia. 6 pp.

Leary, T. (1990). Survey of Wildlife Management in Solomon Islands. *Final Report*. SPREP project PA 17. A joint project of Solomon Islands Government, South Pacific Regional Environment Programme and Traffic (Oceania). SPREP, Noumea, New Caledonia. 74 pp.

Lees, A. (1991). *A representative protected forests system for the Solomon Islands*. Report on behalf of Australian National Parks and Wildlife Service. Maruia Society, Nelson, New Zealand. 185 pp.

Ministry of Natural Resources (1989). *Forest policy statement August 1989*. Forestry Division. 23 pp.

Sloth, B. (1988). *Nature legislation and nature conservation as part of tourism development in the island Pacific*. Pacific Regional Tourism Development Programme. Tourism Council of the South Pacific, Suva, Fiji. 82 pp.

SPREP (1985a). *Action strategy for protected areas in the South Pacific region*. South Pacific Commission, Noumea, New Caledonia. 24 pp.

SPREP (1985b). *Country report: Solomon Islands* In: SPC (1985). *Third South Pacific National Parks and Reserves Conference*. Volume 3. South Pacific Commission, Noumea. Pp. 115-124.

UNEP/IUCN (1988). *Coral reefs of the world. Volume 3. Central and Eastern Pacific*. UNEP Regional Seas Directories and Bibliographies. IUCN, Gland, Switzerland and Cambridge, UK/UNEP, Nairobi, Kenya. 378 pp.

Venkatesh, S, Va'ai, S. and Pulea, M. (1983). An overview of environmental protection legislation in the South Pacific countries. SPREP *Topic Review* No. 13. South Pacific Commission, Noumea, New Caledonia. 63 pp.

Wenzel, L. (1989). Environment and change in the Pacific: a survey of resource use and policy in Fiji, Tonga, W. Samoa, Vanuatu and Solomon Islands. *INR Environmental Studies Report* No. 43. Institute of Natural Resources, University of the South Pacific, Suva, Fiji. 95 pp.

Whitmore, T.C. (1969). The vegetation of the Solomon Islands. *Philosophical Transactions of the Royal Society of London* B 255: 259-270.

ANNEX
Definitions of protected area designations, as legislated, together with authorities responsible for their administration

Title : The National Parks Act

Date: 1954

Brief description: Makes provision for the establishment of strictly protected natural areas as national parks.

Administrative authority: Conservation Officer, The Environment and Conservation Division, Ministry of Lands, Energy and Natural Resources

Designations:

National park

Title : The Wild Birds Protection Act

Date: 1914

Brief description: Enabled the Minister, under Section 14, to declare any island or islands, or any part or parts of any island or any district, as a bird sanctuary. The Act provides protection for bird species only, and makes no provision for the conservation of habitat.

Administrative authority: Unknown

Designations:

Bird sanctuary Only interference with birds is an offence within sanctuaries.

Title: An Ordinance to Consolidate and Amend the Law on Forests and to Control and Regulate the Timber Industry and for Matters Incidental Thereto and Connected (Forests and Timber Ordinance, 1969)

Date: 1969 (amended 1977)

Administrative authority: The Conservator of Forests

Designations:

State forest Established under the provisions of S.9, on any land that is public land, land in which the government holds a freehold interest in land or a leasehold interest in land by or on behalf of the Government. The Commissioner shall not grant any interest in a state forest without the Conservator's consent (S.10). Prohibited activities (S.11) include felling, cutting etc trees; damaging vegetation; clearing land; residing or erecting buildings; grazing livestock; possessing any equipment for cutting or working forest produce; constructing roads, saw pits or workplaces.

Controlled forest Established, in either forests or other vegetation types, for water catchment protection. Rights which may be exercised in a controlled forest are specified in the notification of establishment. Activities specifically prohibited under S.16 comprise felling, cutting etc. forest produce except for personal or domestic use; clearing or breaking land for cultivation or any other purposes; residing, or erecting any building etc.; grazing livestock.

NB: Repealed Forests Ordinance (Cap. 71).

Title: The Town and Country Planning Act

Date: 1979

Brief description: Provides for tree preservation orders for "any tree, groups of trees or woodlands ... in the interests of amenity"

Administrative authority: Unknown

Designation: Unknown

Title: The Lands and Titles Act

Date: 1968 (amended 1970)

Brief description: Makes provision for preservation orders to be applied to land of "historic or religious" value, and permits the establishment of nature reserves.

Administrative authority: Unknown

Designation: Unknown

TOKELAU
(NEW ZEALAND OVERSEAS TERRITORY)

Area 12.25 sq. km

Population 3,000 (1982); 1,700 (1988)
Natural increase: No information

Economic Indicators
GNP: No information

Policy and Legislation Tokelau is a non-self-governing territory of New Zealand. There is no protected areas legislation, and apparently no government conservation policy relating to native species and habitats. Instead, there is a long-standing system of resource management based on traditional custom, although this is now starting to lose its effectiveness (Toloa and Gillett, 1989).

Traditional marine conservation measures can be considered in two categories: those that are specifically designed for conservation and those aspects of the traditional system which indirectly result in a reduced amount of fishing effort on particular species. The most important explicit conservation measure is the "lafu" system, whereby all types of fishing are banned in specific areas of the main reef. The decision to establish a "lafu" is made by the Council of Elders. Another specific conservation measure is the rejection of undersized fish and the prohibition of fish poisoning. In addition to these specific measures, there is a wide variety of traditional practices which effectively reduce fishing effort. For example, only certain individuals may capture sea turtles, and the elevated community status of pelagic fishing relieves pressure from inshore fisheries (Toloa and Gillett, 1989).

International Activities Tokelau is not yet party to the Unesco Man and the Biosphere Programme, the Convention on Wetlands of International Importance especially as Waterfowl Habitat (Ramsar Convention) nor the Convention concerning the Protection of the World Cultural and Natural Heritage (World Heritage Convention). It is not known if New Zealand's active participation in the latter two conventions has any significance for Tokelau.

At a regional level, Tokelau has neither signed nor ratified the Convention on the Conservation of Nature in the South Pacific, 1976. Known as the Apia Convention, it entered into force during 1990. The Convention is coordinated by the South Pacific Commission and represents the first attempt within the region to cooperate on environmental matters. Among other measures, it encourages the creation of protected areas to preserve indigenous flora and fauna. Tokelau has not signed the Convention for the Protection of the Natural Resources and Environment of the South Pacific Region, 1986 (SPREP Convention). However, the Convention has been signed (25 November 1986) and ratified (3 May

1990) by New Zealand, although again it is not clear if this places any obligation upon Tokelau itself. The Convention entered into force during August 1990. Article 14 calls upon the parties to take all appropriate measures to protect rare or fragile ecosystems and threatened or endangered flora and fauna through the establishment of protected areas and the regulation of activities likely to have an adverse effect on the species, ecosystems and biological processes being protected. However, as this provision only applies to the Convention areas, which by definition is open ocean, it is most likely to assist with the establishment of marine reserves and the conservation of marine species.

Administration and Management Responsibility for conservation of natural resources rests with the Council of Elders, a body comprising most males over 60 years of age and totalling some 25 individuals from each island. All land is under customary ownership. The Department of Agriculture and Fisheries has facilitated resource surveys and provided scientific information to support conservation and resource management (Anon., 1989).

Systems Reviews Tokelau consists of three small islands set on a north-west to south-east axis from 8°S to 10°S and 171°W to 173°W. Nukunonu is the biggest atoll at 4.7 sq. km, Fakaofo is 4.0 sq. km and Afafu is 3.5 sq. km. Each atoll consists of a number of reef-bound islets encircling lagoons. These islets vary in length from 90m to 6km and up to 200m in width. At no point do they rise more 5m above sea level. The atolls comprise coral rubble and sand mixed with a thin layer of humus and are of generally low fertility.

In general the vegetation has not been greatly modified. Nukunonu includes areas of *Cordia, Pisonia* and *Guerttarda* woodland which is dense and may reach heights of 10m to 20m. Fakaofo supports a beach scrub of *Scaevola* and *Tournefortia* (Davis *et al.*, 1986). An account of the extent and composition of the vegetation is given in Parham (1971).

There have been few studies of the coral reefs (UNEP/IUCN, 1988). Hinds (1969/1971) carried out a brief survey and reported that coral growth was limited to the upper portions of old coral growth massifs rising from the floor, the main portions of which were barren. Laboute (in press) briefly surveyed reefs on all three atolls in 1987 to assess damage caused by Hurricane Tusi. The degree of destruction was variable, reaching 90% in some places. A significant area of Nukunonu lagoon reef was damaged by a pesticide spill in 1969: all corals in a 2km section of the lagoon along Motu Te Kakai died with the exception of *Porites*. There was little recovery by 1976 (Marshall, 1976). Rhinoceros beetle *Oryctes rhinoceros* was accidentally introduced from

Western Samoa to Nukunonu where it spread to all islets, creating a serious risk to coconuts (Wodzicki, 1971). Control measures have been taken (SPREP, 1980). Turtles are now rare and there has been serious depletion of giant clam *Tridacna* spp. (Wodzicki, 1973), but there is no evidence of depletion of other marine resources (Hooper, 1985). The potentially most threatening environmental hazard is that posed by increasing sea-levels due to global warming. Threats to freshwater supplies, land loss and episodic destruction through hurricanes may make the country uninhabitable, if the worst case scenarios are realised (Pernetta, 1988).

Because of the small size of the island, and the number of people it supports, most conservation measures reflect the need to protect the "human" environment. For example, tree planting is one of the principal aims of an annual conservation week. It has recently been reported that some 47ha of land on Nukunonu has been designated a protected area by the local Council of Elders, with adjoining reef areas to be added at a later date (B. Lear, pers. comm., 1990). More detailed information is not available. However, this development is in line with previous recommendations (Dahl, 1986) that remaining forest on Nukunonu should be protected. Dahl (1986) also suggests that the establishment marine reserves might become necessary if traditional conservation measures cease to be effectively applied.

Addresses

Director of Agriculture and Fisheries, Office of Tokelau Affairs, Apia, Western Samoa

References

Dahl, A.L. (1986). *Review of the protected areas system in Oceania*. IUCN, Gland, Switzerland and Cambridge, UK/UNEP, Nairobi, Kenya. 328 pp.

Davis, S.D., Droop, S.J.M., Gregerson, P., Henson, L., Leon, C.J., Lamlein Villa-Lobos, J., Synge, H. and Zantovska, J. (1986). *Plants in danger: what do we know?* IUCN, Gland, Switzerland and Cambridge, UK. 488 pp.

Hinds, V.J. (1969/71). A rapid fisheries reconnaissance in the Tokelau Islands, August 18-25, 1971. South Pacific Commission, Noumea, New Caledonia. (Unseen)

Hooper, A. (1985). Tokelau fishing in traditional and modern contexts. In: Ruddle, K. and Johannes, R.E. (Eds), *The traditional knowledge and management of coastal systems in Asia and the Pacific*. Unesco/ROSTEA, Jakarta, Indonesia. Pp. 7-38.

Laboute, P. (in press). Mission to the Tokelau Islands to evaluate cyclone damage to coral reefs. SPREP *Topic Review* 31. South Pacific Commission, Noumea, New Caledonia.

Marshall, K.J. (1976). Critical marine habitats and insect control in the South Pacific. *Proceedings of the SPC and IUCN Second Regional Symposium on Conservation of Nature*, Apia, Western Samoa, June. (Unseen)

Parham, B.E.V. (1971). The vegetation of the Tokelau Islands with special reference to plants of Nukunonu Atoll. *New Zealand Journal of Botany* 9: 576-609.

Pernetta, J.C. (1988). Projected climate change and sea level rise: a relative impact rating for countries of the South Pacific Basin. In: MEDU joint meeting of the task team on the implications of climatic change in the Mediterranean. Split, Yuogslavia, 3-7 October. Pp. 1-11.

SPREP (1980). Tokelau. *Country Report* 12. South Pacific Commission, Noumea, New Caledonia. 4 pp.

SPREP (1989). Tokelau. Paper presented at the Fourth South Pacific Conference on Nature Conservation and Protected Areas. Port Vila, Vanuatu, 4-12 September. 2 pp.

Toloa, F. and Gillett, R. (1989). Aspects of traditional marine conservation on Tokelau. *Case Study* 31. Fourth South Pacific Conference on Nature Conservation and Protected Areas, Port Vila, Vanuatu, 4-12 September. 5 pp.

UNEP/IUCN (1988). *Coral reefs of the world. Volume 3. Central and Western Pacific*. UNEP Regional Seas Directories and Bibliographies. IUCN, Gland, Switzerland and Cambridge, UK/UNEP, Nairobi, Kenya. 378 pp.

Wodzicki, K. (1973) The Tokelau Islands – environment, natural history and special conservation problems. Paper 10, Section 3. *Proceedings and Papers, Regional Symposium on Conservation of Nature – Reefs and Lagoons*. South Pacific Commission, Noumea, New Caledonia. Pp. 63-68.

KINGDOM OF TONGA

Area Total land area of 747 sq. km, spread over 347,282 sq. km of sea (IDEC, 1990)

Population Approximately 101,000 (1990) (IDEC, 1990) Natural increase: 0.84% per annum

Economic Indicators
GNP: US$ 800 per capita

Policy and Legislation The role of parks, reserves and protected areas, according to IDEC (1990), is to prevent depletion or extinction of valuable wildlife communities and to enrich and improve production of land and marine resources; and to protect areas or items of importance for Tongan cultural heritage and provide the people of Tonga and visitors with places of recreational, educational and scientific importance.

The legal and land tenure system gives the government considerable powers to acquire and reserve land for public purposes. All land is ultimately owned by the Crown, but is divided into four categories: the King's hereditary estates, Royal family hereditary estates, hereditary estates for the nobles and matapule, and government. The latter two are subdivided by law to provide town and tax (garden) allotments which can only be held by Tongans. The 19th century land reforms have meant that there is no customary tenure of the type found in many other countries of the South Pacific and thus the establishment of protected areas is generally easier than in many Pacific nations because much of the land is owned by the state. The foreshore is the property of the Crown and is defined in the Land Act as "land adjacent to the sea, alternately covered and left dry by the ordinary ebb and flow of the tides and all land adjoining thereunto lying within fifty feet of the high water mark." All territorial seas and internal waters are the property of the Crown and may be subject to government restrictions and regulations. Every Tongan has the right to fish in these waters. There are no traditional fishing rights giving villages or individuals exclusive rights to fish or gather shells in certain areas and such rights may have never existed in the past (Eaton, 1985).

The 1976 Parks and Reserves Act provides for "the establishment of a Parks and Reserves Authority and for the establishment, preservation and administration of parks and reserves". It states that every park "shall be administered for the benefit and enjoyment of the people of Tonga and there shall be freedom of entry and recreation therein by all persons. Every reserve, subject to any conditions and restrictions which the Authority may impose, shall be administered for the protection, preservation and maintenance of any valuable feature of such reserve and activities therein and entry thereto shall be strictly in accordance with any such conditions and restrictions" (SPREP, 1985a).

The 1961 Forest Act provides for the establishment of forest reserves, the conservation of important "culture trees" used in traditional crafts, and the protection of water catchments (Eaton, 1985).

The 1915/1974 Birds and Fish Preservation Act limits or prohibits the catching or injuring of certain species of fish, birds and turtles and establishes the legal authority to fine and imprison offenders, and confiscate equipment used to catch protected animals. Under the provisions of the Act, the two major lagoons of Tongatapu, Fanga'uta and Fanga Kakau, are protected as areas of environmental importance. The trapping of fish without permit, damaging mangroves, drilling, dredging, or discharging any effluent is prohibited (SPREP, 1985a), although some of these activities continue (P. Thomas, pers. comm., 1989). The Birds and Fish Preservation Act is weak and it is acknowledged that it requires updating (SPREP, 1980).

The Preservation of Objects of Archaeological Interest Act (1969) provides for the protection of a number of historical, cultural and archaeological sites, many of which are also protected by traditional law.

Other legislation with provision for environmental protection is reviewed by Venkatesh et al. (1983). An Environmental Protection Act and a Fisheries Act were under consideration in 1987 (U.F. Samani, pers. comm., 1987).

International Activities Tonga is not yet party to any of the international conventions or programmes that directly promote the conservation of natural areas, namely the Convention concerning the Protection of the World Cultural and Natural Heritage (World Heritage Convention), the Unesco Man and the Biosphere Programme and the Convention on Wetlands of International Importance especially as Waterfowl Habitat (Ramsar Convention).

The Convention on the Conservation of Nature in the South Pacific (1976) has been neither signed nor ratified. Known as the Apia Convention, it entered into force during 1990. The Convention is coordinated by the South Pacific Commission and represents the first attempt within the region to cooperate on environmental matters. Among other measures, it encourages the creation of protected areas to preserve indigenous flora and fauna.

Although Tonga is party to the South Pacific Regional Environment Programme (SPREP), the 1986 Convention for the Protection of the Natural Resources and Environment of the South Pacific Region (SPREP Convention) has not yet been signed or ratified. The convention entered into force during August 1990. Article 14 calls upon the parties to take all appropriate measures to protect rare or fragile ecosystems and threatened or endangered flora and fauna through the establishment of protected areas and the regulation of activities likely to have an adverse effect on the species, ecosystems and biological processes being protected. However, as this provision only applies to the convention area, which by definition is open ocean, it is most likely to assist with the establishment of marine reserves and the conservation of marine species.

Other international and regional conventions concerning environmental protection to which Tonga is party are reviewed by Venkatesh *et al.* (1983).

Administration and Management Under the provisions of the 1976 Parks and Reserves Acts, a Parks and Reserves Authority was established in 1989 within the Ministry of Lands, Survey and Natural Resources to protect, manage and develop natural areas in the Kingdom. The authority is headed by an ecologist and environmentalist who is directly responsible to the Secretary of Lands, Survey and Natural Resources. There are two park rangers directly responsible to the head of the Authority. In addition to the general administration of parks and reserves, the authority is also responsible for environmental impact assessment of all physical developments, physical planning and environmental education.

Park management is hampered by shortages of funds and personnel, and there has been only limited development of protected areas. Signposts advertising rules prohibiting the destruction or removal of marine life have been erected. However, there are no means of enforcing these rules and people are not always aware of them; shellfish are still collected from the reef in Ha'atafu Reserve. No biological surveys or inventories of reserves have been carried out (U.F. Samani, pers. comm., 1987). R.H. Chesher (pers. comm., 1987) reported virtually complete neglect of park boundaries. The marine parks on the small islands adjacent to Tongatapu are regularly and destructively fished. The advent of new fishing technology and increasing depletion of fisheries resources indicate that these parks are likely to come under greater threat in the future (IDEC, 1990). There is an urgent need for improved management of existing protected areas.

No baseline information or inventory has been conducted at any of the marine park sites other than a brief overview in 1984 (Chesher, 1985), which revealed extensive damage due to destructive fishing methods in the Pangaimotu reef area, long-term coral damage in the Monuafe area and extensive coral damage of unknown cause in the Malinoa Island area. A recent review (Pernetta, 1988) suggests that the potential impact of climate change and sea-level rise could have severe impacts with economic and social disruption, with inter-island movement of populations, and emigration. Clearly marine and coastal protected are also under considerable threat.

Systems Reviews Tonga, the last remaining Kingdom in the South Pacific, lies between 15°30'-22°20'S and 173°00'-176°15'W. There are 17 main islands forming three major groups, namely the Vava'u Group (143 sq. km) to the north, the central Ha'apai Group (119 sq. km) and the southerly Tongatapu Group (256 sq. km). The islands are mainly flat, elevated coral reefs which cap the peaks of two parallel submarine ridges, although some are high and volcanic in origin.

Brief summaries of the vegetation and coral reefs of most islands are given in Douglas (1969) and UNEP/IUCN (1988), respectively. The original vegetation on limestone islands comprised lowland rain forest. However, on the larger islands this has been entirely cleared (A.L. Dahl, pers. comm., 1989) for settlements and cultivation. According to a 1989 report (Wenzel, 1989), Tonga's exploitable indigenous forest was expected to be exhausted within 3-5 years, although whether this has occurred is not known. The best examples of volcanic island lowland rain forest are found on 'Eua in the Tongatapu Group (Dahl, 1980). In 1982 some 3,779ha of forest reserve existed on 'Eau (Larsen and Upcott, 1982) along the eastern ridge of the island. Subdivision of the area into tax 'apis (leases for agriculture), and deforestation by government and private logging have reduced the indigenous forest of these reserves (IDEC, 1990).

Moss forest occurs on the summit of Kao, and on Tafahi, to the north of the Vava'u Group. Coastal scrub is found on most islands, and recent lava flows support *Casuarina*. The islands of Tongatapu, 'Eua and Uta Vava'u in the Vava'u Group all have extensive areas of secondary vegetation, including scrub and grasslands. Mangroves and *Cyperus* reed swamp are present, the latter especially around the crater lake on Ninafo'ou, an isolated volcanic island to the north-west of the Vava'u Group. The crater zones of most volcanic islands have distinct, but sparse herbaceous flora (Dahl, 1980; IUCN, 1986).

The existing protected areas network covers 4.5% of the terrestrial area of Tonga, but nearly 90% of the network is accounted for by just one site, namely Fanga'uta and Fanga Kakau Lagoons Marine Reserve. Thus, protected habitats are largely marine or littoral and, according to Dahl (1980), include mangrove forest, possibly non-tidal salt marshes in Fanga'uta Lagoon, sea turtle nesting areas, sea grass beds, animals in sediments, algal, barrier, fringing and lagoon reef formations, beach, and open lagoon.

Habitats that are omitted from the protected area system are largely terrestrial and include lowland rain forest, savanna, grasslands, freshwater marsh, rock desert, reed swamp, permanent lake, seabird rookeries, as well as marine lake, offshore environments and submarine trenches (Dahl, 1980). Efforts to incorporate remaining fragments of lowland tropical rain forest in the protected areas system included an attempt to establish Vaomapa Park on Tongatapu, although the lease securing the site has lapsed and the site is no longer protected (SPREP, 1989). A long-standing proposal to gazette a 1,400ha protected area on 'Eua has not yet been implemented, although funding for preliminary surveys and facilities has been sought from the New Zealand government. The proposed park area includes fringing reefs, coastal regions, eastern ridge and terrace, and ridge summit, with rain forest and important bird habitats (Dahl, 1980). The park would also contain some of the last stands of sandalwood remaining in the Pacific, a number of plant species with restricted distributions and Matalanga-'a'Maui natural limestone bridge (SPREP, 1985a).

An Action Strategy for Protected Areas in the South Pacific Region (SPREP, 1985b) has been prepared. The principal goals of the strategy cover conservation education, conservation policy development, establishment of protected areas, effective protected area management and regional and international cooperation. Priority recommendations for Tonga are as follows: develop and implement intensive public education and training programmes, including radio broadcasts on environmental conservation, and grants for overseas study on environment-related disciplines; survey all potential protected areas, followed by preparation of management plans and provision of funds for management; develop a national environmental strategy; prepare resource inventories to help identify areas of critical importance for resource conservation; and exchange between countries in the region of environmental expert staff for short periods, to cooperate in addressing specific problems and in exchanging ideas (SPREP, 1985b).

A number of these recommendations have been acted upon. A government department of public information has been set up in the Prime Minister's Office to coordinate public education and the spread of environmental information from government ministries. Radio broadcasting, however, is seen as less effective than video, itself a very popular medium, for disseminating environmental information. An aerial photography survey has been completed which may provide the basis for a nationwide review of potential protected areas (R.H. Chesher, pers. comm., 1990).

As one response to the Action Strategy, a comprehensive Environmental Management Plan (IDEC, 1990) was published in 1990, in a cooperative effort between the United Nations Economic and Social Commission (UN/ESCAP) and the Government of Tonga. The plan was formally introduced during a three-day symposium in August 1990 in Nuku'alofa. Responsibility for the compilation of the plan was undertaken by the Interdepartmental Environment Committee (IDEC), comprising Lands, Survey and Natural Resources (Chairman); Health; Foreign Affairs; Agriculture, Fisheries and Forestry; Works; Labour, Commerce and Industries, Central Planning; and Tonga Visitors Bureau. The objectives of the environmental management plan are to: examine the existing state of the environment in

Tonga and summarise the relevant existing information in a single document; determine the environmental participants in the Kingdom and their resources needs; discover environmental resource needs that are not being met and identify these as problems; and recommend a plan of action to deal with existing and projected environmental problems. The plan includes coverage on demography and economics, geomorphology, mineral resources and coastal processes, fisheries, tourism, wildlife conservation and management, the institutional and legal framework, and a strategy for implementing the recommendations in the plan.

A number of recommendations concerning the development of protected areas are made in the plan. Community participation in the site selection, research, monitoring and protection of parks and reserves will assist in long-term park management, including establishment of local park or reserve committees to maintain or assist in the management of local resources. Within the framework of community parks, new marine and land parks should be gazetted, including Vava'u Coral Gardens, Neiafu Harbour Wreck and Swallows cave, all on Vava'u. The plan identifies a number of opportunities for further protected areas, namely microparks, resource parks, mangrove parks, botanical gardens, steep-slope forest, turtle and wildlife resources islands, coastal littoral forest and water reserves parks. At present there are no officially designated forest sanctuaries, although some forests on steep coastal slopes and on some volcanic islands with difficult or dangerous access are unofficially protected. The remaining indigenous forests should be divided into parks, reserves and sanctuaries depending on their present condition and status. The proposed national park on 'Eua, designed to protect forest on the steep slopes of south-east 'Eua and to prevent erosion, should be set up in accordance with the recommendations of the Ministry of Lands, Survey and Natural Resources. The proposed 'Ohonua water supply reserve on 'Eua should be implemented by insisting on the maintenance of a 50m border of forest adjacent to all streams and tributaries in the area. Compliance by landowners should be enforced and seedling trees made available to replant areas which have already been cleared. Mount Talau National Park in Vava'u should be established. The plan also makes a series of recommendations aimed at increasing public involvement, monitoring and research. A series of policy recommendations covers institutional, legal and regulatory issues, monitoring and assessment, and research.

Dahl (1980) recommends gazetting the following reserve types: forest reserves on Tafahi or Niuatoputapu, for endemic birds, and perhaps other volcanic islands, viz. Tofua, Kao, Lalo, 'Ata or Toku; other areas of 'Eua of botanical interest; samples of other terrestrial biomes not yet protected; marsh, lake and lagoon habitats, namely Niuafo'ou, Kao and 'Uta Vava'u; and further marine areas to include a full range of marine biomes, especially outside Tongatapu. Much of this is reiterated by Dahl (1986) with the recommendation to gazette 'Eua National Park, a protected area on 'Ata and some protected areas on Niuafo'ou and Kao. Other small reserves should be considered for particular features, and there should be an extended role for coral reef reserves in managing coastal fisheries with the support of the local population. It is intended that the Parks and Reserves Authority will review proposals for terrestrial and marine parks, with a view to implementation.

Addresses

Parks and Reserves Authority, Ministry of Lands, Survey and Natural Resources, PO Box 5, Nuku'alofa

References

Chesher, R. H. (1985). Practical problems in coral reef utilization and management: A Tongan study. In: *Proceedings of the Fifth Coral Reef Congress*, Tahiti. Pp. 213-217.

Dahl, A.L. (1980). Regional ecosystem surveys of the South Pacific Area. *SPC/IUCN Technical Paper* 179. South Pacific Commission, Noumea, New Caledonia. 99 pp.

Dahl, A.L. (1986). *Review of the protected areas system in Oceania.* IUCN, Gland, Switzerland and Cambridge UK/UNEP, Nairobi, Kenya. 328 pp.

Davis, S.D., Droop, S.J.M., Gregerson, P., Henson, L., Leon, C.J., Lamlein Villa-Lobos, J., Synge, H. and Zantovska, J. (1986). *Plants in danger: what do we know?* IUCN, Gland, Switzerland and Cambridge, UK. 488 pp.

Douglas, G. (1969). Draft checklist of Pacific Oceanic Islands. *Micronesica* 5(2): 327-463.

Eaton, P. (1985). Land tenure and conservation in the South Pacific. SPREP *Topic Review* 17. South Pacific Commission, Noumea, New Caledonia. 103 pp.

Interdepartmental Environment Committee. (1990). *Environmental Management Plan for the Kingdom of Tonga.* Economic and Social Committee for Asia and Pacific. Bangkok, Thailand. 197 pp.

Larsen, A. and Upcott, A. (1982). *A study of the forest resources of 'Eua Island and their potential for future management.* Chandler, Fraser And Larsen, Rotorua, New Zealand. 125 pp. (Unseen)

Pernetta, J.C. (1988). Projected climate change and sea level rise: a relative impact rating for countries of the South Pacific Basin. In: MEDU joint meeting of the task team on the implications of climatic change in the Mediterranean. Split, Yugoslavia, 3-7 October. Pp. 1-11.

SPREP (1980). Tonga. *Country Report* 13. South Pacific Commission, Noumea, New Caledonia. 60 pp.

SPREP (1985a). Tonga. In: Thomas, P.E.J. (Ed.), *Report of the Third South Pacific National Parks and Reserves Conference.* Volume III. *Country reviews.* South Pacific Commission, Noumea, New Caledonia. Pp. 211-216.

SPREP (1985b). *Action strategy for protected areas in the South Pacific Region.* South Pacific Commission, Noumea, New Caledonia. 24 pp.

SPREP (1989). Tonga. Paper presented at the Fourth South Pacific Conference on Nature Conservation and Protected Areas, Port Vila, Vanuatu, 4-12 September. 3 pp.

UNEP/IUCN (1988). *Coral reefs of the world. Volume 3: Central and Eastern Pacific.* UNEP Regional Seas Directories and Bibliographies. IUCN, Gland, Switzerland and Cambridge, UK/UNEP, Nairobi, Kenya. 378 pp.

Venkatesh, S., Va'ai, S. and Pulea, M. (1983). An overview of environmental protection legislation in the South Pacific Countries. SPREP *Topic Review* 13. South Pacific Commission, Noumea, New Caledonia. 63 pp.

Wenzel, L. (1989). Environment and change in the Pacific: a survey of resource use in Fiji, Solomon Islands, Tonga, Vanuatu and Western Samoa. *INR Environmental Studies Report* No. 43. Institute of Natural Resources, The University of the South Pacific, Suva, Fiji. 95 pp.

ANNEX
Definitions of protected area designations, as legislated,
together with authorities responsible for their administration

Title: Parks and Reserves Act

Date: 21 September 1976 (Royal Assent: 22 January 1977)

Brief description: Provides for the establishment of a Parks and Reserves Authority and for the establishment, preservation and administration of parks and reserves. The Authority may, with consent of the Privy Council, declare any area of land or sea to be a park or reserve and in the same manner declare any park or reserve to cease to be such.

Administrative authority: Parks and Reserves Authority, established in 1989 within the Ministry of Lands, Survey and Natural Resources.

Designation:

Park Subject to any conditions and restrictions which the Authority may impose, shall be administered for the benefit and enjoyment of the people of Tonga and there shall be freedom of entry and recreation therein by all persons (s.7).

Reserve Subject to any conditions and restrictions which the Authority may impose, shall be administered for the protection, preservation and maintenance of any valuable feature of such reserve, and activities therein and entry thereto shall be strictly in accordance with any such conditions and restrictions (s.8).

Marine reserve Administered for the protection, preservation and control of any aquatic form of life and any organic or inorganic matter therein (s.10). (Defined as an area of sea).

All parks and reserves shall be registered and recorded in accordance with the provisions of the Land Act (s.4.3). Prohibited activities include: alteration, disturbance, destruction or removal etc. of any feature whether organic or inorganic in any park or reserve; damage etc. to any notice, fence or building; depositing rubbish in any park or reserve; obstructing, disobeying etc. any instruction issued by the Authority (s.11).

Title: The Birds and Fish Preservation Act

Date: 1915 (amended in 1916, 1934 and 17 September 1974, gaining Royal Assent on 26 June 1975)

Administrative authority: Enforcement officers include police, fisheries and agricultural officers

Description: Makes provision for the preservation of wild birds and fish

Designation: Not applicable

NB: See The Birds and Fish Preservation (Amendment) Act, 1974

Title: The Birds and Fish Preservation (Amendment) Act, 1974)

Date: 17 September 1974 (Royal Assent 26 June 1975)

Administrative authority: Enforcement officers include police, fisheries and agricultural officers

Description: Amends the Birds and Fish Preservation Act. The Act is to be read and construed as one with The Birds and Fish Preservation Act ("the Principle Act").

Section 2 of the Principle Act is amended by defining "protected area".

Designation:

Protected area Any area comprising land, or water, as specified in the Third Schedule of the Act. Section 7 prohibits discharge of effluent etc.; erecting buildings etc.; cutting, damaging, removing or destroying any mangrove; erecting fish fences or fish traps, fish trawling or commercial fishing; or carrying out any boring, drilling or dredging operations within a protected area.

The two major lagoons of Tongatapu, Fanga'uta and Fanga Kakau, are declared protected in the Third Schedule.

Title: Forests Act

Date: 2 November 1961

Administrative authority: Ministry of Agriculture, Fisheries and Forestry

Description: Provides for the setting aside of areas as forest areas or reserved areas and for the control and regulation of such areas and of forest produce and related matters (the King in Council may declare any unalienated land to be a forest reserve or reserved area (s.3) or may revoke such declarations (s.5)).

Designation:

Forest reserve Any demarcated forest or proclaimed forest reserve but shall not include a village forest reserve.

Reserved area Any demarcated area of land or proclaimed area of land which may be under grass or

243

scrub which may be needed for afforestation in the future.

Village forests Governed by such regulations concerning the protection, control and management of forest produce as the Minister may prescribe.

Section 4 defines activities prohibited within forest reserves and reserved areas, including damage to forest produce, grazing, clearing land, killing, taking or injuring animals, setting fires etc.

Title: Land (Timber Cutting) Act

Date: No information

Administrative authority: No information

Brief description: Regulates cutting and taking of trees. Land (Timber Cutting) Regulations (G33/43) state that, where permission is granted to cut timber

on Crown land, no timber can be cut within 50 feet (c.15m) of the high-water mark.

Designation: No information

Title: The Preservation of Objects of Archaeological Interest Act

Date: 1969

Administrative authority: Committee on Tongan Traditions

Brief description: Provides for the preservation of objects of archaeological interest.

Designation: Conditions can be specified to protect an object of archaeological interest from injury or removal.

Miscellany: An Environmental Protection Act and a Fisheries Act were under consideration in 1987.

SUMMARY OF PROTECTED AREAS

Map ref.	*National/international designation* Name of area	IUCN management category	Area (ha)	Year notified
	Marine Reserve			
1	Fanga'uta and Fanga Kakau Lagoons	VIII	2,835	1974
	Reserves			
2	Ha'atafu Beach	IV	8	1979
3	Hakaumama'o Reef	IV	260	1979
4	Malinoa Island	IV	73	1979
5	Monuafe Island	IV	33	1979
6	Mui Hopo Hoponga	V		1972
7	Pangaimotu Reef	IV	49	1979

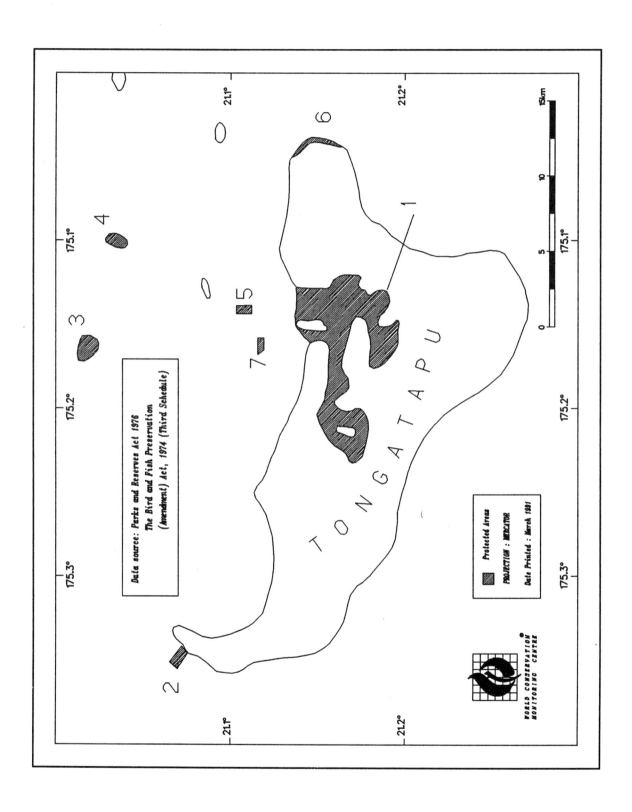

Protected Areas of Tonga

TUVALU

Area 26 sq. km land area in approximately 1.3 million sq. km of territorial waters

Population 8,229 (1985)
Natural increase: No information

Economic Indicators
GNP: No information

Policy and Legislation Prior to independence on 1 October 1978, Tuvalu was part of the Gilbert and Ellice Islands, and therefore subject to the 1938 Gilbert and Ellice Islands Colony Wild Birds Protection Ordinance and the 1975 Wildlife Protection Ordinance; the latter was not enforced.

The Fourth Development Plan 1987-1991 contains a chapter entitled Land Management, Environment and Conservation. This states that the government's general aim is to bring about improved "environmental control through better utilisation of the country's very meagre land and environmental resources".

The Prohibited Areas Ordinance and Wildlife Ordinance (1975) provides for the establishment of wildlife reserves, although none has yet been gazetted, and the protection of seabirds (UNEP/IUCN, 1988).

The Ordinance to Provide for the Conservation of Wildlife (1975, revised 1978) lists 32 fully protected bird species and a number with partial protection. Provision is also made for the declaration of wildlife sanctuaries (Sloth, 1988). There is, however, no legislation for the protection of vegetation (Hay, 1986).

The enactment of legislation to better integrate existing laws and to provide for Environmental Impact Assessments has been recommended (ESCAP, 1988).

International Activities Tuvalu is not yet party to any of the international conventions or programmes that directly promote the conservation of natural areas, namely the Convention concerning the Protection of the World Cultural and Natural Heritage (World Heritage Convention), the Unesco Man and the Biosphere Programme and the Convention on Wetlands of International Importance especially as Waterfowl Habitat (Ramsar Convention).

The Convention on the Conservation of Nature in the South Pacific (1976) has been neither signed nor ratified. Known as the Apia Convention, it entered into force during 1990. The Convention is coordinated by the South Pacific Commission and represents the first attempt within the region to cooperate on environmental matters. Among other measures, it encourages the creation of protected areas to preserve indigenous flora and fauna.

Although Tonga is party to the South Pacific Regional Environment Programme (SPREP), the 1986 Convention for the Protection of the Natural Resources and Environment of the South Pacific Region (SPREP Convention) has not yet been signed or ratified. The convention entered into force during August 1990. Article 14 calls upon the parties to take all appropriate measures to protect rare or fragile ecosystems and threatened or endangered flora and fauna through the establishment of protected areas and the regulation of activities likely to have an adverse effect on the species, ecosystems and biological processes being protected. However, as this provision only applies to the Convention area, which by definition is open ocean, it is most likely to assist with the establishment of marine reserves and the conservation of marine species.

Other international and regional conventions concerning environmental protection to which Tuvalu is party are reviewed by Venkatesh *et al.* (1983).

Administration and Management No information

Systems Reviews Tuvalu comprises nine islands of fossil coral and other calcareous marine materials, rarely exceeding 4m in elevation, with a total land area of 2,511ha. The islands form a 570km chain which describes an arc between 176°E and 180°E. Funafuti, Nanumea, Nui, Nukufetau and Nukulaelae are atolls, generally with narrow strips of land on the east and reefs with scattered islets on the west. Nanumanga, Niulakita and Niutao are reef islands consisting of single islets with brackish internal lakes. Vaitupu is intermediate in type, with a large, but virtually land-locked central lagoon (Douglas, 1969; UNEP/IUCN, 1988; Zann, 1980). Most of the natural vegetation has been cleared for cultivation, especially for copra production. *Rhizophora mucronata* mangrove is found in the central swamp on Funafuti. Niutao is densely covered by coconut, and Nukunono supports small areas of atoll/beach forest (Buckley, 1985; Davis *et al.*, 1986; Douglas, 1969).

Tuvalu's extreme remoteness, small land area and lack of industrial and natural resources pose special problems for development (Zann, 1980). In common with other Pacific states, budget deficits recur. The country is thus dependent upon foreign aid and remittances from expatriates (Paxton, 1985; SPREP, 1981). The major export is postage stamps, earning four times as much as copra, which is the only natural resource exported at present.

Major environmental problems are both anthropogenic and natural. The only urban area is on Funafuti and at present this is the only area where human activity is having a noticeable impact (UNEP/IUCN, 1988). Water may be contaminated by both saltwater and sewage, and in short supply during dry spells. Soils are extremely poor and difficult to manage and improve (SPREP,

1981). The impact of new fishing technology, for example outboard motors, is not yet noticeable, but new management controls and fisheries extension work may be needed (UNEP/IUCN, 1988). There is great concern about the erosion of land areas, but these changes may be natural events and beyond the control of any reasonable coastal engineering projects. Efforts are being made to control coastal erosion with gabion sea wall construction on almost all islands, but there is little understanding of the currents and coastal processes involved (Baines, 1982).

Some of these concerns are reflected in a recent SPREP/ESCAP review of environmental legislation and issues in Tuvalu (ESCAP, 1988). The major environmental issues were identified as coastal erosion; detrimental impacts of land reclamation activities; degradation of fisheries resources; inefficient land use practices; pollution, mainly caused by the absence of a sewerage system and lack of proper garbage disposal sites; and uncontrolled exploitation of flora and fauna, particularly of turtles and coconut crabs. Increased mean sea level, due to global climatic warming, may pose the most serious threat. In the event of severe sea level rises, the loss of freshwater resources, erosion and increased episodic destruction through hurricanes may force the evacuation of the country (Pernetta, 1988).

Dahl (1980) recommends the establishment of reserves for small areas of native vegetation, appropriate series of reef and lagoon environments, perhaps including Kosciusko Bank, and seabird and turtle breeding areas, if any. Dahl (1986) recommends marine protected areas to help control overfishing around Funafuti and Vaitupu. A joint Tourism Council of the South Pacific/Pacific Regional Tourism Development Programme attempted to identify sites of potential tourist interest; suggest development strategies for nature management/ conservation programmes; advise on information for visits to nature sites; and carry out background studies for the establishment of training programmes under nature conservation legislation (Sloth, 1988).

The recent ESCAP (1988) review recommends the establishment of coordinating mechanisms between those government departments principally charged with responsibility for environmental issues to ensure more effective resource management. The extent to which these recommendations have been implemented is not known.

Addresses

Ministry of Commerce and Natural Resources, Vaiku, Funafuti

References

Baines, G.B.K. (1982). Pacific islands: development of coastal marine resources of selected islands. In: Soysa, C., Chia, L.S. and Coulter, W. L. (Eds), *Man, land and sea: coastal resource use and management in Asia and the Pacific*. The Agricultural Development Council, Bangkok, Thailand. Pp. 189-198.

Buckley, R. (1985). Environmental survey of Funafuti Atoll. *Proceedings of the Fifth International Coral Reef Congress, Tahiti* 6: 305-310

Dahl, A.L. (1980). Regional ecosystem survey of the South Pacific Area. *SPC/IUCN Technical Paper* 179. South Pacific Commission, Noumea, New Caledonia. 99 pp.

Dahl, A.L. (1986). *Review of the protected areas system in Oceania*. IUCN, Gland, Switzerland and Cambridge, UK/UNEP, Nairobi, Kenya. 328 pp.

Davis, S.D., Droop, S.J.M., Gregerson, P., Henson, L., Leon, C.J., Lamlein Villa-Lobos, J., Synge, H. and Zantovska, J. (1986). *Plants in danger: what do we know?* IUCN, Gland, Switzerland and Cambridge, UK. 488 pp.

Douglas, G. (1969). Draft checklist of Pacific Oceanic Islands. *Micronesica* 5(2): 327-463.

ESCAP (1988). ESCAP environment mission to Tuvalu, 1988. (Unseen)

Hay, R. (1986). Bird conservation in the Pacific. ICBP *Study Report* No. 7. International Council for Bird Preservation, Cambridge, UK. 102 pp.

Paxton, J. (Ed.). (1989). *The Statesman's yearbook 1989-90*. The Macmillan Press Ltd, London. 1691 pp.

Pernetta, J.C. (1988). Projected climate change and sea level rise: a relative impact rating for countries of the South Pacific Basin. In: MEDU joint meeting of the task team on the implications of climatic change in the Mediterranean. Split, Yugoslavia, 3-7 October. Pp. 1-11.

Sloth, B. (1988). *Nature legislation and nature conservation as part of tourism development in the Island Pacific*. Pacific Regional Tourism Development Programme. Tourism Council of the South Pacific, Suva, Fiji. 82 pp.

SPREP (1981). *Tuvalu*. Country Report No. 18. South Pacific Commission, Noumea, New Caledonia. 13 pp.

UNEP/IUCN (1988). *Coral reefs of the world. Volume 3: Central and Western Pacific*. UNEP Regional Seas Directories and Bibliographies. IUCN, Gland, Switzerland and Cambridge, UK/UNEP, Nairobi, Kenya. 378 pp.

Venkatesh, S., Va'ai, S. and Pulea, M. (1983). An overview of environmental protection legislation in the South Pacific countries. SPREP *Topic Review* 13. South Pacific Commission, Noumea, New Caledonia. 63 pp.

Zann, L.P. (1980). *Tuvalu's subsistence fisheries. Effects of energy crisis on small craft and fisheries in the South Pacific*. Report 4. Institute of Marine Resources, University of the South Pacific, Suva, Fiji.

ANNEX
Definitions of protected area designations, as legislated, together with authorities responsible for their administration

Title: The Prohibited Areas Ordinance and Wildlife Ordinance

Date: 1975

Brief description: Provides for the establishment of wildlife reserves, although none has yet been gazetted, and the protection of seabirds.

Administrative authority: No information

Designation: No information

Title: Ordinance to Provide for the Conservation of Wildlife

Date: 1975, revised 1978

Brief description: Lists 32 fully protected bird species and a number with partial protection. Provision is also made for the declaration of wildlife sanctuaries

Administrative authority: No information

Designation: No information

Area 202 sq. km

Population 36,000 (NPS and ASG, 1988)
Natural increase: No information

Economic Indicators
GNP: No information

Policy and Legislation American Samoa is a semi-autonomous, unincorporated Territory of the United States in which most, but not all, of the Articles of the US Constitution apply. A constitution was adopted in 1960 under which the legislature, executive and judicial branches of government are created. The traditional form of government, the matai system, primarily comprises the high chiefs. In matters concerning the use of land the traditional system plays a pivotal role and decisions regarding land are still controlled by the matai (NPS and ASG, 1988).

Communal tenure exists in American Samoa, with rights to use resources being held in common. The land tenure system is derived from the structure of the family organisation, usually referred to as the matai system. The most important social unit is the "aiga", a large extended family headed by a matai, or chief, who holds the traditional title of that family. Each village has matai titles and land rights belong to the aiga in perpetuity. Land may be passed from one member of the aiga to another, or to another family when a new matai is created. Under American Samoa law, a matai is prohibited from selling, giving, exchanging or in any way disposing of communal lands to a non-Samoan without the written permission of the Governor. While the American Samoan constitution protects Samoans against alienation of their land, communal land may be conveyed or transferred for public purposes to the US government or the government of American Samoa, although this has never been done on a sufficient scale to establish a protected area.

The various protected areas in the country have been established under a diversity of legal acts. Some United States Federal legislation applies, including the 1972 Marine Protection, Research and Sanctuaries Act (P.L. 92-532; 16 USC 1431-1434), which authorises the Secretary of Commerce, with Presidential approval, to designate ocean water as marine sanctuaries for the purpose of preserving or restoring conservation, recreational, ecological or aesthetic values. Such marine sanctuaries are built around existing, distinctive resources where protection and beneficial use require comprehensive planning and management. A site can be nominated by a private party, a state or a state agency. Once a site has been nominated a public hearing must be held in the area most directly affected (Weiting *et al.*, 1975). Fagatele Bay National Marine Sanctuary was established under the provisions of this Act.

Rose Atoll is established under provisions of the National Wildlife Refuge System Administration Act 1966. The basic statutory protection for wildlife refuges prohibits entry, disturbance or exploitation of the site, although there are provisions for exceptions to be made.

A number of national natural landmarks have been designated. The objective of the National Natural Landmark Program, administratively created and not defined in any legislation, is to assist in the preservation of sites illustrating the geological and ecological character of the United States, to enhance the educational and scientific value of sites thus preserved, to strengthen cultural appreciation of natural history, and to foster a wider interest and concern in the nation's natural heritage (NPS, 1977).

American Samoa National Park has been authorised as a national park by the US Congress, under the provisions of federal legislation.

There is an American Samoa Parks and Recreation Act (1980), with provision for the establishment of natural preserves, conservation preserves and territorial parks, although none has been established.

International Activities American Samoa is not yet directly party to any of the international conventions or programmes that promote the conservation of natural areas, namely the Convention concerning the Protection of the World Cultural and Natural Heritage (World Heritage Convention), the Unesco Man and the Biosphere Programme and the Convention on Wetlands of International Importance especially as Waterfowl Habitat (Ramsar Convention). The United States is an active participant in all three conventions, although it is not known if this involves American Samoa in any way.

The Convention on the Conservation of Nature in the South Pacific (1976) has been neither signed nor ratified by either the United States or American Samoa. Known as the Apia Convention, it entered into force during 1990. The Convention is coordinated by the South Pacific Commission and represents the first attempt within the region to cooperate on environmental matters. Among other measures, it encourages the creation of protected areas to preserve indigenous flora and fauna.

The United States is party to the South Pacific Regional Environment Programme (SPREP), and the 1986 Convention for the Protection of the Natural Resources and Environment of the South Pacific Region (SPREP Convention) has been signed (25 November 1986) but not ratified. The Convention entered into force during August 1990. Article 14 calls upon the parties to take all appropriate measures to protect rare or fragile ecosystems and threatened or endangered flora and fauna through the establishment of protected areas and the regulation of activities likely to have an adverse effect on the species, ecosystems and biological processes being protected. However, as this provision only applies to the Convention area, which by definition is open ocean, it is most likely to assist with the establishment of marine reserves and the conservation of marine species.

Administration and Management Administration of protected areas falls into two categories: those that are the responsibility of the territorial government and those that are mandated under federal legislation. The American Samoa Department of Parks and Recreation is principally concerned with development of recreation, and to date most of the areas designated under the Parks and Recreation Act are for recreational purposes. Fagatele Bay is administered and funded by the National Oceanic and Atmospheric Administration (NOAA); it is locally administered by the Economic and Development Planning Office. Rose Atoll is administered by the US Fish and Wildlife Service and the American Samoa Department of Marine Resources. American Samoa National Park is the responsibility of the National Park Service. The national natural landmarks are part of a programme within the National Park Service, although the precise status of each site is not known.

In 1975 an American Samoa Recreation Area Development Plan was issued which provided for recreation facilities and also identified national historic landmarks and national natural landmarks.

Le Vaomatua ("tropical rain forest" in the Samoan language) is an active non-governmental organisation established in 1985 by local residents concerned with protecting forests, reefs and marshes. Recent successful activities have included advocacy for the establishment of American Samoa National Park and cancellation of the dredging and construction of a harbour in Leone Bay.

Systems Reviews American Samoa comprises five volcanic islands and two coral atolls, and is the only United States territory south of the Equator. Principal vegetation types include lowland tropical evergreen rain forest up to 300m altitude and montane forest at 300-700m. Cloud forest is found only on Tau and Olosega at 500-930m and small areas of montane scrub occur on Tutuila. Mangroves, and coastal swamps, largely consisting of sedges and ferns, are found near the coast, but have been drastically reduced in coverage, being felled for housing and other uses. About two-thirds of the native vegetation has been disturbed or cleared for settlements or agriculture and, for example, littoral forest has been almost entirely eliminated from sandy coastal areas. Most of the interior of Tutuila, the main island, is covered by a patchwork of secondary forest mixed with primary or primary-like forest on the steeper, more inaccessible slopes. Most of the remaining primary forest on Tutuila is on the central portion of the north coast (Amerson *et al.*, 1982).

Although much of the forested interior remains inaccessible due to the steep terrain, an increasing

proportion of marginal areas is being cleared. Furthermore, an estimated 1% of rain forest cover is lost annually to plantation cover. One of the consequences of this vegetation clearance has been the increase in erosion, flooding and silting of coastal areas.

Coral reefs, which occur extensively throughout the country, are described in UNEP/IUCN (1988). Tutuila and Manu'a both have fringing reefs rarely more than 50-100m in width which are particularly vulnerable to sedimentation and the effect of run-off carrying agricultural fertiliser, herbicides and pesticides, and animal wastes particularly from piggeries.

An Action Strategy for Protected Areas in the South Pacific Region (SPREP, 1985b) has been prepared. The principal goals of the strategy cover conservation education, conservation policy development, establishment of protected areas, effective protected area management and regional and international cooperation. Priority recommendations for American Samoa are as follows: implement recommendations regarding new areas for reservation and taking fish; possible reservation of reef area of Ofa; possible reservation of forest on Mount Lata; and implementation of a management programme at the Fagatele Bay National Marine Sanctuary. In response to both the general and specific recommendations of the Action Plan, fruit bat surveys have been undertaken by the US-Fish and Wildlife Service. A recent feasibility study (NPS and ASG, 1988) identified remaining areas critical for protection, leading to the establishment of American Samoa National Park. Management activities have been initiated in Fagatele Bay National Marine Sanctuary.

Dahl (1980) identifies a number of habitat types that were excluded from the protected areas system prior to the establishment of American Samoa National Park, and provides a list of potential reserves to remedy these omissions. In a 1986 review, Dahl (1986) states that there is an urgent need for protected areas on Tau, Tutuila and Ofu. However, the new national park has included coral reefs, beaches, coastal, lowland, montane and cloud forest and so probably includes most ecosystems in the country.

The Territorial Comprehensive Outdoor Recreation Plan 1980-85, and the American Samoa Coastal Management Plan (ASCMP) set out a comprehensive system of parks, forest and nature reserves (SPREP, 1980). The ASCMP, which is funded under the National and Oceanic and Atmospheric Administration and operates out of the Economic Development Planning Office, has undertaken some conservation activities. The programme has jurisdiction over all land areas and to the limit of territorial seas. The ASCMP has also initiated a land use permitting system, known as the Project Notification and Review System (PNRS). The PNRS provides for land-use reviews for any planned construction, which include environmental considerations. The ASCMP annually sponsors a three-week coastal awareness programme, highlighting coastal and marine conservation issues.

Addresses

Department of Parks and Recreation, US Department of the Interior, PO Box 3809, Pago Pago 96799

National Oceanic and Atmospheric Adminstration, 3300 Whitehaven NW, Washington DC 20235

National Park Service, Pacific Area Office, 300 Ala Moana Blvd., Box 50165, Room 6305, Honolulu, HI 96850

References

Amerson, A.B., Whistler, W.A. and Schwaner, T.D. (1982). *Wildlife and wildlife habitat of American Samoa: Environment and ecology.* US Fish and Wildlife Service, Washington, DC. 119 pp.

Dahl, A.L. (1980). Regional ecosystem surveys of the South Pacific Area. *SPC/IUCN Technical Paper* 179. South Pacific Commission, Noumea, New Caledonia. 99 pp.

Dahl, A.L. (1986). *Review of the protected areas system in Oceania.* IUCN, Gland, Switzerland and Cambridge UK/UNEP, Nairobi, Kenya. 328 pp.

Davis, S.D., Droop, S.J.M., Gregerson, P., Henson, L., Leon, C.J., Lamlein Villa-Lobos, J., Synge, H. and Zantovska, J. (1986). *Plants in danger: what do we know?* IUCN, Gland, Switzerland and Cambridge, UK. 488 pp.

Eaton, P. (1985). Land Tenure and Conservation: Protected Areas in the South Pacific. Report to the South Pacific Commission.

Fai'ai, P and Daschbach, N. (1989). American Samoa. *Country Review* 3. Fourth South Pacific Conference on Nature Conservation and Protected Areas, Port Vila, Vanuatu, 4-12 September. 10 pp.

Government of American Samoa (1975). American Samoa recreation area development plan: 1975-1980. American Samoa Park and Recreation Board, Pago Pago. 52 pp.

National Park Service and American Samoa Government (1988). National Park feasibility study. Draft 01/88. National Park Service and American Samoa Government. 139 pp.

SPREP (1980). American Samoa. *Country Report* 1. South Pacific Commission, Noumea.

UNEP/IUCN (1988). *Coral reefs of the world. Volume 3. Central and Western Pacific.* UNEP Regional Seas Directories and Bibliographies. IUCN, Gland, Switzerland and Cambridge, UK/UNEP, Nairobi, Kenya. 378 pp.

US Department of Commerce (1984). Final environmental impact statement and management plan for the proposed Fagatele Bay National Marine Sanctuary. Sanctuary Programme Division of the Office of Ocean and Coastal Resource Management, National Oceanic and Atmospheric Administration, US Department of Commerce, Washington DC.

Weiting, H., Lukowski, S., Buckingham, N., Schamis, M and Humke, J. (1975). *Preserving our natural heritage.* Volume 1. *Federal activities.* The Nature Conservancy and US Man and the Biosphere Program. 323 pp.

ANNEX
Definitions of protected area designations, as legislated,
together with authorities responsible for their administration

Title: Parks and Recreation Act

Date: 1980

Brief description: There is created the American Samoa parks system. The department shall inventory all properties belonging to the government and with the Governor's approval determine which properties are included in the park system. The department keeps a list of all areas in the park system according to the classification (see "designations"), with correct and accurate descriptions, and provide the Legislature with a current copy of the list (18.0204).

Administrative authority: Department of Parks and Recreation

Designation:

Natural preserves which are to remain unimproved;

Conservation preserves which may be improved for the purpose of making them accessible to the public in a manner consistent with the preservation of their natural features;

Territorial park or community parks which may be improved for the purpose of providing public recreational facilities in a manner consistent with the preservation and enhancement of the natural features;

Territorial recreation facilities or community recreation facilities which may be improved for the purpose of providing public recreation facilities;

Historical and prehistoric objects and sites which are administered in accordance with federal guidelines as set by the Department of the Interior;

Seashore reserves which include underwater land and water areas of the Territory of American Samoa extending from the mean high water line seaward to 10 fathoms is included within the park system and administered by the Director in accordance with 18.0204.

SUMMARY OF PROTECTED AREAS

Map ref.	National/international designation Name of area	IUCN management category	Area (ha)	Year notified
1	*National Park* American Samoa National Park	II	3,725	1988
2	*National Wildlife Refuge* Rose Atoll	I	653	1973
3	*National Marine Sanctuary* Fagatele Bay	IV	64	1985

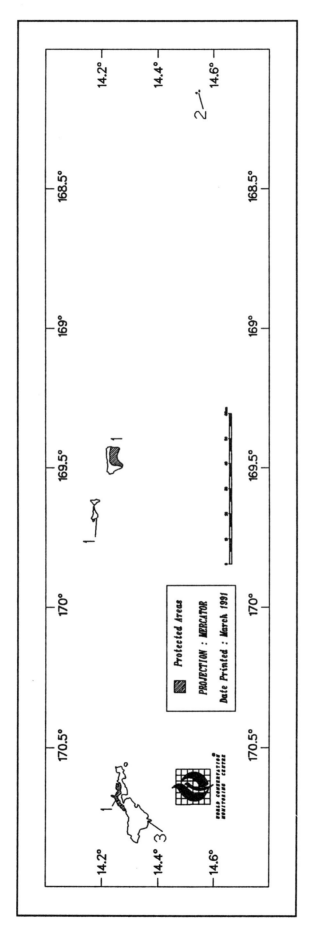

Protected Areas of American Samoa

USA – GUAM

Area 541 sq. km (marine area 218,000 sq. km)

Population 120,000 (1990 estimate)
Natural increase: No information

Economic Indicators
GNP: US$ 19,197 per capita (USA)

Policy and Legislation Guam's constitutional status is that of an unincorporated territory of the United States of America; military authorities control about one-third of the island.

The policy on natural resources is that areas important for recreation, and critical marine and wildlife habitats, "shall be protected through policies and programmes affecting such resources" (Anon., 1985). Areas managed by the government of Guam are protected by the provisions of Public Law No. 12-209 as either natural preserves or as conservation reserves. Natural preserves are intended to remain unimproved, while conservation reserves may be managed for the purpose of making them accessible to the public "in a manner consistent with the perpetuation of their natural features". Multiple-use management or recreation areas (national historic parks and seashore parks) are created to protect outstanding marine life, terrestrial wildlife, oceanic resources, scenery and cultural heritage, including landscapes resulting from World War II (Anon., 1985). New protected areas of government of Guam ownership are added by legislative or administrative action of the Department of Parks and Recreation.

The federal government also administers a number of ecological reserve areas (research natural areas), which were established on 13 March 1984 by the Chief of Naval Operations, United States Navy as physical or biological units in which current natural conditions are maintained. There is no direct legislative protection afforded to such areas, which are established and managed by virtue of administrative procedures of the appropriate land-holding agency. The US Air Force has also established a natural area, in 1973, and is responsible for its management. In addition, Federal Public Law No. 95-348 (1978) created War in the Pacific National Historical Park.

Other relevant legislation specific to conservation in Guam includes the Forestry and Conservation Laws, Territorial Seashore Act, as well as a number of US federal legislative acts. There are several relevant Executive Orders signed by the Governor of Guam, including one on the Protection of Wetlands (Anon., 1985) and Executive Order No. 87-36, which established Anao Conservation Area.

International Activities At the international level, the United States has entered into a number of obligations and cooperative agreements related to conservation. It is party to the Convention concerning the Protection of the World Cultural and Natural Heritage (World Heritage Convention) which it accepted on 7 December 1973. The United States became a contracting party to the Convention on Wetlands of International Importance especially as Waterfowl Habitat (Ramsar Convention) on 18 December 1986, although no sites in Guam are included in the List of Wetlands of International Importance established under the terms of the Convention. The United States participates in the Unesco Man and the Biosphere Programme. To date, 46 biosphere reserves have been declared as part of the international biosphere reserve network, but none is located in Guam.

The Convention on the Conservation of Nature in the South Pacific (1976) has been neither signed nor ratified. Known as the Apia Convention, it entered into force during 1990. The Convention is coordinated by the South Pacific Commission and represents the first attempt within the region to cooperate on environmental matters. Among other measures, it encourages the creation of protected areas to preserve indigenous flora and fauna.

The United States is party to the South Pacific Regional Environment Programme (SPREP) and the 1986 Convention for the Protection of the Natural Resources and Environment of the South Pacific Region (SPREP Convention) has been signed (25 November 1986) but not ratified. The convention entered into force during August 1990. Article 14 calls upon the parties to take all appropriate measures to protect rare or fragile ecosystems and threatened or endangered flora and fauna through the establishment of protected areas and the regulation of activities likely to have an adverse effect on the species, ecosystems and biological processes being protected. However, as this provision only applies to the Convention area, which by definition is open ocean, it is most likely to assist with the establishment of marine reserves and the conservation of marine species.

Administration and Management The Department of Parks and Recreation is responsible for the Guam Territorial Park System, which includes three protected areas (Guam Territorial Seashore Park, Masso River Reservoir Area and Anao Conservation Area).

The development and management of Guam Territorial Seashore Park is divided among several agencies: the Department of Parks and Recreation is responsible for coordination, planning, facility maintenance, outdoor recreation, historic preservation and scenic resources; the Department of Agriculture is responsible for wildlife, marine resources, forestry, fire prevention and soil resources; the Department of Land Management is responsible for leases and land registration and the Guam

Environmental Protection Agency (GEPA) is responsible for water, air pollution and solid waste disposal (Anon., 1985). The Division of Forestry and Soil Resources has responsibility for development, management and protection of forests and watershed resource lands, whilst the US National Park Service is responsible for the management and development of National Park Service areas which are federal protected lands. Federal protected areas are also administered by the US Navy and US Air Force (Anon., 1985).

Systems Reviews Guam is the westernmost territory of the United States and has the finest deep water harbour between Hawaii and the Philippines. With its strategic location and harbour, it can be expected to assume a growing importance for American military and trade activities in the western Pacific (Anon., 1986).

Guam is also the largest and southernmost island of the Marianas Archipelago in the Pacific Ocean, and has been ranked as one of the most important islands for conservation in the South Pacific (Dahl, 1986). Rain forest originally covered most of the island but much has been logged and cleared for development. The mixed forests on old volcanic soils have been completely destroyed, whilst ravine forests survive along river valleys and on some volcanic and limestone hillslopes. There are also still some small areas of mangrove (Davis *et al.*, 1986). Extensive urban development around the island, clearance of land for tourism-related facilities and centuries of slash and burn agriculture have also been responsible for the destruction of natural vegetation cover (Dahl, 1986; Maragos, 1985). Burning of vegetation is also a major issue, particularly in the south where ensuing erosion exacerbates the problem. Guam has the highest incidence of wildfire arson in the US. Other fires are deliberately set for refuse disposal, game flushing and to stimulate grass regrowth to attract game. Public laws Nos. 18-29 and 19-34 Section 28 specifically address the problem of fires and their prevention (Anon., 1989) but enforcement is weak.

Most coral reefs are fringing, although there are two barrier reef lagoons, namely Apra Harbour, which is now extensively modified, and Cocos Lagoon in the south-west. These, and other reefs, are discussed in more detail in UNEP/IUCN (1988). The principal threats to coral reefs are dredging and sedimentation, fertiliser run-off, thermal and other types of water pollution.

The introduced brown tree snake *Boiga irregularis* has been implicated in the widespread and severe decline of native bird species, a number of which are now restricted to the northern part of Guam (Arnett, 1983; Carey, 1988). Consequently, some of the most important areas for conservation are the northern cliffs and north-western plateau habitats.

The biggest threat to the environment is unregulated development, most of which is geared towards tourism and its related infrastructure. Guam is presently undergoing a heavy construction phase, almost all of

which is funded by capital originating from outside the territory. There is some concern amongst those public agencies with responsibilities for environmental issues that this expansion is damaging natural resources. Permits are required for earthmoving and tree felling and mandatory mitigative measures may be imposed by the government, but rarely enforced (SPREP, 1989). Furthermore, various "development", "master" and "management" plans have been developed for all or part of Guam, and several of these have included provision for nature conservation (SPREP, 1985a). However, none of these is followed and most are an exercise in paperwork (J.E. Miculka, pers. comm., 1990).

An action strategy for the development of protected areas in the South Pacific Region (SPREP, 1985b), developed by field managers in the region at the technical session of the Third South Pacific National Parks and Reserves Conference, provides a work programme to implement conservation and protected areas objectives. Priority recommendations for Guam are: to establish the Hilaan area as a protected area; to conduct a multimedia public education programme on the need for further protected areas; to develop tourism programmes to give emphasis to parks and protected areas; to complete a survey of proposed protected areas; and to facilitate the exchange of management and resources information, particularly with the Northern Marianas. All these measures have been initiated to some degree, although very little progress has been achieved (D. Lotz, pers. comm., 1990).

Addresses

Department of Parks and Recreation, 490 Chalan Palasyo Road, Agana Heights, Guam 96919
US National Park Service, War in the Pacific NHP, Marine Drive, Asan, PO Box FA, Agana, Guam 96910
Guam Aquatic and Resource Wildlife Division, Fish and Wildlife Service, PO Box 23367, GAMF, Guam MI 96921
Department of the Navy, US Naval Station, FPO San Francisco 96630-100
Department of the Air Force, Headquarters 633D, Air Base Wing (PACAF), APO San Francisco 96334-5000

References

Anon. (1986). Guam Annual Economic Review, Guam Department of Commerce. (Unseen)
Arnett, G.R. (1983). Proposed endangered status for seven birds and two mammals from the Mariana Islands. *Federal Register* 48 (230): 53729-53733.
Carey, J. (1988). Massacre on Guam. *National Wildlife* 26(5): 13-15.
Dahl, A.L. (1986). *Review of the protected areas system in Oceania.* IUCN, Gland, Switzerland and Cambridge, UK/UNEP, Nairobi, Kenya. 328 pp.
Davis, S.D., Droop, S.J.M., Gregerson, P., Henson, L., Leon, C.J., Lamlein Villa-Lobos, J., Synge, H. and

Zantovska, J. (1986). *Plants in danger: what do we know?* IUCN, Gland, Switzerland and Cambridge, UK. 488 pp.

Maragos, J.E. (1985). Coastal resources development and management in the US Pacific Islands. Report to US Congress Office of Technology Assessment. Unpublished. 130 pp.

Randall, R.H. and Holloman, J. (1974). *Coastal Survey of Guam.* University of Guam Marine Technical Report 14. 404 pp.

SPREP (1980). *Country Report* 6: Guam. South Pacific Commission, Noumea.

SPREP (1985a). Guam. In: Thomas, P.E.J. (Ed.), *Report of the Third South Pacific National Parks and Reserves Conference.* Volume III. *Country reviews.*

South Pacific Commission, Noumea, New Caledonia. Pp. 93-114.

SPREP (1985b). *Action strategy for protected areas in the South Pacific Region.* South Pacific Commission, Noumea, New Caledonia. 24 pp.

SPREP (1989). Guam. Paper presented at the Fourth South Pacific Conference on Nature Conservation and Protected Areas, Port Vila, Vanuatu, 4-12 September. 5 pp.

UNEP/IUCN (1988). *Coral reefs of the world. Volume 3: Central and Western Pacific.* UNEP Regional Seas Directories and Bibliographies. IUCN, Gland, Switzerland and Cambridge, UK/UNEP, Nairobi, Kenya. 378 pp.

ANNEX
Definitions of protected area designations, as legislated, together with authorities responsible for their administration

Title: Public Law No. 12-209

Date: No information

Brief description: Provides for the establishment of protected areas by the Government of Guam

Administrative authority: Department of Parks and Recreation

Designations:

Natural preserve
Conservation reserve
Territorial park
Community park

Recreational facilities (= national historic park, seashore park)

Title: Public Law No. 95-348

Date: 18 August 1978

Brief description: Provided for the establishment of the War in the Pacific National Historical Park

Administrative authority: US National Park Service

Designations: War in the Pacific National Historical Park

Title: Public Law No. 16-62

Date: 9 February 1982

Brief description: An act to amend sections 26003(g), 26007, 26009(b), to add Articles VI, VII, VIII and IX of Chapter IV, Title XIII all of Government Code, relative to forestry and conservation programs of the Department of Agriculture.

Administrative authority: Department of Parks and Recreation

Designations:

Conservation reserve The Department, in cooperation with the Department of Parks and Recreation, shall control and manage lands and water areas that have been set aside by the government of Guam as conservation reserves. Such control and management shall have as its objectives, the wise use of the soil, water, plants and animals of the reserves. Consistent with this objective, the Director, with the concurrence of the Director of the Department of Parks and Recreation, may establish and enforce rules for economic use.

SUMMARY OF PROTECTED AREAS

Map[†] ref.	*National/international designation* Name of area	IUCN management category	Area (ha)	Year notified
	Ecological Reserve Areas			
1	Haputo	IV	102	1984
2	Orote Peninsula	IV	66	1984
	Natural Reserve			
3	Masso River Reservoir Area	IV	67	1976
	Natural Area			
4	Pati Point	IV	112	1973
	Conservation Reserves			
5	Anao	IV	263	1953
6	Bolanos (Chalan Palii)	VIII	365	
	Cotal	VIII	223	
8	Schroeder	VIII	552	
9	Y-Piga	VIII	45	
	Territorial Seashore Park			
10	Guam Territorial Seashore Park	VIII	6,135	1978
	National Historic Park			
11	War in the Pacific	V	779	1978

[†]Locations of most protected areas are shown on the accompanying map.

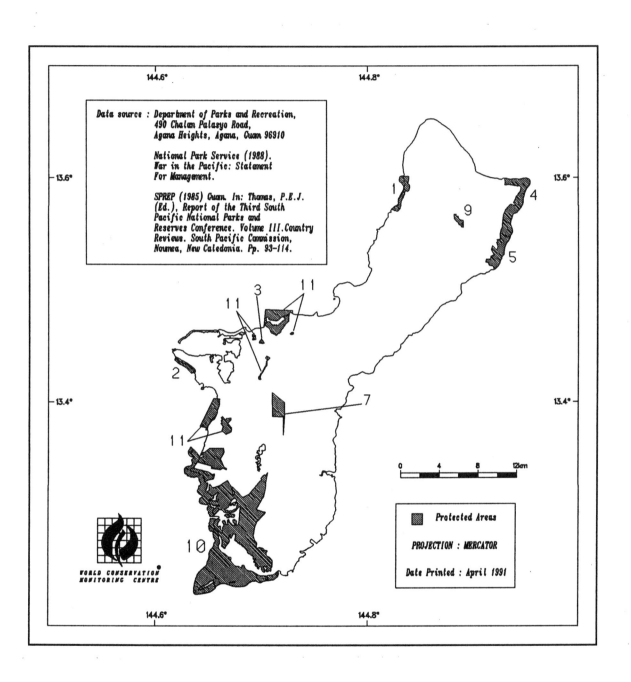

Protected Areas of Guam

USA – HAWAII

Area 16,760 sq. km (land area)

Population 1,062,000 (1986)
Natural increase: No information

Economic Indicators
GNP: No information

Policy and Legislation Legislation exists at both the state and federal levels. The major texts covering protected areas include the Federal Land Policy and Management Act of 1976 (which requires the inventory, assessment and planning of all federal lands); the Fish and Wildlife Act of 1956; the Coastal Zone Management Act of 1972; the Endangered Species Act of 1973 (streamlined in 1982 and supplemented in 1983 by the International Environment Protection Act); the National Policy Act of 1969; and the Fish and Wildlife Improvement Act of 1978. More specific acts of the US Congress which affect the conservation of areas include the Wilderness Act (16 USC 1131), Water Resources Planning Act of 1965 (42 USC 1962), Concessions Policy Act of 1965 (16 USC 20), and the Solid Waste Disposal Act of 20 October 1965 (PL 89-272). Other acts include the Marine Protection, Research and Sanctuaries Act which authorises the Secretary of Commerce to designate ocean waters as marine sanctuaries. In the case of wetland protection, the main legislative provisions are the Migratory Bird Conservation Act of 1929, the Migratory Bird Hunting and Conservation Stamp Act of 1934, the Pittman-Robertson Act of 1937 and the Land and Water Conservation Fund Act of 1965. Regulations published by the Department of Agriculture in 1983 called for the maintenance of habitats in order to sustain viable populations of flora and fauna. The controlling legislation for the activities of the Fish and Wildlife Service is the Fish and Wildlife Coordination Act of 1958 (16 USC 1531) (a supplement to the 1956 Act) and a number of international treaties.

The State of Hawaii controls land by three principal means: the Land Use Law, the Hawaii Environmental Impact Statement Law and the Coastal Zone Management Act. The Land Use Law provides for the creation of a Land Use Commission which has classified some 46% of land area as conservation area, the use of which is controlled by the governing Board of the Department of Land and Natural Resources (the Land Board). Land with a Conservation District classification includes national and state parks, lands with a slope of more than 20° or more, land in existing forestry or water reserves, and marine waters and offshore islands. In reality, only lands within the Protective "P" Subzone of the Conservation District are protected for "conservation" reasons, for example all north-western Hawaiian islands with the exception of Midway Atoll (Callies, 1989).

The first modern conservation measures in Hawaii were taken in 1903 when the US Forestry Service established a professional forestry programme. The early focus was on soil and water conservation rather than natural habitats and included removal of feral cattle from forest areas and the establishment of fenced forest reserves. Nevertheless, these measures are believed to have been critical in helping to preserves much of the biological diversity found today (Holt, 1989).

The establishment of national parks in Hawaii was the earliest major project expressly for habitat conservation (Holt, 1989). Hawaii National Park was created on 1 August 1916 by Act of the US Congress (39 Stat. 432) and consisted of two units each on different islands; one on Hawaii Island and the other on Maui Island. The area of the park was more than doubled as a result of Congressional authorisation in 1922 (45 Stat. 503), in 1928 (45 Stat. 424) and in 1938 (52 Stat. 781). The 'Ola'a Forest Tract was donated in 1951 and 1953 (Executive Order #1640). The park was split into Hawaii Volcanoes National Park (on Hawaii Island) and Haleakala National Park (on Maui Island) in 1961 (75 Stat. 577). Hawaii Volcanoes National Park is protected under 16 USC 1 (National Park Service Organic Act) and under the terms establishing the park as set out in 16 USC 395b, and under several sections of 16 USC 391-396a, which specifically pertain to Haleakala.

The Hawaii Natural Area Reserves System was established under the provisions of Hawaii Revised Statute s. 195-1 *et seq.* in 1970, to "preserve in perpetuity specific land and water areas which support communities, as relatively unmodified as possible, of the natural flora and fauna, as well as geological sites of Hawaii".

Marine life conservation districts are created by administrative authority of the Division of Fish and Game of the Department of Land and Natural Resources, and are established under the provisions of Hawaii Revised Statute s.190-1. The same Division takes responsibility for state wildlife sanctuaries, also established by administrative action.

International Activities The US ratified the Convention on Wetlands of International Importance especially as Waterfowl Habitats (Ramsar Convention) on 18 December 1986, although no sites have yet been listed in Hawaii. The Convention concerning the Protection of the World Cultural and Natural Heritage (World Heritage Convention) was ratified on 7 December 1973 and Hawaii Volcanoes National Parks was inscribed on the World Heritage List in 1987. The US is also party to the Unesco Man and the Biosphere Programme, and Hawaii Islands Biosphere Reserve, comprising Haleakala and Hawaii Volcanoes national parks, has been accepted as part of the international biosphere network.

The Convention on the Conservation of Nature in the South Pacific has been neither signed nor ratified. Known as the Apia Convention, it entered into force during 1990. The Convention is coordinated by the South Pacific Commission and represents the first attempt within the region to cooperate on environmental matters. among other measures, it encourages the creation of protected areas to preserve indigenous flora and fauna.

The United States is party to the South Pacific Regional Environment Programme, and the 1986 Convention for the Protection of the Natural Resources and Environment of the South Pacific Region (SPREP Convention) has been signed (25 November 1986) but not yet ratified. The Convention entered into force during August 1990. Article 14 calls upon the parties to take all appropriate measures to protect rare or fragile ecosystems and threatened or endangered flora and fauna through the establishment of protected areas and the regulation of activities likely to have an adverse effect on the species, ecosystems and biological processes being protected. However, as this provision only applies to the Convention area, which by definition is open ocean, it is most likely to assist with the establishment of marine reserves and the conservation of marine resources.

Administration and Management The National Park Service (NPS) was established by the Act of 25 August 1916 (39 Stat. 535) under the Federal Department of the Interior. National parks and other categories of lands within the national park system are established by individual acts of Congress, except national monuments which can be created by the President on Federal lands, by proclamation under the authority of the Antiquities Act of 8 June 1906. The Service manages Hawaii Volcanoes National Park and Haleakala National Park to preserve outstanding scenic, geological, and biological values, and to ensure their availability for public use and enjoyment to the extent compatible with resource preservation. NPS historic areas have different primary management goals, although natural resource conservation is required.

The responsibility for management and administration of the public lands of the state of Hawaii rests with the Board of Land and Natural Resources, appointed by the Governor. Actions of the Board are carried out by the Department of Land and Natural Resources' six divisions: Land Management, Conveyances, Forestry, Fish and Game, State Parks, Outdoor Recreation and Historic Sites, and Water and Land Development (HRS s.171-3). The Division of State Parks, Outdoor Recreation and Historic Sites carries the responsibilities for preserving state parks "in their natural condition so far as may be consistent with their use and safety", while making improvements for the use and enjoyment of the public. The Division of Fish and Game, amongst other

responsibilities, manages marine life conservation districts, wildlife refuges and bird sanctuaries under authority granted by law through regulations of the Board of Land and Natural Resources. The Division also acts in advisory capacity to the State Natural Area Reserves Systems Commission. The Division of Forestry manages the state's forest resources in a series of forest reserves, under the principle of multiple use, namely: protection and management of water resources; forest recreation; forest management, including timbering, fire and pest control and reforestation; grazing; and protection and conservation of wildlife habitat and natural forest ecosystems. In the area of flora and fauna preservation, the goal of the Forestry Division is to "protect and preserves unique native plant and animal species, and examples of relatively unmodified native forest ecosystems for their productive value to science, education and the cultural or scientific enrichment or satisfaction of future generations".

The Natural Area Reserves System Commission, appointed by the Governor of Hawaii, and part of the Hawaii Department of Land and Natural Resources, the major state land management agency, recommends, to the Department and the Governor, areas for inclusion; recommends policies on controls and uses of areas within the system; advises on preservation of natural areas; and develops ways to extend and strengthen established natural areas. The Governor must approve the nominations by Executive Order. Once an area is included in the system, removal, except through trade for another area, is also through Executive Order. Natural area reserves are selected on the basis of diversity, rarity and viability (TNCH and DFW, 1989).

The Fish and Wildlife Service administers the National Wildlife Refuge System, including Hawaiian Islands NWR, which, according to statute, aims to preserve and manage the habitats of waterfowl, threatened species, big game and other fauna and flora.

The Nature Conservancy of Hawaii (TNCH) was founded in 1980 as an affiliate of a nation-wide, non-profit conservation organisation which had its beginnings in 1917 and is the leading non-governmental organisation in the field of nature conservation in Hawaii. The mission of both the national and Hawaii groups is "to find, protect, and maintain the best examples of communities and ecosystems, and endangered species in the natural world". Through conventional real estate negotiations and innovative practices such as "conservation easements" (paying landowners for allowing use of their land for conservation), TNCH has acquired, and is continuing to acquire, key natural areas. TNCH makes extensive use of volunteer workers, and includes a Board of Trustees (largely comprising influential business people) and a Scientific Advisory Committee, comprising University, State, private and Federal scientists. TNCH has developed the Hawaii Heritage Program (HHP), a sophisticated inventory and database to optimise the selection of further acquisitions (Newman, 1989; Stone and Holt, 1987).

Systems Reviews The Hawaiian Archipelago is the longest and most isolated chain of tropical islands in the world, stretching about 2,300km from Hawaii in the south-east to Kure Atoll in the north-west. Considerable geophysical evidence suggests that the entire chain formed as a result of tectonic motion of the Pacific plate over a relatively stationary hotspot in the mantle of the earth. Hawaii, the youngest island at about 0.8 million years old, is situated over the hotspot and contains the most active volcano in the world, Kilauea, included within Hawaii Volcanoes National Park. Steady crustal movement is transporting each island land mass north-westwards, eventually breaking connection with the hotspot. As a result, the islands to the north-west of Hawaii become progressively older and more effected by subsidence and erosion. All of the islands north-west of Gardner Pinnacles, the last island in the chain with subaerial basalts, are either atolls, coral islands or reefs and shoals of limestone construction. Midway Atoll, the second to last island in the chain, has been dated at 27+/-0.6 million years old. Beyond Kure Atoll the chain continues as a series of drowned atolls or seamounts which extend all the way to Kamchatka (UNEP/IUCN, 1988).

The geography of the Hawaiian Islands is complex. Those islands to the south-east are the summits of some of the largest mountains in the world, measured from their bases, Mauna Loa extending some 9,756m. Some of the older high islands are deeply eroded with spectacular river valleys such as Waipio Valley in Hawaii, Iao Valley on Maui and Halawa Valley on Molokai. All of these are heavily vegetated with many endemic species of flora and fauna. A rise in sea level of approximately 130m during the Holocene has created many bays and estuaries in the drowned heads of river valley systems, particularly on the older islands. The islands to the north-west are much simpler because they are low rocky islets or coral islands at sea level. The climate is mainly tropical but approaches sub-tropical in the extreme north-west. Mild temperatures, moderate humidity and persistent north-easterly tradewinds are typical, whilst rainfall patterns that can vary from very heavy to very light, mainly influenced by altitude, have encouraged the development of a wide variety of terrestrial ecosystems (UNEP/IUCN, 1988).

The native vegetation of Hawaii varies greatly according to altitude, prevailing moisture and substrate. The most recent classification of Hawaiian natural communities recognises nearly 100 native vegetation types categorised in a hierarchy of elevation, moisture and physiognomy. Within these types are numerous island-specific or region-specific associations: an extremely rich array of vegetation types occurs within a very limited geographic area. Major vegetation formations include forests and woodlands, shrublands, grasslands (including savannas, tussock grasslands, and

sedge-dominated associations), herblands, and pioneer associations on lava and cinder substrates (IUCN, 1991).

Forest formations include subalpine, montane and lowland, extending from sea level to above 3,000m on the slopes of the region's highest mountain. Coastal forests are found on one of the north-western Hawaiian Islands, an offshore islet near Molokai, and other main islands of Kaua'i, O'hau, Moloka'i, Maui and Hawaii. Lowland forest occurs below 1,000m on Kaua'i, O'hau, Moloka'i, Maui, Lana'i and Hawaii. Historically such forests were present on the islands of Ni'ihau and Kaho'olawe, but these were lost to development or displaced by alien species. Montane forests are found between 1,000 and 2,000m and are dry to mesic on leeward slopes and mesic to wet on windward slopes. Montane forests are well developed on Kaua'i, Moloka'i, Maui and Hawaii. On the islands of Lana'i and O'hau, the highest forests are similar to montane wet forests on higher islands, but these forests occupy limited area and include lowland elements. Subalpine forests are known only from the two largest and geologically youngest islands: Maui and Hawaii. Shrubland formations are also found, ranging in altitude from sea level to sub-alpine. Most types are in dry and mesic settings that limit forest formations, or on cliffs or slopes too steep to support trees. Eleven native Hawaiian grassland types, including sedge-dominated associations are found in coastal, lowland, montane and subalpine regions. Hawaiian herblands are also found across the full altitudinal range. Pioneer vegetation is well developed in the active and recent volcanic settings of the islands of Maui and Hawaii.

In spite of their low diversity, coral reefs are well developed. Their community structure, succession and development throughout the archipelago has been reviewed by Grigg (1983). Reefs are best developed on leeward (south and south-western) coasts or in bays sheltered from wave action. These include many sites along the Kona Coast and Kealakekua Bay on Hawaii, Molokini "lagoon", the south-east coast of Molokai, Hanauma Bay and some reefs near Barbers Point on Ohau, and the lagoons of the north-western Hawaiian Islands, including Midway and Kure (UNEP/IUCN, 1988). Other important coastal ecosystems include marshes, streams and stream-mouth estuaries, lagoons and beaches.

The Hawaiian biota is characterised by both a high diversity and endemism, caused by 40 million years of isolation and adaptive radiation. The native flora is estimated at some 1,000-1,500 species, whilst arthropods number 6,500-10,000 and native bird species about 100. However, extinction, or the threat of extinction, is also characteristic, and although Hawaii possesses only some 0.2% of the land mass and 14.3% of the native birds and plants of the United States, it contains 27.8% of the threatened birds and plants and 72.1% of extinct birds and plants (DLNR, n.d.). Nine endemic bird species have populations of less than 50 individuals and at least 57 bird species have become extinct, most in the last 200 years. Ayensu and DeFilipps (1978) list 646 candidate "endangered" species, 197 candidate "threatened" and 270 presumed extinct plant species, subspecies and varieties in the Hawaiian islands, which together comprise some 50% of the total indigenous flora. This is largely due to the impact of introduced flora and fauna and introductions can be traced back to the arrival of Polynesians around AD400, who were accompanied by domestic pigs, jungle fowl, dogs, Polynesian rats and various stowaway geckoes, skinks and snails. The Polynesians also actively cultivated the land and as much as 80% of the lowland forest was drastically effected. European visitors, from 1778 onwards, introduced goats (1778), cattle and sheep (1793), horses (1803), axis deer (1867), European pigs and more recently pronghorn antelope, mouflon sheep, brush-tailed rock wallaby, Rocky Mountain mule deer and Columbian black-tailed deer (Degener and Degener, 1961) as well as numerous plants. At least 111 new plant arrivals are recorded between 1778 and 1839 and in total there are some 4,500 exotic plants throughout the islands (Berger, 1975), although only some 2% of these are serious invaders of native ecosystems (Stone and Loope, 1987). As many as 16 insects are accidentally introduced annually (Beardsley, 1962). The most severe documented impacts have been caused by feral goats, through grazing on species that have evolved in the absence of such pressure, and pigs, by encouraging the dispersal of exotic plant species and the direct effect of their digging activities (Atkinson, 1977; Stone and Loope, 1987). One specific example of the impact of introduced species is the severe damage to the fragile ecosystem of Laysan Island by rabbits; as a direct result of this, several endemic terrestrial birds became extinct, namely Laysan millerbird *Acrocephalus familiaris familiaris*, Laysan honeycreeper *Himatione sanguinea freethi*, and Laysan rail *Porzana palmeri* (Anon., 1985)

Terrestrial protected areas management priorities almost invariably include programmes to control non-native species, and activities including fencing, live- and kill-trapping, baiting and snaring. Programmes to eliminate exotic flora have been developed, including the use of selective herbicides. The prospects for long-term survival of native species within protected areas depends on the intensity of long-term management. There is evidence that native flora will survive if pigs and goats are held in check, and that native birds in turn will survive if sufficient habitat remains, as will native invertebrates, although exotic ants, rats and mongooses will have to be controlled (Stone and Loope, 1987). It is also common practice to allow hunting by members of the public, and hunting is frequently carried out by managers of protected areas.

Both national parks were established principally on the basis of their geological interest and while they also contain significant biological resources they do not protect the entire range of habitat for native taxa (R. Kam, pers. comm., 1991). Thus the acquisitions and easements of The Nature Conservancy of Hawaii, the State Natural Areas Reserve System, forest reserves and the national wildlife refuge are key elements in the conservation of biological diversity. The Natural Areas Reserves System currently covers 20 sites which were selected on the basis of biological diversity and relatively undamaged natural conditions. Holt (1989) discusses the effectiveness with which both State and Federal protected areas achieve conservation goals, and also provides a definition of protected natural area which demands lasting legal protection from destructive uses and an active management programme. State forest reserves and conservation districts are judged not to meet the definition; State natural area reserves, sanctuaries and Alaka'i Wilderness Preserve meet the definition in general as do the preserves of The Nature Conservancy; National parks, and the national wildlife refuges are deemed to have effective legal protection and the most comprehensive management programmes of any class of protected area in the State. The existing network, which includes Federal, State and private protected areas, is considered to protect 46 of the 180 natural communities recognised by The Nature Conservancy. Some 88 community types which lie outside the current network are regarded as being in critical need of protection. Holt (1989) makes three recommendations for the further development of the protected areas network: the protection of critically imperilled natural communities must be ensured; active management must be established in existing areas; and conservation techniques for island ecosystems must be developed.

Demand on land around protected areas is increasing, removing buffering areas and reducing still further the extent of near-native ecosystems. Other threats include geothermal and other energy developments, hunting and recreational uses, ranching, agriculture and timber management, housing and heavy industry (Stone and Loope, 1987).

Addresses

National Park Service, Pacific Area Office, 300 Ala Moana Boulevard, Box 50165, Room 6305, Honolulu, HI 96850

Fish and Wildlife Service, National Wildlife Refuge Complex, 300 Ala Moana Boulevard, PO Box 50167, Room 5302, Honolulu, HI 96850

The Nature Conservancy of Hawaii, 1116 Smith Street, Suite 201, Honolulu, HI 96817 (Tel: 808 537 4508; FAX: 808 545 2019)

Department of Land and Natural Resources, 1151 Punchbowl Street, Honolulu

References

Anon. (1985). Plan approved for three songbirds of the northwestern Hawaiian Islands. *Endangered Species Technical Bulletin* 10 (2): 8-10

Atkinson, I.A.E. (1977). A reassessment of factors, particularly *Rattus rattus* L., influencing the decline of endemic forest birds in the Hawaiian Islands. *Pacific Science* 31(2): 109-113.

Ayensu, E.S. and De Filipps, R.A. (1978). *Endangered and threatened plants of the United States.* Smithsonian Institution and World Wildlife Fund-US, Washington. 403 pp.

Berger, A.J. (1975). Hawaii: a dubious distinction. *Defenders* 50(6): 491-496.

Callies, D.L. (1989). Land use planning and priorities in Hawaii. In: Stone, C.P. and Stone, D.B. (Eds.), *Conservation biology in Hawaii.* University of Hawaii Cooperative National Park Resources Study Unit, Honolulu. Pp. 163-167.

Degener, O. and Degener, I. (1961). Green Hawaii: past, present and future of an island flora. *Pacific Discovery* 14(5): 14-17.

DLNR (n.d.). Preserving Hawaii's natural treasures: Hawaii's natural area reserves system. Department of Land and Natural Resources, Honolulu. Brochure.

Davis, S.D., Droop, S.J.M., Gregerson, P., Henson. L., Lamlein Villa-Lobos, J., Synge, H. and Zantovska, J. (1986). *Plants in danger: what do we know?* IUCN, Gland, Switzerland and Cambridge, UK. 488 pp.

Grigg, R.W. (1983). Community structure, succession and development of coral reefs in Hawaii. *Marine Ecology Programme Series* 11:1-14

Holt, A. (1989). Protection of natural habitats. In: Stone, C.P. and Stone, D.B. (Eds), *Conservation biology in Hawaii.* University of Hawaii Cooperative National Park Resources Study Unit, Honolulu. Pp. 168-174.

The Nature Conservancy of Hawaii and Division of Forestry and Wildlife (1989). State of Hawaii, Natural Areas Reserves Systems: Biological Resources and Management Priorities. Summary Report. 26 pp.

Newman, A. (1989). Biological databases for preserve selection. In: Stone, C.P. and Stone, D.B. (Eds.), *Conservation biology in Hawaii.* University of Hawaii Cooperative National Park Resources Study Unit, Honolulu. Pp. 154-157.

Stone, C.P. and Holt, R.A. (1987). Managing the invasions of alien ungulates and plants in Hawaii's natural areas. Unpublished report. 23 pp.

Stone, C.P. and Loope, L.L. (1987). Reducing negative effects of introduced animals on native biotas in Hawaii: what is being done, what needs doing, and the role of the national parks. *Environmental Conservation* 14(3): 245-258.

UNEP/IUCN (1988). *Coral reefs of the world. Volume 3: Central and Western Pacific.* UNEP Regional Seas Directories and Bibliographies. IUCN, Gland, Switzerland and Cambridge, UK/UNEP, Nairobi, Kenya. 378 pp.

ANNEX 1
Definitions of protected area designations, as legislated, together with authorities responsible for their administration

Title: Rules regulating Wildlife Sanctuaries, Department of Land and Natural Resources, Title 13, Subtitle 15 Forestry and Wildlife, Part 2 Wildlife, Chapter 125

Date: No information

Brief description: To conserve, manage and protect indigenous wildlife in sanctuaries

Administrative authority: Board of Land and Natural Resources

Designations:

Wildlife sanctuaries The following activities are prohibited except for agents of the Board and except as authorised by the Board or its authorised representative: to remove, disturb, injure, kill, or possess any form of plant or wildlife; to possess or use any firearm, bow and arrow, or any other weapon, trap, snare, poison, or any device designed to take, capture or kill wildlife; to discharge any weapon on or into a wildlife sanctuary; to possess any explosives or fireworks; to introduce any form of plant or animal life; to start or maintain a fire; to camp or erect any structure; to enter into any area posted "No Trespassing Area"; to remove, damage, or disturb any notice, sign, marker, fence, or structure; to dump, drain, or leave any litter, toxic material, or other waste material except in trash receptacles or areas designated for the deposit of refuse; to enter or remain upon any surface water area; to park, land, or operate any air, water, or land vehicle except on roads and in areas designated for such use.

Title: Rules regulating activities within Natural Area Reserves, Department of Land and Natural Resources, Title 13, Subtitle 9, Chapter 209, Natural Area Reserves System

Date: No information

Brief description: Regulates and prohibits activities within natural area reserves, including provisions for excepted-use activities by permit and penalties for violation of the regulation.

Administrative authority: Board of Land and Natural Resources

Designations:

Natural area reserve Those State lands that have been designated as part of the Hawaii natural area reserves system by the Department pursuant to Section 195-4, Hawaii Revised Statutes.

- Permitted activities include hiking, nature study and bedroll camping without a tent or other temporary structure. Hunting is a permitted activity pursuant to hunting rules of the department.

- It is prohibited: to remove, injure, or kill any form of plant or animal life, except game mammals and birds hunted according to Department rules; to introduce any form of plant or animal life, except dogs when permitted by hunting rules of the Department; to remove, damage, or disturb any

historic or prehistoric remains; to remove, damage, or disturb any notice, marker, or structure; to engage in any construction or improvement; to engage in any camping activity that involves the erecting of a tent or other temporary structure; to start or maintain a fire; to litter, or to deposit refuse or any other substance; to operate any motorised or unmotorised land vehicle or air conveyance of any shape or form in any area, including roads or trails, not designated for its use; to operate any motorised water vehicle of any shape or form in freshwater environments, including bogs, ponds and streams, or marine waters, except as otherwise provided in the boating rules of the Department of Transportation, State of Hawaii; to enter into, place any vessel or material in or on, or otherwise disturb a lake or pond.

SUMMARY OF PROTECTED AREAS

Map[†] ref.	National/international designation Name of area	IUCN management category	Area (ha)	Year notified
	National Parks			
1	Haleakala	II	11,728	1916
2	Hawaii Volcanoes	II	91,960	1916
	Preserves of The Nature Conservancy of Hawaii			
3	Honouliuli	Private	1,495	1990
4	'Ihi'ihilauakea	Private	12	1987
5	Kaluahonu	Private	86	1982
6	Kamakou	Private	1,123	1982
7	Kanepu'u	Private	187	1990
8	Kapunakea	Private	486	1990
9	Maui Lava Tubes	Private	43	1987
10	Mo'omomi	Private	372	1988
11	Pelekunu	Private	2,332	1987
12	Pu'u Kukui Watershed Management Area	Private	3,239	1988
13	Waikamoi	Private	2,117	1983
	Marine Life Conservation Districts			
14	Hanauma Bay	IV	41	1967
15	Honolua-Mokuleia Bay	IV	18	1978
16	Kealakakua Bay	IV	128	1969
17	Lapakahi	IV	59	1979
18	Manele-Hulopoe	IV	125	1976
19	Molokini Shoal	IV		1981
20	Pupukea	IV		1983
21	Wailea Bay	IV	14	1985
	Natural Reserve			
22	Ahihi-Kinau	IV		1973
	National Wildlife Refuges			
23	Hakalau	IV	16,706	1985
24	Hanalei	IV	393	1972
25	Hawaiian Islands (8 sites)	I	103,068	1945

Map[†] ref.	*National/international designation* Name of area	IUCN management category	Area (ha)	Year notified
26	Huleia	IV	97	1973
27	James C. Campbell	IV	63	
28	Kakahaia	IV	18	
29	Kilauea Point	IV	65	1974
30	Pearl Harbour	IV	49	
	Marine Reserve			
31	Little Coconut Island	IV		
	National Estuarine Research Reserve			
32	Waimanu Valley	IV	1,763	1976
	State Natural Area Reserves			
33	Ahihi-Kinau	IV	828	1973
34	Hanawi	IV	3,036	1986
35	Hono O Na Pali	IV	1,275	1983
36	Kaena Point	IV	5	1983
37	Kahaualea	IV	6,772	1987
38	Kipahoehoe	IV	2,260	1983
39	Kuia	IV	662	1981
40	Laupahoehoe	IV	3,196	1983
41	Manuka	IV	10,344	1983
42	Mauna Kea Ice Age	IV	1,577	1981
43	Mount Kaala	IV	445	1981
44	Olokui	IV	662	1985
45	Pahole	IV	266	1981
46	Puu Alii	IV	538	1985
47	Puu Makaala	IV	4,901	1981
48	Puu O Umi	IV	4,106	1987
49	Waiakea 1942 Lava Flow	IV	259	1974
50	West Maui (4 sites)	IV	2,714	1986
	National Historic Parks			
51	Kalaupapa	II	4,343	1980
52	Kaloko-Honokohau	II	470	1978
53	Pu'uhonua o Honaunau	V	73	1961
	State Wildlife Sanctuaries			
54	Hawaii State (39 islets)	I	58	1981
55	Kure Atoll	I	96	1981

	Biosphere Reserve Hawaii Islands (comprises Haleakala and Hawaii Volcanoes national parks)	IX	99,545	1980
	World Heritage Site Hawaii Volcanoes National Park	X	92,964	1987

[†]Locations of some protected areas are shown on the accompanying map.

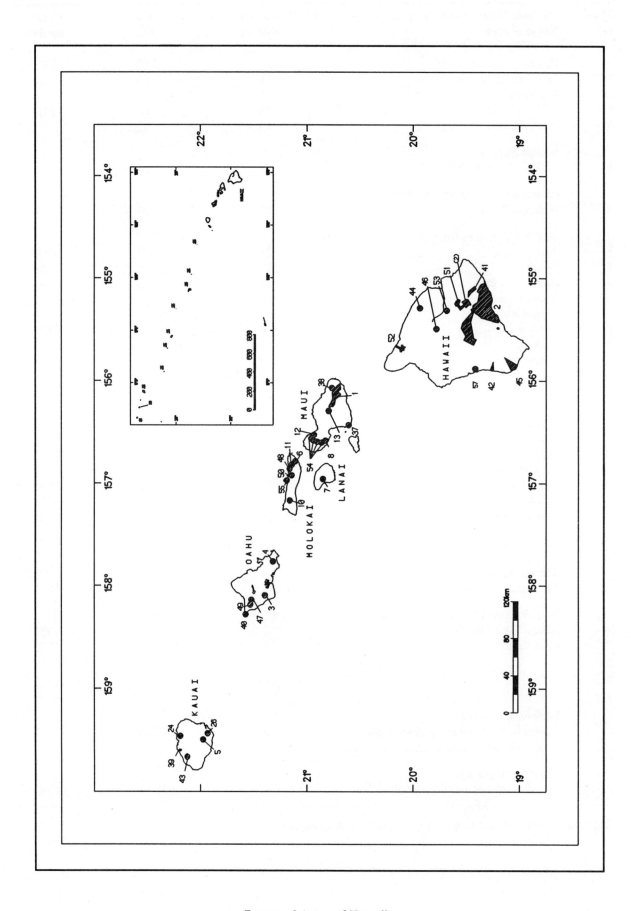

Protected Areas of Hawaii

VANUATU

Area 12,189 sq. km

Population 151,000 (1988) (World Bank, 1990)
Natural increase: 4.2% (urban) and 3.1% (rural) per annum
(Dahl, 1986 and IUCN, 1988)

Economic Indicators
GNP: US$ 820 per capita (1988) (World Bank, 1990)

Policy and Legislation The post-independence Constitution devotes a chapter to land issues, and provides the basis upon which all legislation and policy concerning land tenure in Vanuatu rests. The salient points are that: all land in the Republic belongs to the indigenous custom owners and their descendants; the rules of custom form the basis of ownership and use of land; only indigenous citizens who have acquired land in accordance with a recognised system of land tenure may have perpetual land ownership; and government may own land in the public interest (Nalo *et al.*, 1987). Most land is thus in customary ownership and there is little likelihood of Government acquiring land for protected areas. Such areas would have to be leased or managed jointly by the government and custom owners (M.R. Chambers, pers. comm., 1989). Declaration of a site as public land could well meet with opposition from custom owners, and could involve expensive compensation (Nalo *et al.*, 1987).

The issue of land tenure must be seen in the light of customary practice, or strong traditions concerning land. Although there are certain nationwide principles, such as the important distinction between use rights and ownership rights, the system is highly fragmented and different practices may apply amongst the approximately 40 different language groups in the country. Land tenure, customary tenure, protected areas and conservation are discussed further by Nalo *et al.* (1987) and with particular reference to the establishment of Erromango Kauri Reserve (Barrance, 1989).

Environmental conservation is embodied in the Constitution, which states in Article 7(d) that every person is obliged "to protect Vanuatu and to safeguard the national wealth, resources and environment in the interests of the present generation and of future generations". Two of the six objectives of the Second National Development Plan (1987-1991) specifically address the attainment of sustainable development. Thus, the third objective is to "increase productive utilization of the country's natural resource base as a means of generating viable and sustained economic growth", and the sixth is to "ensure that Vanuatu's unique environmental and cultural heritage is not damaged in the process of economic development and change".

Legislation for the establishment of national parks and reserves and the protection of land and water is generally lacking (Nalo *et al.*, 1987). Laws on Vanuatu which were enacted prior to independence were made by the Resident Commissioners under the provisions of the Anglo-French Protocol of 1914. Some of these are still in force and several include conservation measures. The Forestry Joint Regulation (No. 30 of 1964) provides for the establishment of forest areas, forest lands, and forest reserves (although none have been established), and is incidentally also used for the protection of wrecks. The intention of this legislation is to ensure that forest resources are protected and developed in accordance with good forestry principles.

Under Forestry Act No. 14 (1982), the Director of Forests is authorised to enter into forest plantation agreements with owners to assist reafforestation or to protect threatened forests. Such areas are subject to state control, and clearance without a permit is prohibited. Agreements may be formulated in order to prevent soil erosion, preserve the ecology of an area, conserve land of particular scenic, cultural, historic or national interest, or for recreational use by the public (Sloth, 1988).

Other environmental legislation includes the Fisheries Act No. 37 (1982) which provides for the declaration of marine reserves (Section 20) within which no living organism, sand, coral or part of a wreck may be taken or removed.

Improved legislation for national parks and reserves was under consideration in 1989 (Leaver and Spriggs, 1989) and a number of potential protected areas have been identified. The National Conservation Strategy prospectus (IUCN, 1988) states that a project to review the legislation concerning conservation and environment has been implemented, with the object of making recommendations for modifications or additions to existing laws. One major piece of legislation needed is a heritage conservation act that would provide for both cultural and natural conservation. The Act should provide mechanisms whereby custom leaders and land owners could have sites and objects of special importance registered for protection under the act. Ownership and management would rest with the custom owners, but with the additional support of the law for their protection and specified penalties for the violation of that protection.

The contribution made by traditional custom and practice towards protected areas and resource conservation is significant. For example, seasonal custom taboos are applied in certain coastal waters for the conservation of dugong, fish and turtle (SPREP, 1989). During colonial times land in Vanuatu was registered under freehold titles and were mainly held by non-indigenous interests. Land held by the French, British and Condominium governments prior to independence was vested in the Vanuatu government for the benefit of the Republic, under Land Reform Regulation No. 31 of 1980. A significant number of these land parcels comprised long distances of coastal land about 100m wide. Much of this land has been returned to customary owners, but some has been retained for the establishment of protected areas. The Land Reform Regulation also empowers the Minister responsible for lands to declare any land to be public. This is only done on the advice of the Council of Ministers and after consultation with customary owners. Following requests by the Local Government Council, proposals to declare parts of Espiritu Santo as public land for recreational purposes have been made.

International Activities Vanuatu is not yet party to any of the international conventions or programmes that directly promote the conservation of natural areas, namely the Convention concerning the Protection of the World Cultural and Natural Heritage (World Heritage Convention), the Unesco Man and the Biosphere Programme and the Convention on Wetlands of International Importance especially as Waterfowl Habitat (Ramsar Convention).

The Convention on the Conservation of Nature in the South Pacific (1976) has been neither signed nor ratified. Known as the Apia Convention, it entered into force during 1990. The Convention is coordinated by the South Pacific Commission and represents the first attempt within the region to cooperate on environmental matters. Among other measures, it encourages the creation of protected areas to preserve indigenous flora and fauna.

Although Vanuatu is party to the South Pacific Regional Environment Programme (SPREP), the 1986 Convention for the Protection of the Natural Resources and Environment of the South Pacific Region (SPREP Convention) has not yet been signed or ratified. The Convention entered into force during August 1990. Article 14 calls upon the parties to take all appropriate measures to protect rare or fragile ecosystems and threatened or endangered flora and fauna through the establishment of protected areas and the regulation of activities likely to have an adverse effect on the species, ecosystems and biological processes being protected. However, as this provision only applies to the Convention area, which by definition is open ocean, it is most likely to assist with the establishment of marine reserves and the conservation of marine species.

Other international and regional conventions concerning environmental protection to which Vanuatu is party are reviewed by Venkatesh *et al.* (1983).

Administration and Management The Department of Forestry of the Ministry of Agriculture, Forestry and Fisheries is responsible for the implementation of the Forestry Act No. 14 of 1982. The Department of Fisheries of the same Ministry is responsible for the Fisheries Act No. 37 of 1982, and is consequently responsible for marine reserves, whilst the Ministry is charged with generally overseeing issues concerned with environment and conservation. The administrative status of currently gazetted recreational reserves rests with a local government council, although the legal basis for this is not clear.

In September 1986 an Environment Unit was established in the Ministry of Lands, Geology and Rural Water Supply, with responsibility for coordinating environmental and conservation issues. The Environment Unit essentially has a single objective: to develop policies and programmes such that natural resources are managed so that they are not severely degraded by development activities. Major projects include preparation of a national conservation strategy and comprehensive environmental legislation; resource surveys of fringing reefs, freshwaters, Espiritu Santo Island and biological, geological, scenic and cultural heritage sites; dugong surveys; and surveys of estuarine crocodiles of Vanua Lava (Chambers and Bani, 1987, 1989). The Environment Unit will also be made responsible for the management of any parks and reserves gazetted in the future (SPREP, 1989).

The protected areas system is too small and fails to protect most ecosystems. Existing protected areas are threatened by inadequate management, whilst the options for selecting new protected areas must be declining with time. There is an increasingly urgent need to gazette a protected area on Erromango as there have been recent (1988/1989) proposals to log parts of the remaining kauri stands. Approval has been given for major logging schemes for parts of Malekula despite opposition from Government advisors and technical staff from the Environment, Forestry, Finance, Culture and Labour departments (M.R. Chambers, pers. comm., 1989).

Systems Reviews The Republic of Vanuatu, formerly the Anglo-French Condominium of the New Hebrides, consists of the central and southern part of an archipelago which forms one of the numerous seismic arcs found in the Western Pacific. The Santa Cruz Islands, politically part of the Solomon Islands, constitute the northern part of the archipelago (UNEP/IUCN, 1988).

The archipelago of about 80 islands forms a bifurcating chain; the larger islands are found in the west and are made up of extinct volcanoes covered with fossil or modern coral reefs. The island arc is young and associated with considerable volcanic and seismic activity (Cheney, 1987). The islands are mountainous by Pacific standards, many island interiors being uninhabited (Anon., 1989a), and Tabwemasana Peak on Espiritu Santo attains 1,879m. Brief summaries of the physical characteristics are given by Douglas (1969) and UNEP/IUCN (1988) for most of the islands.

Some 9,000 sq. km, 74% of total land area, is under natural vegetation (Neill, 1987). Principal formations are tropical lowland evergreen rain forest, small areas of broad-leaved deciduous forest, closed conifer forest, montane rain forest between 1,000m and 1,500m; cloud forest above 1,500m, extensive coastal forest, swamp forest on Efate; and scattered mangrove forests covering between 2,500 and 3,500ha, of which 2,000ha occur on Malakula (Beveridge, 1975; David, 1985; Davis *et al.*, 1986).

Although lowland formations have largely been cleared and replaced by anthropogenic vegetation, forest remains the dominant landscape element on most islands. Surveys conducted in the mid-1960s indicate that some 180 sq. km of Erromango were occupied by closed climax forest, including 50 sq. km of kauri pine stands (Johnson, 1981). According to Quantin's (1976) maps, high forests are restricted on most of the islands, especially those that are densely populated (Pentecost, Aoba, Tanna and Shepherd) or have active volcanoes (Ambrym). However, the low montane forests are generally well preserved, and occupy large areas; dense, secondary woody formations, often with a thicket *Hibiscus* community, are extensive.

Forest resources are poor compared with neighbouring countries, due to geological activity, geographic isolation, hurricanes and shifting agriculture. It is estimated that current exploitable, natural forest resources will supply domestic needs for at least 20 years. Logging of indigenous forest has been sporadic, wasteful, largely unprofitable and concentrated on Efate, Espiritu Santo, Erromango, Aneityum and, to a lesser extent, Malekula and Hui. On Erromango, intensive exploitation of kauri pine commenced in 1967. Here, the risk of wind damage to residual vegetation, invasion by weed species and land slips has subsequently increased (Beveridge, 1975), and major stands have been reduced to a single tract. However, a ban on the export of whole logs was due to be in place by 1990. Similarly, a sandalwood export moratorium is to be imposed by 1992.

Coral reefs occur throughout the archipelago, encircling some islands, but discontinuous around active or recently active land masses such as Espiritu Santo, Malakula and Ambrym. Summaries of vegetation and coral reefs on most islands are given by Douglas (1969) and UNEP/IUCN (1988), respectively. Details given in UNEP/IUCN (1988) include particular reference to the reefs of President Coolidge and Million Dollar Point Reserve and Reef Island.

In general, Vanuatu's environment is relatively undisturbed (Chambers and Bani, 1987), due to low population densities (about 10 per sq. km in rural areas), and the limited degree of development (M.R. Chambers, pers. comm., 1989). Nevertheless, this could change due to the high rate of human population increase and the high proportion (85%) of the population engaged in slash-and-burn agriculture and subsistence reef fishing. Further, the concentration of the population in coastal districts, coupled with a reduced fallow period in the slash-and-burn cycle, has led to soil erosion (Anon., 1989). A broad-ranging discussion of environment,

resources and development is given by Baines (1981), covering topics such as tenure, population, administration, financial costs and aspects of natural resources including forests, mangroves, reefs, minerals, wildlife etc. Large-scale agricultural developments are leading to environmental problems, principally accelerated by forest clearance, soil erosion through poor pasture management and increasingly heavy use of pesticides and herbicides (IUCN, 1988).

Protected areas are currently restricted to President Coolidge and Million Dollar Point Reserve, which only includes marine components, and four recreational reserves. The area within the system amounts to less than 0.1% of the total national area, and much of that is accounted for by President Coolidge Reserve.

Dahl (1980) lists 37 habitats, including several forest types, scrub, marsh, grassland, volcanic desert, freshwater and littoral features and various reef formations. With the exception of fringing reefs, none of these is effectively protected.

An Action Strategy for Protected Areas in the South Pacific Region (SPREP, 1985) has been prepared. The principal goals of the strategy cover conservation education, conservation policy development, establishment of protected areas, effective protected area management and regional and international cooperation. Priority recommendations for Vanuatu are as follows: develop a national conservation strategy; establish a kauri reserve on Erromango Island; establish recreational reserves on Espiritu Santo.

The development of a national conservation strategy was delayed by a lack of funds, but has since been financed with funds from Australia, channelled through SPREP. An outline draft National Parks Act has been compiled (Leaver and Spriggs, 1989), although it is not known if this will be integrated into the national conservation strategy or implemented in any other way. Negotiations are underway with landowners on the island of Erromango to secure a long-term lease for the proposed 500ha Kauri Reserve (L. Bule, pers. comm., 1990). SPREP has funded an inter-departmental team to carry out a survey of three potential protected areas on Ambrym, Efate and Malakula and it is hoped that these, along with the proposed site on Erromango, will form the nucleus of a protected areas system (Anon., 1989).

Dahl (1980) has recommended that the protected areas network be consolidated through the establishment of reserves to protect examples of major forest types, grasslands, swamps, lakes and marine habitats; forest reserves on each of the main islands for vegetation and birds, such as 2,000ha of forest in southern Erromango and 2,000-3,000ha in central Efate; a cloud forest reserve on Espiritu Santo, which would be of value for the conservation of Santo mountain starling (Hay, 1986); and reserves along the north-west coast of Malekula or Santo, where reefs were elevated over 6m in 1965. Less specific recommendations are made by Dahl (1986) and include establishing a major protected area on Espiritu Santo, smaller reserves at least on Tanna, Aneityum and Erromango, a recreation and tourism reserve on Efate, and protection of saltwater crocodiles on Vanua Lava (Dahl, 1986).

Addresses

Recreational Reserves
Department of Lands, Ministry of Lands, Geology and Rural Water Supply, Private Mail Bag 007, Port Vila
Environmental Unit, Ministry of Lands, Geology and Rural Water Supply, Private Mail Bag 007, Port Vila
Department of Local Government, Ministry of Home Affairs, Private Mail Bag 036, Port Vila

Marine Parks
Department of Fisheries, Ministry of Agriculture, Forestry and Fisheries, Private Mail Bag 064, Port Vila

Forestry Reserves
Department of Forestry, Ministry of Agriculture, Forestry and Fisheries, Private Mail Bag 064, Port Vila

References

Anon. (1989). Progress with the action strategy for protected areas in the South Pacific. Information Paper 3. Fourth South Pacific Conference on Nature Conservation and Protected Areas. Port Vila, Vanuatu, 4-12 September. 19 pp.

Baines, G.B.K. (1981). Environmental resources and development in Vanuatu. Report to the Government of Vanuatu with support of UNDAT (United Nations Development Advisory Team for the Pacific). Unpublished. 26 pp.

Barrance, A.J. (1989). Erromango kauri reserve – a case study in environmental protection on customary land. Case Study No. 7. Fourth South Pacific Conference on Nature Conservation and Protected Areas, Port Vila, Vanuatu, 4-12 September. 6 pp.

Beveridge, A.E. (1975). Kauri forests in the New Hebrides. *Philosophical Transactions of the Royal Society of London* B 272: 369-383.

Chambers, M.R. and Bani, E. (1987). Wildlife and heritage conservation in Vanuatu. *Resources development and environment*. ESCAP, Port Vila, Vanuatu. Pp. 124-133.

Chambers, M.R. and Bani, E. (1989). Vanuatu – safe haven for the dugong. *The Pilot*. September. Pp. 13-14.

Cheney, C. (1987). Geology and the environment. In: Chambers, M.R. and Bani, E., *Resources development and environment*. ESCAP, Port Vila, Vanuatu. Pp. 1-16.

Dahl, A.L. (1980). Regional ecosystem survey of the South Pacific Area. *SPC/IUCN Technical Paper* 179. South Pacific Commission, Noumea, New Caledonia. 99 pp.

Dahl, A.L. (1986). *Review of the protected areas system in Oceania*. IUCN, Gland, Switzerland and Cambridge, UK/UNEP, Nairobi, Kenya. 328 pp.

David, G. (1985). Les mangroves de Vanuatu: 2ème partie, présentation générale. *Naika* 19: 13-16.

Davis, S.D., Droop, S.J.M., Gregerson, P., Henson, L., Leon, C.J., Lamlein Villa-Lobos, J., Synge, H. and Zantovska, J. (1986). *Plants in danger: what do we know?* IUCN, Gland, Switzerland and Cambridge, UK. 488 pp.

Douglas, G. (1969). Draft checklist of Pacific Oceanic Islands. *Micronesica* 5: 327-463.

Hay, R. (1986). Bird conservation in the Pacific. *ICBP Study Report* No. 7. International Council for Bird Preservation, Cambridge, UK. 102 pp.

IUCN (1988). National Conservation Strategy: Vanuatu. Phase I: Prospectus. IUCN, Gland, Switzerland. 39 pp.

Johnson, M.S. (1971). New Hebrides Condominium, Erromango forest inventory. *Land Resources Study* No. 10. Overseas Development Administration, Land Resources Division, Surbiton, UK. 91 pp.

Leaver, B. and Spriggs, M. Erromango kauri reserve. *Working Paper* No. 1. TCP/VAN/6755. FAO, Rome. 28 pp.

Nalo, C., Hunt, L. and Boote, D. (1977). Land tenure in Vanuatu today. In: Chambers, M.R. and Bani, E., *Resources development and environment.*, ESCAP, Port Vila, Vanuatu. Pp. 78-92.

Neill, P. (1987). Forestry resources and policies in Vanuatu. In: Chambers, M.R. and Bani, E., *Resources development and environment*. ESCAP, Port Vila, Vanuatu. Pp. 59-62.

Quantin, P. (1976). *Archipel des Nouvelles Hébrides: sols et quelques données du milieu naturel, Santo*. Office de la Recherche Scientifique et Technique de Outre-Mer, Paris. 37 pp.

Schmid, M. (1978). The Melanesian forest ecosystem (New Caledonia, New Hebrides, Fiji Islands and Solomon Islands). In: Unesco/UNEP/FAO, *Tropical forest ecosystems*. Unesco, Paris. Pp. 654-683.

Sloth, B. (1988). *Nature legislation and nature conservation as part of tourism development in the island Pacific*. Pacific Regional Tourism Development Programme. Tourism Council of the South Pacific, Suva, Fiji. 82 pp.

SPREP (1985). *Action strategy for protected areas in the South Pacific Region*. South Pacific Commission, Noumea, New Caledonia. 24 pp.

SPREP (1989). Vanuatu. Paper presented at the Fourth South Pacific Conference on Nature Conservation and Protected Areas, Port Vila, Vanuatu, 4-12 September. 6 pp.

UNEP/IUCN (1988). *Coral reefs of the world. Volume 3: Central and Western Pacific*. UNEP Regional Seas Directories and Bibliographies. IUCN, Gland, Switzerland and Cambridge, UK/UNEP, Nairobi, Kenya. 378 pp.

Venkatesh, S, Va'ai, S. and Pulea, M. (1983). An overview of environmental protection legislation in the South Pacific countries. SPREP *Topic Review* No. 13. South Pacific Commission, Noumea, New Caledonia. 63 pp.

World Bank (1990). *World Tables*. 1989-90 Edition. The John Hopkins University Press, Baltimore. 646 pp.

ANNEX
Definitions of protected area designations, as legislated, together with authorities responsible for their administration

Title: The Forestry Joint Regulation (No. 30)

Date: 1964

Brief description: Enacted prior to independence under the provisions of the Anglo-French Protocol of 1914.

Administrative authority: Director of Forests

Designation:

Forest areas
Forest lands
Forest reserves

Title: Forestry Act No. 14

Date: 1982

Brief description: The Director of Forests is authorised to enter into forest plantation agreements with owners to assist reafforestation or to protect threatened forests. Such areas are subject to state control and clearance without a permit is prohibited. Agreements may be formulated in order to prevent soil erosion, preserve the ecology of an area, conserve land of particular scenic, cultural, historic or national interest, or for recreational use by the public.

Administrative authority: Director of Forests

Designation: No information

Title: Fisheries Act No. 37

Date: 1982

Brief description: Provides, *inter alia*, for the declaration of marine reserves (Section 20).

Administrative authority: Department of Fisheries

Designation:

Marine reserves No living organism, sand, coral or part of a wreck may be taken or removed.

Area 2,830 sq. km land area in an Exclusive Economic Zone of 680,000 sq. km

Population 168,000 (1988) (World Bank, 1990)
Natural increase: 0.7% per annum (SPREP, 1989a)

Economic Indicators
GNP: US$ 610 per capita (1988) (World Bank, 1990)

Policy and Legislation The Fifth Development Plan (1985-87) stated, *inter alia*, that the protection of the environment and conservation of natural resources was a principal national goal, although overriding priority was to be given to an increase in production, particularly for export (Firth and Darby, 1988). The Constitution (1960) stipulates that "all land in Western Samoa is either customary land, private freehold, or public land". The relative proportions of these different types of tenure are customary land (80.5% of land area), private freehold (3.7%), Western Samoa Trust Estates Corporation (4.5%) and public government land (11.3%) (Eaton, 1985). All land below the high water mark is defined as public land (Pearsall, 1988). The Constitution also states that customary land may only be acquired compulsorily for public purposes. Customary land may be leased for an authorised purpose if the lease is in accordance with Samoan custom and usage, the desires and interests of the owners, or the public interest (Eaton, 1985).

Legislation which enables government to acquire customary land includes the Taking of Lands Act (1964) and Article 102 of the Constitution. The first gives the government the power to obtain land for public purposes by negotiation or compulsory processes, although the latter is rarely used. The 1964 Act could be used as a legal instrument for obtaining land for parks and reserves (Tiavolo, 1985) and has provision for land to be taken to protect catchment areas, especially where agricultural development is a threat (Venkatesh *et al.*, 1983).

The principal current legislation for the establishment of protected areas is the 1974 National Parks and Reserves Act. Section 4 enables the Head of State, acting on the advice of the Cabinet, to declare any public land to be a national park, provided it is not set aside for any other public purpose and is not less than 607.5ha (1,500 acres), except in the case of an island. As defined in Section 5, every national park shall be preserved in perpetuity for the benefit and enjoyment of the people of Western Samoa and shall be administered so that: it is preserved as far as practical in its natural state; flora and fauna are preserved as far as possible; its value as a soil, water and forest conservation area is maintained; and, subject to a number of provisos, the public has freedom of access. Section 6 provides for the establishment of nature reserves within which either named species are protected, or all taxa within a specified area are protected. Access may be restricted, except where a nature reserve is declared in a marine area, in which case

customary fishing rights remain unaffected. Sections 7, 8 and 9, respectively, make provision for the establishment of recreation reserves, historic reserves, and reserves for other purposes. A site fully gazetted under the Act can only be degazetted by Parliament. The principal weakness of the 1974 Act is that it only enables national parks and reserves to be established on public land. There is no legal mechanism for establishing areas on customary land. For example, a recently established reserve in Falealupo District, comprising 1,200ha of pristine lowland rain forest, has no legal basis (E. Bishop, pers. comm., 1988). Environmental and conservation legislation is currently being reviewed, and the amendment of the National Parks and Reserves Act has again been recommended with an expectation that it may be completed by the end of 1991 (I. Reti, pers. comm., 1991).

Habitat protection is also provided for under the 1959 Agricultural, Forests and Fisheries Ordinance and the 1967 Forests Act. These enable the Forestry Division within the Department of Agriculture, Forests and Fisheries to "conserve, protect, and develop the resources of the country especially soil, water and forest" and to establish forest reserves for water, soil and climate protection and a sustained timber harvest. The Forests Act allows for protection of forest and water catchment areas as "protected land". Large areas of indigenous forest have been designated as protection forests, although in practice logging has continued (Firth and Darby, 1988).

Other legislation which incorporates measures relating to protected areas includes the 1965 Water Act, under which watersheds and riparian vegetation may be protected.

A summary review of legislation covering aspects of environmental protection, planning and tourism is given in Firth and Darby (1988).

At the present time, no protected area in Western Samoa receives full legal protection. It is a legal requirement that all proposed protected areas are surveyed and that the proposed boundary, with a written submission, be presented to the Land Board in the Ministry of Lands. Public comment is duly invited, and, if this is positive, the proposal is submitted to the Head of State for signature. Subsequently, the Ministry of Lands will amend its land use maps to indicate a protected area. This process has not yet been completed in any instance. The lack of due process fails to invoke the full legal protection of the available legislation, and other Government departments are technically at liberty to propose and implement changes in land use within nominal protected areas (E. Bishop, pers. comm., 1989).

International Activities Western Samoa is not yet party to any of the international conventions or programmes that directly promote the conservation of natural areas, namely the Convention concerning the Protection of the World Cultural and Natural Heritage (World Heritage Convention), the Unesco Man and the Biosphere Programme and the Convention on Wetlands of International Importance especially as Waterfowl Habitat (Ramsar Convention).

At a regional level, Western Samoa has signed (12 June 1976) and ratified (20 July 1990) the Convention on the Conservation of Nature in the South Pacific, 1976. Known as the Apia Convention, it entered into force during 1990. The Convention is coordinated by the South Pacific Commission and represents the first attempt within the region to cooperate on environmental matters. Among other measures, it encourages the creation of protected areas to preserve indigenous flora and fauna.

Western Samoa is also party to the South Pacific Regional Environment Programme (SPREP) and has signed (25 November 1986) and ratified (19 July 1990) the Convention for the Protection of the Natural Resources and Environment of the South Pacific Region, 1986 (SPREP Convention). The Convention entered into force during August 1990. Article 14 calls upon the parties to take all appropriate measures to protect rare or fragile ecosystems and threatened or endangered flora and fauna through the establishment of protected areas and the regulation of activities likely to have an adverse effect on the species, ecosystems and biological processes being protected. However, as this provision only applies to the Convention area, which by definition is open ocean, it is most likely to assist with the establishment of marine reserves and the conservation of marine species.

Other international and regional conventions concerning environmental protection to which Western Samoa is party are reviewed by Venkatesh *et al.* (1983).

Administration and Management The National Parks and Reserves Act (1974) is currently administered by the Department of Agriculture, Forests and Fisheries, although the Department of Lands and Environment undertakes all responsibilities (I. Reti, pers. comm., 1991). The Assistant Director for Forests and Conservation has the overall responsibility for forestry and national parks (I. Reti, pers. comm., 1989). Management aims are to: establish examples of each type of reserve, ensuring that as many of the different types of vegetation and wildlife as possible are conserved; improve and develop appropriate facilities, to enable the full enjoyment and appreciation of the reserves; and promote public awareness, understanding and appreciation of these areas. Government departments were being restructured during 1989 and it was intended that the National Parks and Reserves Section would be transferred from its current ministry to the Ministry of Lands on 1 January 1990 (E. Bishop, pers. comm., 1990).

The National Parks and Reserves section operating budget was reduced in real terms by 93% between 1979 and 1987, while staff numbers were reduced by 42% in the same period. This trend is now being reversed and the 1988 budget was 28,000 WS Tala (US$ 14,000), twice that of 1987. The 1989 budget was 56,000 WS Tala, the proposed 1990 budget was WS Tala 104,000. Similarly, staff numbers have increased and were proposed to increase from 19 in 1988 to 27 in 1990. Most of these funds were to pay for additional labourers at O' Le Pupu Pu'e National Park, installation of a water pipe in Vailima Botanical Garden and the preparation of a poster depicting Samoan birds (E. Bishop, pers. comm., 1988). Staff levels and annual budgets previously reached a peak of 30 staff in 1981 and a budget of 66,300 WS Tala in 1979, a period during which substantial external financial and management assistance was available (SPREP, 1985b). The Department of Lands and Environment has four staff (1991) and a current six-month budget of WS Tala 87,610.

A Division of Environment and Conservation has been established within the Department of Lands and Environment. Its principal function is the provision of advice on: policies influencing the management of natural and physical resources; impact of private or public development; means of ensuring public participation in environmental planning and policy formulation; the control and management of hazardous substances; establishment and naming of national parks; control of pollution of air, water and land and the control of litter; and research and training relevant to these functions (I. Reti, pers. comm., 1991).

Systems Reviews Western Samoa, lying between 13°-15'S and 171°-173'W, is situated approximately 1,000km north-east of Fiji and a similar distance north-north-east of Tonga, and includes nine islands, the largest being Savai'i (1,709 sq. km) and Upolu (1,118 sq. km). The biologically related, but politically distinct, American Samoa lies some 100km to the east.

Most of the original lowland tropical forest on Savai'i and Upolu has been cleared or highly modified. Only two major forested areas, on customary land, persist, namely Tafua and Mt Silisili on Savai'i (E. Bishop, pers. comm. 1988). Figures given by Firth and Darby (1988) indicate that approximately 1,500 sq. km remain forested, comprising 550 sq. km of protection forest or proposed national parks and reserves, and 950 sq. km of commercial forest. However, no firm commitment has been made to the environmental protection role of protection forests and there is virtually no primary forest remaining (Firth and Darby, 1988). Montane forests are less damaged and still contain a rich endemic flora. Cloud forests, montane lava flow scrub and montane meadows are found in the upland regions. The Aleipata Islands include a number of littoral communities, as well as *Diospyros* coastal forests and *Dysoxylum* lowland forests, which are otherwise rare in Western Samoa (Davis *et al.*, 1986; Whistler, 1983). Mangroves cover less than 1,000ha (Bell, 1985).

Forest resources, particularly lowland and sub-montane formations, are threatened by both agricultural encroachment and commercial logging (Pearsall, 1988). In 1977 responsibility for parks and reserves was transferred to a department which included indigenous forest logging amongst its primary responsibilities and the momentum of an initially dynamic protected areas programme was lost. Within a few years, most of the remaining lowland and foothill forest in Savai'i and Upolu had been destroyed (Firth and Darby, 1988). A recent development has been the risk of timber concessions being sold for cash remuneration, particularly on Savai'i. The destruction of lowland rain forest is largely complete, and regions of higher elevations are threatened by governmental road construction, followed by clearing of large areas in order to expand plantations or to facilitate cattle farming; commercial logging activities by Australian and Japanese concerns, formerly engaged in selective felling but currently clear felling; and uncontrolled clear felling by local families on village-owned land (Beichle and Maelzer, 1985). The cutting and in-filling of mangroves has been largely uncontrolled (Bell, 1985). Although the 1976 Forestry Act provides for the exploitation of forests on a sustained yield basis, the criteria for this to be judged by, for example, annual allowable cut, have never been established and there is a risk that the bulk of indigenous timber resources may be exhausted by as soon as 1995 (Firth and Darby, 1988).

Coastal and peripheral woodlands have been heavily disturbed by cultivation and lower slopes around main settlements are cultivated. Approximately 20% of Savai'i is cultivated, whilst the largest cultivated areas on Upolu are along the north coast around Apia and Satapuala (Douglas, 1969). The coastal district has long been settled and soils around villages have been exhausted (Firth and Darby, 1988). A process of land tenure conversion is underway, from the traditional (and legally constituted) system to a *de facto* system of private control and virtual ownership by converting land to agriculture.

Western Samoa's current park and reserve system consists of one national park and five reserves. A little over 1% of the land area is protected, but O Le Pupu Pu'e National Park alone accounts for 95% of the protected area. The system's growth years were 1978-79, when one national park and three reserves were established. O Le Pupu Pu'e National Park is a key area within the Samoa, Wallis and Futuna biogeographic province identified by Dahl (1980). Dahl (1980) identifies a number of habitats as occurring in Western Samoa, including varieties of lowland rain forest, montane rain forest, cloud, riverine, swamp, mangrove and atoll/beach forest, scrub, grassland, freshwater marsh, rock desert, lakes and streams, and a variety of littoral, coastal and marine habitats, including reefs and lagoons. A significant number are protected within O Le Pupu Pu'e, namely four lowland rain forest formations, three montane rain forest formations, cloud and swamp forest,

Pandanus sp. littoral scrub, crater meadow, crater marsh, caves and rocky coast (Dahl, 1980). It also protects populations of all the major forest bird species of Upolu, including tooth-billed pigeon *Didunculus strigirostris* and ma'o *Gymnomyza samoensis* (Hay, 1986). Nevertheless, a similar reserve is required on Savai'i, in view of deforestation at the western end of the island (Hay, 1986). Two sites, one incorporating coastal forest at Tafua, south-east Savai'i, and another on the slopes of Mt Silisili up to the summit, have been proposed as protected areas (Holloway and Floyd, 1975) and would preserve the greatest range of habitats (Hay, 1986). These two sites still support lowland rain forest, but are in the process of being cleared and are on customary land, thus precluding their designation as national parks or reserves under current legislation (E. Bishop, pers. comm., 1988). A coastal reserve is particularly important, as are examples of the high altitude habitats of the island. Such reserves would assist in and may ensure the protection of species such as Samoan white-eye, tooth-billed pigeon and ma'o, possibly Samoan wood-rail if its existence is re-confirmed, and other endemic Samoan species (Hay, 1986). A recent development has been the agreement between a private group from the USA and villagers in the Falealupo District in Savai'i, whereby some 1,200ha of lowland rain forest is protected in return for financial assistance and restricted exploitation rights (SPREP, 1989). However, as the site is on customary land it will not be possible to gazette the site under the provisions of existing legislation. It is not known if this arrangement has any basis in protected areas legislation.

An Action Strategy for Protected Areas in the South Pacific Region (SPREP, 1985a) has been prepared. The principal goals of the strategy cover conservation education, conservation policy development, establishment of protected areas, protected area management and regional and international cooperation. Priority recommendations for Western Samoa are as follows: develop a national conservation strategy; investigate the remaining traditional marine fishing rights; train researchers and managers in the marine field; assess conservation status and availability of reef and lagoon areas; include environmental concerns in education systems; establish an Environmental Management Unit responsible for environmental assessment, liaison, environmental education, legislation and environmental reporting; map all national ecosystems; make an inventory of flora and fauna, particularly threatened species; prepare management plans for all protected areas; develop a national environmental public awareness campaign to be implemented through schools, mass media and Pulenu'us; set aside areas which include ecosystems not covered by existing parks and reserves; complete a review of marine resources; and send the Superintendent of Parks and Reserves to the 1987 Parks Management Training Course in New Zealand.

Regional progress on the implementation of the action strategy has been reviewed (SPREP, 1987, 1989b). Western Samoa has participated in a sub-regional course in coastal resource management and planning, emphasising the role of protected areas. A SPREP project to assess the potential environmental impact of tourism, with emphasis on the protection of critical habitats, species and potential protected areas, has been completed (Firth and Darby, 1988). An ecosystem survey was scheduled to commence in 1989, using aerial photographs taken during 1987 (Pearsall, 1989). This is intended to identify a representative system of natural areas. In conjunction with the Land Use Resource Programme of the Ministry of Lands, it will provide a powerful resource planning and management tool (Anon., 1989). The compilation of management plans for each of the five reserves and updating that for O Le Pupu Pu'e was planned 1988-90 (E. Bishop, pers. comm., 1988). An Asian Development Bank funded project on land use planning commenced in 1989 (I. Reti, pers. comm., 1989).

A major UNDAT survey in the mid-1970s (Holloway and Floyd, 1975) identified 6 potential national parks, 24 nature reserves, 11 historical sites and 7 archaeological sites. Of these, O Le Pupu Pu'e, Palolo Deep Marine Reserve and Mount Vaea have been gazetted. Recommended sites considered to be of particular value include Mt Silisili and Tafua (Hay, 1986). However, the proposed areas occur on customary land and gazettement under the current 1974 legislation is precluded. Three methods of protecting the areas were mooted: government purchase of the land, which would only be possible with outside financial assistance; leasing, which might not provide the necessary security of tenure; and the dedication of customary land (Holloway and Floyd, 1975). Dahl's (1980) proposed reserves reiterate those of Holloway and Floyd (1975), with the addition of a recommended reserve type aimed at protection of Palolo worm *Eunice viridis* (K). Dahl (1986) identifies as highest priority the protection of a major park in central Savai'i, possibly also including a sample of lowland forest. Protection of both land and marine areas in the Aleipata Islands should also be considered and a survey of the latter site has been completed (Andrews and Holthus, 1989).

The preparation and adoption of a national conservation strategy is recommended by Firth and Darby (1988) to ensure the sustainable exploitation of natural resources, a call reiterated in SPREP (1989a). Strengthening the national parks and reserves systems is perhaps more urgent, and should be addressed as a high priority. This would principally entail; amending the 1974 National Parks and Reserves Act, such that reserves may be established on customary land (not necessarily with government ownership), provision of adequate funding, the establishment of some 30 protected areas, the principal one being Mount Silisili National Park on Savai'i, and the imposition of a moratorium on any further logging in Tafua Forest (Firth and Darby, 1988).

Despite repeated recommendations for the establishment of protected areas, especially Mount Silisili and the Aleipata Islands, there has been no progress in gazetting either site. Nevertheless, the establishment of the Division of Environment and Conservation, increased funding and the adoption of the Lands and Environment Act 1989 indicate a greater willingness on the behalf of government to give the environment equal priority with other pressing issues (I. Reti, pers. comm., 1991).

Addresses

National Parks and Reserves, Department of Agriculture, Forests and Fisheries, PO Box L1874, Apia

Division of Environment and Conservation, Department of Lands and Environment, PO Private Bag, Apia

References

Andrews, G.J. and Holthus, P.F. (1989). *Marine environment survey: proposed Aleipata Islands National Park, Western Samoa.* South Pacific Commission, Noumea, New Caledonia. 67 pp.

Beichle, U. and Maelzer, M. (1985). A conservation programme for Western Samoa. *ICBP Technical Publication* No. 4. International Council for Bird Preservation, Cambridge, UK. Pp. 1-3.

Bell, L.A.J. (1985). Coastal zone management in Western Samoa. In: Thomas, P.E.J. (Ed.), *Report of the Third South Pacific National Parks and Reserves Conference.* Volume II. South Pacific Commission, Noumea, New Caldenonia. Pp. 57-73.

Dahl, A.L. (1980). Regional ecosystem surveys of the South Pacific Area. SPC/IUCN *Technical Paper* 179. South Pacific Commission, Noumea, New Caledonia. 99 pp.

Dahl, A.L. (1986). *Review of the protected areas system in Oceania.* IUCN, Gland, Switzerland and Cambridge, UK/UNEP, Nairobi, Kenya. 328 pp.

Davis, S.D., Droop, S.J.M., Gregerson, P., Henson, L., Leon, C.J., Lamlein Villa-Lobos, J., Synge, H. and Zantovska, J. (1986). *Plants in danger: what do we know?* IUCN, Gland, Switzerland and Cambridge, UK. 488 pp.

Douglas, G. (1969). Draft checklist of Pacific Oceanic Islands. *Micronesica* 5: 327-463.

Eaton, P. (1985). Land tenure and conservation: protected areas in the South Pacific. SPREP *Topic Review* No. 17. South Pacific Commission, Noumea, New Caledonia. 103 pp.

Firth, N. and Darby, d'E.C. (1988). *Environmental planning for tourism in Western Samoa.* A report to the Government of Western Samoa and the South Pacific Regional Environment Programme. KRTA Limited, Auckland, New Zealand. 124 pp.

Hay, R. (1986). Bird conservation in the Pacific. *ICBP Study Report* No. 7. International Council for Bird Preservation, Cambridge, UK. 102 pp.

Holloway, C.W. and Floyd, C.H. (1975). *A national parks system for Western Samoa*. United Nations Development Advisory Team for the South Pacific (UNDAT), Suva, Fiji. 71 pp.

Pearsall, S.H. (1988). Western Samoa. Country Report. The Nature Conservancy, Honolulu, Hawaii. Unpublished. 75 pp.

Pearsall, S.H. (1989). A system of representative natural areas for Western Samoa. Case Study 29. Fourth South Pacific Conference on Nature Conservation and Protected Areas, Port Vila, Vanuatu, 4-12 September. 12 pp.

SPREP (1985a). *Action strategy for protected areas in the South Pacific Region*. South Pacific Commission, Noumea, New Caledonia. 24 pp.

SPREP (1985b). Western Samoa. In: Thomas, P.E.J. (Ed.), *Report of the Third South Pacific National Parks and Reserves Conference. Volume III. Country reviews*. South Pacific Commission, Noumea, New Caledonia. Pp. 232-269.

SPREP (1987). SPREP makes considerable progress with implementation of action strategy for protected areas. *Environment Newsletter*. South Pacific Regional Environment Programme. July-September. Pp. 15-22.

SPREP (1989a). Western Samoa. Paper presented at the Fourth South Pacific Conference on Nature Conservation and Protected Areas, Port Vila, Vanuatu, 4-12 September. 13 pp.

SPREP (1989b). Progress with the Action Strategy for protected areas in the South Pacific Region. Information Paper 3. Fourth South Pacific Conference on Nature Conservation and Protected Areas, Port Vila, Vanuatu, 4-12 September. 19 pp.

Tiavolo, A. (1985). Land tenure system in Western Samoa. South Pacific Commission, Noumea, New Caledonia. Unpublished report. 3 pp.

Trotman, I.G. (1979). Western Samoa launches a national park program. *Parks* 3(4): 5-8.

Venkatesh, S., Va'ai, S. and Pulea, M. (1983). *An overview of environmental protection legislation in the South Pacific countries*. South Pacific Commission, Noumea, New Caledonia. 63 pp.

Whistler, W.A. (1983). Vegetation and flora of the Aleipata Islands, Western Samoa. *Pacific Science* 37(3): 227-249. (Unseen)

World Bank (1990). *World Tables*. 1989-90 Edition. The John Hopkins University, Baltimore. 646 pp.

ANNEX
Definitions of protected area designations, as legislated, together with authorities responsible for their administration

Title: National Parks and Reserves Act

Date: 30 December 1974

Brief description: Provides for the establishment, preservation and administration of national parks and reserves for the benefit of the people of Western Samoa.

Administrative authority: As stated in the Act: Department of Lands and Survey, Minister of Lands (National Parks and Reserves, Ministry of Agriculture, Forestry and Fisheries)

Designation:

National park As defined in Section 5, every national park shall be preserved in perpetuity for the benefit and enjoyment of the people of Western Samoa and shall be administered so that: it is preserved as far as practical in its natural state; flora and fauna are preserved as far as possible; its value as a soil, water and forest conservation area is maintained; and, subject to a number of provisos, the public has freedom of access.

Nature reserves Within which either specified species of flora and fauna are protected, or all taxa within a specified area are protected. Access may be restricted, except where a nature reserve is declared in a marine area, in which case customary fishing rights remain unaffected.

Recreation reserves (S.7)
Historic reserves (S.8)
Reserves for other purposes (S.9)

NB: A site fully gazetted under the Act can only be degazetted by Parliament. The principal weakness of the 1974 Act is that it only enables national parks and reserves to be established on public land.

Title: Forests Act

Date: 1967

Brief description: Consolidates the law relating to conservation, protection and development of natural resources, especially soil, water and forests.

Administrative authority: Department of Agriculture, Forests and Fisheries

Designation:

State forests on customary or freehold land

Protected forestry
Historic, cultural and archaeological sites

NB Source: Firth and Darby (1988); original act not seen

SUMMARY OF PROTECTED AREAS

Map[†] ref.	*National/international designation* Name of area	IUCN management category	Area (ha)	Year notified
	National Park			
1	O Le Pupu Pu'e	Unassigned	2,857	1978
	Historic and Nature Reserve			
2	Tusitala (comprises 3 parts as follows)	Unassigned	64	1958
	Mount Vaea Scenic Reserve	Unassigned	51	1958
	Stevenson's Historic Site	Unassigned	1	1958
	Vailima Botanic Garden	Unassigned	12	1978
	Reserves			
3	Palolo Deep	IV	22	1979
4	Togitogiga Recreation	V	3	1978
	Unclassified			
5	Falealupo Forest	Private	1,215	1989

[†]Locations of most protected areas are shown on the accompanying map.

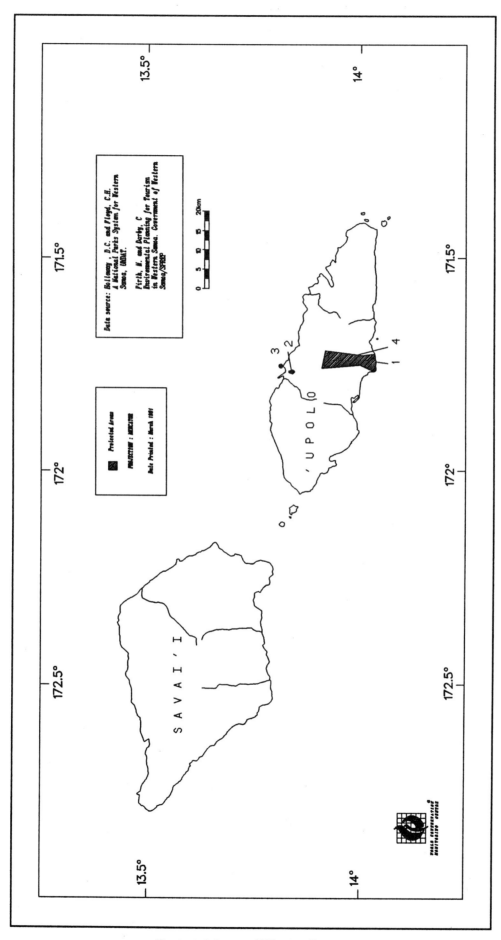

Protected Areas of Western Samoa

Australasia

AUSTRALIA

Area 7,682,427.85 sq. km (Terrestrial area excluding External Territories)

Population 16,873,000 (1990)
Natural increase: 1.18% per annum

Economic Indicators
GNP: US$ 11,203 (1989)

Policy and Legislation

Politically, Australia is a federation. There are six states (New South Wales, Queensland, South Australia, Tasmania, Victoria and Western Australia) and two self-governing territories (Australian Capital Territory and Northern Territory). The eight states and territories have jurisdiction over the management of their own land and territorial sea, including nature conservation management. The Australian federal government is referred to as the Commonwealth of Australia. It does not have direct jurisdiction over area management or nature conservation in the eight states and territories, but is responsible for these matters in six external territories (Australian Antarctic Territory, Ashmore and Cartier Islands, Christmas Island, Cocos (Keeling) Islands, Coral Sea Islands, and Heard Island and the McDonald Islands), and also in the Territory of Jervis Bay, an enclave on the coast of New South Wales. In addition, it has some specific responsibilities in these matters in the External Territory of Norfolk Island, the Australian Capital Territory and the Northern Territory.

Commonwealth government responsibilities for nature conservation in Australia as a whole stem from its constitutional powers in relation to international treaties and foreign affairs generally, Aboriginal affairs, exports and imports, and corporations. In addition, the Commonwealth fosters cooperation and liaison among the states and territories in relation to nature conservation. The governments of the Commonwealth and the eight states and territories each has an agency which has primary responsibility for nature conservation. Development of common policies on nature conservation among the Commonwealth, state and territory governments occurs mainly through the Council of Nature Conservation Ministers (CONCOM), its Standing Committee and working groups. Membership of CONCOM consists of the government ministers responsible for nature conservation in the various jurisdictions. From July 1989 it also includes the corresponding Minister from New Zealand. The ministers meet annually. The CONCOM Standing Committee comprises the heads of the government nature conservation agencies, who meet twice a year. Working groups are established, as the need arises, to address specific issues. Officers of the various nature conservation agencies with specialist knowledge participate in the working groups, which cover topics

such as: management of national parks, staff training and development, community education and interpretation, endangered fauna, endangered flora, international agreements and fire management.

Under the jurisdiction of the Commonwealth, the Australian National Parks and Wildlife Service administers the National Parks and Wildlife Conservation Act 1980. The Commonwealth and the eight states and territories all have legislation to achieve their aims in nature conservation. The major laws administered by the respective nature conservation agencies are shown in the Annex. In most cases the basic legislation was put in place during the early 1970s. In recent years there have been some significant new developments in legislation.

Commonwealth The Australian National Parks and Wildlife Service has responsibility for administering the National Parks and Wildlife Conservation Act 1980, the Whale Protection Act 1980 and the Wildlife Protection (Regulation of Exports and Imports) Act 1982.

Australian Capital Territory Under the Nature Conservation Ordinance 1980, the Minister responsible for the ACT may declare unleased Commonwealth (federal government) land in Australian Capital Territory and Jervis Bay Territory to be a reserved area. Similar powers under the Public Parks Ordinance 1928-66 provide for the declaration of public parks or recreation reserves and some of these are managed as natural areas.

Under the Fishing Ordinance 1967, 800ha of the waters of Jervis Bay Territory are "managed in sympathy" with Jervis Bay Nature Reserve. Certain activities including dredging, trawling, seaweed harvesting and commercial fishing without a licence are prohibited.

External Territories and Commonwealth Waters The National Parks and Wildlife Conservation Act 1975 provides for the establishment of parks and reserves over land or sea areas where constitutionally there is a basis for Commonwealth interest. These areas may be designated national park or some other designation and may only be revoked by a resolution of both Houses of Federal Parliament. Plans of management are required to be prepared and, after being subject to public comment and amendment, are considered by the responsible Minister and laid before both houses of federal parliament.

Norfolk Island National Park has been proclaimed under the Norfolk Island National Park and Norfolk Island Botanic Garden Act 1984 and also under the National Parks and Wildlife Conservation Act.

Several historic shipwrecks in waters along the coast of Australia have had "protected zones" designated around them under the Historic Shipwrecks Act 1976. While these areas are declared primarily to protect the enclosed shipwrecks, the legislation prohibits ships entering and mooring, trawling, and diving or any other underwater activity, effectively protecting all marine life in the designated area. A permit is required to enter the area for any purpose.

New South Wales The Wilderness Act was passed in 1987 and provides for: the permanent protection and proper management of wilderness areas in New South Wales; and promoting the education of the public and the appreciation, protection and management of wilderness. Under the Act, the Director of the New South Wales National Parks and Wildlife Service is required to report on the status of areas identified as wilderness and to investigate proposals for additional wilderness areas.

The National Parks and Wildlife Act 1974 provides for parks and nature reserves to be created by proclamation. Plans of management are prepared with regard to the objectives of conservation, study and appreciation of wildlife and natural features, and to promote appreciation and enjoyment of the natural values of these areas. They are subject to public comment before adoption by the Minister for the Environment.

The Act establishes a National Parks and Wildlife Advisory Council to advise the Minister on the control and management of national parks and nature reserves. Advisory committees may also be established for one or more national parks, nature reserves and historic sites or any combination thereof, to make recommendations to the above Council, the Director or Superintendent of the respective national park. A recent change to the Act provides for assistance to private property holders to conserve sites of natural or cultural importance on their land through a voluntary conservation agreement. Conditions to be negotiated in a conservation agreement may include restrictions on use of the land, such as limitations to grazing, clearing and the use of pesticides.

The Forestry Act 1916 provides that land within state forest and on certain other Crown lands may be dedicated, by notice in the Gazette, as a flora reserve for the preservation of native flora. To date all such reserves have been located within state forest, and while the preservation of native flora remains a keystone in all flora reserves, the establishment and management of the reserves has always been based on a broader objective of preserving samples of the natural environment. A flora reserve can only be revoked by Act of Parliament, and for each reserve the Forestry Commission is required to prepare a working plan for approval by the Minister for Natural Resources. Advisory Committees have been set up under the working plans to assist in the care of those flora reserves which receive appreciable public use.

The Fisheries and Oyster Farms (Amendment) Act 1979 provides for the creation and management of aquatic reserves in New South Wales. At present, the legislation specifies that aquatic reserves may only be declared over Crown lands and that the regulations may prohibit or regulate the taking of fish and other marine fauna as specified in the Act. The regulations also provide for the management, protection and development of any aquatic reserve.

Northern Territory The Conservation Commission Act 1980 established the Conservation Commission (CCNT) of the Northern Territory to promote the conservation and protection of the natural environment and establish and manage parks, reserves and sanctuaries and undertake other functions relating to soil and environmental conservation. The Commission is a corporation of nine members, two of whom are the Director of Conservation and his or her Deputy, and another two members nominated by the Aboriginal Land Councils.

Land is reserved under the provisions of the Territory Parks and Wildlife Conservation Act 1980 by the Administrator of the Northern Territory, following receipt of a report from the Conservation Commission. This Act refers only to the creation of "parks and reserves", allowing considerable flexibility in the interpretation of these terms. Accordingly, areas declared may range from small sites of specific interest, to major national parks. Land is held by the Conservation Land Corporation as constituted under the Conservation Commission Act. Revocation of reserved land is by declaration by the Administrator following a resolution by the Northern Territory Legislative Assembly. An amendment to this Act provides for the establishment of local management committees for any park, reserve, sanctuary or protected area. This facilitates the participation of local people in the park planning and management process.

Management plans are prepared by the Conservation Commission of the Northern Territory for areas under its control in accordance with the Territory Parks and Wildlife Conservation Act.

The Cobourg Peninsula Aboriginal Land Sanctuary Act 1981 established a major national park on Cobourg Peninsula, owned and controlled by Aborigines but managed by the CCNT on their behalf. It is anticipated that Katherine Gorge (Nitmiluk) National Park, recently awarded to the Jawoyn Aboriginal people under the Land Claims Act, will be managed by the CCNT on behalf of the traditional owners under mutually agreed guidelines.

Kakadu National Park and Uluru (Ayers Rock-Mount Olga) National Park are proclaimed under the National Parks and Wildlife Conservation Act 1975. Much of Kakadu National Park, and all of Uluru (Ayers Rock-Mount Olga) National Park, is Aboriginal land leased to the Director, Australian National Parks and Wildlife Service (ANPWS). Management plans are prepared by the ANPWS in consultation with the traditional Aboriginal owners. Within Kakadu National

Park is a conservation zone, a temporary reservation to protect the area until a decision is made whether to declare all or part of the zone as an addition to the park.

The Fisheries Act 1988 allows the Minister to declare an area, place or any waters to be a fishery management area and requires that the management of each management area be subject to general regulations to the Act and to the provisions of a management plan prepared for the purpose. Two aquatic life reserves were created under the now repealed Fish and Fisheries Act 1980.

Queensland National parks are established under the National Parks and Wildlife Act 1975-84 to conserve areas of scenic, scientific or historic interest. Vacant Crown lands unencumbered in any way are permanently reserved and can be revoked only with the authority of Parliament, though under certain circumstances land can be excluded by Order in Council for tourist purposes or for roads.

Oil exploration activities may be carried out on national parks under conditions set by the Director of National Parks and Wildlife.

The National Parks and Wildlife Act prescribes that "the cardinal principle to be observed in the management of national parks shall be the permanent preservation to the greatest possible extent, of their natural condition". Provision is made for declaration within national parks of special management zones, including primitive areas, primitive and recreation areas, recreation areas, scientific areas and historic areas.

The Director is empowered to carry out whatever works or make provisions as he considers necessary or desirable for the preservation, proper management, or public enjoyment of national parks.

The Fauna Conservation Act 1974-85, provides for fauna reserves and fauna refuges – the latter may be declared over land of any tenure with the agreement of the landholder. Fauna sanctuaries are also established under this Act to protect fauna but not habitat, though in general, a sanctuary is declared only if habitat protection is otherwise assured in the state of Queensland. All national parks and all islands off the coast are fauna sanctuaries.

Provision for the establishment of environmental parks and scientific purpose reserves (department and official purposes reserves) is contained in the Land Act 1962-86 and in particular in the Land Amendment Act 1973.

The Department of Forestry is responsible for the management of State Forests under the Forestry Act 1959-1987. Section 33 of the Forestry Act sets out the cardinal principles of management of state forests as being:

– permanent reservation;
– production of timber and associated products in perpetuity;
– protection of watersheds therein;

having due regard to:

– benefits of permitting grazing;
– desirability of conservation of soil and environment and protection of water quality; and
– the possibility of applying the area to recreational purposes.

The Conservator of Forests may recommend to the Governor in Council that the whole or part of a state forest be declared a scientific area or feature protection area so as to preserve it as a sample of the natural environment of the state forest concerned.

Reserves for Fisheries Purposes are declared under the Fisheries Act 1976-1989 as part of the management of fisheries resources within Queensland. The lands of these reserves and sanctuaries are important fish nursery or feeding areas and in the case of reserves, often contain important commercial and recreational fishing grounds and act as protection measures to assist the ongoing viability of fisheries of a region.

Great Barrier Reef Marine Park (GBRMP) was established under the Great Barrier Reef Marine Park Act 1975 (Cwlth). The Act provides for the conservation of the Great Barrier Reef by the establishment, control, care and development of the marine park in the Great Barrier Reef Region – an area of approximately 345,000 sq. km off the northern coast of Queensland. It is a multiple-use region which allows for reasonable human use whilst still ensuring the protection of the reef. GBRMP extends from low water, outwards to beyond the outer reefs, and from Cape York in the north to approximately Bundaberg in the south.

Human use is controlled mainly by regulations, public education and by the development of zoning plans. Zones separate conflicting activities and allow for uses such as diving, reef-walking, recreational and commercial fishing, and for general tourist activities. In other zones, use is restricted to permitted scientific research whilst still other zones are preservation zones to be left always undisturbed by people. The development of zoning plans is accomplished, in part, by extensive public participation. Certain activities such as mining or oil exploration, littering, spearfishing with scuba, and the taking of certain protected animal species are prohibited in all of the Great Barrier Reef Region.

South Australia The National Parks and Wildlife Act 1972 provides for the establishment and management of reserves for public benefit and for the conservation of wildlife in a natural environment. Recent changes to the Act have created a new category of multiple-use reserve where conservation of wildlife and use of natural resources can occur alongside one another in areas set aside as regional reserves.

The reserves comprise national parks, conservation parks, game reserves, recreation parks and regional reserves. They may be abolished or their boundaries altered by a proclamation of the Governor, subject to a resolution passed by both Houses of Parliament.

Management plans for each reserve are prepared by the Minister, in conjunction with comments and suggestions of the National Parks and Wildlife Reserves Advisory Committee and representations from the public. Objectives in the management of reserves include, the preservation and management of wildlife, the preservation of features of geographical, natural or scenic interest, and the encouragement of public use and enjoyment of reserves. The management plan may also provide for the division of a reserve into zones which shall be kept and maintained under the conditions declared by the plan.

The National Parks and Wildlife Reserves Advisory Committee, at the request of the Minister, can investigate and advise the Minister upon any matter referred to the Committee for advice. The Committee may also refer any matter affecting the administration of the Act, to the Minister for consideration. Five members are appointed to the Committee by the Governor.

The Fisheries Act 1982 provides, inter alia, for the protection of the aquatic habitat. Thirteen aquatic reserves have been proclaimed in South Australia pursuant to the Act. In addition, the Fisheries Act (Aquatic Reserves) Regulations 1984 provide for limited access and/or limited fishing activities within the waters of most of the reserves. Other reserves are designated as non-entry, thereby affording complete protection to marine life within the reserve.

The Forestry Act 1950-81 dedicates specific areas of Crown land exclusively for the purpose of forestry. These areas are called forest reserves. Under the same Act, areas of forest within forest reserves can be proclaimed and dedicated as native forest reserves specifically for the conservation of flora and fauna.

Native forest reserves are required to have a "statement of purpose" for which they are established and are then protected under the Forestry Act from "operations inconsistent with this purpose". Proclamation occurs after approval from state Parliament.

The Historic Shipwrecks Act 1981 (SA) provides for the declaration of marine waters to protect historic shipwrecks. The legislation effectively gives protection to marine life in the designated area by prohibiting shipping, and by prohibiting diving or other underwater activities except in accordance with a permit issued by the Department of Environment and Planning.

Tasmania The National Parks and Wildlife Act 1970 provides for the establishment of conservation areas by the Governor's proclamation. Conservation areas may include privately owned lands subject to the consent of the owners. Conservation areas that are Crown land may be declared state reserves by Governor's proclamation but may not be revoked unless the Governor's draft proclamation is first approved by each House of Parliament. This proclamation may give a name to the state reserve including that of state reserve, national park, nature reserve, historic site or Aboriginal site. Other statutory powers for example, to grant mining leases or forestry rights, do not apply in state reserves unless a management plan approved by both Houses of Parliament so provides. The sub-Antarctic Macquarie Island has been declared a nature reserve under this Act (Clark and Dingwall, 1985).

In conservation areas (some of which are named wildlife sanctuaries) wildlife and habitat are protected by regulations, but other activities such as mining and forestry are permitted. Management plans can provide additional protection. Section 22 A-G of the Forestry Act 1920 authorises preparation and approval of Forest Management Plans in respect of Crown land that is both reserve land within the meaning of the National Parks and Wildlife Act 1970, i.e. set aside for a conservation purpose, and land in a state forest or timber reserve. Forestry operations may proceed in such areas with concurrence of both the Department of Lands, Parks and Wildlife, and Forestry Commission, subject to those operations not jeopardising the conservation values for which the area was declared.

National parks are generally outstanding natural areas greater than 4,000ha. Nature reserves comprise areas of significant natural features reserved for nature conservation and scientific study. Crown land conservation areas may also be declared game reserves where management is aimed at producing native or introduced game species which may be hunted in season.

Management plans are required to be prepared in respect of all areas proclaimed under the National Parks and Wildlife Act. These are required to be publicly displayed and comment sought before being approved by the Governor. Where provision is made for use of a state reserve other than as provided for in the Act, relevant parts of the management plan require the approval of both Houses of Parliament.

Under the Forestry Act, the Forestry Commission may, by notification in the Gazette, set aside land within a state forest as a forest reserve.

Under the Sea Fisheries Act 1959, the Crayfish Point Marine Reserve has been established to provide an undisturbed area for rock lobster research.

Victoria Under the National Parks Act 1975, provision is made for the establishment of national parks and other parks for the preservation and protection of the natural environment including wilderness areas, indigenous flora and fauna, and features of scenic, archaeological, geological, historical or other scientific interest. National parks and other parks may be zoned by publication of the Governor in Council proclamation to that effect. New national parks and other parks are established by parliamentary amendment of the appropriate schedules to the National Parks Act to include the name and description of the lands included in the park. Provision is also made in the Act for the National Parks Service to manage land not reserved in the Schedules to the Act. National parks comprise predominantly unspoilt landscapes and are of extensive size whereas other parks are usually of relatively less size or significance. The Act makes statutory provision for special works to be conducted in certain parks and provides the Director with authority to manage parks generally.

Under the Crown Land (Reserves) Act 1978, state wildlife reserves may be established for the preservation or management of wildlife or wildlife habitat. Flora and fauna reserves and flora reserves are also reserved under the Crown Land (Reserves) Act.

Under the Wildlife Act 1975, the Director is required to prepare as soon as practical a plan of management for each wildlife reserve. The Minister may adopt or vary such plans.

The Reference Areas Act 1978 provides for reference areas to be proclaimed by the Governor in Council, and for the Minister to issue directives for their protection, control and management.

Two historic shipwrecks have been declared under the Historic Shipwrecks Act 1981 (Vic). The legislation prohibits interference of the shipwrecks by boating, fishing, diving or other underwater activities, effectively protecting all marine life in the designated area.

The Flora and Fauna Guarantee Act was passed by Parliament in May 1988. The Act provides for the protection of the habitat of endangered plant and animal species and includes measures to prevent further deterioration in the conservation status of all native species. It also establishes a system for effective consultation with landholders and the community and places great importance on preventive action so that species which currently appear secure do not become endangered. Under the Act, a Scientific Advisory Committee is established to advise the Minister on technical matters.

Western Australia Under the Land Act 1933, the Governor may reserve land for public purposes. Notice of such reservations is published in the Gazette. The Governor may also proclaim reserved areas as Class A, B or C. Class A reserves remain dedicated for the purpose declared in the proclamation until revoked by Act of Parliament. Class B reserves may be revoked by the Governor by notice in the Gazette subject to the Minister for Lands presenting a report to both Houses of Parliament explaining the reasons for any revocation or alteration. Class C reserves may be revoked or altered by Gazettal or a Ministerial Notice to that effect.

The Land Act provides that the Governor may vest reserves in a private, semi-government or government authority for specific purposes. In Western Australia, most conservation areas are vested in the National Parks and Nature Conservation Authority as Class A, B or C

reserves for purposes such as conservation of flora and fauna, or national park. Reserves vested in the Authority for a purpose that includes flora and fauna are termed nature reserves.

Under the Conservation and Land Management Act 1984, the Governor may also reserve marine or inland waters as a marine nature reserve or marine park. Such areas are also vested in the National Parks and Nature Conservation Authority, and may be of Class A.

Under the Fisheries Act 1905, the Governor may reserve any part of Western Australian waters vested in the Crown and the land at any time covered by those waters as aquatic reserves.

International Activities The Unesco Convention concerning the Protection of the World Cultural and Natural Heritage (World Heritage Convention) was ratified by Australia in 1974. Eight properties have been inscribed on the World Heritage List as of July 1991. To date, twelve sites have been nominated as biosphere reserves under the Unesco Man and the Biosphere (MAB) Convention. Australia signed the Convention on Wetlands of International Importance Especially as Waterfowl Habitat (Ramsar Convention) in May 1974 and forty wetlands are included currently on the list.

At a regional level, Australia ratified the 1976 Convention on the Conservation of Nature in the South Pacific on 28 March 1990. Known as the Apia Convention, it entered into force during 1990. The Convention is coordinated by the South Pacific Commission and represents the first attempt within the region to cooperate on environmental matters. Among other measures, it encourages the creation of protected areas to preserve indigenous flora and fauna.

Australia is also party to the South Pacific Regional Environment Programme (SPREP) and the Convention for the Protection of the Natural Resources and Environment of the South Pacific Region, 1986 (SPREP Convention) was signed on 25 November 1986 and ratified on 9 September 1987. The Convention entered into force during August 1990. Article 14 calls upon the parties to take all appropriate measures to protect rare or fragile ecosystems and threatened or endangered flora and fauna through the establishment of protected areas and the regulation of activities likely to have an adverse effect on the species, ecosystems and biological processes being protected. However, as this provision only applies to the convention areas, which by definition is open ocean, it is most likely to assist with the establishment of marine reserves and the conservation of marine species.

Australia recognises also the importance of protecting migratory species. A bilateral agreement has been signed with Japan under which both countries agree to protect the birds that migrate between Australia and Japan and also the endangered birds in each country. Recently, a similar agreement between Australia and the People's Republic of China came into force.

Administration and Management

Australian Capital Territory All unleased open space throughout the ACT, including Namadgi National Park and several nature reserves, are managed by the ACT Parks and Conservation Service, ACT Administration. The ACT Parks and Conservation Consultative Committee advises the Minister and the Department on management.

Overall management objectives are:

— to maintain natural ecosystems and landscapes and to protect sites of prehistoric and historic significance for present and future generations of Australians;
— to provide opportunities for recreational, scientific and educational use and enjoyment of these resources consistent with their protection.

The Minister may make regulations to protect reserves, govern their use and the conduct of the public in them, define the powers of rangers and impose penalties.

External Territories The Director of the Australian National Park and Wildlife Service (ANPWS) is responsible for the areas proclaimed under the National Parks and Wildlife Conservation Act.

Norfolk Island National Park and Norfolk Island Botanic Garden are managed by the ANPWS. Advice is provided by the Norfolk Island National Park Advisory Committee.

The management of the Historic Shipwrecks Act has been delegated to the State or Territory museums in some instances.

New South Wales National parks, nature reserves, state recreation areas, historic sites and Aboriginal areas are managed by the New South Wales National Parks and Wildlife Service established under the Act and responsible to the Minister for the Environment.

Flora reserves are managed by the Forestry Commission of New South Wales under the Forestry Act 1916.

Aquatic reserves are managed by New South Wales Agriculture and Fisheries on behalf of the Minister for Agriculture and Rural Affairs, under the Fisheries and Oyster Farms (Amendment) Act 1979. As at 31 December 1988 there were seven aquatic reserves gazetted and managed within New South Wales. These are additional to numerous closures throughout all waters as protective measures for fish habitats and their fauna.

Northern Territory The Director of Conservation, a Deputy Director and staff are public servants employed for the purpose of carrying out the functions of the Commission. The Commission administers the Territory Parks and Wildlife Conservation Act as well as legislation relating to forestry, bushfires, soil conservation and environmental assessment.

Kakadu National Park and Uluru (Ayers Rock-Mount Olga) National Park are managed by the Australian National Parks and Wildlife Service.

The Territory's aquatic life reserves are managed by the Department of Primary Industry and Fisheries under the Fisheries Act 1988. It is intended that in 1989 the two existing reserves will be declared as fishery management areas and management plans for each will be prepared.

Queensland The National Parks and Wildlife Act provides for the appointment of a Director of National Parks and Wildlife to administer the Act and Regulations, the environmental park provisions of the Land Act, the Fauna Conservation Act and Regulations, The Native Plants Protection Act 1930, and the Queensland Marine Parks Act 1982.

The Forestry Act provides for the appointment of a Conservator of Forests to administer the Act and Regulations.

The Fisheries Act 1976-1989 is administered by the Queensland Department of Primary Industries with day-to-day management by the Fisheries Branch of that Department.

The Great Barrier Reef Marine Park Authority (GBRMPA) is responsible for the management of the marine park. To accomplish this, the Authority delegates some responsibilities for day-to-day management to other Commonwealth and Queensland agencies, in particular the Queensland National Parks and Wildlife Service (QNPWS). Most of the day-to-day management of the GBRMP is carried out by the QNPWS under special agreement with the Commonwealth.

The Great Barrier Reef Marine Park Authority, established under the Great Barrier Reef Marine Park Act, has three members who reflect the cooperative arrangements between the Commonwealth and Queensland. The Chairman is a Commonwealth nominee, one member is nominated by the Queensland Government and the third represents non-government interests. The administration of most islands in the region is the responsibility of the Queensland Government.

South Australia The Director of the National Parks and Wildlife Service is responsible to the Director-General of the Department of Environment and Planning. Both are responsible to the Minister for the management of reserves. The National Parks and Wildlife Service constitutes a division of the Department of Environment and Planning. All staff, including the Director, are public servants employed to carry out the functions of the National Parks and Wildlife Act.

The Department of Fisheries actively fosters public awareness of the existence of the aquatic reserves. This is being achieved by publication of relevant information, signposting of reserves, Fisheries Officers (enforcement staff) passing on information to people who may be fishing or intending to fish in or near the reserves, and by the Publicity Officer who spends part of his or her time conducting discussions with schools and interested groups. The Department of Fisheries works closely with the National Parks and Wildlife Service (NPWS) when establishing management arrangements for aquatic reserves. Some aquatic reserves abut terrestrial conservation reserves managed by the NPWS.

The Director of the Woods and Forests Department is responsible to the Minister of Forests for the management of all forest reserves. Softwood plantation forestry is the major activity conducted on land managed by the Woods and Forests Department. Forest reserves are under the direct control of district managers and occur in four major geographical regions of the State: South East, Central, Murraylands and Northern. The regions are under the control of two regional managers. Flora and fauna on all forest reserves in currently protected under the National Parks and Wildlife Act. Permits must be obtained from the South Australian National Parks and Wildlife Service then from the Woods and Forests Department before other than observational activities involving flora and fauna can be undertaken on forest reserves.

The Department of Environment and Planning administers the Historic Shipwrecks Act.

Tasmania State reserves, game reserves and conservation areas are administered by the Department of Lands, Parks and Wildlife under the National Parks and Wildlife Act. Whereas some conservation areas are administered directly by the Department, there are, in addition, approximately 35 conservation areas where other government authorities or the owners are the managing authorities. The Department also administers protected areas, state recreation areas, coastal reserves etc., reserved under the Crown Lands Act.

The Forestry Commission manages forest reserves under the Forestry Act.

The Tasmanian Sea Fisheries Development Authority is responsible for the management of Crayfish Point Marine Reserve.

Victoria Under administrative arrangements ordered by the Governor in Council in 1987, a Director of National Parks and Wildlife administers the National Parks Act and the Wildlife Act. Detailed management in the field of parks and wildlife is by regional managers of the Department of Conservation, Forests and Lands.

The Victorian Archaeological Survey administers the Historic Shipwrecks Act.

Western Australia Under the Conservation and Land Management Act 1984, the National Parks and Nature Conservation Authority is established to develop policies in preserving the natural environment and in encouraging public appreciation of nature. On the Authority's recommendation, the Minister for

Conservation and Land Management may classify land vested in the Authority for such purposes as prohibited or restricted access or, in the case of a national park, for the purpose of a wilderness area.

The Department of Conservation and Land Management manages areas vested in the Authority, provides and maintains facilities for the enjoyment of natural areas by the public, and undertakes other necessary functions for the management of national parks, nature reserves, marine parks and marine nature reserves. The Department administers the Conservation and Land Management Act, which covers the management of land with which the Department is involved, and the Wildlife Conservation Act 1950-80, which covers the protection of flora and fauna.

The Department of Fisheries manages areas of water for the conservation and protection of marine flora and fauna through declaration of aquatic reserves and proclamation of fishing zones and closed waters.

Systems Reviews The main land masses of Australia are situated between latitudes 11°S and 44°S and longitudes 113°E and 154°E, and separate the Indian and Pacific oceans. In addition, there are scattered islands in the surrounding seas and oceans, extending from the tropics to the subantarctic. The land area of Australia is almost 770 million hectares and the coastline is approximately 40,000km long. There is an extensive Continental Shelf and a 200 nautical mile Australian Fishing Zone has been declared. While a large proportion of interior Australia has an arid or semi-arid climate, the coastal fringes are generally well-watered, except along parts of the west and south coasts. The major mountain ranges along the east coast of the country attract high rainfall and, at the highest altitudes in the south-eastern corner (up to 2,160m), snow.

Because Australia covers a large area and encompasses a wide range of ecosystems, the task of identifying, classifying and naming all species of its biota is a major undertaking. The Commonwealth Government has established the Australian Biological Resources Study with the objective of documenting the totality of the Australian flora and fauna. The Study funds grants for research into taxonomy and distribution. As research on major taxonomic groups is completed, volumes of the *Flora of Australia* and the *Fauna of Australia* are published. The latter includes a general discussion of the Australian environment (Dyne and Walton, 1987).

Forests cover a very small part of Australia, about 5% of total land mass, restricted to coastal and near coastal areas with high rainfall and suitable soils. Of the forests that existed at the time of the arrival of the first white settlers, only one-third remains, and most of this loss has occurred since 1900. Rain forests are scattered unevenly along the eastern fringe of the continent and some pockets exist in Northern Territory. Tropical rain forests, however, are only found in Queensland. In New South Wales the rain forests range from subtropical, such as the

Border Ranges, to warm temperate rain forests such as are found on the Dorrigo Plateau, and cool temperate rain forests further south at Mount Boss and Barrington Tops. There are also some 4,000 sq. km of temperate rain forest in Tasmania, much of it in the north-western part of the state (AHC, 1985).

In New South Wales the main wetland areas are in the coastal zone and the drainage basins of the Murray-Darling and their tributaries; alpine bogs and fresh water lakes are a feature of the ranges and tablelands. In Victoria, the main wetland areas are the billabongs, anabranch systems and flood plain swamps of the River Murray, and the Gippsland Lakes and adjoining estuaries and coastal lagoons. South Australia, the driest state, has relatively few wetlands but those of the Murray River, the coastal saline lagoons, swamps in the south-eastern region of the state and those along Cooper Creek and Diamantina River are the most important. Most of South Australia's wetlands have either been drained years ago, or are within protected areas. In Western Australia, the main wetland areas are the seasonally flooded plains in the Kimberleys. In Queensland the greatest concentration of wetlands is in the Gulf Country, Cape York Peninsula and along the lower regions of the inland-draining western rivers. Loss of wetlands through draining is relatively small except in coastal areas and in the flood plains of the eastward draining rivers. Tasmania has a high percentage of wetland, alpine in particular; however, large areas have been modified. In Northern Territory vast areas of coastal plain are seasonally wet (Paijams, 1978). The arid lands of Australia occupy some three-quarters of the land surface of the continent, an area exceeded only by the North African and Middle-East region. Although they are often perceived as being of low biological value, they contain highly specialised and adapted organisms (AHC, 1985).

Australian alpine areas include unique elements, with two main regions occurring: the Great Dividing Range and the Central Highlands of Tasmania, and both areas are well represented in protected areas. Four principal environments are found in alpine areas, *viz.*, lowland, below 1,000m; montane, up to 1,370m; sub-alpine, from 1,370m to 1,800m; and alpine, over 1,800m. This wide range supports a rich flora and fauna (AHC, 1985).

The first national park in Australia, Royal National Park, south of Sydney, was established in 1879. The century that followed saw the progressive development of a comprehensive system of national parks and reserves. As at 31 December 1988 approximately 40,780,930ha, or about 5.3% of the total land surface, had been reserved under different categories. In addition, an area of 38,354,763ha was reserved as marine and estuarine protected areas (MEPAs), some of which is reserved as a part of, or contiguous with terrestrial reserves.

Addresses

Australian Capital Territory

Director, ACT Parks and Conservation Service, ACT Administration, GPO Box 158, CANBERRA ACT 2601 (Tel: 6 246 2308)

National

Director, Australian National Parks and Wildlife Service, GPO Box 636, CANBERRA ACT 2601 (Tel: 6 246 6211; FAX: 6 250 0339/0228; Tlx: 62971)

New South Wales

Director, National Parks and Wildlife Service, PO Box 1967, HURSTVILLE NSW 2220(Tel: 2 585 6300; FAX: 2 585 6555; Tlx: 26034)

Commissioner, Forestry Commission of New South Wales, GPO Box 2667, SYDNEY NSW 2001 (Tel: 2 234 1567)

Director, New South Wales Agriculture and Fisheries, PO Box K220, HAYMARKET NSW 2000 (Tel: 2 217 6666)

Northern Territory

Director, National Parks and Wildlife Service, PO Box 1260, DARWIN NT 0800 (Tel: 89 508211; Tlx: 85130)

Director, Conservation Commission of the Northern Territory, PO Box 496, PALMERSTON NT 0831 (Tel: 89 895511)

Secretary, Department of Primary Industry and Fisheries, GPO Box 990, DARWIN NT 0801 (Tel: 89 894211)

Queensland

Director, National Parks and Wildlife Service, PO Box 190, NORTH QUAY, Queensland 4002 (Tel: 7 227 4850; FAX: 7 221 5718)

Under Secretary, Department of Environment and Conservation, PO Box 155, NORTH QUAY QLD 4002 (Tel: 7 227 4111)

Conservator of Forests, Department of Forestry, GPO Box 944, BRISBANE QLD 4001 (Tel: 7 234 0111)

Director, Fisheries Branch, Department of Primary Industries, GPO Box 46, BRISBANE QLD 4001 (Tel: 7 227 4111)

Chairman, Great Barrier Reef Marine Park Authority, PO Box 1379, TOWNSVILLE QLD 4810 (Tel: 77 818811)

South Australia

Director, National Parks and Wildlife Service, PO Box 1782, ADELAIDE SA 5001 (Tel: 8 216 7777)

Director, Woods and Forests Department, GPO Box 1604, ADELAIDE SA 5001 (Tel: 8 226 9900)

Director, Department of Fisheries, GPO Box 1625, ADELAIDE SA 5001 (Tel: 8 226 0600)

Tasmania

Director, Department of Parks, Heritage and Wildlife, GPO Box 44A, HOBART TAS 7001 (Tel: 2 302336; FAX: 2 238765)

The Secretary, Department of Lands, Parks and Wildlife, GPO Box 44A, HOBART TAS 7001 (Tel: 2 308033)

Chief Commissioner, Forestry Commission of Tasmania, GPO Box 207B, HOBART TAS 7000

Victoria

Director, National Parks & Wildlife Division, PO Box 41, EAST MELBOURNE, VIC 3002 (Tel: 3 412 4011)

Director-General, Department of Conservation, Forests and Lands, PO Box 41, EAST MELBOURNE VIC 3002 (Tel: 3 651 4011)

Western Australia

Director, Parks, Recreation and Planning, Department of Conservation and Land Management, PO Box 104, COMO WA 6152 (Tel: 9 386 8811; FAX 9 386 1578; Tlx: 94585)

Director, Fisheries Department of Western Australia, 108 Adelaide Terrace, PERTH WA 6000 (Tel: 9 325 5988)

References

AHC (1985). Australia's National Estate. *Special Australian Heritage Publication Series* No. 1. Australian Heritage Commission. Australian Governmant Publishing Service, Canberra. 225 pp.

ANPWS (1989). Nature Conservation Reserves in Australia (1988). *Occasional Paper* No. 19. Australian National Parks and Wildlife Service, Canberra. 69 pp.

Dyne, G.R. and Walton, D.W. (Eds) (1987). *Fauna of Australia* General Articles. Volume 1A. Australian Government Publishing Service, Canberra. 339 pp.

Clark, M.R. and Dingwall, P.R. (1985). *Conservation of islands in the Southern Ocean*. A discussion document prepard for IUCN's Commission on National Parks and Protected Areas. IUCN, Gland, Switzerland and Cambridge, UK. 188 pp.

Paijams, K.I. (1978). *Feasibility Report on a National Wetland Inventory*. CSIRO Division of Land Use Research, Canberra. Pp. 2-3. (Unseen)

SPREP (1989). Australia. Paper presented at the Fourth South Pacific Conference on Nature Conservation and Protected Areas. Port Vila, Vanuatu, 4-12 September. 12 pp.

ANNEX
Definitions of protected areas designations, as legislated

Australian Capital Territory

National parks Extensive areas for the conservation of natural ecosystems, enjoyment and study of the natural environment and public recreation.

Nature reserves Land set aside primarily for conservation and also for compatible recreational use.

Reserves Land set aside for both conservation and compatible recreational use.

Waters managed in sympathy An area of water in Jervis Bay Territory managed for conservation, recreation and education.

External Territories

National parks Relatively large areas which contain representative samples of major natural regions, features or scenery of national or international significance where plant and animal species, geomorphological sites, and habitats are of special scientific, education, and recreational interest.

National nature reserves Nationally significant areas set aside primarily for nature conservation.

Marine parks Nationally significant areas set aside primarily for protection of the marine environment and its biota. The area is zoned to allow for various activities.

New South Wales

National parks Relatively large areas set aside for their features of predominantly unspoiled natural landscape, flora and fauna, permanently dedicated for public enjoyment, education and inspiration, and protected from all interference other than essential management practices, so that their natural attributes are preserved.

Nature reserves Areas of special scientific interest containing wildlife or natural phenomena where management practices aim at maximising the value of the area for scientific investigation and education purposes.

State recreation areas Permanent reservations in the form of large regional parks established to provide recreational opportunities in an outdoor environment.

Historic sites Areas preserved as the sites of building, objects, monuments or landscapes of national importance.

Aboriginal areas Places of significance to Aborigines or sites containing relics of Aboriginal culture.

Flora reserves Land set aside for the preservation of native flora and the natural environment.

Aquatic reserves Aquatic environments requiring protection and management to ensure future fisheries are maintained for all users. They can vary from very small units of two to three hectares, representing a particularly sensitive area, through to more extensive areas of significant habitats and faunal assemblages, covering 200-400ha. The larger units are generally declared such that the area covered provides a comprehensive management unit within which there is a zoning scheme to provide degrees of protection and reasonable levels of use which are consistent with the conservation values.

Northern territory

National parks Large areas of unspoiled landscape reserved for conservation, public enjoyment, education and inspiration.

Conservation reserves Areas set aside primarily for conservation of anthropological, natural or scientific values.

Nature parks Land reserved primarily for public recreation and enjoyment in a fairly natural environment.

Hunting reserves Areas set aside primarily for maintenance of game which can be harvested under permit.

Historical reserves Areas set aside primarily for their historical significance even though they may be used for other purposes such as recreation.

Other areas managed by the Conservation Commission Areas which have been purchased by the Northern Territory Government for conservation and recreation related purposes but which have not yet been declared parks. Areas declared as reserves under Section 103 of the Crown Lands Act and subject to Aboriginal land claim. Areas designated as protected areas under Section 22 of the Territory Parks and Wildlife Conservation Act but which are not owned by the Northern Territory Government.

Aquatic life reserves Areas established to conserve, enhance and protect the marine environment. These areas may be used for compatible purposes such as education or recreation.

Queensland

National parks Relatively large areas of natural landscape above high water mark with a high level of diversity of flora and fauna and which may be of historic interest. They are permanently dedicated for public enjoyment and education and protected from all interference other than essential management practices to ensure that the natural attributes are preserved.

Environmental parks Natural or near natural areas, less outstanding in size or natural attributes than national parks, totally protected for public enjoyment.

Fauna reserves Areas of land held permanently in their natural state. They are undisturbed, other than by naturally occurring processes, and are closed to the public.

Fauna refuges Land declared to preserve habitat and protect fauna.

Scientific purpose reserves Land declared for specific scientific studies relating to flora and fauna conservation.

Scientific areas Areas of native forest selected and managed to preserve significant natural ecosystems and to provide for their scientific investigation.

Feature protection areas Areas that may possess one or more of the following qualities:

— outstanding natural beauty;
— spectacular biological or geological features;
— unique or unusual qualities;
— representative examples of landscape of high scenic quality in readily accessible or visually sensitive locations;
— significant stimulating or aesthetic sensory qualities, other than visual.

Marine parks (Queensland State) Tidal lands and tidal waters declared for conservation, research, and public use and enjoyment.

Marine park (Great Barrier Reef Marine Park) Multiple-use marine area extending from low water mark declared for the protection of plant and animal life while allowing for reasonable use compatible with the conservation needs of the Reef. The GBRMP includes over 2,900 individual reefs, every one is included in zoning plans which permit particular activities. Regulations also cover the air space above the ocean and affords protection to bird species as well as marine organisms. For details, refer to zoning plans obtainable from GBRMPA or QNPWS offices.

Reserves for fisheries purposes:

Fish habitat reserves Are declared over tidal lands of significant importance to fisheries. This type of reserve is specifically designed to protect and preserve the lands of the reserve for a public benefit and to not allow private development. Normal access, boating and most forms of commercial and recreational fishing are not restricted. Revocation of a fish habitat reserve in whole or part is by an Order-in-Council following parliamentary assessment.

Wetland reserves Are declared over tidal habitat lands of importance to fisheries but which do not meet the criteria for declaration as fish habitat reserves. The reserves may also be declared as a buffer between a fish habitat reserve and some form of disturbance. This type of reserve is designed to preserve the area for a public enjoyment but to also accommodate under permit and guidelines private disturbances of minimal impact on the existing habitat features. Normal access, boating and fishing activities are not restricted. Revocation of a wetland reserve in whole or in part is by Order-in-Council.

Fish sanctuaries Are declared over tidal lands of significant importance to fish and crustaceans within certain phases of their life cycles or where these are particularly vulnerable to disturbance. A fish sanctuary is designed to protect the fish and marine products only in a given area, does not protect the habitat and does not allow any form of fishing. Normal access is not restricted. Revocation of a fish sanctuary in whole or part is by Order-in-Council.

Where habitat and fish and marine products are both to be protected, a fish habitat reserve or wetland reserve may be declared simultaneously with a fish sanctuary over the one area of tidal lands.

South Australia

National parks Protected areas "of national significance by reason of the wildlife or natural features of those lands". Generally they are contiguous areas of substantial size, preferably tens of thousands of hectares, with controlled provision for public visitation and enjoyment. They are reserves encompassing many natural values including scenic beauty, wildlife, history and inspiration to visitors.

Conservation parks Lands that should "be protected or preserved for the purpose of conserving any wildlife or the natural or historic features of those lands". Although these areas may contain all or some of the features represented in national parks, they tend to be subject to less visitation by the public, and subsequently are developed to a minimal extent.

Game reserves Lands which should "be preserved for the conservation of wildlife and management of game". Game reserves may be "set aside for the purpose of fishing or hunting". These areas have an important conservation role and may be declared open at prescribed times for strictly controlled hunting.

Recreation parks Lands that should "be conserved and managed for public recreation and enjoyment". These areas protect natural values, landscape, and historic sites but may also provide facilities for public recreation in a natural setting.

Regional reserves Lands that should be preserved for the purpose of conserving any wildlife or natural or historic features of the area while at the same time permitting the utilisation of the natural resources of the land.

Native forests reserves Areas of native vegetation that are a significant size and/or have important ecological features.

Aquatic reserves Any waters, including the bed of such waters, or land and waters proclaimed pursuant to the Fisheries Act 1982 to be an aquatic reserve. The objectives of the aquatic reserves are to:

- preserve examples of different marine habitats;
- protect endangered species;
- conserve nursery areas for economically important species;
- serve as sites to learn about marine ecosystems.

Tasmania

National parks Extensive areas for the conservation of natural ecosystems, enjoyment and study of the natural environment and public recreation/tourism.

State reserves Generally small reserves set aside for scenic and recreational reasons and/or to protect geological sites.

Nature reserves Areas set aside because of the significance for nature conservation. Public use is not encouraged where this might be detrimental, although provision may be made for appropriate tourism and recreational activities.

Aboriginal sites Areas containing relics of Aboriginal people or known to be of significance to them. Degree of public use will depend on needs of site for protection.

Historic sites Areas of significance in terms of European exploration, settlement or use, with encouragement of tourism and recreational use.

Game reserves Essentially the same as nature reserves except that specific provisions are made for hunting and the maintenance of game populations.

Conservation areas Large multiple use reserves set aside primarily to protect animals and their habitats and to provide for recreation and controlled use of resources.

Muttonbird reserves Reserves where special provision is made for private and commercial muttonbirding.

Forest reserves Land set aside within a state forest for recreational purposes; preservation or protection of any features of land of aesthetic, scientific, or other value; or for the preservation of the fauna or flora.

Marine reserve Area set aside for scientific research on commercially important species.

Victoria

National parks Certain Crown land characterised by its predominantly unspoilt landscape, and its flora, fauna or other features, should be reserved and protected permanently for the benefit of the public.

Other parks Certain areas of Crown land with landscape or other features of particular interest or suitability for the enjoyment, recreation and education of the public, of or in matters appertaining to the countryside, should be reserved permanently and made available for the benefit of the public. Other parks are generally classified into coastal parks, historic parks, state parks and parks. Other areas include some flora and fauna reserves, and reserves.

Wildlife reserves Areas of land that may:

- have ecological significance for a particular species or faunal association;
- contain the habitat of endangered species;
- contain favoured breeding grounds;
- have a high species diversity;
- be of educational, recreational or scientific interest;

- provide for the conservation of species, such as wild ducks, which may be harvested by the community.

Reference areas Areas of public land containing viable samples of one or more land types that are relatively undisturbed and that are reserved in perpetuity. These areas must not be tampered with, and natural processes should be allowed to continue undisturbed.

Flora and fauna reserves Areas set aside for the protection of particular groups of flora and fauna.

Marine parks Areas set aside for the purposes of conservation, recreation and scientific research.

Marine reserves Areas set aside for the purpose of conservation, and various exploitive and non-exploitive activities.

Western Australia

National parks Established to preserve for all times scenic beauty, wilderness, native wildlife, indigenous plant life and areas of scientific importance and to provide for the appreciation and enjoyment of those things by the public in such a manner and by such means as will leave them unimpaired for the future.

Marine parks Established to preserve a selection of marine and estuarine habitats for recreation and nature conservation purposes.

Conservation/recreation reserves Land reserved for recreation and conservation purposes.

Nature reserves Land reserved for the conservation of flora and fauna.

Aquatic reserves Set aside for preservation, culture or propagation of marine or freshwater fauna and flora.

SUMMARY OF PROTECTED AREAS

Map[†] ref.	National/international designation Name of area	IUCN management category	Area (ha)	Year notified
	CAPITAL TERRITORY			
	National Park			
1	Namadgi	II	94,000	
	Nature Reserves			
2	Gudgenby	II	62,000	1979
3	Jervis Bay	II	4,921	1971
4	Tidbinbilla	II	5,500	1964
	Other areas			
5	Lanyon	II	1,300	
6	Other reserves	II	6,000	
	NEW SOUTH WALES			
	National Parks			
1	Bald Rock	II	5,451	1969
2	Barrington Tops	II	39,121	1969
3	Ben Boyd	II	9,455	1971
4	Blue Mountains	II	245,716	1959
5	Boonoo Boonoo	II	2,692	
6	Border Ranges	II	31,368	1979
7	Bouddi	II	1,167	1937
8	Brisbane Water	II	11,369	1959
9	Broadwater	II	3,737	1974
10	Budawang	II	16,102	
11	Budderoo	II	5,700	
12	Bundjalung	II	17,545	
13	Cathedral Rock	II	6,529	
14	Cocoparra	II	8,358	1969
15	Conimbla	II	7,590	
16	Crowdy Bay	II	8,005	1972
17	Deua	II	81,625	
18	Dharug	II	14,834	1967
19	Dorrigo	II	7,885	1967
20	Gibraltar Range	II	17,273	1963
21	Goulburn River	II	67,897	
22	Guy Fawkes River	II	35,630	1972
23	Hat Head	II	6,445	1973
24	Heathcote	II	2,251	1963
25	Kanangra-Boyd	II	68,276	1969
26	Kingchega	II	44,182	1969
27	Kings Plains	II	3,140	
28	Kosciusko	II	646,911	1944
29	Ku-Ring-Gai Chase	II	14,614	1894
30	Macquarie Pass	II	1,064	1969
31	Mallee Cliffs	II	57,969	
32	Marramarra	II	11,727	1979
33	Mimosa Rocks	II	5,181	
34	Mootwingee	II	68,912	
35	Morton	II	154,195	1938
36	Mount Imlay	II	3,808	1972
37	Mount Kaputar	II	36,817	1960
38	Mount Warning	II	2,210	
39	Mungo	II	27,847	
40	Murramarang	II	1,609	

Map[†] ref.	National/international designation Name of area	IUCN management category	Area (ha)	Year notified
41	Myall Lakes	II	31,493	1972
42	Nalbaugh	II	3,764	1972
43	Nangar	II	3,492	
44	New England	II	29,881	1931
45	Nightcap	II	4,945	1983
46	Nungatta	II	6,100	
47	Nymboida	II	1,368	
48	Oxley Wild Rivers	II	38,890	
49	Royal	II	15,020	1879
50	Sturt	II	310,634	1972
51	Tarlo River	II	6,759	
52	Wadbilliga	II	76,675	
53	Wallaga Lake	II	1,237	1972
54	Warrabah	II	2,635	
55	Warrumbungle	II	20,914	1961
56	Washpool	II	27,715	1983
57	Weddin Mountains	II	8,361	1971
58	Werrikimbe	II	35,178	1975
59	Willandra	II	19,386	1972
60	Woko	II	8,285	
61	Wollemi	II	487,289	1979
62	Yengo	II	140,000	
63	Yuraygir	II	18,285	1973
	State Recreation Areas			
64	Booti Booti	V	1,488	
65	Bournda	V	2,305	
66	Bungonia	V	3,836	
67	Burrendong	V	1,227	
68	Burrinjuck	V	1,714	
69	Davidson Park	V	1,200	
70	Illawarra Escarpment	V	1,266	
71	Munmorah	V	1,007	
72	Parr	V	38,000	
73	Wyangala	V	2,034	
	Nature Reserves			
74	Avisford	IV	2,437	
75	Banyabba	IV	12,560	1969
76	Barren Grounds	IV	2,024	1956
77	Bimberi	IV	7,100	
78	Binnaway	IV	3,699	
79	Bournda	IV	5,831	1972
80	Burrinjuck	IV	1,300	
81	Camerons Gorge	IV	1,280	
82	Cocoparra	IV	4,647	1963
83	Coolbaggie	IV	1,793	1963
84	Copperhannia	IV	3,494	1972
85	Coturaundee	IV	6,688	
86	Curumbenya	IV	9,380	1964
87	Dananbilla	IV	1,855	
88	Egan Peakes	IV	2,145	1972
89	Georges Creek	IV	1,190	1967
90	Guy Fawkes River	IV	1,534	1970
91	Ingalba	IV	4,012	1970
92	Ironbark	IV	1,604	
93	Kajuligah	IV	13,660	
94	Kemendok	IV	1,043	

Map[†] ref.	National/international designation Name of area	IUCN management category	Area (ha)	Year notified
95	Kooragang	IV	2,926	1983
96	Lake Innes	IV	3,510	
97	Limeburners Creek	IV	8,892	1971
98	Limpinwood	IV	2,443	1963
99	Macquarie Marshes	IV	18,211	1971
100	Mann River	IV	5,640	
101	Manobalai	IV	3,733	1967
102	Mount Hyland	IV	1,636	1984
103	Mount Neville	IV	2,666	
104	Mount Seaview	IV	1,704	1974
105	Mundoonen	IV	1,374	1970
106	Munghorn Gap	IV	5,934	1961
107	Muogamarra	IV	2,274	1960
108	Nadgee	IV	17,116	1957
109	Narran Lake	IV	4,527	
110	Nearie Lake	IV	4,347	1973
111	Nocoleche	IV	74,000	1979
112	Nombinnie	IV	70,000	
113	Pantoneys Crown	IV	3,230	1977
114	Pilliga	IV	69,595	1968
115	Razorback	IV	2,595	
116	Round Hill	IV	13,630	1960
117	Rowleys Creek Gulf	IV	1,659	1962
118	Scabby Range	IV	3,449	
119	Severn River	IV	1,947	1968
120	Sherwood	IV	2,444	1966
121	The Basin	IV	2,318	1964
122	The Hole Creek	IV	5,587	
123	Tinderry	IV	11,559	
124	Tollingo	IV	3,232	
125	Ulandra	IV	3,931	
126	Wallabadah	IV	1,132	1971
127	Watsons Creek	IV	1,260	
128	Winburndale	IV	9,396	1967
129	Wingen Maid	IV	1,077	
130	Woggoon	IV	6,565	1972
131	Yanga	IV	1,773	1972
132	Yathong	IV	107,241	1971
	Flora Reserves			
133	Banda Banda	IV	1,400	1984
134	Gilgai	IV	2,400	
135	Moira Lakes	IV	1,435	
136	Mt Dromedary	IV	1,259	
137	Nunnock Swamp	IV	1,020	
138	The Castles	IV	2,360	
139	Toolum Scrub	IV	1,665	
140	Waihou	IV	1,800	
	NORTHERN TERRITORY			
	National Parks			
1	Finke Gorge	II	45,856	1967
2	Gurig	II	220,700	
3	Kakadu	II	1,755,200	1979
4	Katherine Gorge (Nitmiluk)	II	180,352	1963
5	Keep River	II	59,700	
6	Ormiston Gorge and Pound	II	4,655	
7	Simpsons Gap	II	30,950	1970

Map[†] ref.	National/international designation Name of area	IUCN management category	Area (ha)	Year notified
8	Uluru (Ayers Rock-Mount Olga)	II	132,538	1974
	Conservation Reserves			
9	Connells Lagoon	IV	25,890	
10	Devils Marbles	IV	1,828	
11	Fogg Dam	IV	1,569	
12	Mac Clark (Acacia place)	IV	3,042	
	Nature Parks			
13	Cutta Cutta Caves	V	1,499	
14	Douglas Hot Springs	V	3,107	
15	Ellery Creek Big Hole	V	1,766	
16	Red Bank	V	1,295	
17	Ruby Gap	V	9,257	
18	Trephina Gorge	V	1,771	
	Marine Park			
19	Cobourg	II	229,000	1983
	Hunting Reserve			
20	Howard Springs	VII	1,605	
	Other areas			
21	Alice Springs Waterfowl Protected Area	VI	1,725	
22	Arnhem Highway Protected Area	VI	30,400	
23	Black Jungle & Lambells Lagoon	VI	4,052	
24	Cape Hotham Conservation Reserve	VI	12,900	
25	Cape Hotham Forestry Reserve	VI	10,830	
26	Casuarina Coastal Reserve	VI	1,180	
27	Dulcie Ranges	VI	13,700	
28	Gosse Bluff	VI	4,759	
29	Gregory National Park (proposed)	VI	1,104,340	
30	Indian Island Forest Reserve	VI	2,648	
31	Junction Reserve	VI	19,930	
32	Keep River Extensions	VI	31,340	
33	Kings Canyon	VI	71,720	
34	Litchfield Park	VI	65,700	
35	Longreach Waterhole	VI	8,000	
36	Marrakai Flora & Fauna Reserve	VI	35,000	
37	Mary River Conservation Reserve	VI	27,000	
38	Mary River Crossing	VI	2,590	
39	Melacca Swamp	VI	6,367	
40	North Island	VI	5,500	
41	Point Stuart	VI	6,279	
42	Rainbow Valley	VI	2,483	
43	Tennant Creek Telegraph Station	VI	1,797	
44	Upper Roper River	VI	13,840	
45	Vernon Islands	VI	7,659	
46	West MacDonalds	VI	16,890	
47	Wildman River	VI	22,940	
	QUEENSLAND			
	National Parks			
1	Archer Bend	II	166,000	1977
2	Barron Gorge	II	2,784	1940
3	Bellenden Ker	II	31,000	1921
4	Blackdown Tableland	II	23,800	1979
5	Bladensburg	II	33,700	
6	Bowling Green Bay	II	55,300	1950

Map† ref.	National/international designation Name of area	IUCN management category	Area (ha)	Year notified
7	Bunya Mountains	II	11,700	1908
8	Burrum River	II	1,618	
9	Byfield	II	4,090	
10	Camooweal Caves	II	13,800	
11	Cania Gorge	II	2,000	
12	Cape Melville	II	36,000	1977
13	Cape Palmerston	II	7,160	1977
14	Cape Tribulation	II	16,965	
15	Cape Upstart	II	5,620	1967
16	Capricorn Coast	II	1,067	1938
17	Capricorn-Bunker Group	II	167	1937
18	Carnarvon	II	223,000	1979
19	Castle Tower	II	4,980	1975
20	Cedar Bay	II	5,650	1967
21	Chillagoe-Mungana Caves	II	1,932	1940
22	Conondale (2)	II	1,740	1931
23	Conway	II	23,800	
24	Cooloola	II	40,900	1975
25	D'Aguilar	II	1,328	1938
26	Dagmar Range	II	1,585	
27	Daintree	II	56,450	1977
28	Deepwater	II	4,090	
29	Dipperu	II	11,100	1967
30	Dunk Island	II	730	1936
31	Edmund Kennedy	II	6,200	1977
32	Ella Bay	II	3,430	
33	Endeavour River	II	1,840	1970
34	Epping Forest	II	3,160	1971
35	Eubenangee Swamp	II	1,520	
36	Eungella	II	50,800	1950
37	Eurimbula	II	7,830	1977
38	Flinders Group	II	2,962	
39	Forty Mile Scrub	II	4,500	1970
40	Girraween	II	11,399	1932
41	Gloucester and Middle Islands	II	3,970	
42	Graham Range	II	2,930	
43	Great Basalt Wall	II	30,500	
44	Great Sandy	II	52,400	1977
45	Haslewood Island Group	II	1,210	
46	Herbert River Falls	II	2,428	
47	Herbert River Gorge	II	18,900	1963
48	Hinchinbrook Channel	II	5,585	
49	Hinchinbrook Island	II	39,350	1968
50	Hook Island	II	5,180	
51	Hull River	II	1,250	1968
52	Iron Range	II	34,600	1977
53	Isla Gorge	II	7,800	1964
54	Jardine River	II	235,000	1977
55	Jourama Falls	II	1,070	
56	Kroombit Tops	II	2,360	
57	Lakefield	II	537,000	1979
58	Lamington	II	20,200	1915
59	Lawn Hill	II	12,200	
60	Littabella	II	2,420	
61	Lizard Island	II	990	1939
62	Lonesome	II	3,367	
63	Magnetic Island	II	2,720	1954

Map[†] ref.	National/international designation Name of area	IUCN management category	Area (ha)	Year notified
64	Maiala	II	1,140	
65	Main Range	II	11,500	
66	Mazeppa	II	4,126	1972
67	Mitchell & Alice Rivers	II	37,100	1977
68	Moreton Island	II	15,400	
69	Mount Aberdeen (1)	II	1,667	
70	Mount Aberdeen (2)	II	1,242	1952
71	Mount Barney	II	11,400	1947
72	Mount Blackwood	II	1,060	
73	Mount Maria	II	3,430	1969
74	Mount Mistake	II	5,560	
75	Mount Spec	II	7,224	1952
76	Mount Tempest	II	9,360	1966
77	Mount Walsh	II	2,987	1947
78	Northumberland Group	II	3,702	1937
79	Orpheus Island	II	1,300	1960
80	Palmerston	II	14,200	1941
81	Porcupine Gorge	II	2,938	1970
82	Possession Island	II	510	1977
83	Pumicestone	II	1,930	
84	Robinson Gorge	II	77,300	1953
85	Rokeby	II	291,000	
86	Royal Arch Caves	II	1,514	
87	Scawfell	II	1,090	
88	Shaw Island	II	1,659	
89	Simpson Desert	II	555,000	1967
90	Snake Range	II	1,209	1972
91	South Island	II	1,619	
92	Southwood	II	7,120	1970
93	Springbrook	II	2,159	1956
94	Staaten River	II	470,000	1977
95	Starcke	II	7,960	1977
96	Sundown	II	11,200	1941
97	Upper Tully	II	1,586	1962
98	West Hill Island	II	398	1938
99	Whitsunday Island	II	10,930	1977
100	Wild Duck Island	II	207	
101	Woodgate	II	5,490	1974
102	Woody Island	II	660	
103	Yamanie Falls	II	9,712	
	Scientific Reserves			
104	Mariala	IV	27,300	
105	Palm Grove	IV	2,550	
106	Taunton	IV	11,470	1980
	Scientific Areas			
107	No 33 (Hurdle Gully Scrub)	IV	1,674	
108	No 39 (North Bargoo Creek)	IV	1,000	
109	No 40 (West Spencer Creek)	IV	3,700	
110	No 46 (Platypus Creek)	IV	1,200	
	Faunal Reserve			
111	Palmgrove	I	25,617	1967
	Faunal Refuge			
112	Taunton	IV	5,346	

Map[†] ref.	*National/international designation* Name of area	IUCN management category	Area (ha)	Year notified
	Marine Parks			
113	Great Barrier Reef	VII	34,500,000	1979
114	Green Island Reef	II	3,000	1974
115	National Park sections within GBRMP	II	4,700,314	1979
116	Pumicestone Passage	II	8,000	
	Environmental Parks			
117	Goneaway	V	24,800	1974
118	Lake Broadwater	V	1,220	
119	Mount Archer	V	1,990	
120	Mount Zamia	V	1,140	
121	Townsville Town Common (1)	V	2,920	
122	Wilandspey	V	5,200	1977
	SOUTH AUSTRALIA			
	National Parks			
1	Canunda	II	9,358	1966
2	Coffin Bay	II	30,380	
3	Coorong	II	39,904	1966
4	Flinders Chase	II	73,662	1919
5	Flinders Ranges	II	94,908	1970
6	Gammon Ranges	II	128,228	1970
7	Innes	II	9,141	1970
8	Lake Eyre	II	1,228,000	
9	Lincoln	II	29,060	1941
10	Mount Remarkable	II	8,649	1965
11	Nullarbor	II	231,900	
12	Witjira	II	776,900	
	Game Reserves			
13	Bool Lagoon	VII	2,690	
14	Coorong	VII	6,841	1966
15	Katarapko	VII	8,905	
16	Loch Luna	VII	1,905	
17	Moorook	VII	1,236	
	Aquatic Reserves			
18	American River	IV	1,525	1971
19	Barker Inlet-St Kilda	IV	2,055	1973
20	Blanche Harbour-Douglas Bank	IV	3,160	1980
21	Seal Beach-Bales Bay	IV	1,140	1971
22	Whyalla-Cowleds Landing	IV	3,230	1980
23	Yatala Harbour	IV	1,426	1980
	Native Forest Reserves			
24	Headquarters (Penola Forest Reserve)	IV	1,700	
25	Mount Gawler North (Mt Crawford Forest Reserve)	IV	1,013	
26	Murtho Forest Reserve	IV	1,910	
27	The Bluff Range (Wirrabara Forest Reserve)	IV	2,633	
	Conservation Parks			
28	Bakara	II	1,022	
29	Barwell	II	4,561	
30	Bascombe Well	II	32,200	1970
31	Big Heath	II	2,351	1964
32	Billiatt	II	59,148	1963
33	Brookfield	II	6,333	
34	Calpatanna Waterhole	II	3,630	1974
35	Cape Gantheaume	II	21,254	1971

Map† ref.	National/international designation Name of area	IUCN management category	Area (ha)	Year notified
36	Carcuma	II	2,881	1969
37	Clinton	II	1,964	1970
38	Cocata	II	6,876	
39	Danggali	II	253,480	1976
40	Deep Creek	II	4,184	1971
41	Dudley	II	1,122	1970
42	Dutchmans Stern	II	3,532	
43	Elliot Price	II	64,570	1967
44	Fairview	II	1,398	1960
45	Franklin Harbor	II	1,334	1976
46	Gum Lagoon	II	6,589	1970
47	Hambidge	II	37,992	1962
48	Hincks	II	66,285	1962
49	Isles of St Francis	II	1,320	1967
50	Karte	II	3,565	1969
51	Kellidie Bay	II	1,780	1962
52	Kelly Hill	II	7,374	1971
53	Kulliparu	II	13,536	
54	Lake Gilles	II	45,114	1971
55	Lathami	II	1,190	
56	Little Dip	II	1,977	1975
57	Martin Washpool	II	1,883	
58	Messent	II	12,246	1964
59	Mount Boothby	II	4,045	1967
60	Mount Rescue	II	28,385	1962
61	Mount Scott	II	1,238	1972
62	Mount Shaugh	II	3,460	1971
63	Munyaroo	II	12,385	1977
64	Ngarkat	II	207,941	1979
65	Nuyts Archipelago	II	5,420	1967
66	Pandappa	II	1,057	1973
67	Peebinga	II	3,371	1962
68	Pinkawillinie	II	127,164	1970
69	Pooginook	II	2,852	1970
70	Scorpion Springs	II	30,366	1970
71	Simpson Desert	II	692,680	1967
72	Swan Reach	II	2,016	1970
73	Telowie Gorge	II	1,946	1970
74	Unnamed	II	2,132,600	1970
75	Venus Bay	II	1,460	1976
76	Warrenben	II	4,061	1969
77	Western River	II	2,364	1971
78	Whyalla	II	1,011	
79	Winnanowie	II	4,318	1990
80	Yumbarra	II	106,189	1968
	Regional Reserves			
81	Innamincka	V	1,381,765	
82	Simpson Desert	V	2,964,200	
	Recreation Parks			
83	Onkaparinga River	V	1,680	
84	Para Wirra	V	1,409	
	TASMANIA			
	National Parks			
1	Asbestos Range	II	4,281	1976
2	Ben Lomond	II	16,527	1947

Map[†] ref.	*National/international designation* Name of area	IUCN management category	Area (ha)	Year notified
3	Cradle Mountain-Lake St Clair	II	161,108	1922
4	Douglas-Apsley	II	16,080	1989
5	Franklin-Lower Gordon Wild Rivers	II	440,120	1981
6	Freycinet	II	10,010	1916
7	Hartz Mountains	II	7,226	1939
8	Maria Island	II	9,672	1964
9	Mount Field	II	16,265	1916
10	Mount William	II	13,899	1973
11	Rocky Cape	II	3,064	1967
12	Southwest	II	605,000	1968
13	Strzelecki	II	4,215	1967
14	Walls of Jerusalem	II	51,800	1981
	Nature Reserves			
15	Betsey Island	IV	181	1928
16	Big Green Island	IV	270	1983
17	Chappell Islands	IV	1,350	1975
18	East Kangaroo Island	IV	200	1984
19	Lavinia	IV	6,800	1975
20	Lime Bay	IV	1,310	1976
21	Three Hummock Island	IV	7,284	1977
	Wildlife Sanctuaries			
22	Ben Lomond	VI	2,665	1946
23	Southport Lagoon	VI	3,556	1976
	Game Reserves			
24	Bruny Island Neck	IV	1,450	1979
25	New Year Island	IV	112	1981
	Muttonbird Reserves			
26	Babel Island	VII	445	1957
27	Hunter Island	VII	7,365	1957
28	Outer and Inner Sister Islands	VII	1,012	
	Conservation Areas			
29	Central Plateau	VII	23,250	1978
30	Derwent River Wildlife Sanctuary	VII	1,568	1941
31	Egg Islands	VII	128	1978
32	Logan Lagoon	VII	2,256	1968
33	Southport Lagoon Wildlife Santuary	VII	1,068	
34	Southwest	VI	777,151	1966
35	Tamar River	VII	4,600	1978
	State Reserves			
36	Alum Cliffs	III	1,540	1979
37	Cape Pillar	II	3,200	1974
38	Cape Raoul	II	2,089	1978
39	Labillardiere	II	2,332	1975
40	Pieman River	II	3,314	1940
	VICTORIA			
	National Parks			
1	Alfred	II	3,050	1925
2	Baw Baw	II	13,300	1979
3	Bogong	II	81,200	1981
4	Brisbane Ranges	II	7,517	1975
5	Burrowa-Pine Mountain	II	17,600	1978
6	Cobberas-Tingaringy	II	116,600	

Map[†] ref.	National/international designation Name of area	IUCN management category	Area (ha)	Year notified
7	Coopracambra	II	35,100	1979
8	Croajingolong	II	87,500	1979
9	Dandenong Ranges	II	1,920	
10	Errinundra	II	25,100	
11	Fraser	II	3,750	1957
12	Grampians	II	167,000	
13	Hattah-Kulkyne	II	48,000	1960
14	Kinglake	II	11,430	1928
15	Lind	II	1,365	1926
16	Little Desert	II	132,000	1968
17	Lower Glenelg	II	27,300	1969
18	Mitchell River	II	11,900	
19	Mount Buffalo	II	31,000	1898
20	Mount Eccles	II	5,470	
21	Mount Richmond	II	1,733	1960
22	Otway	II	12,750	1981
23	Point Nepean	II	2,200	
24	Port Campbell	II	1,750	1965
25	Snowy River	II	95,400	1979
26	Tarra-Bulga	II	1,230	
27	The Lakes	II	2,390	1927
28	Wilsons Promontory	II	49,000	1898
29	Wonnangatta-Moroka	II	107,000	
30	Wyperfeld	II	100,000	1909
	Reference Areas			
31	Benedore River (Croajingolong NP)	I	1,200	
32	Bungil (Mt Lawson SP)	I	1,300	
33	Burnside	I	1,190	
34	Cudgewa Creek	I	1,130	
35	Disappointment	I	1,090	
36	Little Desert West (Little Desert NP)	I	2,240	
37	Roseneath	I	2,226	
38	Ryan Creek	I	1,570	
39	Seal Creek	I	1,000	
40	Sunset	I	8,400	
41	Toorour	I	1,750	
	Wildlife Reserves			
42	Bronzewing	IV	11,200	
43	Dowd Morass	IV	1,501	
44	Ewing Morass	IV	7,300	
45	Jack Smith Lake	IV	2,781	
46	Kings Billabong	IV	2,166	
47	Koorangie	IV	2,853	
48	Lake Coleman	IV	2,055	
49	Lake Connewarre	IV	3,300	
50	Lake Murdeduke	IV	1,500	
51	Lake Timboram	IV	2,060	
52	Nooramunga	IV	9,996	1964
53	Red Bluff	IV	8,800	
54	Reedy Lake	IV	1,400	
55	Rocky Range	IV	4,453	
56	Tooloy-Lake Mundi	IV	2,416	
57	Wandella Forest	IV	1,060	
58	Wandown	IV	1,591	
59	Wathe	IV	5,763	
60	Westernport	IV	1,650	1979

Map[†] ref.	National/international designation Name of area	IUCN management category	Area (ha)	Year notified
	Flora and Fauna Reserves			
61	Bull Beef Creek	IV	1,490	
62	Crinoline Creek	IV	1,550	
63	Deep Lead	IV	1,240	
64	Fryers Ridge	IV	2,000	
65	Jilpanger	IV	8,290	
66	Lake Timboram	IV	2,060	
67	Lansborough	IV	1,800	
68	Mount Bolangum	IV	2,930	
69	Mullungdung	IV	1,520	1979
70	Providence Ponds	IV	1,650	
71	Stradbroke	IV	2,660	
72	Sweetwater Creek	IV	1,240	
73	Timberoo	IV	1,230	
74	Turtons Track	IV	1,525	
75	Wilkin	IV	3,600	
76	Wychitella	V	3,780	
77	Yarrara	IV	2,200	
	Coastal Parks			
78	Discovery Bay	V	8,590	1979
79	Gippsland Lakes	V	17,200	1979
	Marine and Coastal Parks			
80	Corner Inlet	V	18,000	
81	Nooramunga	V	15,000	
82	Shallow Inlet	V	2,000	
	Parks			
83	Reef Hills	V	2,040	
84	Tyers	V	1,810	
	Historic Park			
85	Beechworth HP	V	1,130	
	State Parks			
86	Angahook-Lorne	V	21,000	
87	Barmah	V	7,900	
88	Black Range	V	11,700	
89	Carlisle	V	5,600	
90	Cathedral Range	V	3,577	1979
91	Chiltern Park	V	4,255	
92	Eildon	V	24,000	
93	French Island	V	8,300	1979
94	Holey Plains	V	10,576	1978
95	Kamarooka	V	6,300	
96	Kara Kara	V	3,840	
97	Kooyoora	V	3,593	
98	Lake Albacutya	V	10,700	
99	Lerderberg	V	13,340	
100	Moondarra	V	6,470	
101	Mount Arapiles-Tooan	V	5,050	
102	Mount Lawson	V	13,150	
103	Mount Napier	V	2,800	
104	Mount Samaria	V	7,600	1979
105	Mount Worth	V	1,040	
106	Pink Lakes	V	50,700	1979
107	Terrick Terrick	V	2,493	
108	Wabonga Plateau	V	21,200	1980

Map[†] ref.	National/international designation Name of area	IUCN management category	Area (ha)	Year notified
109	Warby Ranges	V	3,540	
110	Whipstick	V	2,300	
	Other areas			
111	Avon Wilderness	V	40,000	
112	Big Desert Wilderness	V	113,500	1979
113	Langi Ghiran	V	2,695	
114	Murray-Kulkyne	V	1,550	
	WESTERN AUSTRALIA			
	National Parks			
1	Alexander Morrison	II	8,501	1970
2	Avon Valley	II	4,366	1970
3	Badgingarra	II	13,121	1973
4	Beedelup	II	1,530	
5	Boorabbin	II	26,000	1977
6	Bungle Bungle	II	208,723	
7	Cape Arid	II	279,415	1969
8	Cape Le Grand	II	31,390	1948
9	Cape Range	II	50,581	1965
10	Collier Range	II	277,841	
11	D'Entrecasteaux	II	57,722	1967
12	Drovers Cave	II	2,681	1972
13	Drysdale River	II	435,906	1974
14	Eucla	II	3,342	1979
15	Fitzgerald River	II	242,804	1954
16	Frank Hann	II	61,420	1970
17	Geikie Gorge	II	3,136	1967
18	Goongarrie	II	60,356	
19	Hamersley Range	II	617,602	1969
20	Hassell	II	1,265	1971
21	Hidden Valley	II	1,817	
22	John Forrest	II	1,508	1957
23	Kalbarri	II	186,071	1963
24	Leeuwin-Naturaliste	II	16,172	1970
25	Millstream-Chichester	II	199,730	1969
26	Moore River	II	17,543	1969
27	Mount Frankland	II	30,830	
28	Nambung	II	17,491	1968
29	Neerabup	II	1,082	1945
30	Peak Charles	II	39,959	1979
31	Pemberton	II	3,141	1977
32	Porongurup	II	2,572	1957
33	Rudall River	II	1,569,459	1977
34	Scott	II	3,273	1959
35	Shannon	II	52,598	
36	Sir James Mitchell	II	1,087	1969
37	Stirling Range	II	115,661	1913
38	Stockyard Gully	II	1,406	
39	Stokes	II	9,509	1974
40	Tathra	II	4,322	1970
41	Torndirrup	II	3,919	1968
42	Tuart	II	1,785	
43	Walpole-Nornalup	II	15,877	1957
44	Walyunga	II	1,812	1972
45	Warren	II	1,355	
46	Watheroo	II	44,512	1969
47	West Cape Howe	II	3,517	

Map[†] ref.	National/international designation Name of area	IUCN management category	Area (ha)	Year notified
48	William Bay	II	1,739	1971
49	Windjana Gorge	II	2,134	1971
50	Wolf Creek Crater	II	1,460	
51	Yalgorup	II	11,819	1968
52	Yanchep	II	2,799	1905
	Nature Reserves			
53	25 Mile Brook	IV	1,020	
54	Arthur River	IV	3,234	
55	Ascot	IV	1,861	
56	Austin Bay Reserve	IV	1,305	
57	Bakers Junction	IV	1,090	
58	Barlee Range	IV	104,544	1963
59	Barrow Island	IV	23,483	1910
60	Basil Road	IV	1,162	
61	Bendering	IV	1,602	1970
62	Bernier and Dorre Islands	IV	9,720	1970
63	Beynon	IV	3,323	
64	Billyacatting Hill	IV	2,063	
65	Boyagin	IV	4,804	1960
66	Buntine	IV	3,289	1963
67	Burngup	IV	1,289	1970
68	Burrma Road	IV	6,890	
69	Cairlocup	IV	1,577	
70	Camel Lake	IV	3,215	1962
71	Capamauro	IV	3,588	
72	Carlyarn	IV	2,723	1974
73	Cheadanup	IV	6,813	
74	Chiddarcooping	IV	5,217	
75	Chinocup	IV	19,821	
76	Coblinine	IV	4,033	1958
77	Cooloomia	IV	50,350	
78	Corackerup	IV	4,334	1970
79	Corneecup	IV	1,952	
80	Dobaderry	IV	4,005	
81	Dolphin Island	IV	3,203	
82	Dongolocking	IV	1,312	
83	Dragon Rocks	IV	32,219	
84	Duladgin	IV	1,619	
85	Dumbleyung Lake	IV	3,958	
86	Dunn Rock	IV	24,819	
87	Durokoppin	IV	1,030	1971
88	Gibson Desert	IV	1,859,286	1977
89	Gingilup Swamps	IV	2,807	
90	Goodlands	IV	1,349	
91	Great Victoria Desert	IV	2,495,777	1970
92	Haddleton	IV	1,161	
93	Harris	IV	3,610	
94	Jebarijup Lake	IV	1,016	1962
95	Jilbadji	IV	208,866	1972
96	Joverdine	IV	1,754	
97	Kathleen	IV	1,190	
98	Kondinin Salt Marsh	IV	2,208	
99	Kooljerrenup	IV	1,050	
100	Kundip	IV	2,170	
101	Lacepede Islands	IV	160	
102	Lake Ace	IV	2,392	
103	Lake Bryde	IV	1,454	

Map[†] ref.	National/international designation Name of area	IUCN management category	Area (ha)	Year notified
104	Lake Campion	IV	10,752	
105	Lake Cronin	IV	1,016	
106	Lake Gounter	IV	3,328	1955
107	Lake Hurlstone	IV	5,002	
108	Lake Liddelow	IV	1,133	
109	Lake Logue	IV	4,835	
110	Lake Magenta	IV	94,170	1958
111	Lake Muir	IV	11,310	
112	Lake Shaster	IV	10,756	
113	Lake Varley	IV	2,197	1970
114	Lakeland	IV	3,315	
115	Mill Brook	IV	1,484	
116	Moondyne	IV	1,991	
117	Mt Manypeaks	IV	1,328	
118	Mungaroona Range	IV	105,842	1972
119	Namming	IV	5,432	
120	Nilgen	IV	5,507	
121	No 01058	IV	1,036	
122	No 01059	IV	1,861	
123	No 07634	IV	1,926	
124	No 08029	IV	1,020	
125	No 08434	IV	2,259	
126	No 10129	IV	2,509	
127	No 14429	IV	6,637	
128	No 16305	IV	1,235	
129	No 18583	IV	1,156	
130	No 19210	IV	5,262	
131	No 19881	IV	1,072	
132	No 20262	IV	1,019	
133	No 23825	IV	1,917	
134	No 24486	IV	12,622	
135	No 24496	IV	69,066	
136	No 26442	IV	1,400	
137	No 26792	IV	1,039	
138	No 26885	IV	5,200	
139	No 27386	IV	1,417	
140	No 27388	IV	4,467	
141	No 27487	IV	1,468	
142	No 27768	IV	1,106	
143	No 27872	IV	2,075	
144	No 27888	IV	4,341	
145	No 27985	IV	6,065	
146	No 28323	IV	1,180	
147	No 28940	IV	4,377	
148	No 29012	IV	1,403	
149	No 29020	IV	1,528	
150	No 29027	IV	1,252	
151	No 29184	IV	1,309	
152	No 29920	IV	1,036	
153	No 30583	IV	5,418	
154	No 31424	IV	2,936	
155	No 31742	IV	1,651	
156	No 31799	IV	3,618	
157	No 31967	IV	23,945	
158	No 32129	IV	1,752	
159	No 32130	IV	2,481	
160	No 32131	IV	1,058	

Map[†] ref.	National/international designation Name of area	IUCN management category	Area (ha)	Year notified
161	No 32776	IV	4,732	
162	No 32777	IV	8,551	
163	No 32779	IV	1,046	
164	No 32780	IV	1,485	
165	No 32783	IV	7,082	
166	No 32784	IV	1,709	
167	No 32864	IV	1,437	
168	No 32995	IV	1,886	
169	No 33113	IV	8,860	
170	No 33466	IV	5,131	
171	No 33475	IV	1,735	
172	No 34604	IV	3,636	
173	No 34605	IV	308,990	
174	No 34720	IV	723,072	
175	No 34776	IV	2,249	
176	No 35659	IV	1,049	
177	No 35752	IV	20,925	
178	No 35918	IV	14,182	
179	No 36003	IV	1,113	
180	No 36053	IV	10,850	
181	No 36203	IV	6,612	
182	No 36208	IV	153,293	
183	No 36271	IV	321,946	
184	No 36419	IV	1,406	
185	No 36913	IV	6,090	
186	No 36915	IV	4,435	
187	No 36918	IV	13,750	
188	No 36936	IV	309,678	
189	No 36957	IV	780,883	
190	No 37083	IV	1,099	
191	No 38450	IV	1,009	
192	No 38545	IV	1,671	
193	No 39422	IV	40,105	
194	No 40156	IV	6,620	
195	No 40161	IV	1,170	
196	No 40628	IV	405,424	
197	North Karlgarin	IV	5,186	
198	North Sister	IV	1,008	
199	North Tarin Rock	IV	1,416	
200	Nuytsland	IV	625,343	1965
201	Pallarup	IV	4,191	
202	Palm Springs	IV	2,154	
203	Parry Lagoons	IV	12,370	
204	Pinjarrega	IV	18,221	
205	Point Coulomb	IV	28,676	1969
206	Prince Regent	IV	634,952	1964
207	Quarram	IV	3,825	
208	Queen Victoria Spring	IV	272,598	
209	Rock View	IV	1,733	
210	Roe	IV	1,246	
211	Seagroatt	IV	1,149	
212	Sheepwash Creek	IV	1,111	
213	Silver Wattle Hill	IV	1,660	
214	South Buniche	IV	1,298	
215	South Eneabba	IV	5,979	
216	South Stirling	IV	1,710	
217	Taarblin Lake	IV	1,285	

Map[†] ref.	National/international designation Name of area	IUCN management category	Area (ha)	Year notified
218	Tarin Rock	IV	2,391	1960
219	Tutanning	IV	2,090	1970
220	Two Peoples Bay	IV	4,745	1966
221	Unicup Lake	IV	3,290	1960
222	Wanagarran	IV	11,069	
223	Wandana	IV	25,976	
224	Wanjarri	IV	53,248	1971
225	Welsh	IV	1,717	
226	Wotto	IV	2,892	
227	Yenyening Lakes	IV	2,435	
	Marine Parks			
228	Marmion	II	10,500	
229	Ningaloo (Commonwealth Waters)	II	200,000	1987
230	Ningaloo (State waters)	II	230,000	1987
	ASHMORE AND CARTIER ISLANDS			
	National Nature Reserve			
1	Ashmore Reef	I	58,300	1983
	CHRISTMAS ISLAND			
	National Park			
2	Christmas Island	II	8,700	1990
	CORAL SEA ISLANDS TERRITORY			
	National Nature Reserves			
3	Coringa-Herald	I	885,600	1982
4	Lihou Reef	I	843,600	1982
	NEW SOUTH WALES – LORD HOWE ISLAND			
	National Park			
5	Lord Howe Island Park Preserve	II	1,176	1981
	TASMANIA – MACQUARIE ISLAND			
	Nature Reserve			
6	Macquarie Island	I	12,785	1972

	Biosphere Reserves			
	Croajingolong	IX	101,000	1977
	Danggali Conservation Park	IX	253,230	1977
	Fitzgerald River National Park	IX	242,804	1978
	Hattah-Kulkyne National Park and Murray-Kulkyne Park	IX	49,500	1981
	Kosciusko National Park	IX	625,525	1977
	Macquarie Island Nature Reserve	IX	12,785	1977
	Prince Regent River Nature Reserve	IX	633,825	1977
	Southwest National Park	IX	403,240	1977
	Uluru (Ayers Rock-Mount Olga) National Park	IX	132,550	1977
	Unnamed Conservation Park of South Australia	IX	2,132,600	1977
	Wilson's Promontory National Park	IX	49,000	1981
	Yathong Nature Reserve	IX	107,241	1977
	Ramsar Wetlands			
	Apsley Marshes	R	940	1982
	Barmah Forest	R	28,500	1982

Map[†] ref.	National/international designation Name of area	IUCN management category	Area (ha)	Year notified
	Bool and Hacks Lagoons	R	3,200	1985
	Cape Barren Is. East Coast Lagoons	R	4,230	1982
	Cobourg Peninsula	R	191,660	1974
	Coongie Lakes	R	1,980,000	1987
	The Coorong and Lakes Alexandrina and Albert	R	140,500	1985
	Corner Inlet	R	51,500	1982
	Eighty-mile Beach	R	125,000	1990
	Forrestdale and Thomsons Lakes	R	754	1990
	Gippsland Lakes	R	43,046	1982
	Gunbower Forest	R	19,450	1982
	Hattah-Kulkyne Lakes	R	1,018	1982
	Hosnie's Spring	R	1	1990
	Jocks Lagoon	R	70	1982
	Kakadu (Stage II)	R	692,940	1989
	Kakadu (Stage I)	R	667,000	1980
	Kerang Wetlands	R	9,172	1982
	Kooragang Nature Reserve	R	2,206	1984
	Lake Albacutya	R	10,700	1982
	Lake Warden System	R	2,300	1990
	Lake Crescent (NW corner)	R	270	1982
	Lake Toolibin	R	437	1990
	Lakes Argyle and Kununurra	R	150,000	1990
	Little Waterhouse Lake	R	90	1982
	Logan Lagoon	R	2,320	1982
	Lower Ringarooma River	R	1,650	1982
	Macquarie Marshes Nature Reserve	R	18,200	1986
	Moulting Lagoon	R	3,930	1982
	Ord River Floodplain	R	130,000	1990
	Peel-Yalgorup System	R	21,000	1990
	Pittwater-Orielton Lagoon	R	2,920	1982
	Port Phillip Bay (western shoreline) and Bellarine Peninsula	R	7,000	1982
	Riverland	R	30,600	1987
	Roebuck Bay	R	50,000	1990
	Sea Elephant Conservation Area	R	1,730	1982
	Towra Point Nature Reserve	R	281	1984
	Vasse-Wonnerup System	R	740	1990
	Western Port	R	52,325	1982
	Western District Lakes	R	30,182	1982
	World Heritage Sites			
	East Coast Rain Forest Parks	X	203,564	1986
	Great Barrier Reef	X	4,870,000	1981
	Kakadu National Park	X	1,307,300	1981
	Lord Howe Island Group	X	1,540	1982
	Uluru (Ayers Rock) National Park	X	132,566	1977
	Tasmanian Wilderness	X	1,081,348	1982
	Wet Tropics	X	920,000	1988
	Willandra Lakes	X	600,000	1981

[†]Locations of some protected areas are shown on the accompanying maps.

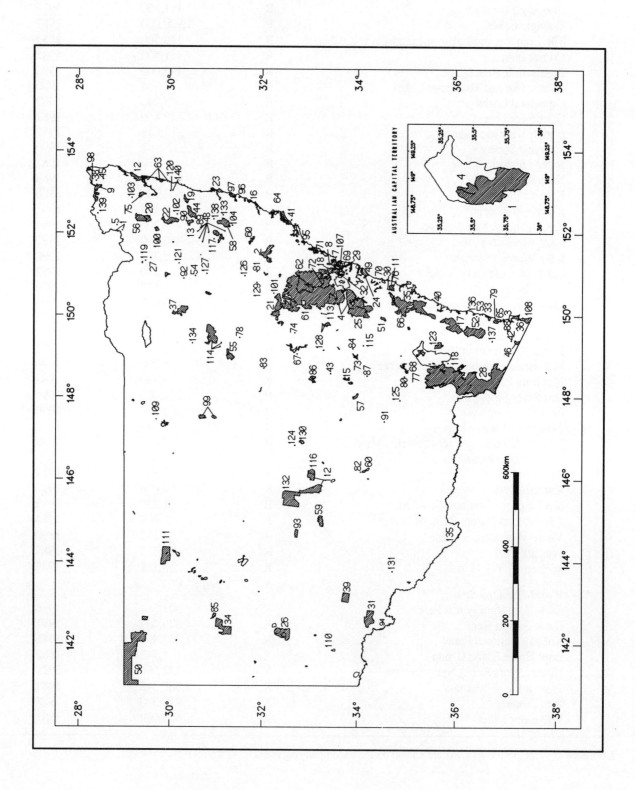

Protected Areas of Australia – New South Wales

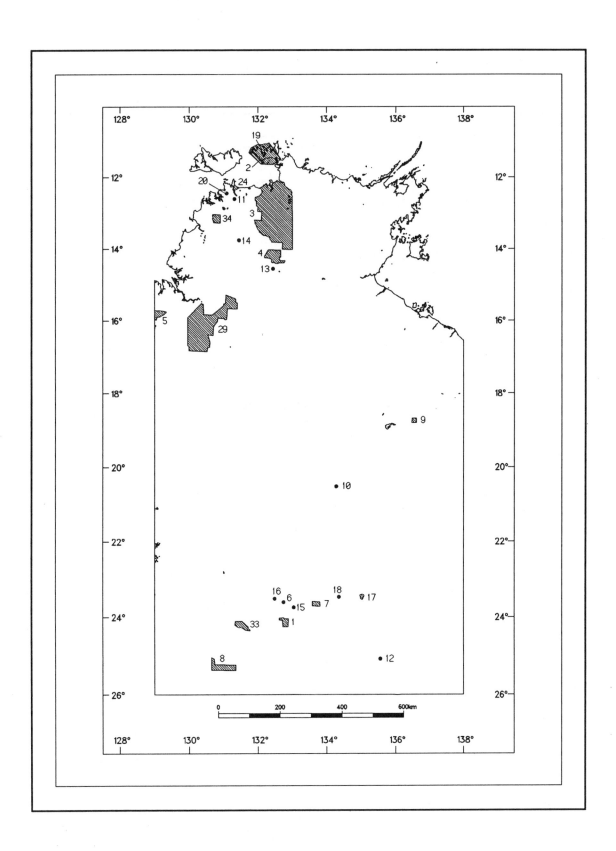

Protected Areas of Australia – Northern Territory

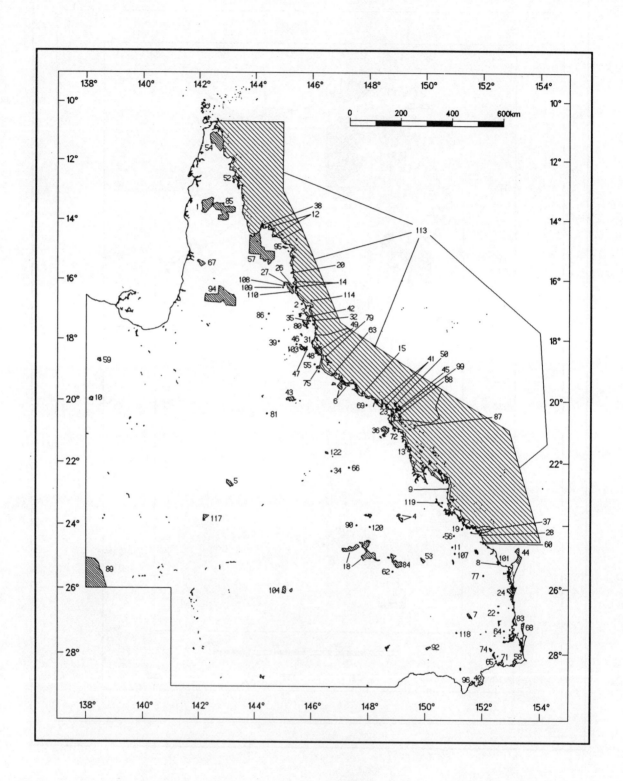

Protected Areas of Australia – Queensland

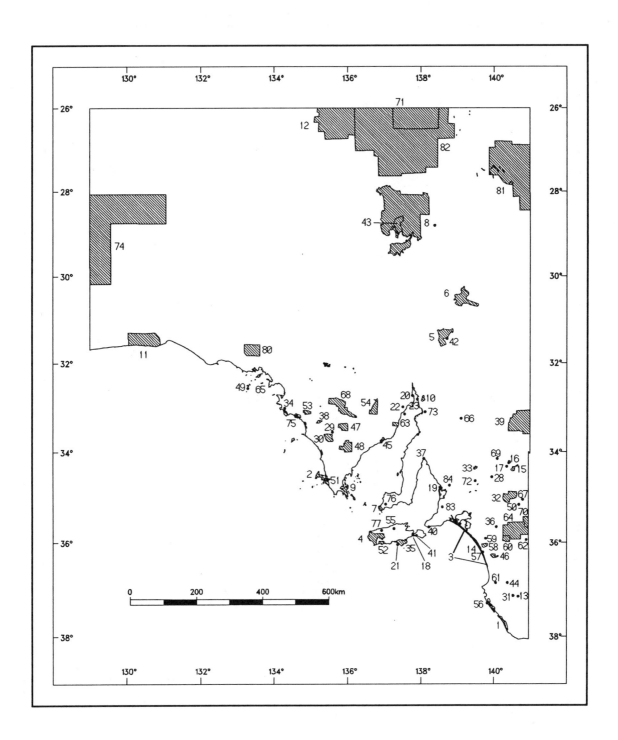

Protected Areas of Australia – South Australia

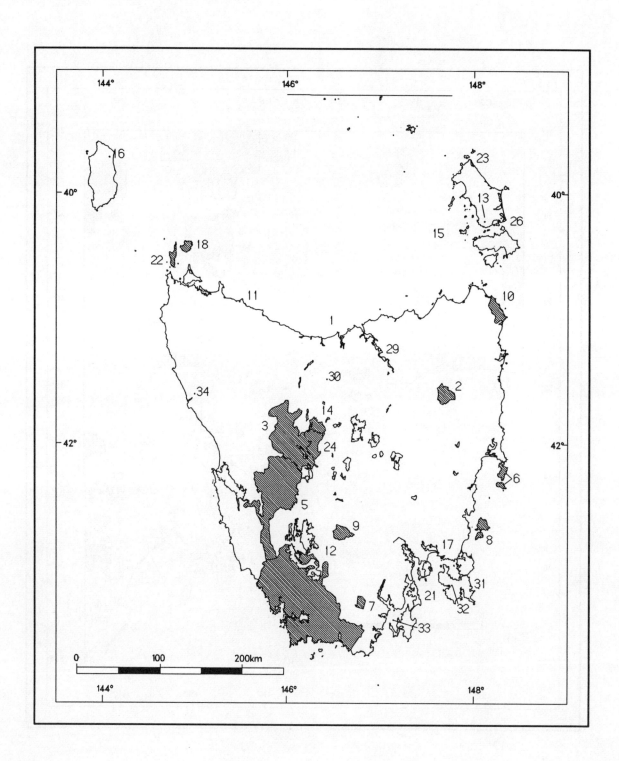

Protected Areas of Australia – Tasmania

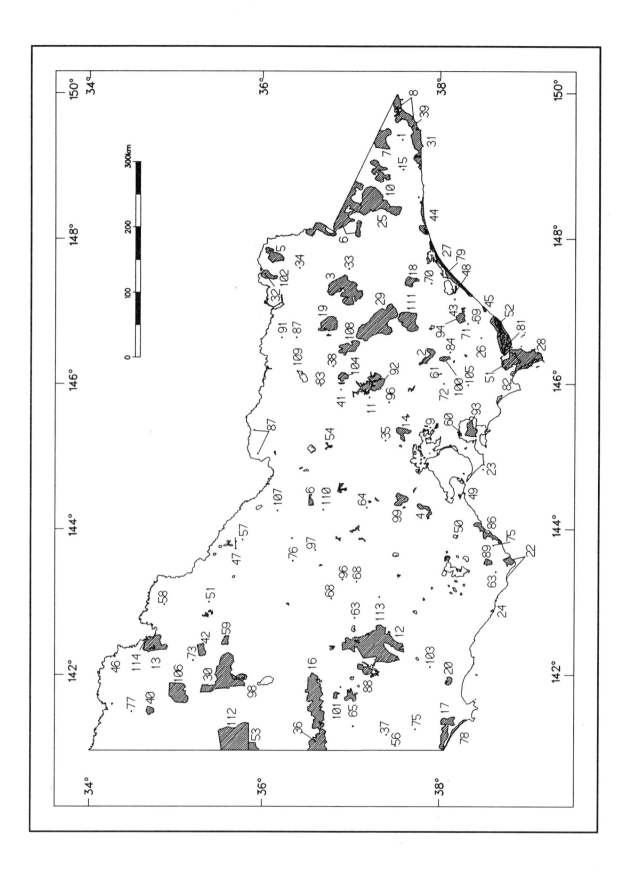

Protected Areas of Australia – Victoria

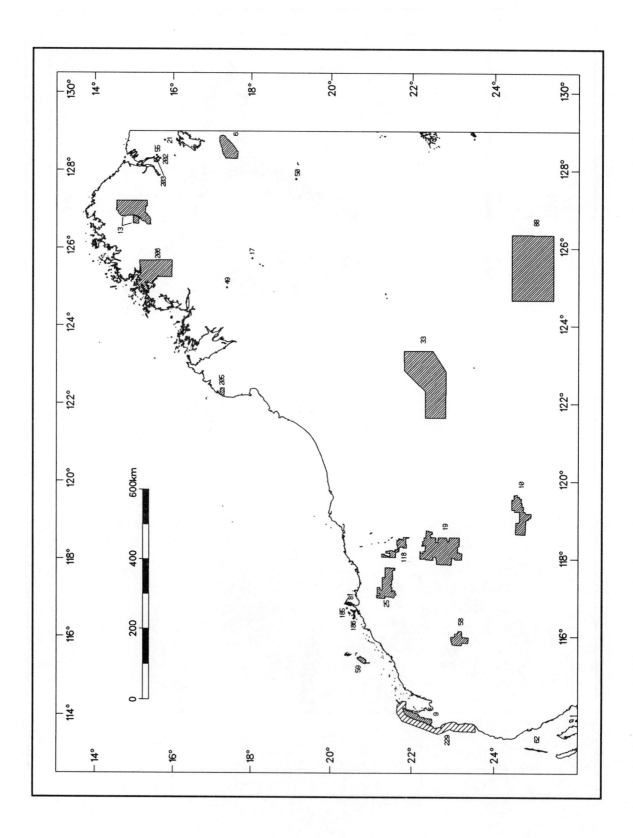

Protected Areas of Australia – Western Australia (North)

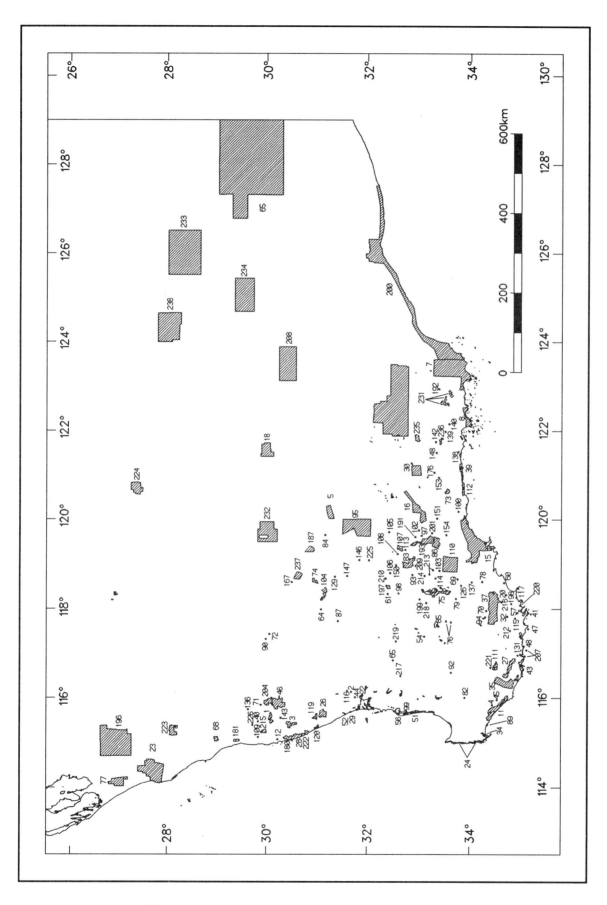

Protected Areas of Australia – Western Australia (South)

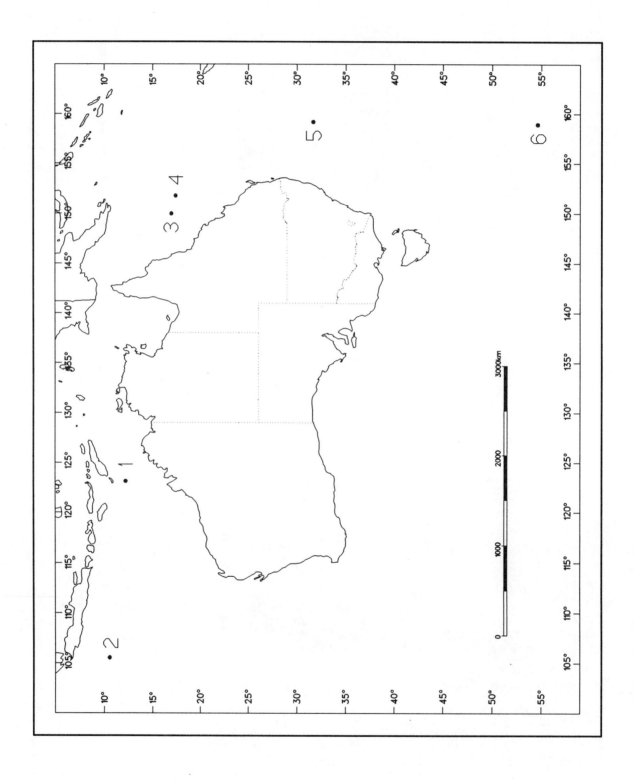

Protected Areas of Australia – Australian Islands

NEW ZEALAND

Area 268,684 sq. km

Population 3,392,000
Natural increase: 0.82% per annum

Economic Indicators
GNP: US$ 9,860 per capita (1988)

Policy and Legislation Following the 1987 reorganisation of government environmental administration, the protected area system comprised 20 principal categories established under six separate statutes, viz. Conservation Act 1987; Reserves Act 1977; National Parks Act 1980; Wildlife Act 1953; Queen Elizabeth II National Trust Act 1977; and Marine Reserves Act 1971. It became necessary to rationalise the structure of quasi-autonomous government organisations (quangos) which play an important role in conservation administration. A major review initiated in October 1987, including widespread public consultation, culminated in the Conservation Law Reform Act 1989, which rationalised the structure and responsibilities of quangos and reformed the law relating to freshwater fish and game management, marginal strips, and conservation management planning. At that stage, the existing National Parks and Reserves Authority and the twelve Regional National Parks and Reserves Boards that covered the country remained operative under the National Parks Act, and the Reserves Act, in order to provide a "bridge" during the first years of the new Department of Conservation. Early in 1990, under the Conservation Law Reform Act, the national parks and reserves bodies were replaced by the New Zealand Conservation Authority and seventeen district conservation boards.

These assume the responsibilities previously exercised by the National Parks and Reserves Boards, Nature Conservation Council, Forest Park Advisory Committees, Walkways Commission and Committees, Marine Reserves Management Committees, National Recreational Hunting Advisory Committee, and Hauraki Gulf Maritime Park Board. The Act also gives the Guardians of lakes Manapouri and Te Anau statutory recognition. In order to carry out these functions, the new bodies have the status of local government, the right to appear before tribunals, and a substantial independence (for example, the right to advocate and publish). They are, in effect, citizen partners of the department, able to reach out into the whole community.

Further consolidation of the law relating to protected areas is considered desirable, to simplify the approach and reduce the number of categories involved. These reforms will be achieved by amendments to the principal acts.

International Activities The Convention concerning the Protection of the World Cultural and Natural Heritage (World Heritage Convention) came into force on 22 November 1984, and two sites have been inscribed on the World Heritage List: Tongariro National Park, and South-West New Zealand, the latter encompassing 10% of the country. The Convention on Wetlands of International Importance Especially as Waterfowl Habitat (Ramsar Convention) was signed without reservation as to ratification on 13 August 1976, and five sites are included in the List of Wetlands of International Importance. New Zealand participates in the Unesco Man and the Biosphere Programme (MAB), but has not designated any Biosphere Reserves.

New Zealand is party to the South Pacific Regional Environment Programme (SPREP) and the Convention for the Protection of the Natural Resources and Environment of the South Pacific Region, 1986 (SPREP Convention) was signed on 25 November 1986 and ratified on 3 May 1990. The Convention entered into force during August 1990. Article 14 calls upon the parties to take all appropriate measures to protect rare or fragile ecosystems and threatened or endangered flora and fauna through the establishment of protected areas and the regulation of activities likely to have an adverse effect on the species, ecosystems and biological processes being protected. However, as this provision only applies to the convention areas, which by definition is open ocean, it is most likely to assist with the establishment of marine reserves and the conservation of marine species.

Administration And Management Recent years have witnessed revolutionary change in environmental administration in New Zealand. On 1 April 1987, the Department of Lands and Survey, New Zealand Forest Service, Wildlife Service and Commission for the Environment were disestablished and replaced by the Department of Conservation and a number of smaller policy agencies to continue with specific functions. Establishment of the Department created for the first time in New Zealand a single agency of central government with a statutory conservation mandate – with conservation defined as the protection and preservation of natural and historic resources, held in stewardship for present and future generations.

The new organisation comprises a Head Office in Wellington and 14 regional conservancies, with boundaries established according to geographical, ecological, local government and management considerations. The conservancies are designed to conduct all management operations through a regional office and network of field stations. Head Office handles policy, planning, audit, resource allocation, and ministerial servicing functions at a national level. It is made up of five policy divisions (protected species, estate protection, resource use and recreation, advocacy and information, historic resources); four servicing divisions (personnel and administration, finance, science and research and information systems); a Maori perspectives and Iwi liaison unit; a ministerial servicing/corporate planning/revenue generation unit; and a legal services unit.

Systems Reviews New Zealand is a mountainous archipelago set in the temperate latitudes of the south-west Pacific region. The total land area is 268,684 sq. km, made up of the main North and South islands, the smaller Stewart Island, and more than 600 offshore islands extending from the sub-tropical Kermadec Group to sub-antarctic Campbell Island in the Southern Ocean, and including the Chatham Group some 800km to the east of the mainland.

Mountains dominate the landscape. About 60% of the country is more than 300m above sea level, and the Southern Alps contain 29 peaks rising to above 3,000m. New Zealand straddles the "hot rim" of the Pacific Basin, so has a long volcanic history. Ruapehu and Ngaruhoe volcanoes, which dominate Tongariro National Park, are the most spectacular andesitic volcanoes in the south-west Pacific and both are highly active, the former having erupted 45 times in the past 125 years. Associated with the North Island volcanoes is an extensive complex of hot springs and geysers, constituting one of the five major hydrothermal regions in the world.

Human occupation over the past millennium has produced massive and widespread ecological change. This has been particularly so in the period of planned European settlement beginning 150 years ago. Some 60% of the original vegetation cover has been destroyed, especially the forests which are now reduced to only 23% of the total land area, mainly in the mountains. The introduction of plants and animals has been likened to a biological invasion, and the number of introduced plant species (c. 2,000) now rivals that of the higher native plants, while among the successfully introduced animals are at least 30 species of birds and 20 species of mammals.

The New Zealand protected area system is among the most comprehensive in the world and was one of the earliest to develop, having its origins in the days of European settlement in the latter half of the last century. In 1989 the system contained more than 2,000 individual areas, covering approximately 5 million hectares, or almost 20% of the total land area. This represents a considerable increase in the network since 1983 when there were 1,660 areas covering 4.5 million hectares or 17% of the country. However, protection of the marine environment has lagged well behind that on land. Seaward of the coast there is less than 1% of the country's 11,000km of coastline within protected areas. The protected areas system also includes the sub-Antarctic Antipodes Islands, Auckland Islands and Bounty Islands which were established as nature reserves under the 1977 Reserves Act (Clark and Dingwall, 1985).The Department of Conservation is committed to the establishment of an extensive network of marine protected areas. In 1990, proposals and/or investigations for additional marine reserves were under action at 33 sites, including extensions to two national parks.

The Reserves Act 1977 includes a commitment to work toward establishment of a protected area system which is representative of the full range of ecological diversity in the country. The bias in ecological representation at present toward higher altitude and forested landscapes, and the ongoing widespread loss of indigenous ecosystems make this task imperative. To meet this commitment, the Protected Natural Areas Programme was developed to undertake a rapid survey of remaining natural areas, identify those meriting protection and protect the most important. The country was sub-divided into 268 ecological districts as a framework for the survey, and 133 of these were singled out for urgent attention. As of 1989, 46 ecological districts had been surveyed in the period since 1983 when fieldwork began. Priority is being given to protection of areas on pastoral leasehold land in the South Island high country, where protection will be effected primarily by conservation covenant. So far, 26 areas have been accorded protection, by purchase as reserve, transfer to the Department, or inter-agency agreement. A further 35 areas are under negotiation; those in Central Otago and northern Southland covering some 40,000ha. In North Island five protected private land agreements have been completed in the Taranati region, and one area purchased for reserve in eastern Coromandel. Survey and protection of lowland and coastal forest in the East Cape region, much of it on Maori land, have been given priority for future work. However, despite support from the Department of Conservation, other government agencies, and public and local authorities, there is mounting concern that progress with the survey and protection of remaining natural areas is insufficient to combat losses of habitat throughout the country.

The Government took important initiatives, starting in 1990, to protect forests on private land. A Forest Heritage Fund was established with an annual vote of $5 million to protect forests either by legal covenant agreements or purchase. To protect forests on private land owned by Maori another fund with $2.5 million has been created.

A number of issues are seen as threatening the protected areas system. These are the spread of possum throughout Northland with threats to coastal forests of outstanding importance, in one of the few remaining areas of the country free from possum; increased damage from goat browsing, especially following accidental releases from growing numbers of commercial herds; increases in number of red deer as a consequence of reductions in live recovery operations by commercial hunters; proliferation of exotic plants, for example *Clematis vitalba*, which is steadily invading reserves in the lower and western parts of North Island and in Nelson and Marlborough; and the inexorable spread of wild conifers which is a long-standing problem in areas of natural grassland and shrubland, particularly in Central Otago, and in Tongariro National Park.

Addresses

Department of Conservation (Director General), PO Box 10420, WELLINGTON (Tel: 4 471 0726; Fax: 4 471 1082)

References

Clark, M.R. and Dingwall, P.R. (1985). *Conservation of islands in the Southern Ocean.* A discussion document prepared for IUCN's Commission on National Parks and Protected Areas. IUCN, Gland, Switzerland and Cambridge, UK. 188 pp.

Dingwall, P.R. (1989). New Zealand. *Country Review 8.* Fourth South Pacific Conference on Nature Conservation and Protected Areas, Port Vila, Vanuatu, 4-12 September. South Pacific Commission, Noumea, New Caledonia. 15 pp.

ANNEX
Definitions of protected area designations, as legislated, together with authorities responsible for their administration

Title: Conservation Law Reform Act

Date: 1989

Brief description: Text not available

SUMMARY OF PROTECTED AREAS

Map† ref.	National/international designation Name of area	IUCN management category	Area (ha)	Year notified
	National Parks			
1	Abel Tasman	II	22,543	1942
2	Arthur's Pass	II	94,422	1929
3	Egmont	II	33,540	1900
4	Fiordland	II	1,023,186	1904
5	Mount Aspiring	II	355,518	1964
6	Mount Cook	II	69,923	1953
7	Nelson Lakes	II	96,112	1956
8	Tongariro	II	76,504	1894
9	Urewera	II	207,462	1954
10	Westland	II	117,547	1960
11	Whanganui	II	74,231	1986
	National Park Special Areas			
12	Fiordland (Sinbad Gully Stream)	I	2,160	1974
13	Fiordland (Takahe)	I	177,252	1953
14	Mount Aspiring	I	1,722	1973
15	Secretary Island	I	8,980	1973
	Scientific Reserve			
16	Waituna Wetlands	I	3,557	1983
	Nature Reserves			
17	Anglem	I	16,977	1907
18	Farewell Spit	I	11,388	1938
19	Kapiti Island	I	1,970	1975
20	Little Barrier Island	I	2,817	1895
21	Mount Uwerau	I	1,012	1966

Map[†] ref.	National/international designation Name of area	IUCN management category	Area (ha)	Year notified
22	Pegasus	I	67,441	1907
23	Waitangiroto	I	1,214	1957
	Wildlife Refuges			
24	Lake Alexandrina/McGregor's Lagoon	IV	2,200	1957
25	Pouto Point	IV	6,789	1957
26	Wairau River Lagoons	IV	1,040	1959
	Marine Reserve			
27	Poor Knights Islands	I	2,410	1981
	State Forest Ecological Areas			
28	Big River	IV	6,733	1980
29	Blackwater	IV	9,150	1980
30	Coal Creek	IV	3,025	1980
31	Diggers Ridge	IV	4,235	1982
32	Flatstaff	IV	1,622	1980
33	Greenstone	IV	1,144	1980
34	Kapowai	IV	1,400	1982
35	Lake Christabel	IV	10,648	1981
36	Lake Hochstetter	IV	1,803	1981
37	Lillburn	IV	2,670	1982
38	Manganuiowae	IV	1,760	1981
39	Mangatutu	IV	2,533	1980
40	Moehau	IV	3,634	1977
41	Onekura	IV	2,351	1981
42	Papakai	IV	3,366	1982
43	Pororari	IV	6,448	1980
44	Pukepoto	IV	1,906	1980
45	Roaring Meg	IV	3,600	1980
46	Saltwater	IV	1,438	1981
47	Saxton	IV	4,120	1980
48	Tiropahi	IV	3,451	1980
49	Waikoau	IV	2,800	1982
50	Waipapa	IV	1,830	1979
51	Waipuna	IV	1,910	1980
	Forest Sanctuaries			
52	Hihitahi	I	2,170	1973
53	Ngatukituki	I	1,600	1973
54	Waipoua	I	9,105	1952
	Reserve			
55	Lake Whangape Government Purpose	IV	1,450	
	Ecological Area			
56	Fletchers Creek	IV	2,586	1980
	Scenic Reserves			
57	Arapawa Island	IV	1,035	1973
58	Chance, Penguin and Fairy Bays	IV	1,599	1903
59	Chaslands	IV	1,334	1937
60	Codfish Island	I	1,396	1915
61	D'Urville Island	IV	4,072	1912
62	Glen Allen	IV	1,000	1914
63	Glenhope	IV	5,936	1907
64	Glory Cove	IV	1,297	1903
65	Gordon Park	IV	1,817	1938
66	Gouland Downs	IV	6,564	1917

Map[†] ref.	National/international designation Name of area	IUCN management category	Area (ha)	Year notified
67	Hakarimata	IV	1,795	1905
68	Isolated Hill	IV	2,160	1924
69	Jordan Stream	IV	1,151	1916
70	Karamea Bluff	IV	1,445	1910
71	Kenepuru Sound	IV	1,687	1895
72	Lake Ianthe	IV	1,308	1905
73	Lake Kaniere	III	7,252	1906
74	Lake Okareka	III	1,143	1930
75	Lake Okataina	III	4,388	1974
76	Lake Tarawera	III	5,819	1973
77	Leithen Bush	IV	1,342	1978
78	Lewis Pass	IV	13,737	1907
79	Lower Buller Gorge	IV	5,941	1907
80	Mangamuka Gorge	IV	2,832	1927
81	Matahuru	IV	1,336	1905
82	Maungatautari	III	2,389	1927
83	Maurihoro	IV	1,797	1936
84	Meremere Hill	IV	1,368	1982
85	Moeatoa	IV	1,212	1927
86	Mokau River	IV	2,273	1920
87	Mount Stokes	IV	4,396	1977
88	Mt Courtney	IV	1,772	1912
89	Mt Hercules	IV	8,024	1911
90	Mt Te Kinga	IV	3,747	1905
91	Nydia Bay	IV	1,408	1938
92	Paradise Bay	IV	2,743	1977
93	Pelorus Bridge	IV	1,010	1906
94	Pokaka	IV	8,068	1982
95	Pryse Peak	IV	3,646	1903
96	Pukeamaru Range	IV	3,265	1907
97	Punakaiki	IV	2,037	1914
98	Rahu	IV	2,132	1936
99	Rakeahua	IV	6,463	1903
100	Rangitoto Island	IV	2,333	1980
101	Robertson Range	IV	3,689	1912
102	Saltwater Lagoon	IV	1,359	1928
103	Ship Cove	IV	1,093	1896
104	South Cape	IV	5,077	1903
105	Tahuakai	IV	1,561	1983
106	Tangarakau	IV	2,640	1918
107	Tapuaenuku	IV	2,226	1962
108	Te Arowhenua	IV	1,705	1981
109	Te Kopia	IV	1,408	1911
110	Te Tapui	IV	2,382	1925
111	Tennyson Inlet	IV	5,596	1896
112	Toatoa	IV	2,847	1982
113	Upper Buller Gorge	IV	5,920	1979
114	Waioeka Gorge	IV	18,645	1933
115	Waipapa	IV	2,528	1974
116	Waipori Falls	IV	1,352	1913
117	Waituhui Kuratau	IV	1,319	1953
118	Wanganui River	IV	35,858	1915
119	Warbeck River	IV	1,283	1931
120	Whangamumu	IV	2,154	1981
	State Forest Parks			
121	Catlins	VIII	57,881	1974
122	Coromandel	VIII	68,693	1972

Map[†] ref.	National/international designation Name of area	IUCN management category	Area (ha)	Year notified
123	Craigieburn	VIII	35,735	1967
124	Hanmer	VIII	16,843	1978
125	Haurangi	VIII	16,198	1974
126	Kaimai Mamaku	VIII	36,965	1975
127	Kaimanawa	VIII	74,965	1969
128	Kaweka	VIII	65,549	1974
129	Lake Sumner	VIII	102,296	1974
130	Mt Richmond	VIII	180,271	1977
131	North West Nelson	VIII	376,328	1970
132	Pirongia	VIII	13,324	1971
133	Pureora	VIII	69,532	1978
134	Raukumara	VIII	115,102	1979
135	Rimutaka	VIII	14,217	1972
136	Ruahine	VIII	93,056	1976
137	Tararua	VIII	117,226	1967
138	Victoria	VIII	180,592	1981
139	Whakarewarewa	VIII	3,830	1975
	Other area			
140	Waipoua Kauri Management Area	IV	3,747	1976
	ISLANDS			
	Nature Reserves			
141	Auckland Islands	I	62,564	1934
142	Campbell Islands	I	11,331	1954
143	Raoul Island and Kermadec Group	I	3,089	1934
	Ramsar Wetlands			
	Waituna Lagoon	R	3,556	1976
	Farewell Spit	R	11,388	1976
	Whangarmarino	R	5,690	1989
	Kopuatai Peat Dome	R	9,665	1989
	Firth of Thames	R	7,800	1990
	World Heritage Sites			
	South West New Zealand (Te Wahipounamu)	X	2,600,000	1990
	Tongariro National Park	X	76,504	1990

[†]Locations of most protected areas are shown on the accompanying maps.

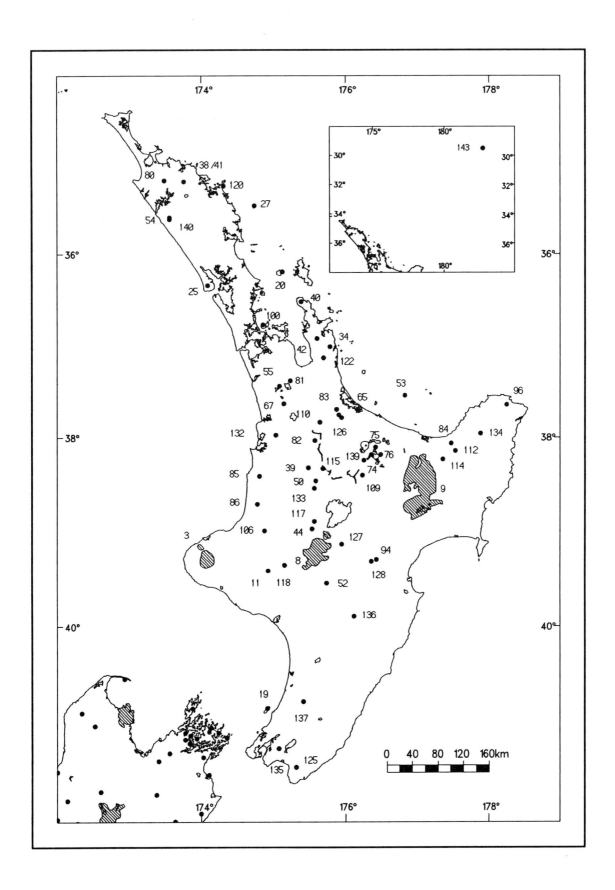

Protected Areas of New Zealand – North Island

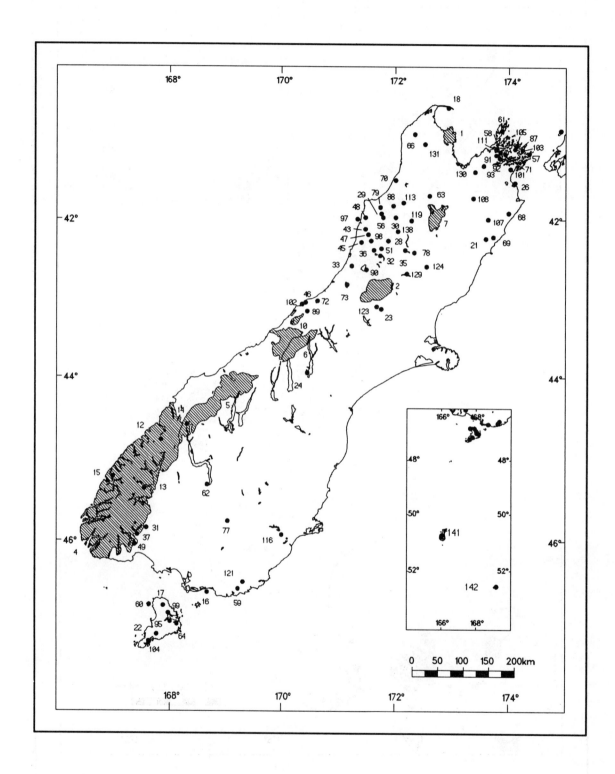

Protected Areas of New Zealand – South Island

The Antarctic and
Southern Ocean Islands

ANTARCTIC TREATY TERRITORY

Area 14,245,000 sq. km approximately

Population No permanently resident population (transient population of about 2,500 in winter and 4,500 in summer)

Economic Indicators
GNP: Not applicable
GDP: Not applicable

Policy and Legislation A series of international agreements, with measures adopted pursuant to them as guidelines for national legislation, as well as national instruments themselves, comprise the existing system of governance for the Antarctic and the activities conducted in the territory.

The 1959 Antarctic Treaty was negotiated in order to ensure that Antarctica be used for peaceful purposes only and to further international cooperation and scientific research. This treaty entered into force in June 1961, and is subject to review, if requested by one of its States Parties with decision-making status (the Antarctic Treaty Consultative Parties) after June 1991; that is, thirty years after entry into force.

The Treaty would not have achieved its objectives without Article IV which preserves all previously asserted rights or claims to territorial sovereignty in Antarctica and the position of those who recognise no claim or basis of claim in Antarctica, and nullifies any basis of claim during its operation. This balance between the claimant and non-claimant states is reflected in other components of the Treaty System. Article IX of the Treaty requires that the Consultative Parties meet periodically and upon consultation recommend to their governments measures in furtherance of the principles and objectives of the Treaty. These "recommendations", which come into effect when approved by all the Consultative Parties, operate as with the principles of international law, do not constitute laws enforceable in any court but are guidelines, viewed as moral obligations by the Parties which could, if breached, do no more than render a state liable to pressure and political sanctions.

In furtherance of the requirement in Article IX (f) that the Parties formulate measures on the conservation of living resources, the Agreed Measures for Conservation of Antarctic Fauna and Flora were negotiated in 1964 and several other measures followed. Although as an annex to Recommendations III-VIII, they are actually voluntary guidelines. By virtue of Articles III and XIII, the Consultative Parties have since 1982 been legally required to implement the provisions of the Agreed Measures. The conservation provisions prohibit the taking of native species, except in accordance with a permit, and establish criteria for the issuance of permits with regard to specially protected species, as well as for the specially protected areas (SPAs). At the Antarctic Treaty Consultative Meeting (ATCM) VII and VIII the Treaty Parties incorporated recommendations to review SPAs, to include the following categories: representative examples of the major Antarctic land and freshwater systems; areas with unique complexes of species; areas which are the type locality or only known breeding area of any plant or invertebrate species; areas which contain especially interesting breeding colonies of birds or mammals; and areas which should be kept inviolate so that they may be used for comparison with disturbed localities. The seventh meeting also made provision for the establishment of sites of special scientific interest (SSSIs), for specified periods of time, to cover areas where scientific investigation was being carried out, or where there were sites of exceptional scientific interest that required long-term protection (Bonner and Lewis Smith, 1985). The legal regime of the Antarctic Treaty System does not apply to territory and territorial waters of many of the Subantarctic islands, most of which are under undisputed national jurisdiction and all of which have some form of related conservation legislation applying to them. South Orkney and South Shetland Islands, Peter I Øy, Scott Island and Balleny Islands are exceptional, lying within the 60°S latitude, and are covered by the Agreed Measures of the Antarctic Treaty.

SSSIs differ from SPAs in a number of respects. It was recognised at an early stage that the designation of an area as an SPA virtually isolated it from active scientific research programmes. However, there was a good case for providing sites with a measure of protection while research was in progress, or was likely to be continued. Furthermore, the Agreed Measures did not allow sites of non-biological interest to be designated as SPAs. It was to rectify this that the concept of SSSIs was formulated at the Seventh Antarctic Treaty Consultative Meeting. It has been stipulated that SSSIs should only be proposed when either scientific research was being, or was likely to be, carried out and that there was a demonstrable risk that interference would jeopardise such investigations, or that they were of exceptional scientific interest and therefore required long-term protection (Bonner and Lewis Smith, 1985).

An attempt to enhance the limited range of protected area designations was made at the XIV and XV ATCMs, in which two new protected area categories were established. Specially reserved areas are intended for the protection of representative examples of major geological, glaciological and geomorphological features and representative examples of outstanding aesthetic, scenic and wilderness values. Multiple-use planning areas are designed to promote cooperative planning to minimise environmental impacts (IUCN, 1990).

The 1972 Convention for the Conservation of Antarctic Seals entered into force in 1978. The Convention contains comprehensive measures to regulate the taking of seals, including specifying permissible catch levels, protected species, and the opening and closure of sealing seasons and zones, and it establishes three Seal Reserves. The Convention is subject to regular review. The latest review was conducted in September 1988.

The 1980 Convention on the Conservation of Antarctic Marine Living Resources (CCAMLR), which provides objectives and means for such conservation, entered into force in 1982. It established an Executive Commission and an Advisory Scientific Committee which meet annually at their headquarters in Hobart, Tasmania, Australia.

The Convention on the Regulation of Antarctic Mineral Resource Activities (CRAMRA) was completed on 2 June 1988 and opened for signature on 25 November 1988. It establishes objectives and institutions for the regulation of possible minerals development activities. The Convention requires ratification by the Parties before it can come into force and has largely been superceded by the environmental profile of 1991.

In addition to the legal instruments and measures of the Antarctic Treaty System, activities in Antarctica are also subject to a variety of legal obligations that stem from treaties that are more broadly applicable, such as the International Convention for the Regulation of Whaling and several International Maritime Organisation sponsored treaties on vessel safety and pollution control.

The scope of these measures notwithstanding, it is becoming increasingly apparent that there are deficiencies that require remedying, notable among which are the following (IUCN, 1990):

(i) The intermittent development of measures over a period of almost 30 years has introduced differing standards and procedures, reflecting the situation existing at the time they were written.

(ii) There are gaps in the series of measures which require filling, especially to deal with activities such as tourism that have recently increased.

(iii) The implementation of some of the instruments has not always been consistent or up to desired standards.

(iv) The various disparate elements of the system for environmental protection require better coordination.

(v) While there are principles which span the various legal and management frameworks (e.g. peaceful use, freedom of scientific information, international cooperation), it is also clear that the aspirations and goals of individual governments and institutions involved in Antarctica vary considerably.

International Activities The Antarctic Treaty was signed in Washington on 1 December 1959 by 12 states and entered into force on 23 June 1961. These Original Signatories comprise Argentina, Australia, Belgium, Chile, France, Japan, New Zealand, Norway, South Africa, Soviet Union, United Kingdom and United States of America, all of which are also Consultative Parties. The following are Acceding States only: Austria, Bulgaria, Canada, Colombia, Cuba, Czechoslovakia, Democratic People's Republic of Korea, Denmark, Greece, Guatemala, Hungary, Papua New Guinea, Romania and Switzerland. States that are both Acceding States and Consultative Parties, but are not amongst the Original Signatories, are Brasil, Ecuador, Finland, Germany, India, Italy, Netherlands, People's Republic of China, Peru, Poland, Republic of Korea, Spain, and Uruguay.

Administration and Management The CCAMLR provides for a Commission and Secretariat with an associated Scientific Advisory committee and CRAMRA makes provision for regulatory and administrative bodies. However, there is no overall administrative framework for conducting Antarctic activities. Unlike most international conventions, there is no Antarctic Treaty Secretariat, responsibility being taken by the Parties that host Consultative Meetings. Some non-governmental groups have called for the establishment of an Environmental Protection Agency to rectify this situation. The Scientific Committee on Antarctic Research (SCAR), a body of the International Council of Scientific Unions (ICSU), was established in 1958 with a brief to initiate, coordinate and promote Antarctic Science. The Group of Specialists on Environmental Affairs and Conservation has been established to allow SCAR to advise on protective measures.

Most of the principles promoted by SCAR for the protection of the Antarctic environment have been adopted under the Agreed Measures for the Conservation of Antarctic Fauna and Flora, including the establishment of protected areas and many other recommendations of the Antarctic Treaty Consultative Meetings.

Systems Reviews There are several terms used to refer to the South Polar regions; the generally accepted nomenclature of the SCAR is as follows. "Antarctica" is used to refer to the Antarctic Continent, including the offshore islands of the continental shelf. "The Antarctic" includes all the region, sea, ice shelves and land to the south of the Antarctic convergence (the Antarctic Convergence zone). The oceanic part of the Antarctic is called the Southern Ocean. Because the Antarctic convergence is not a fixed feature, the CCAMLR used a series of coordinates to define a geographical boundary which more or less follows the position of the Antarctic Convergence (Bonner and Lewis Smith, 1985). This boundary is wider than the Antarctic Treaty, which is limited by latitude 60°S.

Antarctica forms a hub about the South Pole, with its coasts mostly lying south of latitude 70°S. Two deep indentations – the Weddell Sea south of the Atlantic and the Ross Sea south of the Pacific break into the near-circular mass, with open water in summer to over 77°S and floating ice shelves at the heads of these seas extending far south of 80°S. The Antarctic Peninsula projects far to the north towards South America, its tip reaching 63°S and its flanking islands almost to 60°S.

Antarctica, geologically, is two continents. The larger, Greater (or East) Antarctica, is an almost circular block of land with most of its rock surface near sea level, but with fringing mountains around the greater part of its margin, including the Transantarctic Mountains, rising to some 4,500m, which traverse the modern continent and form the western margin of the Ross and the eastern margin of the Weddell Sea. Lesser (or West) Antarctica is smaller, and consists essentially of a high inland mountain massif, the Ellsworth Mountains, which contain the highest summit in Antarctica at 4,897m, extended northwards by the Antarctic Peninsula and westward by the Crary and Executive Committee ranges. Between Greater and Lesser Antarctica there are some deep sub-ice channels which would sub-divide the continent in the event of complete melting of the ice cap.

The shores of Antarctica are fringed by seas that are uniformly ice-covered in winter and even in summer bear immense expanses of pack ice. At summer minimum sea ice covers some 3 million sq. km: at winter maximum the cover expands to some 20 million sq. km. This fluctuating ice fringe expands the effective boundary of the Antarctic, but the cooling of the sea surface by the melting of this ice and of the immense icebergs that break from the ice shelves of the Ross and Weddell seas and some other coastal margins has an even wider influence.

Ice and snow support some life: snow algae which stain the surfaces yellow, green or reddish colours. Certain crystalline rocks contain remarkable miniature "endolithic" ecosystems of bacteria, algae and fungi. Ice-free land supports a considerable diversity of lichens, with a total of some 150 Antarctic species, and in the more mild coastal habitats there are numerous different vegetation types dominated by mosses (with around 100 Antarctic species) and liverworts (with perhaps 25 species). In such places, varied mats and carpets of moss inject a range of green and yellow tints into the landscape and greatly increase its beauty as well as its scientific interest. Only two species of vascular plant, a grass and a low-growing cushion plant, grow in Antarctica and they are confined to the maritime region of the northern Antarctic Peninsula and its fringing islands. The Subantarctic islands are much more richly vegetated, with flowering plant assemblages dominant near the coasts, and areas of wetland and feldmark. The subantarctic islands are relatively rich in both terrestrial and freshwater fauna.

Freshwater lakes of Antarctica have dense mats of aquatic moss and blue-green algae on their beds, planktonic algae in their surface waters, and fauna of small crustaceans, worms, and rotifers. Some algal populations have also been recorded recently in streams, freshwater seeps and wetland habitats outside lakes.

A review by the SCAR Sub-Committee on Conservation, published in 1985, and considering the 17 SPAs and 11 SSSIs existing at that time, concluded that most types of maritime Antarctic vegetation were included within the system, but that there was inadequate representation of freshwater bodies. No sites had been defined to safeguard the inshore marine ecosystem, although some such sites have since been included. The protected areas system, as of 1990, comprised 32 SSSIs totalling 93,000ha, and 16 SPAs covering 12,000ha, in addition to 55 historic sites and monuments and three seal reserves (Dingwall, 1990).

Until the formulation of the specially reserved area designation, no provision had been made for protection of extensive landscapes. In particular, SPAs are by definition established at the smallest possible size to fulfil their function, thus failing to include extensive regions of high scenic value (IUCN, 1990).

A series of recommendations to strengthen the protected areas system is made in IUCN (1990), focusing on the protection of ecological interest and biological diversity, protection of scientific research sites, compilation of protection management plans, recognition and protection of outstanding Antarctic landscapes, integration of protection and other activities, protection of historic and cultural sites, and further protection of Subantarctic islands.

Interest remains high, especially amongst some non-governmental organisations, in assigning world park (or an equivalent) status on the Antarctic region, and there is some support for this amongst the Treaty Parties. The governments of Australia and France, in refusing to sign the Minerals Convention, advanced in 1989 a proposal for negotiation of a new convention on environmental protection, which would establish Antarctica as a "nature reserve – land of science". Widespread support exists amongst the Antarctic Treaty Contracting Parties for a comprehensive and integrated environmental protection regime, and interest has emerged in negotiating a protocol to the Antarctic Treaty on environmental protection (Dingwall, 1990). These proposals were discussed at a special ATCM in Chile in 1990 during which the continuation of a temporary ban on mineral exploration was agreed.

Other Relevant Information Five island groups lie within the Antarctic Treaty Area, and are covered by the Agreed Measures of the Convention, namely South Orkney and South Shetland Islands, Peter I Øy, Scott Island and Balleny Islands. The British include South Orkney and South Shetland island groups in the British Antarctic Territory, designated in 1962, which is administered by a High Commissioner in Council, resident in London. Peter I Øy is a Norwegian dependency. Scott Island and Balleny Islands are part of the Ross Dependency, New Zealand territory. There is no formal administration of protected areas, although the British Antarctic Survey station leaders are responsible for local permits.

References

Bonner, W.N. and Lewis Smith, R.I. (1985). *Conservation areas in the Antarctic*. A review prepared by the Sub-Committee of Conservation Working Group on Biology. Scientific Committee on Antarctic/International Council of Scientific Unions. 299 pp.

Dingwall, P. (1990). Recent progress in developing the protected areas network in the Antarctic Realm. Report to the 34th Working Session of the IUCN Commission on National Parks and Protected Areas, Perth, Australia, 26-27 November.

IUCN (1990). A strategy for Antarctic conservation. General Assembly Paper GA/19/90/19. 18th Session of the IUCN General Assembly, Perth, Australia, 28 November-5 December. 111 pp.

SUMMARY OF PROTECTED AREAS

Map[†] ref.	National/international designation Name of area	IUCN management category	Area (ha)	Year notified
	Specially Protected Areas			
1	Ardery and Odbert Islands	I	220	1966
2	Beaufort Island	I	1,865	1966
3	Cape Hallett	I	?	1966
4	Coppermine Peninsula	I	65	1970
5	Dion Islands	I	17	1966
6	Green Island	I	20	1966
7	Lagotellerie Island	I	?	1985
8	Litchfield Island	I	250	1975
9	Lynch Island	I	8	1966
10	Moe Island	I	117	1966
11	New College Valley	I	?	1985
12	North Coronation Island	I	?	1985
13	Rookery Islands	I	65	1966
14	Sabrina Island	I	60	1966
15	Southern Powell & adjacent islands	I	610	1966
16	Taylor Rookery	I	?	1966
	Sites of Special Scientific Interest			
17	Ablation Point, Ganymede Heights, Alexander Is.	I	18,000	1990
18	Arrival Heights	I	?	1975
19	Avian Island, North-West Marguerite Bay	I	100	1990
20	Barwick Valley	I	29,120	1975
21	Biscoe Point	I	?	1985
22	Byers Peninsular	I	3,027	1975
23	Canada Glacier	I	100	1985
24	Cape Crozier	I	462	1975
25	Cape Royds	I	2	1975
26	Cape Shirreff	I	265	1989
27	Caughley Beach	I	?	1985
28	Chile Bay (Discovery Bay)	I	?	1987
29	Cierva Point	I	?	1985
30	Clark Peninsula	I	?	1985
31	Fildes Peninsula	I	154	1975
32	Harmony Point	I	?	1985
33	Haswell Island	I	80	1975
34	Linnaeus Terrace	I	?	1985
35	Marine Plain, Mule Peninsula	I	2,340	1987
36	Mount Flora, Hope Bay, Antarctic Peninsula	I	?	1990
37	North-east Bailey Peninsula	I	?	1985

Map[†] ref.	*National/international designation* Name of area	IUCN management category	Area (ha)	Year notified
38	North-west White Island	I	?	1985
39	Parts of Deception Island	I	?	1985
40	Port Foster, Deception Island	I	?	1987
41	Potter Peninsula	I	?	1985
42	Rothera Point	I	?	1985
43	South Bay, Doumer Island	I	?	1987
44	Summit of Mt Melbourne	I	?	1987
45	Svarthamaren	I	390	1987
46	Tramway Ridge	I	1	1985
47	Western Shore, Admiralty Bay	I	160,000	1979
48	Yukidori Valley	I	300	1987
	Specially Reserved Area			
49	Dufek Massif			

[†]Locations of most protected areas are shown on the accompanying map.

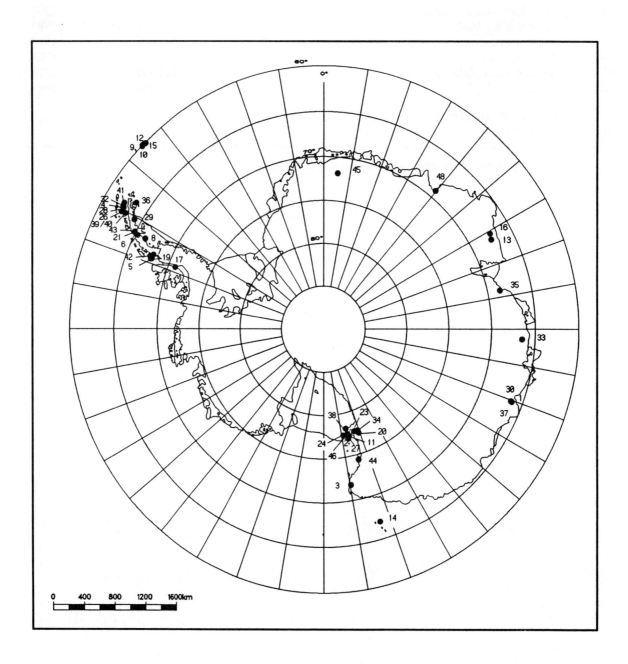

Protected Areas of the Antarctic Treaty Territory

SOUTHERN OCEAN ISLANDS

FRANCE – ILES CROZET

Area 500 sq. km

Population No permanent inhabitants

Economic Indicators
GDP: No information.
GNP: No information

Policy and Legislation The islands are a part of the Territoire des Terres Australes et Antarctiques Françaises (TAAF).

Most of the islands (a total area of 22,500ha), with the exception of Ile de la Possession and Ile de l'Est, form a part of the "Parc National Antarctique Français". This is thought to approximate IUCN category II (National Park). It was first established under an individual decree of 30 December 1924 which was replaced by a more comprehensive decree in 1938. Numbers of seals and elephant seals which may be killed, and time of year when this is permissible, can be regulated. Birds are also protected in some areas.

The Nature Conservation Act (No. 76 629) of 10 July 1976, which, among other things, provides the legislative framework for defining, designating and establishing nature reserves in France, was extended to include the TAAF in 1982.

Antarctic and Subantarctic Lands Act (Décret 1966) regulates activities near breeding colonies of birds, to minimise human disturbance. There is strict control on fishing within a 200-mile Exclusive Economic Zone around the islands, while the exploitation of whales in the 200-mile zone is prohibited (Arrêté, 1981).

Most of the Crozet Islands have been proposed (under five separate proposals) as specially protected areas, to be awarded similar status to those areas covered under the Antarctic Treaty (see Antarctic Treaty Territory information sheet) (Bonner and Lewis Smith, 1985).

International Activities Iles Crozet are considered to be a part of the Antarctic for the purposes of the Scientific Committee on Antarctic Research (SCAR). SCAR members are national academies of sciences, and SCAR has been closely involved with conservation issues in the region. The islands do not, however, fall within the limits of the Antarctic Treaty and hence do not fall under the obligations associated with it, including those related to protected areas.

Iles Crozet lie within the area covered by the Convention for the Conservation of Antarctic Marine Living Resources (CCAMLR) which regulates exploitation of marine animals, but local French management of fisheries occurs.

Administration and Management The TAAF is divided into four districts, administered from Paris by the Head of Territory assisted by an Advisory Council. Local responsibility for the District of Crozet is assumed by a District Head, based at Port Alfred-Faure. The Comité de l'Environnement (Committee for the Environment), established in 1982, advises the Head of the Territory on matters relating to environmental and wildlife protection, protected areas, and management problems. The administration of TAAF has prepared guidelines for protecting the fauna of the islands. Killing or removal of animals is forbidden unless specifically permitted.

Research in TAAF is conducted under the auspices of a scientific council in Paris, chaired by the Head of the Territory. The council is assisted by the Direction Generale de la Recherche Scientifique et Technique (General Directorate for Scientific and Technical Research) and the Comité National Français des Recherches Antarctiques (French National Committee on Antarctic Research). In recent years, the level of research carried out in the Crozet group has increased, with emphasis on biology and ecology of avifauna, and the effects of alien animals on the native biota.

Systems Reviews Iles Crozet comprise an archipelago in the southern Indian Ocean, 2,400km south-east of South Africa. It consists of five main volcanic islands, in two groups: a western group of Ile aux Cochons, Ile des Pingouins, and Ilôts des Apôtres; and an eastern group comprising Ile de la Possession, and Ile de l'Est.

Vegetation of the three main islands is relatively uniform. Herbs are common on the coasts. Tussock grassland dominates to about 50m. Grass is important from 50-100m, with Kerguelen cabbage, dwarf shrub, and fern. Above 100m, feldmark vegetation predominates, comprising cushion herbs and grass. Above 200-300m, only sparse mosses and lichens are found.

The islands have a very rich and relatively diverse avifauna, with the total number of birds estimated at 25 million. The invertebrate fauna is extensive, with numerous endemic insect species. A number of seal species breed and notothenid fishes and lithodid crabs are common in coastal waters.

The islands were discovered in 1772 by Marion Dufresne, and were visited sporadically by sealers in the 19th century. A meteorological station, Port Alfred-Faure, was established by the French on Ile de la Possession in 1962. It is manned continually, by 20 men on average, and relieved annually.

Alien mammals have been introduced to several islands and Ile aux Cochons is particularly badly affected. Rats and mice are the only mammals currently on Ile de la Possession and there are no mammals on Ile de l'Est at present. Salmonid fishes have also been introduced into the group, but are not well established; the fishes may have a localised impact on the native invertebrate fauna. Alien plants are widespread on some of the islands, especially Ile de la Possession where about 50 alien vascular species are established, at least 12 being naturalised with widespread distributions.

Iles Crozet are considered to be of high biological value, with a very rich avifauna. The impact of man in the past has been variable rather than heavy. Several islands in the group are largely unmodified with intact native fauna and flora; islands such as Ilôts des Apôtres, Ile des Pingouins, and to a lesser extent Ile de l'Est. The former two islands are covered under the national park classification, but Ile de l'Est is not. Clark and Dingwall (1985) consider that protection equivalent to IUCN Category I (Scientific/Strict Nature Reserve) is warranted.

Ile de la Possession is the most modified island in the group, and is the only one with human occupation, but still retains some notable fauna and flora. In 1982 it was proposed to construct a major road between the coastal landing area and the meteorological base, which would seriously disturb a major king penguin *Aptenodytes patagonicus* rookery. The current status of this proposal is unknown.

Present fishing operations in the area of Iles Crozet are light, following over-fishing in the early 1970s. However, recent surveys have indicated that several economically important species are abundant around the islands. The possibility of renewed development of a major fishery raises questions concerning the effects of fisheries on the bird life of the islands.

Other Relevant Information Further information concerning protected areas in a wider French context may be obtained from the main body of information on France (see Volume 2).

Addresses

Terres Australes et Antarctiques Françaises (Director), 27 rue Oudinot, 75700 Paris, France

References

Bonner, W.N. and Lewis Smith, R.I. (1985). *Conservation areas in the Antarctic*. Scientific Committee on Antarctic Research, International Council of Scientific Unions. 299 pp.

Clark, M.R. and Dingwall, P.R. (1985). *Conservation of islands in the Southern Ocean: a review of the protected areas of Insulantarctica*. IUCN, Gland, Switzerland and Cambridge, UK. 188 pp.

FRANCE – ILES AMSTERDAM AND ST PAUL

Area 55 sq. km (Ile Amsterdam); 7 sq. km (Ile St Paul)

Population No permanent inhabitants

Economic Indicators
GDP: No information.
GNP: No information.

Policy and Legislation The islands are a part of the Territoire des Terres Australes et Antarctiques Françaises (TAAF).

The entire area of both islands forms a part of the "Parc National Antarctique Français", which is thought to approximate IUCN Category II (National Park). It was first established under an individual decree of 30 December 1924 which was replaced by a more comprehensive decree in 1938. Numbers of seals and elephant seals which may be killed, and time of year when this is permissible, can be regulated. Birds are also protected in some areas.

The Nature Conservation Act (No. 76 629) of 10 July 1976, which, among other things, provides the legislative framework for defining, designating and establishing nature reserves in France, was extended to include the Territory of TAAF in 1982.

The Antarctic and Subantarctic Lands Act (Décret, 1966) covers and restricts the activities of French personnel. Exploitation of whales in the 200-mile Exclusive Economic Zone around the island is prohibited.

Roche Quille, a small island off the coast of Ile St Paul, and a 20 sq. km area of the central plateau and south-western cliffs of Ile Amsterdam, have been proposed as specially protected areas (under two separate proposals), to be awarded similar status to those areas covered under the Antarctic Treaty (see Antarctic Treaty Territory information sheet) (Bonner and Lewis Smith, 1985).

International Activities Iles Amsterdam and St Paul are considered to be a part of the Antarctic for the purposes of the Scientific Committee on Antarctic Research (SCAR). SCAR members are national academies of sciences, and SCAR has been closely involved with conservation issues in the region. The islands do not however fall within the limits of the Antarctic Treaty and hence they do not fall under the obligations associated with it, including those related to protected areas. Furthermore, they lie outside the boundaries of the Convention for the Conservation of Antarctic Marine Living Resources.

Administration and Management The TAAF is administered from Paris, as four districts. The District of Amsterdam and Saint-Paul is the responsibility of the District Head, based at La Roche Godon.

The Comité de l'Environnement (Committee for the Environment), established in 1982, advises the Head of the Territory on matters relating to environmental and wildlife protection, protected areas, and management problems.

Specific details of policy are not known. Guidelines for conduct of personnel on the island, and visitors, have been prepared by TAAF. Personnel are periodically informed of environmental issues and conservation regulations.

Systems Reviews Both islands (which are some 100km apart) lie on the same raised plateau in the southern Indian Ocean, 1,300km north-east of Iles Kerguelen. Ile Amsterdam has steep coastal cliffs, especially on the western side, rising to a central plateau at around 600m which forms the floor of a crater. Highest points on the island are the remnants of the crater wall, with Mont de la Dives reaching 881m. Numerous craters and vents occur on the plateau and flanks of the central cone. There is no present volcanic activity.

Ile St Paul is the summit of another volcano, with steep cliffs along its eastern coast: the crater associated with the volcano forms a deep bay along this coast. It has only a very narrow entrance to the sea and this is protected by a shallow sill. The highest point lies on the crater rim and is some 268m; the southern and eastern flanks of the island are more gently sloping. The islands lie to the north of the Subtropical Convergence. Lowland slopes to 250m are dominated by meadows of tussock grass. Dense grasslands of sedges, and to a lesser extent *Spartina arundinacea*, occur to about 600m. Above this, vegetation comprises feldmark of dwarf shrub, sphagnum bogs, and mosses. *Phylica arborea* trees, up to 7m high, and ferns were formerly widespread on Ile Amsterdam, but now occur only in sheltered areas free from grazing. They are not found on Ile St Paul.

Only seven bird species are known to breed on Ile Amsterdam and only five on Ile St Paul. The breeding population of yellow-nosed albatross (37,000 pairs) on Ile Amsterdam comprises more than one-half of the world population. Amsterdam Island fur seal *Arctocephalus tropicalis* breeds on both islands, and southern elephant seal *Mirounga leonina* occurs.

Ile Amsterdam was discovered by Sebastian del Cano (on one of Magellan's ships) in 1522, but was not landed on until 1696. It was visited sporadically by explorers, sealers, scientists and crayfishermen in the 1800s, and claimed by France in 1843. Cattle farming was attempted in 1871. A French meteorological station was established in 1949-50 and is manned by 35 men on average, relieved annually.

Research in TAAF is centred on Iles Kerguelen and Iles Crozet, but limited studies are conducted on Ile Amsterdam; very little research is conducted on Ile St

Paul. On both islands, fur seals were heavily exploited in the 1800s, but populations are now recovering. However, southern elephant seal no longer breeds on the island.

Introduced mammals are widespread and cause considerable problems; Norwegian rats, mice and cats are found on both islands, while rabbits are found on Ile St Paul. Vegetation on Ile Amsterdam has been modified considerably by grazing of introduced mammals, in particular cattle, which are still present, but also sheep, goats and pigs which were formerly present. Cattle are widely distributed, and their grazing has destroyed large areas of tree *Phylica arborea* and ferns. General reduction of vegetation cover has contributed to spread of alien vegetation, caused soil erosion, and reduced suitable habitat for some bird species. Vegetation has also been frequently destroyed by fire, the last one in 1973 burning for many months and affecting colonies of birds. Alien plants are widespread on both islands.

Control measures to limit number and distribution of introduced mammals should be considered: control programmes of cats and rats would need to be introduced together and be long-term. Discovery of Amsterdam Island albatross in 1982 (numbering only 25-30 breeding pairs), and the presence of an undescribed species of procellariid highlights the need for a comprehensive biological survey of Ile Amsterdam. Preparation of detailed inventories of fauna and flora is an essential prerequisite for any conservation planning.

Other Relevant Information Further information concerning protected areas in a wider French context may be obtained from the main body of information on France (see Volume 2).

Addresses

Terres Australes et Antarctiques Françaises (Director), 27 rue Oudinot, 75700 Paris, France

References

Bonner, W.N. and Lewis Smith, R.I. (1985). *Conservation areas in the Antarctic*. Scientific Committee on Antarctic Research, International Council of Scientific Unions. 299 pp.

Clark, M.R. and Dingwall, P.R. (1985). *Conservation of islands in the Southern Ocean: a review of the protected areas of Insulantarctica*. IUCN, Gland, Switzerland and Cambridge, UK. 188 pp.

FRANCE – ILES KERGUELEN

Area 7,000 sq. km

Population No permanent inhabitants

Economic Indicators
GDP: No information
GNP: No information

Policy and Legislation The islands were a dependency of Madagascar from 1924 to 1955 (then a French Colony), and then became part of Territoire des Terres Australes et Antarctiques Françaises (TAAF).

Areas of Iles Kerguelen are part of the "Parc National Antarctique Français" which is thought to approximate IUCN Category II (National Park). It was first established under an individual decree of 30 December 1924 which was replaced by a more comprehensive decree in 1938. The latter provides protection for seals on the west coast of Ile Kerguelen. Numbers of seals and elephants seals which may be killed, and time of year when this is permissible, can be regulated. Birds are also protected in some areas.

The Nature Conservation Act (No. 76 629) of 10 July 1976, which, among other things, provides the legislative framework for defining, designating and establishing nature reserves in France, was extended to include the TAAF in 1982.

The Antarctic and Subantarctic Lands Act (Décret 1966) limits human activities near breeding colonies of birds. There is a ban on exploitation of whales within the 200-mile Exclusive Economic Zone around the islands (Arrêté, 1981). Other restrictions in this zone include a ban on fishing in inshore waters, closed seasons and areas where appropriate, minimum size of fish, need to hold a fishing licence, return of detailed catch records, and presence of observers at times.

The Iles Nuageuses, some 14km to the north of the main island, have been proposed as a specially protected area, to be awarded similar status to those areas covered under the Antarctic Treaty (see Antarctic Treaty Territory information sheet) (Bonner and Lewis Smith, 1985).

International Activities Iles Kerguelen are considered to be a part of the Antarctic for the purposes of the Scientific Committee on Antarctic Research (SCAR). SCAR members are national academies of sciences, and SCAR has been closely involved with conservation issues in the region. The islands do not, however, fall within the limits of the Antarctic Treaty and hence do not fall under the obligations associated with it, including those related to protected areas.

The area around Iles Kerguelen is within the boundaries of the Convention for the Conservation of Antarctic Marine Living Resources (CCAMLR), although local French management of fisheries occurs.

Administration and Management TAAF is administered in Paris, as four districts. The district of Kerguelen has its own district head based at Port-aux-Français, assisted by an advisory council.

The Comité de l'Environnement (Committee for the Environment), established in 1982, advises the Head of the Territory on matters relating to environmental and wildlife protection, protected areas, and management problems. The administration of TAAF has prohibited the taking of birds, with the exception of Kerguelen pintail. Personnel in TAAF are periodically informed of conservation regulations and requirements.

A local scientific council, chaired by the head of the Territory, in conjunction with the Direction Générale de la Recherche Scientifique et Technique (General Directorate for Scientific and Technical Research), coordinates research programmes. Research is conducted in association with the Comité National Français des Recherches Antarctiques (French National Committee on Antarctic Research), a member of the international Scientific Committee on Antarctic Research (SCAR).

Systems Reviews Iles Kerguelen is an archipelago comprising one large island, Ile Kerguelen (Grande Terre) of 660,000ha, with about 300 off-lying islands, islets and rocks lying in the southern Indian Ocean, 1,800km south-west of Australia. The islands lie just south of the Antarctic Convergence. Ile Kerguelen (Grande Terre) itself is 120km by 140km, its rocky shoreline deeply dissected by inlets into a number of peninsulas. The islands are of volcanic origin, and rocks are mainly of basaltic composition. High mountains occur in the west, with the highest peak, Mont Ross, reaching 1,850m. There are extensive plateaux in the east, deep glacial valleys, and numerous rivers and lakes. In the west there is a large ice cap, Calotte Glaciaire Cook, some 50km by 20km, the remains of an ice sheet which formerly covered the entire island. Numerous other glaciers occur.

Mires are widespread, dominated by bryophytes, low-lying shrubs, cushion- forming flowering plants, and tussock-forming grasses. Vegetation varies considerably to 500m. Above the plateaux grasses, ferns, mosses, and lichens can occur, but many high areas are devoid of vegetation. The islands host 30 breeding species of bird, including penguins, albatrosses, petrels, gulls and terns. Southern elephant seal *Mirounga leonina* and Antarctic fur seal *Arctocephalus gazella* breed on the islands. In nearshore coastal waters, several fish species are abundant, in particular notothenid cods.

The islands were discovered by Captain Yves Joseph de Kerguelen-Tremarec in 1772. Sealing, whaling, and several scientific expeditions visited the islands in the 19th century. In 1908 a sheep station and whaling station were established by Bossiere at Port Jeanne d'Arc continuing until 1932. A meteorological station was established at Port-aux-Français, in the Golfe du Morbihan, in 1949. There is now a permanent research station at this site. Research undertaken is mostly atmospheric and geophysical, although some biological research is also carried out. Research on introduced mammals has increased under the sponsorship of TAAF, with the objective of conducting control programmes. Staff number 90 on average (1980), and are relieved annually.

Exploitation of marine mammals and seabirds occurred throughout much of the 19th century, and elephant seals were killed up until the 1960s. Stocks of Antarctic fur seal and elephant seal have only recently begun to recover.

Introduced mammals have had considerable impact on the islands' native biota. Rabbits were introduced in 1874, and grazing has decimated certain important plant species, while less palatable species have become dominant over large areas, in contrast to off-lying islands without alien mammals. Grazing has also caused the loss of much invertebrate fauna closely associated with the original flora. Feral populations of sheep, mouflon and reindeer have added to the effect of rabbits, causing considerable erosion, and reducing nesting habitat for seabirds. It has been estimated that there are 3,500 cats on Ile Kerguelen (1977) killing more than 1 million birds annually. Cat control was attempted in 1972-75, with some success, but the programme was discontinued. The salmonid species, intentionally introduced as game fish, have reportedly had a marked effect on the invertebrate fauna which previously was not subject to predation by fishes. Over much of Ile Kerguelen modification of biota is too advanced, and populations of mammals too well-established for any possibility of restoration to its original state; a systematic programme of control would reduce pressures on vegetation and seabirds. Alien plants are also widespread over Ile Kerguelen. Despite the impact on the indigenous biota, plant and animal introductions have continued until recently (for example, the introduction of conifer trees in 1976).

There is a need for a full biological survey of Iles Kerguelen, in particular of the off-lying islands which are thought to be largely free of introduced mammals and to have intact native biota. Species which are rare on the main island are known to be abundant on some offshore islands. Identification of priority areas for conservation is of immediate concern. To this end, an increase in permanent scientific staff attached to TAAF would be beneficial, enabling more long-term investigations to be undertaken. An increase in logistical support for biological research is also possibly necessary. A greater level of legal protection is needed for areas of the main island, and total protection of off-lying island environments and biota would help ensure conservation of the diverse fauna of Iles Kerguelen.

Other Relevant Information Further information concerning protected areas in a wider French context may be obtained from the main body of information on France (see Volume 2).

Addresses

Terres Australes et Antarctiques Françaises (Director), 27 rue Oudinot, 75700 Paris, France

References

Bonner, W.N. and Lewis Smith, R.I. (1985). *Conservation areas in the Antarctic*. Scientific Committee on Antarctic Research, International Council of Scientific Unions. 299 pp.

Clark, M.R. and Dingwall, P.R. (1985). *Conservation of islands in the Southern Ocean: a review of the protected areas of Insulantarctica*. IUCN, Gland, Switzerland and Cambridge, UK. 188 pp.

NORWAY – BOUVETØYA (BOUVET ISLAND)

Area 50 sq. km

Population No permanent inhabitants

Economic Indicators
GDP: No information.
GNP: No information

Policy and Legislation The entire island is owned by the Crown. It is a strict nature reserve and was established under an Order of Council in December 1971, which gives protection to the flora and fauna; prohibits alien introductions, the construction of buildings, roads etc, vehicular transport, and the disposal of waste which could harm flora and fauna; and governs scientific activities by permit. The conditions include territorial waters. Seals are additionally protected by Norwegian Ordinance.

The government's main goals concerning polar regions include "the improvement of the knowledge on which the management of the polar regions is based and taking of the necessary steps for their protection, ensuring the sound management of resources in the Antarctic and the protection of the natural environment and the strengthening of international cooperation under the Antarctic Treaty". One specific aim for the furtherance of research put forward by the government is that regular Antarctic expeditions should be organised every third year, with emphasis on Norway's management responsibilities and environmental issues (Ministry of the Environment, 1989).

International Activities Bouvetøya does not fall within the limits of the Antarctic Treaty and hence does not fall under its obligations, including those related to protected areas. The island is within the boundaries defined by the Convention for the Conservation of Antarctic Marine Living Resources (CCAMLR) which regulates exploitation of marine animals.

Administration And Management The island is administered by the Norwegian Ministry of Environment (*Miljoverndepartementet*) which was created in 1972 in conjunction with the Department of Justice (Polar Division). These bodies are advised by the Norsk Polarinstitutt (Norwegian Polar Research Institute), which is one of five directorates/institutions under the Ministry of Environment. Activities on the island are governed by permit.

There have been occasional visits by scientific expeditions since the first landing of scientists from the "Norvegia" in 1927. The level of scientific research increased has recently with several Norwegian Antarctic Research Expeditions (NARE) expeditions, but further detailed biological surveys appear to be needed. Short-term weather stations have been established on the island since 1977. South Africa has been granted permission to construct a meteorological station on the island, but no building has yet been undertaken.

Systems Reviews Bouvetøya lies in the South Atlantic Ocean, 3,330km south of South Africa. It is roughly oval in outline, 9.5km by 7km, and is a domed plateau reaching a maximum altitude of 800m, with steep cliffs except on the eastern side. A low lava platform (Nyroysa) on the west coast resulted from a landslide in 1957/58. A single rocky islet, Larsoya, occurs to the south-west. The island is probably a complex of volcanic cones, and is of basaltic composition. It is heavily glaciated, being 95% ice-covered. It lies about 500km south of the Antarctic Convergence, and pack ice from the Weddell Sea is occasionally present.

The flora is regarded as impoverished, showing affinities with island groups to the west such as the South Sandwich Islands and South Shetland Islands. Antarctic fur seal *Arctocephalus gazella* is numerous, and southern elephant seal *Mirounga leonina* probably also breeds. Three species of penguins and several other species of seabird are found on the island. There are no introduced animals.

Bouvetøya was discovered in 1739 by the Frenchman Jean-Baptiste Lozier de Bouvet. Sporadic sealing occurred in the 1800s. The island was claimed by Norway in 1928, with formal annexation in 1930. Several scientific expeditions have visited the island this century.

Other Relevant Information Further information concerning protected areas in a wider Norwegian context may be obtained from the main body of information on Norway.

Addresses

Ministry of Environment, PO Box 8013 Dep, Oslo 1, Norway

Justis-og politidepartementet, Polaravdelingen, PO Box 8005 Dep, Oslo 1, Norway

References

Bonner, W.N. and Lewis Smith, R.I. (1985). *Conservation areas in the Antarctic*. Scientific Committee on Antarctic Research, International Council of Scientific Unions. 299 pp.

Clark, M.R. and Dingwall, P.R. (1985). *Conservation of islands in the Southern Ocean: a review of the protected areas of Insulantarctica*. IUCN, Gland, Switzerland and Cambridge, UK. 188 pp.

Ministry of Environment (1989). *Report to the Storting No. 46 (1988-89)*. Environment and Development, programme for Norway's follow-up of the report of the World Commission on Environment and Development. Ministry of Environment. 74 pp.

SOUTH AFRICA – PRINCE EDWARD ISLANDS

Area 300 sq. km (Marion Island); 44 sq. km (Prince Edward Island)

Population No permanent inhabitants

Economic Indicators
GDP: No information
GNP: No information

Policy and Legislation The islands are not legally protected, although the Sea Birds and Seals Protection Act (Act 46) of 1973 gives some specific legal protection to fauna. This Act was expected to be incorporated into the 1989 Environmental Conservation Act (Cooper and Condy, 1988). Despite this limited statutory protection, both islands are effectively managed as a nature reserve.

Like the rest of South Africa the islands have a territorial sea of 12 nautical miles and an exclusive fishing zone of 200 nautical miles.

A conservation policy, including "operational guidelines" for visiting scientists, has been prepared which is equivalent to a management plan. This "Code of Conduct for the Environmental Protection of the Prince Edward Islands" was developed by the South African Scientific Committee for Antarctic Research in 1982 and was being revised by the Department of Environment Affairs in 1988 prior to its formal adoption. It regulates construction and maintenance activities, sets out procedures for the disposal of waste materials and the prevention of pollution, restricts tourism, limits the taking of flora and fauna, and deals with certain aspects of aircraft and ship operation to minimise disturbance to animals. Nine areas on Marion Island, and the entire area of Prince Edward Island have been designated "wilderness areas", while sealers' and shipwreck remains are totally protected. In wilderness areas greater restrictions are placed on access and activities: only essential research may be conducted, while rigid precautions must be taken to prevent the accidental introduction of alien species. Research carried out on Prince Edward Island is very limited: it is only permitted for projects specifically requiring information from that island, and the island is only visited once or twice a year for 5-10 days. The Code of Conduct also provides for the proclamation and deproclamation of permanent and seasonal specially protected areas and sites of special scientific interest, to be equivalent to those declared within the Antarctic Treaty areas, although to date none has been proclaimed on Prince Edward Islands. Environmental impact assessments prior to the implementation of major projects is recommended.

Further recommendations have been made, including the drafting and implementation of a formal and effective management plan and the formal, statutory protection of the islands as a wilderness area or nature reserve (Cooper and Condy, 1988).

International Activities Prince Edward Islands are considered to be a part of the Antarctic for the purposes of the Scientific Committee on Antarctic Research (SCAR). SCAR members are national academies of sciences, and SCAR has been closely involved with conservation issues in the region. The islands do not, however, fall within the limits of the Antarctic Treaty and hence do not fall under the obligations associated with it, including those relating to protected areas.

The Islands lie within the area covered by the Convention for the Conservation of Antarctic Marine Living Resources (CCAMLR) which regulates exploitation of marine animals.

Administration and Management Activities on the islands are administered jointly by the South African Scientific Committee for Antarctic Research (SASCAR), a body within the Council for Scientific and Industrial Research (CSIR) and the Antarctic section of the Department of Transport. Annual scientific expeditions to the island are sponsored by the Department of Transport, which provides transport, food and clothing, but are managed by the CSIR. In 1983 the research budget for the islands was approximately R300,000.

Systems Reviews The islands lie some 22km apart in the South Indian Ocean, some 925km west of Iles Crozet and 2,300km south-east of South Africa. Marion Island is roughly oval, with a central mountainous area rising from a coastal plain: the highest mountain, State President Swart Peak, rises to 1,230m. The island is the summit of a basaltic volcano and there has been some recent activity, with a small fissure eruption on the west coast in 1980. Prince Edward island is also oval in shape, with a coastal plain in the east, but with precipitous cliffs in the north and south and only a narrow coastal plain in the west. There are a several volcanic cones in the high central area.

On the salt-sprayed coastline herbs dominate, inland tussock grass dominates, with ferns in more protected areas and mires and bogs over a fairly wide area. Feldmark communities occur above 300m. A number of seal species are found, and the islands have a very rich avifauna.

The islands were probably both discovered by the Dutchman Barent Ham in 1663, but the first confirmed sighting was in 1772 by Marion Dufresne, while Captain James Cook visited and named the islands in 1775. By 1802 sealers had set up temporary bases on the islands. The islands were annexed by South Africa in 1948 and 1949 (Prince Edward Islands Act No. 43 of 1948). A meteorological base, established on Marion Island in 1949, is now a scientific base with up to 22 staff and is visited by a research and supply vessel twice annually, when staff may be relieved.

There are a number of introduced alien species on Marion Island, including the grass *Poa annua* which has become well established, mice, which are not regarded as a serious problem, and cats. Cats were increasing dramatically in numbers until a control programme, started in 1977, resulted in an immediate population reduction and is continuing to prove effective: since 1986 a major programme of night hunting has been underway and it is thought that this will continue until cats have been completely eradicated. The only major alien species on Prince Edward Island is the grass *Poa annua*.

Other Relevant Information Further information concerning protected areas in a wider South African context may be obtained from the main body of information on South Africa (see Volume 2).

Addresses

SASCAR, c/o Foundation for Research and Development, Council for Scientific and Industrial Research,

Program for the Environment, PO Box 395, Pretoria 0001, South Africa (Tel: 12 841 2911)

Antarctic Section, Department of Transport, Private Bag X193, Pretoria 0001, South Africa

References

Bonner, W.N. and Lewis Smith, R.I. (1985). *Conservation areas in the Antarctic.* Scientific Committee on Antarctic Research, International Council of Scientific Unions. 299 pp.

Clark, M.R. and Dingwall, P.R. (1985). *Conservation of islands in the Southern Ocean: a review of the protected areas of Insulantarctica.* IUCN, Gland, Switzerland and Cambridge, UK. 188 pp.

Cooper, J. and Condy, P.R. (1988). Environmental conservation at the sub-Antarctic Prince Edward Islands: a review and recommendations. *Environmental Conservation* 15(4): 317-326.

UNITED KINGDOM – FALKLAND ISLANDS

Area 13,000 sq. km

Population 1,916 permanent (1986), 2,000 (temporary)
Natural increase: no information

Economic Indicators
GDP: No information
GNP: No information

Policy and Legislation The present constitution dates from 1985. Sovereignty over the Islands is claimed by Argentina, which calls them the Islas Malvinas. Administration is exercised by the Governor, appointed by the UK government, a Legislative Council and an Executive Council.

The major protected areas legislation in the Falklands is the Nature Reserves Ordinance of 1964, which provides for the designation of nature reserves (Annex). Sanctuaries may be declared under the Wild Animals and Birds Protection Ordinance 1964, which also provides for more general protection of animals. Private land may, with the consent of the owner, be made a wild animal and bird sanctuary (Annex). Other conservation legislation includes the Seal Fishery Ordinance of 1953; the Whale Fishery Ordinance of 1964 and the Control of Kelp Ordinance of 1970. The conservation legislation has been recommended for revision for a number of years, and in 1991 a preliminary review of this, as well as the protected areas themselves, was in progress (S. Oldfield, pers. comm., 1991).

In 1987, a 150-mile fishing zone was declared although a number of countries are permitted to fish within the zone (Hills, 1991).

International Activities The Falkland Islands are included in the UK ratification of the Convention for the Protection of Wetlands Especially as Waterfowl Habitat (Ramsar Convention) and the Convention Concerning the World Cultural and Natural Heritage (World Heritage Convention). However, the application of these conventions remains problematic due to the disputed claim over sovereignty: no sites have been declared.

Administration and Management At present there is no government agency specifically responsible for conservation in the Falklands. Conservation advice is provided to the Falkland Islands government by local scientists and naturalists on a voluntary basis. The Falkland Islands Economic Study of 1982 (which updated an earlier study by Lord Shackleton in 1975) recognised the international conservation importance of the Falklands and recommended the employment of a permanent scientific officer for conservation matters (Oldfield, 1987).

The UK Ministry of Defence has an important role in the conservation of the Falklands and takes an active interest in conservation issues. A conservation group has been established in the garrison, an officer has been nominated responsible for conservation matters, and independent consultants have been employed as advisors. All visiting members of the services are briefed on the importance of the islands, and strict controls have been placed on protected sites near the military base (Oldfield, 1987; Oldfield, *in litt.*, 1991).

The Falklands Islands Foundation was a charity registered in the UK which promoted and supported both natural and cultural conservation in the Islands. It was associated with the Falklands Islands Trust, based in the Falklands. The two groups were combined in 1991 and are now known as Falkland Conservation. Among the activities now being undertaken by the new group is the acquisition of land, notably small offshore islands, scientific research and education (Oldfield, 1987). A fully-paid member of Falkland Conservation is based in Stanley (Thompson, pers. comm., 1991). The NGO Forum for Nature Conservation in the UK Dependent Territories is another non-governmental organisation which aims to promote and coordinate conservation in UK Dependent Territories. It liaises with the UK government and acts as a link for voluntary organisations and individuals concerned with conservation issues in these areas.

Systems Reviews The Falklands consist of two main islands, East Falkland and West Falkland, together with over 700 associated islands and islets. They are composed of sedimentary rocks: the landscape is generally rugged and hilly, with the highest peaks being Mount Adam (700m) on West Falkland and Mount Usborne (705m) on East Falkland. East Falkland is divided into two large land blocks, connected by a narrow land bridge the southern block is dominated by the plain of Lafonia, an area of gentle relief. The coastline is deeply indented. Virtually the whole area is covered by blanket peat which is eroded and associated with scree and stone runs at high altitude (Clark and Dingwall, 1985).

On the main islands the vegetation has been severely degraded by burning and overgrazing. Most of the main islands are dominated by oceanic heath, with grass and small shrubs; bush and scrub associations are found inland on well-drained slopes; bog communities are found on poorly drained areas; while feldmark vegetation occurs above 600m, with cushion-forming vascular plants and associated mosses and lichens. Vegetation of the ungrazed areas and the offshore islands is dominated by dense stands of tussock grass. There are no trees. Kelp beds are common around the coasts (Clark and Dingwall, 1985; Oldfield, 1987).

The avifauna includes 63 breeding bird species of which 16 are endemic. Southern elephant seal, southern sea lion and South American fur sea are common, while leopard seal is occasionally present (Clark and Dingwall, 1985).

The Falklands are thought to have been discovered in 1592 by the English Captain, John Davis; the first recorded landing was in 1690 by Captain John Strong who named the islands after Viscount Falkland, then Treasurer of the British navy. Various settlements were established on the islands by the British, French and Spanish during the 1700s. Current British settlement dates from 1833 (Clark and Dingwall, 1985).

Nature reserves, on Crown land, cover about 6,000ha. Privately-owned reserves and sanctuaries, which may be owned by individuals or organisations, cover a further 4,000ha (Oldfield, 1987).

Introduced mammals are an environmental problem. Domesticated pigs, rabbits, goats and cattle introduced onto several of the larger islands have, together with the burning of the tussock grass, led to the destruction of much of the native vegetation. The introduction of sheep farming to the Islands in the 1860s compounded this effect. Rats, cats and mice found on the main islands are also likely to have greatly affected the bird populations. Dogs, foxes and horses are also present (Clark and Dingwall, 1985). Uncontrolled burning on the main islands continues to threaten the natural vegetation and has led to erosion, while there is some concern over the sustainability of fishing levels in the surrounding waters. It has been recommended that a detailed biological and ecological survey of the Islands be undertaken to provide the basis for future conservation planning (S. Oldfield, *in litt.*, 1991).

Addresses

The Executive Council, Government House, Stanley, Falkland Islands

Falkland Conservation, Stanley, Falkland Islands
Falkland Conservation (Secretary), 13 Well Court, Dean Village, Edinburgh EH4 3BE, Scotland, UK (Tel: 031 556 6226)
NGO Forum for Nature Conservation in the UK Dependent Territories, 22 Mandene Gardens, Great Gransden, Sandy, Bedfordshire SG19 3AP, UK (Tel: 07677 558)

References

Clark, M.R. and Dingwall, P.R. (1985). *Conservation of islands in the Southern Ocean: a review of the protected areas of Insulantarctica.* IUCN, Gland, Switzerland and Cambridge, UK. 188 pp.

Hills, A. (1991). Keeping a stiff upper lip. *Geographical Magazine* June. Pp. 22-24.

Oldfield, S. (1987). *Fragments of paradise: a guide for conservation in the UK dependent territories.* British Association of Nature Conservationists, Oxford, UK. 192 pp.

ANNEX
Definitions of protected area designations, as legislated, together with authorities responsible for their administration

Title: Nature Reserves Ordinance

Date: 1964

Brief description: Provides for the designation of nature reserves

Administrative authorities: Governor-in-Council, Falkland Islands

Designations

Nature reserve Land reserved for the purposes of protecting flora and fauna, and for providing opportunities for research on the biota. They may only be designated on Crown land. Entry into, and activities within, reserves are restricted, including the burning and cutting of vegetation and the hunting of animals.

Source: Clark and Dingwall, 1985

Title: Wild Animals and Birds Protection Ordinance

Date: 1964

Brief description: Provides for the designation of animal and bird sanctuaries as well as providing for the more general protection of animals and birds

Administrative authority: Governor-in-Council, Falkland Islands

Designations:

Animal and bird sanctuary Established for the protection of animals and birds, with specified exceptions. The introduction of animals is prohibited. There is provision for scientific research.

Private sanctuary Protected with the consent of the landowner: provisions are as above, although subject to the landowner's stipulations. Limited farming, tourism and research can occur.

Source: Clark and Dingwall, 1985